Whale Sharks

CRC

MARINE BIOLOGY SERIES
The late Peter L. Lutz, Founding Editor
David H. Evans and Stephen Bortone, Series Editors

Methods in Reproductive Aquaculture: Marine and Freshwater Species,
edited by Elsa Cabrita, Vanesa Robles, and Paz Herráez

Sharks and Their Relatives II: Biodiversity, Adaptive Physiology, and Conservation,
edited by Jeffrey C. Carrier, John A. Musick, and Michael R. Heithaus

Artificial Reefs in Fisheries Management,
edited by Stephen A. Bortone, Frederico Pereira Brandini, Gianna Fabi, and Shinya Otake

Biology of Sharks and Their Relatives, Second Edition,
edited by Jeffrey C. Carrier, John A. Musick, and Michael R. Heithaus

The Biology of Sea Turtles, Volume III,
edited by Jeanette Wyneken, Kenneth J. Lohmann, and John A. Musick

The Physiology of Fishes, Fourth Edition,
edited by David H. Evans, James B. Claiborne, and Suzanne Currie

Interrelationships Between Coral Reefs and Fisheries,
edited by Stephen A. Bortone

Impacts of Oil Spill Disasters on Marine Habitats and Fisheries in North America,
edited by J. Brian Alford, PhD, Mark S. Peterson, and Christopher C. Green

Hagfish Biology,
edited by Susan L. Edwards and Gregory G. Goss

Marine Mammal Physiology: Requisites for Ocean Living,
edited by Michael A. Castellini and Jo-Ann Mellish

Shark Research: Emerging Technologies and Applications for the Field and Laboratory,
edited by Jeffrey C. Carrier, Michael R. Heithaus, and Colin A. Simpfendorfer

Red Snapper Biology in a Changing World,
edited by Stephen T. Szedlmayer, Stephen A. Bortone

The Physiology of Fishes,
edited by Suzanne Currie, David H. Evans

Whale Sharks: Biology, Ecology, and Conservation,
edited by Alistair D. M. Dove and Simon J. Pierce

For more information about this series, please visit: https://www.crcpress.com/CRC-Marine-Biology-Series/book-series/CRCMARINEBIO?page=&order=pubdate&size=12&view=list&status=published,forthcoming

Whale Sharks
Biology, Ecology, and Conservation

Edited by
Alistair D. M. Dove and Simon J. Pierce

CRC Press
Taylor & Francis Group
Boca Raton London New York

CRC Press is an imprint of the
Taylor & Francis Group, an **informa** business

First edition published 2022
by CRC Press
6000 Broken Sound Parkway NW, Suite 300, Boca Raton, FL 33487-2742

and by CRC Press
2 Park Square, Milton Park, Abingdon, Oxon, OX14 4RN

ISBN: 978-1-032-04940-3 (hbk)
ISBN: 978-1-138-57129-7 (pbk)
ISBN: 978-0-203-70291-8 (ebk)

Typeset in Times
by codeMantra

Contents

Preface

There are over 28,000 living species of fishes, including more than 500 species of sharks, so it may seem odd to single out just one to be the focus of a whole book. In this case, though, our chosen topic is one of the most iconic living animals. The whale shark *Rhincodon typus* is the largest of all living fish species. Based on the fossil record, the whale shark may even be the largest fish of all time, bigger even than "megalodon", the terrifying whale-eating Miocene predatory shark, *Otodus megalodon*. While the whale shark is a fish of superlatives – the biggest, the heaviest, the deepest-diver – the species also challenges most common shark stereotypes. Rather than being a toothy apex carnivore like megalodon or the contemporary great white shark *Carcharodon carcharias*, the whale shark is a placid, polka-dotted filter-feeder that really only poses a threat to zooplankton. This gentle giant has captured the interest of scientists and the imagination of the public, and gives us plenty to talk about in this volume.

The first whale shark was documented in 1828. Whale shark science had a rather inauspicious beginning, with this initial description published not in some esteemed expert journal, but in the *South African Commercial Advertiser* magazine. To add further insult, apparently the intended species name in Latin was misspelt by the typesetter, leading to more than a century of argument over what they should correctly be called. The eventually accepted genus name, *Rhincodon*, refers to – of all the species' many unique features – the whale shark's "rasp-like teeth." Fortunately, this did not stop anyone else from coming up with their own, more interesting names for the species in the interim. The English name, whale shark, was conferred because the species is as big as a whale and, like the baleen whales, is a filter-feeder. Whale sharks have attracted numerous other common names, in dozens of languages, as befits their vast circumtropical distribution. The whale shark's Japanese name is *jinbei zame*, the "patterned pyjama shark", while in Kenyan Swahili, they are called *papa shillingi* ("the shark covered with coins"). One of our favorites is the Vietnamese *cá Ông*, literally "Sir Fish". If ever there were a Sir Fish, then surely the whale shark would earn that title.

The whale shark is the only oceanic member of the order Orectolobiformes, which is otherwise composed of much smaller sharks that tend to dwell on or near the seafloor. While the fossil record is incomplete, it is thought that the whale shark has been swimming the world's oceans since the late Oligocene period some 23 million years ago. The modern-day whale shark is unambiguously recognized as the only member of its genus and family (Rhincodontidae). There are two other giant filter-feeding planktivorous sharks: the basking shark *Cetorhinus maximus*, which is also the second-largest shark species, and the megamouth shark *Megachasma pelagios*. The filter-feeding giant manta ray *Mobula birostris* also deserves a mention, as the largest of all rays. None of these species are closely related to whale sharks, with each of their feeding habits apparently evolving independently. Clearly, planktivory and filter-feeding are strategies that correlate with huge body size.

The whale shark's huge size is clearly influenced by and in turn influences, the lifestyle and habits of the species. Their massive adult size helps whale sharks to retain heat in their body, even while diving into the near-freezing darkness of the deep ocean, and enables them to swim thousands of kilometers every year as they search for food. Both of these are highly useful traits while navigating the clear, blue, nutrient-poor surface waters of the open ocean. The species' broad movements mean that whale sharks link marine habitats, ecosystems, and nations, with their range incorporating all the tropical and subtropical oceans.

Given their wide distribution, then, and their ability to grow larger than a school bus, it might be surprising to learn how recent most of our knowledge is on the species. There's an apocryphal anecdote that Jacques Cousteau, during his illustrious career exploring the oceans, only ever saw two whale sharks. Indeed, only a few hundred sightings of the species were documented before the mid-1980s. Up to that time, knowledge of whale sharks comprised simple reports of where and when they had been seen, and the occasional opportunistic examination of stranded or fished individuals. Field research on whale sharks began only when a medical doctor, Geoff Taylor, enlisting his family and friends for help, embarked on a pioneering research program along the Ningaloo Reef in Western Australia. This began the longest continuous identification dataset on the species and what, in hindsight, can be considered the start of the "modern" phase of whale shark research. Ningaloo Reef was the first location where whale sharks were determined to gather near to the coast with predictable seasonality, apparently in response to the abundance of prey species. This, in turn, facilitated planned, systematic research on the species. Over 150 years after the species' first description, the scientific literature began to expand as knowledge of their diet, life history, and horizontal and vertical movements were finally amenable to study.

Fast-forwarding to the present, we now know of some very special locations, such as off Yucatan Mexico and Qatar, where hundreds of whale sharks come together in truly awesome instances of biological congress. These gatherings of sharks are typically referred to, rather blandly, as "aggregations" and in most instances seem to be communal feeding events, rather than any kind of socially driven gathering. It is genuinely awesome to see such a tonnage of massive sharks together, and, with that in mind, we think these events deserve a better name. Throughout this book, we generally prefer the term "constellations". First, this term reflects the alternate common names for the whale shark that poetically reference the distinctive patterns on their skin, such as *marokintana* in Madagascar, meaning "many stars". Second, the Groth algorithm that has been used for photographic identification and matching of whale sharks, which provided the initial technical foundation for the online global whale shark photo-identification database, Wildbook for Whale Sharks (www.whaleshark.org), was originally developed by astrophysicists for identifying star patterns. Third, "constellation" works definition-wise, and it's just a much prettier name for what are truly exceptional biological phenomena. Once you've seen one of these events in person, we have no doubt that you'll agree "constellation" to be a far more apt collective noun.

Constellations of whale sharks, similar to the one discovered at Ningaloo Reef, have subsequently been discovered from several countries across the world, including Indonesia, Mexico (both Atlantic and Pacific), Qatar, Tanzania, Philippines, Mozambique and St. Helena as well as several others. This has enabled scientific knowledge to develop rapidly from the 1990s until the present. There have been several major summaries of progress along the way. Geoff Taylor published a book on his observations and experiences in 1994 and updated the text in 2019. Colman (1997) wrote a review of global knowledge on the species, with particular reference to Ningaloo studies. A significant milestone was the first "International Whale Shark Conference", held in Western Australia in 2005, which provided an opportunity for Stevens (2007) and Martin (2007) to summarize the published literature and the behavioral ecology of the species, respectively. That 2005 conference has subsequently been followed by four more (Mexico, Atlanta USA, Qatar, and Australia), with the most recent in 2019. David Rowat, following his publication of a book on the whale sharks of the Seychelles in 2010, led a comprehensive global review of the species in 2012 (Rowat and Brooks 2012). That review documented all that was known of whale sharks and asked, "Are they still mysterious?" (Spoiler alert: Yes.). Leading this book's final chapter, Dr. Rowat has the final word again on that question.

Predictable, seasonal constellations of whale sharks have not only been popular with whale shark scientists, but also with tourists. Public awareness of whale sharks has soared over the past 20 years, and the species has become the focus of intense wildlife tourism in several countries. Whale shark tourism is now worth well over $100 million worldwide each year and provides a fascinating study of both best practice in marine ecotourism and, arguably, wildlife tourism gone wrong. In some countries, too, the discovery of regular constellations or occurrences led to them being identified as a reliable target species in fisheries that, as it turned out, resulted in dramatic, far-reaching population declines.

Whale sharks are now famous, much-loved, and globally endangered. The species has survived for millions of years, but can they survive us? As we enter the third century of, and our modern relationship with, the world's biggest fish (who knows what the ancients made of them!), this book gathers the world's leading experts in whale shark biology, ecology, and conservation to summarize our collective knowledge on the species. Studying an enormous, highly mobile, deep-diving shark poses unique research challenges, and the technologies developed to answer these questions have helped to advance marine science as a whole. Chapters include discussions of satellite-linked tags used to track whale shark movements; genome sequencing to examine evolutionary adaptations; and even the use of underwater ultrasound units to investigate the species' reproduction. We hope that by collating what is known, and what we think we know, and explicitly highlighting how we, or you, could fill some of the remaining knowledge gaps, we can make the job easier for future researchers, conservationists, and resource managers in this field and give you the information you need to join the team.

Whale sharks are an amazing fish to work with. In our time studying this species, we have come to both admire and respect the efficiency of their adaptations to a life spent searching for planktonic food in the tropical pelagic oceans. Their effortless cruising, and the apparently lazy sweep of their tail, belies an elegant efficiency that is rarely matched in nature. Combine that with their peaceful demeanor, as well as those endearing polka dots, and it's hard not to love them. And why try? We hope that, as you work your way through this book, you too will develop a sense of awe and marvel at all of our good fortune to share the ocean, and the planet, with this utterly extraordinary species.

Al Dove, USA
Simon Pierce, New Zealand
December 2020

REFERENCES

Colman, J. G. 1997. A review of the biology and ecology of the whale shark. *Journal of Fish Biology* 51:1219–34.

Martin, R. A. 2007. A review of behavioural ecology of whale sharks (*Rhincodon typus*). *Fisheries Research* 84:10–16.

Norman, B. M., J. A. Holmberg, Z. Arzoumanian, et al. 2017. Undersea constellations: The global biology of an endangered marine megavertebrate further informed through citizen science. *Bioscience* 67:1029–43.

Rowat, D. 2010. *Whale Sharks: An Introduction to the World's Largest Fish from One of the World's Smallest Nations, the Seychelles.* Victoria: Marine Conservation Society Seychelles.

Rowat, D., and K. S. Brooks. 2012. A review of the biology, fisheries and conservation of the whale shark *Rhincodon typus. Journal of Fish Biology* 80:1019–56.

Smith, A. 1828. Description of new, or imperfectly known objects of the animal kingdom, found in the south of Africa. *South African Commercial Advertiser* 145:2.

Stevens, J. D. 2007. Whale shark (*Rhincodon typus*) biology and ecology: A review of the primary literature. *Fisheries Research* 84:4–9.

Taylor, G. 2019. *Whale Sharks: The Giants of Ningaloo Reef.* 2nd Revised Edition. Perth: Scott Print.

Acknowledgments

A contributed chapter scientific book like this places the editors squarely in the role of "cat herders in chief", and so we thank profusely all of the contributors who gave so willingly and generously of their time and expertise to pull together this body of knowledge, much of which happened while the world was in the grip of a global pandemic. We'd like to thank Dr. Jeff Carrier for the original invitation and his advice and support, and John Sulzycki and Alice Oven at Taylor & Francis for their patience with a couple of neophytes trying to spin a few too many plates. We would also like to thank the photographers who contributed their amazing images to help communicate and enhance this book. We managed this process while working from opposite sides of the world, so each of us also has a sphere of people whose support made this work possible.

Al appreciates all those who have contributed to whale shark care and research at Georgia Aquarium since 2005 but who are not contributors to this book. In particular, Dr. Bruce Carlson, Mike Maslanka, Dr. Tim Mullican, Ray Davis, Tim Binder, the late Jeff Swanagan, Dr. Brian Davis, Dr. Julia Kubanek, Dr. Michael Black, Dr. Marc Weissburg, Matt Foretich, Dr. Timothy Read, Dr. Milton Tan, Dr. Al Camus, Dr. David Marancik, Jeff Reid, Jake Levenson, D. Harry Webb, the late Dr. Greg Bossart, and all the current and former staff of the zoological operations, veterinary services, environmental health, dive operations, and research and conservation teams. Abundant thanks also to key field research partners, especially Rafael de la Parra-Venegas, Beti Galvan-Pastoriza, Emilio and Eugenio de la Parra, Dr. Mark Erdmann, Abraham Sianipar, Thom Maughan, Elizabeth Clingham, Annalea Beard, LeeAnn Henry, Sam Cherrett, Rhys Hobbs, Dr. George Leonard, Nick Mallos, Captain Vico Rosaro, Beth Taylor, Kenickie Andrews, Johnny Herne, Craig and Keith Yon, the irrepressible Graeme Sim and everyone at Marine Megafauna Foundation, Wildbook and Okinawa Churaumi Aquarium. Most of all, Al would like to thank his wife Trish and their wonderful kids Lucy and Andrew for their patience, love and support, and for tolerating his lengthy absences on expeditions to the other end of the world. He would like to thank his mother Marj and dedicate his contributions to his late father, Dr. Hugh Dove AM, whose example inspired Al to a career in science.

Simon is hugely indebted to the team at the Marine Megafauna Foundation for their support through the gestation of this book, and to the funders of MMF's global whale shark research and conservation efforts. In particular, the GLC Charitable Trust and another private trust largely enabled his contribution to this book, and some of the research insights herein, through their several years of support. Gemma Oakley's support and expertise was fundamental. Thanks also to Aqua-Firma and their guests, the Shark Foundation and Waterlust for their generous long-term support for our international project work. Dr. Chris Rohner, Dr. Clare Prebble, and Dr. David Robinson contributed help and well beyond the chapters they have authored in this volume. He would also like to thank his other long-term research collaborators at the Galapagos Whale Shark Project, the Large Marine Vertebrate Research Institute in the Philippines, the Georgia Aquarium team, the project team in Tanzania, the Qatar Whale Shark Research Project, the Madagascar Whale Shark Project, and where it all started back in Mozambique, as well as the all-important funders and stakeholders in these projects. The international whale shark research community is a fantastic bunch, and he feels lucky to be a part of it. Finally, a special thanks to Dr. Andrea Marshall, whose friendship enabled this amazing and unexpected career; to his late father, Trevor Pierce, and mother, Sue Pierce, for their unstinting encouragement and support for this unconventional life, not least during the interesting times that were 2020; and to his wife Madeleine for everything.

Editors

Alistair D. M. Dove, PhD, is a broadly trained marine biologist and currently Vice President of Science and Education at Georgia Aquarium in Atlanta, Georgia, USA, where he oversees international research programs on whale sharks, manta rays, coral reefs, sharks, and dolphins. Dove graduated from the University of Queensland in Brisbane, Australia, with a BSc Honours first class in 1995 and a PhD in Microbiology and Parasitology in 1999, for which he was awarded a University Medal and Dean's List commendation. His early research focus was on parasites and diseases in freshwater and marine environments, but after a period studying diseases of lobsters and a move to Georgia, Aquarium in 2006, he began focusing on the biology and ecology of whale sharks. He has led expeditions to study whale sharks in Mexico, Indonesia, the Galapagos and the remote island of St. Helena. He chaired the organizing committee for the third International Whale Shark Conference in 2013 and co-led the Whale Shark Genome Project, culminating in the publication of the complete whale shark genome in 2017. He has published over 60 other peer-reviewed articles and several book chapters, but this is his first book as editor. He is a passionate science communicator, primarily through blogging at the Deep Sea News network, on Twitter as @DrAlistairDove, and in frequent appearances in print and television media. He is an adjunct professor at Georgia Institute of Technology, the University of Georgia and Georgia State University, and a professional fellow of the Association of Zoos and Aquariums. He lives in Atlanta with his wife Patricia, also a marine biologist, their two children, and a very mediocre home aquarium.

Simon J. Pierce, PhD, is a co-founder and Principal Scientist at the Marine Megafauna Foundation, where he leads the global whale shark research and conservation program. Pierce is also a Co-Chair for the Sub-Equatorial Africa region within the IUCN Shark Specialist Group, a Science Advisor to the Wildbook for Whale Sharks global database, and a founding board member of the Sawfish Conservation Society. Pierce holds a BSc in Ecology from Victoria University of Wellington in New Zealand, and a BSc (Hons, First Class) and doctoral degree in marine biology from The University of Queensland in Brisbane, Australia. He began working with whale sharks in Mozambique in 2005 and now leads or collaborates on conservation biology and population ecology research programs across the world. In 2016, he led the most recent global conservation assessment of whale sharks for the IUCN Red List, which classified the species as endangered for the first time, then led the technical proposal that resulted in whale sharks being successfully listed on Appendix I of the United Nation's Convention on Migratory Species in 2017. He has published more than 30 scientific papers on whale sharks with his postgraduate students and collaborators, as well as many others on threatened and poorly known marine megafauna species. Simon is also an award-winning marine wildlife photographer, with his photos and videos being routinely featured by major media outlets and used to support international conservation, education, and advocacy initiatives. His usual summer habitat is in New Zealand and then, as the temperature drops, he migrates to scientific or photographic field projects through the world. Together with his wife Madeleine, he maintains a wildlife travel and photography website at naturetripper.com.

Contributors

Gonzalo Araujo
Large Marine Vertebrates (LAMAVE)
    Research Institute Philippines
Jagna, Bohol, Philippines

Tonya Clauss
Animal and Environmental Health
Georgia Aquarium
Atlanta, Georgia

Christopher Coco
Zoological Operations
Georgia Aquarium
Atlanta, Georgia

Professor Philip Dearden
University of Victoria
Victoria, Canada

Dr. Molly K. Grace
Department of Zoology
University of Oxford
Oxford, United Kingdom

Jonathan R. Green
Galapagos Whale Shark Project
Cumbaya, Quito, Ecuador

Dr. Alex R. Hearn
Universidad San Francisco
    de Quito
Quito, Ecuador

Jason Holmberg
WildMe (Wildbook)
Portland, Oregon

Dr. Lisa Hoopes
Research, Conservation and Nutrition
Georgia Aquarium
Atlanta, Georgia

Dr. Rui Matsumoto
Okinawa Churaumi Aquarium
Motobu, Japan

Dr. Craig R. McClain
Louisiana Universities Marine Sciences
    Consortium
Chauvin, Louisiana

Mark G. Meekan
Australian Institute of Marine Sciences
and
University of Western Australia
Crawley, Australia

Kiyomi Murakumo
Okinawa Churaumi Aquarium
Motobu, Japan

Dr. Bradley M. Norman, AM
ECOCEAN Inc.
The University of Queensland
Brisbane, Australia

Dr. Ryo Nozu
Okinawa Churaumi Aquarium
Motobu, Japan

Dr. Sebástian A. Pardo
Biology Department
Dalhousie University
Halifax, Canada

Emily E. Peele
Department of Biology and Marine Biology
UNCW Center for Marine Science
University of North Carolina Wilmington
Wilmington, North Carolina

Dr. César Peñaherrera-Palma
Pontificia Universidad Católica del Ecuador
Portoviejo, Ecuador

Cameron Perry
Georgia Institute of Technology
Atlanta, Georgia
and
Maldives Whale Shark Research Project
Malé, Maldives

Dr. Clare E. M. Prebble
Marine Megafauna Foundation
Truckee, California
and
National Oceanography Centre
University of Southampton
Southampton, United Kingdom

Samantha Reynolds
ECOCEAN Inc.
The University of Queensland
St Lucia, Australia

Marlon Román
Inter-American Tropical Tuna Commission
La Jolla, California

Dr. David P. Robinson
Sundive Research
Byron Bay, Australia

Dr. Christoph A. Rohner
Marine Megafauna Foundation
Truckee, California

Dr. David Rowat
Marine Conservation Seychelles (ret.)
Victoria, Seychelles

Dr. Keiichi Sato
Okinawa Churashima Research Foundation
Okinawa Churaumi Aquarium
Motobu, Japan

Dr. Jennifer V. Schmidt
Shark Research Institute
Princeton, New Jersey

Christian Schreiber
Zoological Operations
Georgia Aquarium
Atlanta, USA

Dr. Ana M. M. Sequeira
University of Western Australia
Crawley, Australia

Freya Womersley
Marine Biological Association
Plymouth, United Kingdom

Dr. Makio Yanagisawa
Marine Palace Co. Ltd.
Oita City, Japan

Dr. Kara E. Yopak
UNCW Center for Marine Science
University of North Carolina Wilmington
Wilmington, North Carolina

Dr. Jackie A. Ziegler
University of Victoria
Victoria, Canada

CHAPTER 1

How and Why Is the Whale Shark the World's Largest Fish?

Alistair D. M. Dove
Georgia Aquarium

Mark G. Meekan
Australian Institute of Marine Science

Craig R. McClain
Louisiana Universities Marine Sciences Consortium

CONTENTS

1.1 INTRODUCTION

People are inherently interested in extremes, especially when it comes to animals. Exploring and identifying which member of any given animal group is the biggest, the fastest, the smallest, or the heaviest seems to be a natural and irresistible part of human nature, and our drive to classify and measure things. Doubtless, the enormous size of whale sharks is a principal factor promoting public interest in this species, and may even be one reason why you are reading this book. In this brief introductory chapter, we explore fundamental questions about the size of whale sharks, and seek to place this extraordinary animal in a comparative context (Figure 1.1). It is not a trivial issue, not just biologically, but because body size plays a surprisingly important role in conservation. For better or worse, people tend to prioritize large animals in conservation (Ward et al. 1998), so the future plight of whale sharks may be tied to their tremendous size, in much the same way as is the increasing public awareness of this exceptional animal.

1.2 THE WHALE SHARK BODY PLAN

The whale shark is not only enormous but has a very unusual body function, and plan, especially relative to other elasmobranchs. This doubtless is a case of form following function and largely reflects their lifestyle as pelagic filter-feeding planktivores, but it is worth detailing the unique aspects of the whale shark body plan because they contribute significantly to its singular nature.

The unusual features of whale sharks start at the very front: the whale shark has a terminal mouth. In other words, the mouth opens right at the front of the head, which is unlike every other extant shark species, in which the mouth is subterminal – sometimes even ventral – and preceded dorsally by a section of the body called the rostrum or snout. This rostrum varies from pointed (e.g. mako sharks), Isurus spp. to bluntly squared (tiger sharks), Galeocerdo cuvier to rounded (most carcharhinids or "typical" sharks), or to wildly modified, as in the hammerhead, winghead, and bonnethead sharks (Sphyrnidae). This is not a trivial matter, because in many sharks, the underside of the rostrum acts as a planing surface that creates lift at the front of the body. This, combined with additional lift from the planing surfaces of the pectoral fins, counteracts the heterocercal nature (uneven lobes) of the tail, wherein the upper lobe is longer than the lower, which creates lift at the back end of the animal, driving the nose down. Keeping these two forces in dynamic balance allows the shark to maintain level swimming while moving forward. For whale sharks, the loss of the rostrum is offset by the large and broad pectoral fins and the broad, flat expanse of the underside of the head; it's just that most of the planing surface is behind the mouth rather than in front of it.

The main advantage of the terminal mouth is to facilitate filter feeding, especially the active surface suction feeding, in which whale sharks swim along with the body inclined upwards at roughly 25° and the top jaw often out of the water, while they actively suck in mouthfuls of water from the very surface layer, ingesting floating food items (Motta et al. 2010). Interestingly, basking sharks do have a rostrum and do *not* engage in active surface suction feeding. Rather, they are obligate subsurface ram feeders that simply drive the body forward through the water with the mouth wide open (Sims 2000). Whale sharks can feed this way too (see Chapter 8), but it is a less common mode than active surface suction feeding. In this sense, the whale shark appears specialized to feed more on the neuston (those organisms floating on or just under the surface) than either of the other filter feeding sharks.

Contrary to popular belief, the jaws of whale sharks do have up to 300 rows of teeth extending from one corner of the mouth to the other, each row with up to 16 files (layers from anterior to posterior). The teeth are very small and have a simple, single teardrop-shaped cusp directed posteriorly. Of all of the extraordinary anatomical features that Smith (1828) could have used to name the whale shark, it was these seemingly vestigial teeth that gave them the genus name *Rhincodon* (Greek "rasp-tooth"). Given that it is often only the teeth of sharks that are preserved in the fossil record, however, there may have been some method to Smith's madness. Just behind the teeth in both upper and lower jaws are several velums. These are transverse flaps of skin anchored to the floor and roof of the mouth, which, when the shark closes its mouth, act as check valves to seal the mouth and force water on a one-way path out through the gills.

The very wide and square mouth, which can be up to 1.5 m in horizontal diameter, has a side effect, which is to widely separate the nares (analogous to nostrils) of the whale shark. Much is made of the cephalofoil (the "hammer") of hammerhead sharks and how it serves to separate the nares in those species (Mara et al. 2015), allowing them to better determine the direction of origin of smells, a sort of stereo-olfaction, if you will. But the nares of whale sharks are large (see Chapter 3) and far more widely separated than those of even the largest hammerhead, it's just less noticeable because their head lacks that iconic hammer shape. The directional smelling abilities of whale sharks must surely be excellent.

The rostrum of most sharks sets the eyes in a lateral position, but the terminal mouth of whale sharks places the eyes much closer to the front corners of the head. This is a potentially sensitive place, since they will be among the first things to be affected if the shark hits something. To overcome this problem, whale sharks have evolved two unique adaptations: eyes that are both armored and retractable (Tomita et al. 2020). Whale shark eyes do not have nictitating membranes ("3rd eyelids"), and whale sharks are the only species known to have denticles (i.e. scales) on the very surface of the eyeball. Those denticles are different from the ones that cover the rest of the body; they have a bramble-like shape that may provide resistance to abrasion. For more serious insults, whale sharks are able to retract the eyeballs within the head using a unique cartilage bar that connects the eyeball to the lateral parts of the cranium (equivalent to the skull in sharks); this bar can be flexed by attached musculature to retract the eye and rotate it ventrally (Tomita et al. 2020). When this happens, the socket is filled with gelatinous connective tissue that normally occupies the space behind the eye and is displaced into the socket when the retraction mechanism is triggered. Partial ventral rotation of the eye also occurs (without retraction) when whale sharks are active surface suction feeding and the body is inclined upwards at more than 25°, aiding forward vision, although whale sharks do seem to have a blind spot directly in front of the animal.

Behind the jaws lies a truly enormous orobranchial (mouth and gills) chamber that makes the whale shark effectively a "swimming mouth". It extends back as far as the middle of the pectoral fins, accounting for perhaps 30% of the body length. This capacious chamber houses the ten gills and 20 filter pads, of which more is said in Chapter 8. In short, the filtration pads, which derive from the gill rakers, are so modified so as to make it impossible for the animal to open the gills for bulk water flow to pass to the outside. Anything that enters the mouth must be swallowed or must be small enough to pass through the filter pads, or it can be ejected back out of the mouth by a motion that resembles coughing. Whale sharks do have a patent spiracle just posterior to the eye that communicates between the inside of the mouth and the exterior, opening, but this is thought to be largely non-functional, although it serves as a convenient home for small remoras (see Chapter 4).

The first dorsal fin of whale sharks is set well back on the body at roughly the midpoint (compare this to many carcharhinids, in which the dorsal starts in the forebody, before the trailing edge of the pectoral fins). Otherwise, their fins are unremarkable in every respect other than size. The body surface itself is distinguished by longitudinal ridges and valleys starting behind the head and continuous with the keels of the caudal peduncle (base of the tail). These ridges are a prominent feature of the extraordinarily thick skin; both features are shown in Figure 1.2. The skin consists of a tough, thin, pigmented outer layer or epidermis, in which the denticles are embedded, and an extraordinarily thick, rubbery, white dermis of dense and collagenous connective tissue. This remarkable skin may be one of their most extraordinary adaptations, providing structural support (enhanced by the ridges), insulation, and exceptional defense against both predators and well-meaning scientists trying to attach satellite tags. All but the very largest of predatory animals that attempt to bite a whale shark is going to come away with a mouth full of gristle, if anything at all. There has been much speculation about the function of the polka dots on the skin, which are irregular anterior to the pectoral fins but give way to more regular alternating stripes and spots posteriorly. It is noteworthy that the zebra shark, *Stegostoma tigrinum* is one of the whale shark's closest relatives and has both the ridges on the skin and the polka dot pattern, although in that case it is dark spots on a light background, the opposite of the pattern in whale sharks. Perhaps their closest relative, however, the tawny nurse shark *Nebrius ferrugineus*, lacks both ridges and spots.

1.3 JUST HOW BIG IS THE WHALE SHARK?

While the whale shark is the largest extant species of fish, cartilaginous or bony, by a considerable margin, exactly how large it grows is not a simple question to answer. Perhaps the most definitive work in this regard is the critical literature review of body sizes among ocean giants by McClain et al.

(2015), which concluded that 18.8 m in total length (62 ft) is the largest reliable record of a whale shark, and further observed that several other records are also around the 18 m mark (Figure 1.1), with less reliable reports as much as 20 m. This is substantially larger than the next largest shark, the basking shark *Cetorhinus maximus* at 12.3 m maximum length (McClain et al. 2015), which is also a filter feeder. The 18.8 m maximum observed size for whale sharks is close to the theoretical maximum in modeling studies such as 18–20 m in Hsu et al. (2014), 18 m in Perry et al. (2018), 18 m (for females) in Meekan et al. (2020), and 15–20 m range posited by an earlier model (Wintner 2000).

The whale shark is not only the largest extant fish species, but perhaps the largest fish *ever* to have existed. Critical evaluation of this possibility is hampered by the poor fossilization character-istics of sharks, where teeth are often all that is preserved and body sizes must be extrapolated from fossil tooth sizes using scaling relationships derived from extant species (Ehret et al. 2009). The legendary Miocene mega-predatory shark *Otodus (Carcharocles) megalodon* (Agassiz 1835) may seem comparable in size to the whale shark (it is certainly the one we whale shark researchers are asked about the most!). However, Shimada (2019) critically evaluated size estimates of this species and concluded that the maximum size range was likely 14.2–15.3 m, commenting specifically that animals greater than 15 m in length would have been "exceptionally rare". The extinct Jurassic bony fish *Leedsichthys problematicus* Woodward 1889 is another candidate for largest fish of all time, but the five known fossil specimens were estimated to be between 8.0 and 16.5 m in total length (Liston et al. 2013). While the fossil record is, of course, incomplete, no fish skeleton of any species has yet been discovered that was longer than 18.0 m and thus it appears that the whale shark really is the all-time record holder.

The data compiled in McClain et al. (2015) referred to the maximum recorded length of various taxa, but this provides only one perspective on the size of whale sharks. Maximum sizes represent the extreme right tail of histogram distributions that are otherwise approximately normal, so these extreme records are, by definition, very rare. Maybe of more practical significance is how large is the "average" adult whale shark? In other words, what is the central tendency of the size distribution of adult whale sharks? Given that animals of this size should be far more common, this would seem an easier question to answer than the maximum possible size. But answering this question is subject to a bias that is somewhat unique to whale sharks: the majority of research on this species is cur-rently conducted at feeding constellations at coastal sites that tend to be dominated by juvenile, i.e. *smaller*, males in roughly 3:1 ratio (see Chapter 7, and Norman et al. 2017). Relatively few studies examine substantial populations of *adult* whale sharks of both sexes, simply because this happens in so few places. Maturation is thought to occur at approximately 9 m in length, based on the dif-ferentiation of claspers in the male and the development of a pelvic shelf in females (see Chapter 2 and Matsumoto et al. 2019).

Meekan et al. (2020) recently proposed that significantly different growth trajectories may exist for male and female whale sharks. Male growth curves tend to asymptote (i.e. slow down) at the onset of maturity, which occurs at an average adult size of 8.0–9.0 m, whereas female growth curves do not appear to follow an asymptotic pattern, instead continuing to a larger average size of 12–14.5 m. This suggests that when we are considering the body size of the largest whale sharks, we will generally be talking about females. This is certainly consistent with field observations in places like the Galapagos, which is home to the largest known whale sharks and where the popula-tion is almost exclusively female (see Chapter 7). Researchers interested in demography of whale sharks often comment on the apparent lack of really big males, concluding that they must be in some part of the ocean as yet undiscovered. But perhaps we don't find really big males because there aren't any, or many at least. Interestingly, Meekan et al. (2020) proposed that differences in sex-specific growth trajectory may be one explanation for the phenomenon of coastal constel-lations (see Chapter 7) because young males with faster growth trajectories have higher energy demands that can only be met in the more nutrient-rich coastal zone. Notwithstanding the potential differences in growth trajectories of males and females, we can confidently say that the majority of adult whale sharks are probably between 9 and 13 m in length and, from any viewpoint, that's objectively a very big fish.

1.4 HOW BIG IS THE WHALE SHARK RELATIVE TO OTHER LARGE MARINE ANIMALS?

Compared to its closest relatives, which are other members of the shark order Orectolobiformes (carpet sharks), the body size of whale sharks is extremely divergent. The largest tawny nurse shark, which is the next biggest orectolobiform shark and the whale sharks' closest relative, is around 4 m in maximum size, or less than a quarter of the length of the largest whale sharks (Pillans 2003). Not surprisingly, whale sharks are also genetically distinct within the phylogeny of this order (Alam et al. 2014). This evolutionary distinctiveness mirrors an ecological divergence, too: most orectolobiform sharks are sluggish bottom-dwelling predators, whereas the whale shark is the only one that is pelagic and planktivorous (Rowat and Brooks 2012).

In comparison with other shark species outside the Orectolobiformes, only the basking shark approaches the whale shark in size and, even then, not very closely. The largest basking shark documented in McClain et al. (2015) was just over 12 m in length, so the difference in maximum length between these two species (6.8 m) is larger than the maximum size of almost all other shark species! Although the whale shark and basking shark are both filter feeders, the two species are not closely related, and their filtering mechanisms are very different and certainly evolved independently (Paig-Tran and Summers 2014). It is interesting and perhaps a little surprising that whale sharks are significantly larger than basking sharks, for two reasons. First, the closest relatives of basking sharks are among the lamniform sharks, including the mako and white sharks, which are relatively much larger than the closest relatives of whale sharks. So, we might expect basking sharks to have the genomic basis to attain much greater sizes than does the whale shark, whose relatives are so much smaller. Furthermore, whale sharks live in the tropics, which are generally less productive waters than the seasonal seas of the higher latitudes inhabited by basking sharks. For these reasons, we would predict the opposite pattern: that basking sharks should be the larger species.

The third filter feeding shark, the megamouth *Megachasma pelagios*, is also relatively large at around 6 m in total length (Figure 1.1), although only half the size of basking sharks and a third the size of the largest whale shark. Megamouth sharks are more closely related to basking sharks than to whale sharks (Maisey 1985), but their filtration apparatus is different from either of the other two species (Paig-Tran and Summers 2014), suggesting that filter feeding has appeared at least three separate times in the evolutionary history of sharks, four if you include extinct filter-feeding taxa. Among the apex predatory sharks, at just over 7 m in length the largest great white sharks, *Carcharodon carcharias* are smaller than many juvenile whale sharks and roughly a third of the length of the largest recorded whale shark (McClain et al. 2015). So, the white shark may be "great" relative to other apex predatory sharks, but it is relatively puny compared to the whale shark (see Figure 1.1).

Despite their outstanding size relative to other sharks, whale sharks do not approach the size of several mysticete (baleen) whales, and they are smaller than at least one odontocete (toothed) whale: the sperm whale *Physeter microcephalus* (Linnaeus 1758; McClain et al. 2015). Among the mysticetes, the blue whale, fin whale, and sei whale, all have larger maximum sizes than the whale shark, and several other baleen whales like the humpback, bowhead, right, and gray whale are all on par with maximum whale shark lengths, and much heavier. To place the whale shark in better context, the blue whale is the largest animal to have ever existed on this planet, with a maximum length of over 33 m and a mass of over 200 t (McClain et al. 2015, and see Figure 1.1) (Table 1.1).

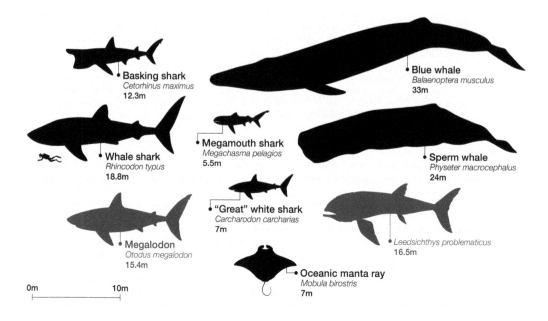

Figure 1.1   Comparative body size among select marine megavertebrates. (Jen Richards Art, commissioned.)

Table 1.1   Comparison of Maximum Body Size Among Select Large Marine Vertebrates

Species	Max Length (m)	Status	References
Blue whale *Balaenoptera musculus*	33	Extant	McClain et al. (2015)
Sperm whale *Physeter macrocephalus*	24	Extant	McClain et al. (2015)
Whale shark *Rhincodon typus*	18.8	Extant	McClain et al. (2015)
Leeds fish *Leedsichthys problematicus*	16.5	*Extinct*	Liston et al. (2013)
Megalodon *Otodus megalodon*	15.4	*Extinct*	Shimada (2019)
Basking shark *Cetorhinus maximus*	12.3	Extant	McClain et al. (2015)
Oceanic manta ray *Mobula birostris*	7[a]	Extant	McClain et al. (2015)
"Great" white shark *Carcharodon carcharias*	7	Extant	McClain et al. (2015)
Megamouth shark *Megachasma pelagios*	5.5	Extant	Compagno (2001)

[a]   Measured as lateral disc width rather than total length.

1.5 WHY IS THE WHALE SHARK SO LARGE?

The anatomical and ecological uniqueness of whale sharks suggests a long, lonely, and distinct evolutionary path, but is their extraordinary body size a result of phylogeny (that is, their evolutionary ancestry) or their ecology? Evidence argues against phylogeny as the only explanation for whale shark gigantism. Whale sharks are the only species in their genus, which is in turn the only extant genus in its family, which is in an order whose other members are all demersal species of unremarkable size relative to other shark species. There is another rhincodontid known from the fossil record, however, *Palaeorhincodon* Herman 1975, the fossils of which are distributed broadly across the globe and includes at least four putative species. As with many sharks, the descriptions of these four species are all are based exclusively on measurements of fossil teeth, which are similar to those of whale sharks except in possessing two lateral cuspids that are absent in those of modern whale sharks (Herman 1975). Although the fossils of *Palaeorhincodon* are from the Eocene at least 166 million years ago, and therefore predate the earliest appearance of fossilized *Rhincodon* teeth in Oligocene formations 23 million years ago, the relationship of these genera to each other and to their most recent common ancestor has not been established. If other gigantic members of the Orectolobiformes ever existed, none of them are represented in the current fossil record. The whale shark is the only remaining gigantic member of this order and does not seem to represent a remnant of some previous speciose radiation of giant carpet sharks, and so we must look elsewhere than the phylogenetic constraints on orectolobiforms for the explanation of the whale shark's extraordinary size.

How might ecology generate a phenotypic extreme like the whale shark over evolutionary timescales? When it comes to extremely large body size in the ocean, there is a clear correlation with planktivorous feeding habits, regardless of taxon (Pyenson and Vermeij 2016; Pimiento et al. 2019). The largest whales are the planktivorous mysticetes, the two largest sharks are also planktivores, and the largest batoid elasmobranchs, the manta rays, are also filter feeders. So, perhaps a better and more generalized question is "are animals able to become planktivores because they are so large or do large sizes result from planktivory?".

Pimiento et al. (2019) discussed gigantism among marine species and evaluated the relationship of this trait to planktivory and partial endothermy, i.e. being slightly warm-blooded or warm-blooded in only part of the body. Partial endothermy is present in a number of large bodied and pelagic species, such as bluefin tuna, lamnid sharks, the opah *Lampris guttatus*, and swordfish (Dickson and Graham 2004). Pimiento et al. (2019) concluded that planktivorous habits are tightly correlated with gigantism, but that partial endothermy and planktivory do not occur together. The latter conclusion is debatable. While we lack direct measurements of body temperature in most gigantic elasmobranchs, evidence of anatomical adaptations to retain body heat *does* exist in mobulid rays (e.g. the *rete mirabile* or blood vessel networks for counter-current exchange of heat) (Schweitzer and Sciara 1986, Alexander 1996). Thus, planktivory and endothermy probably co-occur in at least one lineage of mega-elasmobranchs. Partial endothermy also seems at least plausible in the case of basking sharks, which live in colder water and whose closest relatives among the Lamniformes are partially endothermic (Bernal et al. 2001). In the only direct study of body temperature in whale sharks that we are aware of, Nakamura et al. (2020) found no evidence that muscle activity contributed to elevated temperature in superficial skeletal musculature, although they were not able to obtain measurements from most other parts of the body, so regional endothermy remains a possibility. The red muscle of whale sharks is close to the surface rather than near the body core as in most homeothermic sharks and fish (Figure 1.2), which would argue against it, but Nakamura et al. (2020) did find that whale sharks have the lowest heat transfer coefficient of any fish yet measured, concluding that the very large body size of whale sharks confers great temperature *stability* at least. This may be a significant adaptation allowing for excursions into cold, deep water, which they do with some regularity (see Chapter 6). Upon reaching these depths, the retention of warmth from the surface makes whale sharks effectively warm-blooded until such time as they equilibrate or, more likely, return to the surface to warm up again (Thums et al. 2013). In practical terms, the size of whale sharks means that they can use the heat captured at the surface by basking and from warmer surface waters and retain it within their very large bodies to be able to swim and feed in deep, cold water. This strategy of evolving a very large body size to stabilize and maintain temperatures is known as "gigantothermy" and is an adaptation shared with leatherback turtles today and the extinct sauropod dinosaurs of the Mesozoic (Meekan et al. 2015).

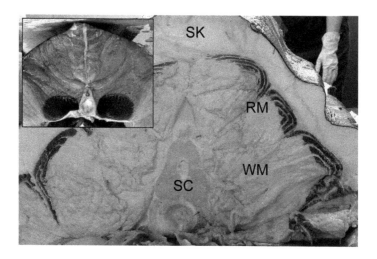

Figure 1.2 Transverse section of skeletal musculature of the whale shark, showing the superficiality of the red muscle layer. SK – skin, RM – red muscle, WM – white muscle, SC – spinal column. Inset: transverse section of porbeagle shark showing two mediolateral cords of red muscle that retain metabolic heat at the core of the body. (Alistair Dove/Georgia Aquarium; inset Wikimedia Commons.)

Another contributing factor to whale shark gigantism may be their metabolism. While direct measurements of whale shark metabolic rates are lacking, the metabolism of most of their relatives is slow compared to other pelagic sharks (e.g. Carcharhiniformes, Lamniformes), reflective of their sluggish demersal lifestyle. Diet consumption in aquarium-held animals supports this notion; whale sharks in captivity eat surprisingly little, given their body size, suggesting either slow metabolism or great efficiency, or both (see Chapters 8 and 9). Another suggestion of slow metabolic rate comes from the size of the red blood cells (erythrocytes) in whale sharks and other orectolobiforms. Larger red cells have lower oxygen diffusion efficiency and tend to occur in animals with slower metabolisms (Kozłowski et al. 2003). In other words, there is a negative correlation between red cell diameter and metabolic rate, corrected for body size. At an average of 22 μm in diameter, the red cells of whale sharks are approximately twice the size of those in carcharhinids (requiem sharks) and sphyrnids (hammerheads), but similar to those of orectolobid wobbegongs (Dove et al. 2011). Revisiting the idea that ancestry could have had a role in the evolution of gigantism, slow orectolobiform metabolism may have allowed whale sharks to live "on the metabolic knife edge" of harvesting plankton in the open ocean (Sims 1999), particularly in the nutrient-poor tropics. In that sense, the extraordinary size of whale sharks may result from *both* phylogeny and ecology.

A final explanation for the great size of whale sharks may come from simple physics. Large animals have a lower surface area-to-volume ratio than do smaller animals, because volume increases as a cube function of length. Relatively lower surface area means relatively less drag and therefore more efficient swimming, even when the absolute drag and absolute energetic costs may be much bigger for the larger animals (Ohlberger et al. 2006). Large whale sharks (for their species) may be able to roam farther in the ceaseless search for patchy food than can smaller animals, because they have a relatively lower energetic cost of transport. This may explain why we tend to observe smaller whale sharks near the coast and larger ones around oceanic islands.

1.6 WHY ISN'T THE WHALE SHARK LARGER?

While it is interesting to explore how whale sharks come to be large, we must also ask "Why isn't the whale shark even *larger*?". Given that whale sharks have adopted the same planktivorous habits as the mysticete or baleen whales, why have whale sharks not achieved the same body sizes seen in that group? We propose at least two possible explanations, operating separately or in concert.

The first explanation concerns the biomechanics of cartilaginous fishes. The robust and truly bony skeletons of marine mammals can support larger body sizes than those that can be supported by the rib-less, flexible, cartilaginous skeleton of sharks and rays. Furthermore, the rigid bony skeletons of whales can withstand the tremendous mechanical stresses of lunge-feeding enormous volumes of water (Potvin et al. 2012). Whale sharks, on the other hand, have a quite different and very efficient feeding mechanism (Motta et al. 2010), which does not rely on lunge feeding and is thus much less mechanically stressful. Interestingly, whale sharks seem to derive a significant amount of structural support from their incredibly thick and dense skin, which, at up to 15 cm, is the thickest in the animal kingdom (Martin 2007 and references therein) and consists mostly of a white, rubbery, dense, collagenous matrix.

The upper limit of body size in whale sharks may also be set by energetic constraints. In mysticete whales, the energy demands of increasingly large body sizes trade off against the ever-higher energetic costs of lunge feeding, which may be many times the basal metabolic rate (Potvin et al. 2012). At the upper size limit, the cost of feeding exceeds the energy that can be harvested from even the densest food patches (Goldbogen et al. 2011). Could a similar upper limitation exist for whale sharks? Obviously, the mode of feeding differs markedly from the mysticetes, and there are few studies of the energetic cost of feeding in whale sharks to compare. Recently, however, Cade et al. (2020) showed that overall dynamic body acceleration (ODBA, a proxy for metabolic rate) in whale sharks approximately doubled during surface feeding periods compared to non-feeding periods. By contrast, the average metabolic rate of mysticetes, which as mammals is already much higher than that of sharks, was 3.7 times higher than the base rate during feeding periods and sometimes peaked at instantaneous values 48 times the base rate during feeding lunges (Potvin et al. 2012). While the cost of feeding is indeed lower in whale sharks than in whales, whale sharks also live and, importantly, feed in the relatively nutrient-poor tropics. So, it remains possible that it becomes difficult for whale sharks to harvest enough energy to sustain their metabolic needs beyond their contemporary body size.

1.7 CAN SIZE SAVE THE WHALE SHARK?

Evidence is accumulating that body size may be an important factor in how humans approach conservation initiatives, and this obviously has significant implications for whale sharks. For example, Ward et al. (1998) showed that larger animals are more popular in zoos, with guests willing to pay more money to symbolically adopt larger animals than smaller. Similarly, Clucas et al. (2008) found that animals featured on the covers of popular nature magazines tend to be large-bodied, regardless of taxon or even conservation status. The attachment of greater cultural value to animals based on their size has given rise to terms like "charismatic megafauna" to capture the idea that very large animals are in some way more appealing, better, or more important than smaller ones. This is a completely arbitrary distinction, of course, and yet true nonetheless. Charismatic megafauna are excellent candidates for "umbrella species", which is a conservation concept wherein protection of the large charismatic species and/or their habitats confers proxy protections on other animals that share those habitats (discussed further in Chapter 12).

The positive bias towards larger animals would seem to bode well for whale sharks, as the largest fish on the planet, but any such benefit may be offset to a degree by increased extinction risk, which also correlates positively and strongly with body size (Payne et al. 2016). Increased extinction risk for larger animals generally results from lower population sizes, slow maturation, and lower intrinsic population growth, in other words, the preponderance of K-selected life history strategies (Macarthur and Wilson 1963). Whale sharks display some of the K-selected characteristics, like large size, delayed maturation, and longevity, but they also have the largest known litter size among sharks and show no parental investment, which are distinctly r-selected traits. The documented population declines of 30%–60% over the last 75 years (Pierce & Norman 2016) would suggest that whale sharks do indeed have predominantly K-selected life history traits that make them susceptible to extinction threats.

Conservation practitioners would do well to model efforts to protect whale sharks after the successful "save the whales" and "save the panda" campaigns of the late 20th century, which have been extraordinarily effective in bringing many mysticetes and the giant panda back from the brink of extinction. There seems little doubt that the large body size and peaceful behavior of these animals contributed to increases in their perceived value in the eyes of the public, once they were made aware of their conservation plight. In the case of whales, this was aided by discoveries concerning their high level of intelligence and social behavior. While whale sharks do not show remotely similar levels of intelligence or social behavior as whales, there is still no reason the whale shark could not be cultivated as a "panda of the ocean", and conservationists should shamelessly take advantage of body size bias, along with their peaceful demeanor and endearing polka dots, to improve the conservation marketing of this species (Chapter 12).

1.8 CONCLUSIONS

There is no doubt that the whale shark is the largest shark and the largest extant fish of any kind, and it seems likely to be the largest fish ever to have swum in the world's oceans. How fortunate we are to live in a time when this species is still around! Some questions remain to be answered about the body size of whale sharks and how it relates to their lifestyle. For example, we still do not know for certain which factors limit the body size of whale sharks. We do not understand the genetic basis of gigantism in whale sharks or indeed most other species. The availability of the complete whale shark genome (Read et al. 2017) may be a useful tool in the search for a genotypic mechanism to explain such a divergent phenotype. Finally, questions remain about the precise relationship between body size and body temperature in this species. Is it just thermal stability, or passive gigantothermy, or do whale sharks have specific adaptations for partial endothermy, as occur in many other pelagic species? While scientists seek the answers to these questions, the sheer mind-blowing size of these ocean giants will continue to captivate the scientific community and the public alike.

ACKNOWLEDGMENTS

We thank Jen Richards for the line art in Figure 1.1 and Georgia Aquarium for permission to reprint Figure 1.2.

REFERENCES

Alam, M. T., Petit, R. A., Read, T. D., and A. D. M. Dove. 2014. The complete mitochondrial genome sequence of the world's largest fish, the whale shark (*Rhincodon typus*), and its comparison with those of related shark species. *Gene* 539(1):44–49.

Alexander, R. L. 1996. Evidence of brain-warming in the mobulid rays, *Mobula tarapacana* and *Manta birostris* (Chondrichthyes: Elasmobranchii: Batoidea: Myliobatiforrnes). *Zoological Journal of the Linnean Society* 118:151–64.

Bernal, D., Dickson, K. A., Shadwick, R. E., and J. B. Graham. 2001. Analysis of the evolutionary convergence for high performance swimming in lamnid sharks and tunas. *Comparative Biochemistry and Physiology* 129A:695–726.

Cade, D. E., Levenson, J. J., Cooper, R., et al. 2020. Whale sharks increase swimming effort while filter feeding, but appear to maintain high foraging efficiencies. *Journal of Experimental Biology* 223. doi:10.1242/jeb.224402.

Clucas, B., McHugh, K., and T. Caro. 2008. Flagship species on covers of US conservation and nature magazines. *Biodiversity Conservation* 17:1517–28.

Compagno, L. J. V. 2001. *Sharks of the World. An annotated and illustrated catalogue of the shark species known to date. Volume 2. Bullhead, mackerel and carpet sharks (Heterodontiformes, Lamniformes and Orectolobiformes).* Rome: FAO.

Dickson, K. A., and J. B. Graham. 2004. Evolution and consequences of endothermy in fishes. *Physiological and Biochemical Zoology* 77(9):998–1018.

Dove, A. D. M., J. Arnold, and T. M. Clauss. 2011. Blood cells and serum chemistry in the world's largest fish: The whale shark *Rhincodon typus*. *Aquatic Biology* 10:177–84.

Ehret, D., Hubbell, G., and B. J. McFadden. 2009. Exceptional preservation of the white shark *Carcharodon* (Lamniformes, Lamnidae) from the early Pliocene of Peru. *Journal of Vertebrate Palaenotology* 29(1):1–23.

Goldbogen, J. A., Calambokidis, J., Oleson, E., et al. 2011. Mechanics, hydrodynamics and energetics of blue whale lunge feeding: Efficiency dependence on krill density. *Journal of Experimental Biology* 214:131–46.

Herman, J. 1975. Les Sélaciens des terrains néocrétacés & paléocènes de Belgique & des contrées limitrophes. Eléments d'une biostratigraphie intercontinentale [The sharks from the Late Cretaceous and Paleocene terrains of Belgium and adjacent lands. Elements of an intercontinental biostratigraphy]. *Mémoires pour servir à l'explication des Cartes géologiques et minières de la Belgique* 15:1–450.

Hsu, H. H., Joung, S. J., Hueter, R. E., and K. M. Liu. 2014. Age and growth of the whale shark (*Rhincodon typus*) in the north-western Pacific. *Marine and Freshwater Research* 65:1145–54.

Kozłowski, J., Konarzewski, M., and T. Gawelczyk. 2003. Cell size as a link between noncoding DNA and metabolic rate scaling. *Proceedings of the National Academy of Sciences* 100:14080–85.

Liston, J., Newbrey, M., Challands, T., and C. Adams. 2013. Growth, age and size of the Jurassic pachycormid *Leedsichthys problematicus* (Osteichthyes: Actinopterygii). In *Mesozoic Fishes 5 – Global Diversity and Evolution*, ed. G. Arratia, H. Schultze, and W. M. Munich, 145–75. Germany: Verlag.

MacArthur R. H., and E. O. Wilson. 1963. An equilibrium theory of insular zoogeography. *Evolution* 17(4):373–87.

Maisey, J. G. 1985. Relationships of the megamouth shark, *Megachasma*. *Copeia* 1985(1):228–31.

Mara, K. R., Motta, P. J., Martin, A.P., and R. E. Hueter. 2015. Constructional morphology within the head of hammerhead sharks (Sphyrnidae). *Journal of Morphology* 276:526–39.

Martin, R. A. 2007. A review of behavioural ecology of whale sharks (*Rhincodon typus*). *Fisheries Research* 84:10–16.

Matsumoto, R., Matsumoto, Y., Ueda, K., et al. 2019. Sexual maturation in a male whale shark (*Rhincodon typus*) based on observations made over 20 years of captivity. *Fishery Bulletin* 117(1):78–86.

McClain, C. R., Balk, M. A., Benfield, M. C., et al. 2015. Sizing ocean giants: Patterns of intraspecific size variation in marine megafauna. *PeerJ* 3:e715.

Meekan, M. G., Fuiman, L. A., Davis, R., et al. 2015. Swimming strategy and body plan of the world's largest fish: Implications for foraging efficiency and thermoregulation. *Frontiers in Marine Science* 2:64.

Meekan, M. G., Taylor, B. M., Lester, E., et al. 2020. Asymptotic growth of whale sharks suggests sex-specific life-history strategies. *Frontiers in Marine Science* 7:774.

Motta, P. J., Maslanka, M., Hueter, R. E., et al. 2010. Feeding anatomy, filter-feeding rate, and diet of whale sharks *Rhincodon typus* during surface ram filter feeding off the Yucatan Peninsula, Mexico. *Zoology* 113(4):199–212.

Nakamura, I., Matsumoto, R., and K. Sato, 2020. Body temperature stability observed in the whale sharks, the world's largest fish. *The Journal of Experimental Biology.* doi:10.1242/jeb.210286.

Norman, B. M., Holmberg, J. A., Arzoumanian, Z., et al. 2017. Undersea constellations: The global biology of an endangered marine megavertebrate further informed through citizen science. *BioScience* 67 (12):1029–43.

Ohlberger, J., Staaks, G., and F. Hölker. 2006. Swimming efficiency and the influence of morphology on swimming costs in fishes. *Journal of Comparative Biochemistry and Physiology B* 176:17–25.

Paig-Tran, E. W., and A. P. Summers. 2014. Comparison of the structure and composition of the branchial filters in suspension feeding elasmobranchs. *The Anatomical Record* 297(4):701–15.

Payne, J. L., Bush, A., Heim, N. A., et al. 2016. Ecological selectivity of the emerging mass extinction in the oceans. *Science* 353(6305):1284–86.

Perry, C. T., Figeiredo, J., Vaudo, J. J., et al. 2018. Comparing length-measurement methods and estimating growth parameters of free-swimming whale sharks (*Rhincodon typus*) near the South Ari Atoll, Maldives. *Marine and Freshwater Research* 69:1487–95.

Pierce, S. J. and B. Norman. 2016. *Rhincodon typus. The IUCN Red List of Threatened Species* 2016:e. T19488A2365291 (accessed 9 September 2020).

Pillans, R. 2003. *Nebrius ferrugineus. The IUCN Red List of Threatened Species* 2003:e.T41835A10576661 (accessed 29 September 2020.)

Pimiento, C., Cantalapiedra, J. L., Shimada, K., et al. 2019. Evolutionary pathways toward gigantism in sharks and rays. *Evolution* 73(3):588–99.

Potvin, J., Goldbogen, J. A., and R. E. Shadwick. 2012. Metabolic expenditures of lunge feeding rorquals across scale: Implications for the evolution of filter feeding and the limits to maximum body size. *PLoS ONE* 7(9):e44854.

Pyenson, N. D., and G. J. Vermeij. 2016. The rise of ocean giants: Maximum body size in Cenozoic marine mammals as an indicator for productivity in the Pacific and Atlantic Oceans. *Biology Letters* 12(7):20160186.

Read, T. D., Petit, R. A., Joseph, S. J., et al. 2017. Draft sequencing and assembly of the genome of the world's largest fish, the whale shark: *Rhincodon typus* Smith 1828. *BMC Genomics* 18:532.

Rowat, D., and K. S. Brooks. 2012. A review of the biology, fisheries and conservation of the whale shark *Rhincodon typus. Journal of Fish Biology* 80(5):1019–56.

Schweitzer, J., and G. N. Sciara. 1986. The *rete mirabile cranica* in the genus *Mobula*: A comparative study. *Journal of Morphology* 188: 167–78.

Shimada, K. 2019. The size of the megatooth shark, *Otodus megalodon* (Lamniformes: Otodontidae), revisited. *Historical Biology.* doi:10.1080/08912963.2019.1666840.

Sims, D. 1999. Threshold foraging behaviour of basking sharks on zooplankton: Life on an energetic knife-edge? *Proceedings of Biological Sciences* 266:1437–43.

Sims, D. 2000. Filter-feeding and cruising swimming speeds of basking sharks compared with optimal models: They filter-feed slower than predicted for their size. *Journal of Experimental Marine Biology and Ecology* 249(1):69–76.

Smith, A. 1828. Description of new, or imperfectly known objects of the animal kingdom, found in the south of Africa. *South African Commercial Advertiser* 145:2.

Thums, M., Meekan, M., Stevens, J., et al. 2013. Evidence for behavioural thermoregulation by the world's largest fish. *Journal of the Royal Society Interface* 10(78):20120477.

Tomita, T., Murakumo, K., Komoto, S., et al. 2020. Armored eyes of the whale shark. *PloS One* 15(6):e0235342.

Ward, P. I., Moseberger, N., Kistler, C., and O. Fischer. 1998. The relationship between popularity and body size in zoo animals. *Conservation Biology* 12(6):1408–11.

Wintner, S. P. 2000. Preliminary study of vertebral growth rings in the whale shark, *Rhincodon typus*, from the east coast of South Africa. *Environmental Biology of Fishes* 59:441–51.

CHAPTER 2

Whale Shark Reproduction, Growth, and Demography

Simon J. Pierce
Marine Megafauna Foundation

Sebástian A. Pardo
Dalhousie University

Christoph A. Rohner
Marine Megafauna Foundation

Rui Matsumoto, Kiyomi Murakumo, and Ryo Nozu
Okinawa Churaumi Aquarium

Alistair D. M. Dove
Georgia Aquarium

Cameron Perry
Georgia Institute of Technology
and
Maldives Whale Shark Research Project

Mark G. Meekan
Australian Institute of Marine Science

CONTENTS

2.1 INTRODUCTION

Large shark species typically have low reproductive output, take many years to reach adulthood, and can live for a long time. As the largest of all sharks, and hence a biological extreme, the whale sharks' reproduction and growth is an inherently interesting topic. For such a well-known animal, though, knowledge of the reproductive biology of whale sharks is surprisingly lacking. Few individual whale sharks have ever had their anatomy examined in detail. Some of these were caught in the "bad old days" when the species was subject to large-scale fisheries, and a few have been recovered after stranding. A small number of whale sharks have been held in aquaria, which has allowed for longer-term observations on these animals. Fortunately, the high degree of public and research interest in the species has helped inspire the development of new technologies and techniques to study live, free-swimming whale sharks in the wild.

New life-history insights in recent years allow us to update the demographic model for whale sharks. Demography is a science that was created to forecast the growth rate of human populations and provides the mathematical basis for the modern life insurance industry. Here, we can use data and inferences from whale shark reproduction, age, and growth studies to estimate population-level characteristics, such as the "intrinsic rate of population increase" (r), calculated on an annual basis. Demography can be thought of like a bank account. *Deposits* are the average number of babies, divided by the frequency in which a species gives birth: for sharks, this can be annually, biennially, or over even longer time scales. Removals from the population caused by natural mortality are the *withdrawals*. If deposits exceed withdrawals, the population will grow. The rate of growth (i.e. r) is then the *interest*, which allows for a compounding balance over time.

Sadly, whale sharks are a globally endangered species, and may still be in decline overall (see Chapter 12). Additional mortality from threats such as fisheries, bycatch, and vessel collisions also impacts their population growth or decline (Chapter 11). Estimating r based on natural mortality alone gives us a metric with which to measure the impact of management and conservation efforts that aim to reduce or eliminate mortality from human activities. When whale shark populations are recovering at the maximum rate that their biology allows, we will know these initiatives are succeeding. In this case, a healthy population – like a healthy bank account – is one that is continually growing.

Understanding the biology of whale sharks is a key component of formulating an effective conservation strategy (Chapter 12). Where we lack knowledge on the biological inputs to this model, we can investigate how different (theoretical) scenarios will influence the estimate of r. That helps to highlight the most important research questions that scientists need to answer going forward. Here, we also provide some pointers on how these answers might be obtained.

2.2 REPRODUCTIVE MODE AND EMBRYONIC DEVELOPMENT

The 500+ contemporary species of sharks have developed a wide variety of reproductive modes to maximize their evolutionary success. At a species level, different habitats and ecological niches provide their own evolutionary pressures, and the range of shark body forms also introduces some practical constraints. The diversity and flexibility of reproductive strategies employed by sharks to prevail against these challenges, ranging from egg-laying (oviparity) in many species to "sibling cannibalism" (adelphophagy), where fetal sand tiger sharks *Carcharias taurus* actively hunt their siblings within the uterus, has made the group an ongoing favorite for study by reproductive and evolutionary biologists (Musick et al. 2005; Parsons et al. 2008; Penfold and Wyffels 2019). Even with this attention, sharks still have the capacity to astonish scientists; multiple shark species have recently been shown to be capable of "virgin births" through parthenogenesis, where females produce viable offspring through asexual reproduction, even in the complete absence of males (Chapman et al. 2007, 2008; Robinson et al. 2011). Sharks do like to keep us guessing.

Most teleost (bony) fishes use external fertilization. Adult teleost fishes typically expel hundreds of thousands of eggs into the water, where they suffer heavy mortality as gametes and then as free-swimming developing larvae. Sharks take a fundamentally different approach, called internal fertilization, where the development of the fetal pups takes place at least partially inside the mother's body. In oviparous species, the embryos continue their development, while encased within tough egg cases, on the seafloor. "Viviparity" is where the embryos develop entirely in the mother, who gives birth (parturition) to free-swimming, independent pups. Yolk-sac viviparity, formerly known as "ovoviviparity", is the term used for species, which include the whale shark, in which the yolk-sac is the main source of fetal nutrition through development until birth. The embryos "hatch" inside the mother early in the development cycle, absorb most or all of this yolk-sac while in the uterus, and are born free-swimming and independent. The yolk must contain all the necessary nutrients to bring the embryos to term, but they produce relatively small young as they are limited to this finite resource.

The whale shark is a member of the order Orectolobiformes, broadly known as carpet sharks. However, the species is alone in the family Rhincodontidae and, as such, it is evolutionarily distinct from other extant sharks. Indeed, the whale shark is thought to have been around in something akin to its current form for at least 23 million years (Cicimurri and Knight 2009; Carrillo-Briceño et al. 2020). Genetic data indicate that the whale shark's closest living relatives are the short-tail nurse shark *Pseudoginglymostoma brevicaudatum* and the zebra shark *Stegostoma tigrinum*, along with the (slightly more distant again) tawny nurse shark *Nebrius ferrugineus* and nurse shark *Ginglymostoma cirratum* (Naylor et al. 2012). Even among these four related species, there are substantial differences in reproductive modes. The zebra shark is oviparous, producing up to around 100 eggs per year in captivity (Watson and Janse 2017), which each hatch around 160 days after deposition (Robinson et al. 2011; Dudgeon et al. 2017). The poorly known short-tail nurse shark is also thought to be oviparous (Ebert et al. 2013). The tawny nurse shark appears to vary in reproductive mode through its distribution. In Australia, the species is yolk-sac viviparous, with large litters of up to 32 pups (Ebert et al. 2013). In Japanese waters, however, the developing pups feed inside the uterus on large infertile yolky eggs, referred to as oophagy, and litter size is only 1–4 pups (Teshima et al. 1995; Tomita et al. 2018). The nurse shark is yolk-sac viviparous, producing 20–50 pups per litter on a biennial basis following a 5–6-month gestation period (Castro 2000; Pratt and Carrier 2001).

Whale shark reproduction was a conundrum until relatively recently. No whale shark pups had been documented at all until 1953, when a large egg case, containing a live 35 cm whale shark embryo, was collected in a trawl in the Gulf of Mexico (Baughman 1955). Unsurprisingly, this led to many researchers thinking the species was oviparous for decades afterward. This embryo still had an attached external yolk-sac, and its gut cavity was full of yolk (Wolfson 1983). The ~30 cm long egg case itself was relatively thin and light in color compared to other oviparous sharks, whose durable eggs must provide some protection against abrasion and potential predators, leading Wolfson (1983) to surmise that it may have been aborted. Resolution of this question had to wait until 1995, when a 10.6 m total length (TL), 16 t pregnant female shark was caught off Taiwan. This "megamamma"

Figure 2.1 The single litter of whale shark pups to be examined by scientists, from a 10.6 m "megamamma" female caught in Taiwan in 1995. (Photo: Hua Hsun Hsu, George Chen Shark Research Center, National Taiwan Ocean University, Taiwan.)

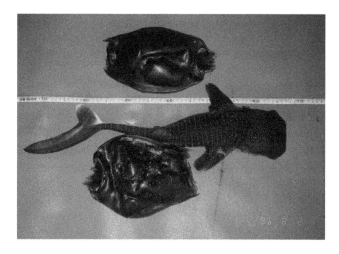

Figure 2.2 Empty egg cases and a fetus from the "megamamma" pregnant female whale shark caught in Taiwan. (Photo: Hua Hsun Hsu, George Chen Shark Research Center, National Taiwan Ocean University, Taiwan.)

shark, as it was described in the resulting paper (Joung et al. 1996), contained around 300 embryos in varying stages of development in its twin uteri (Figure 2.1). Some empty egg cases from embryos that had already hatched were found in the uteri; 24 egg cases were measured and had an average length of 21 cm (Chang et al. 1997). Some other embryos were still present inside their egg cases, with their yolk-sac attached. Most embryos, however, had already left their eggs and were nearly ready for birth (Figure 2.2). The largest size-class of embryos, 58–64 cm TL, had clear vitelline scars at the prior attachment point of their now fully absorbed yolk-sacs (Joung et al. 1996). The sex ratio of all the embryos that were inspected was 123:114 (female:male), which was not significantly different from 1:1. This litter, at slightly above 300 embryos (variably referred to as 304–307 in the initial description), is the largest – by a considerable margin – that has ever been recorded in a shark. The next largest litter in an individual shark appears to be 135 from a blue shark *Prionace glauca*, although the average for that species is only around 30 (Nakano and Stevens 2008).

This singular female remains the only pregnant whale shark that has ever been described in detail by scientists. In 1991, three of the authors on the "megamamma" paper searched Taiwanese official records for additional pregnant females, finding no results. Their personal interviews with six long-time whale shark fishermen (with 14–43 years of experience) confirmed that none had seen a whale shark carrying fetal pups (Joung et al. 1996). Their interviewees did, however, recount having seen more than 200 eggs the size of "ping pong balls and larger" in the "belly" of "very fat females" ranging from 10 to 16.5 t and 3.5 to 10 m TL. Over 100,000 much smaller eggs were reportedly seen in one 15 t female. The authors surmised that these would have been the follicles on the surface of the right ovary (see Section 2.3.1 for context), which would be conspicuous when the shark was cleaned, whereas their uteri may not have been examined. Only one fisherman described what may have been uterine contents. In a 15 t female, he reported over 100 "pale milky eggs encased in a membrane" in cases measuring approximately 12 × 17 cm. There were also many smaller encased eggs that he reported as "deeper yellow", interpreted as being the yolk showing through the outer case (Joung et al. 1996). This female was, therefore, presumed to be carrying developing embryos.

A few other observations are worth mentioning for completeness. In 1910, Thomas Southwell recorded a trawl capture off Sri Lanka of a shark identified as a whale shark with a "very ripe ovary. Oviduct full of eggs, 16 cases counted, same form as in dog-fish". Following a further investigation of the report, Gudger (1933) discounted it as a misidentification of an unrelated species, which does seem likely if it was caught in a trawl fishery. In 1996, 16 juveniles of about 1 m TL were reported to be swimming alongside a 5.5 m TL whale shark off India (Krishna-Pillai 1998, cited in Akhilesh et al. 2013). However, Akhilesh et al. (2013) regarded the original author's assessment that they were the larger sharks young as "highly unlikely" and, in the absence of photographic evidence, viewed this record as more likely to have been teleost associates such as cobia *Rachycentron canadum* (see Chapter 4). The size of the whale shark in question makes it likely to have been immature (see below), further supporting this view, and cobia are often confused with young whale sharks by people unfamiliar with the species.

More credibly, Geoff Taylor, a pioneering whale shark researcher from Western Australia, has reported witnessing a whale shark giving birth while conducting an aerial survey of the Ningaloo Reef (Taylor 2019). He recounted that "We circled down to 200 feet over the water in our Cessna aircraft watching the baby sharks appear from under the mother. We counted 14 newborn whale sharks emerging at the surface alongside the mother. Their distinctive tadpole shape was unmistakable". This appears to be the only record of a whale shark actually giving birth. No postpartum maternal care has been recorded from any shark species to date (Jacoby et al. 2012; Awruch 2015), so any association with the mother is presumed to be fleeting.

2.3 REPRODUCTIVE ANATOMY IN THE WHALE SHARK

2.3.1 Female Reproductive Anatomy and Maturity

As noted earlier, internal fertilization in sharks has required both sexes to develop a number of specialized anatomical adaptations, described in detail by Walker (2020). Female sharks have a pair of reproductive tracts, and either a single or paired ovary located in the anterior body cavity, about level with the anterior end of the liver (see Figure 2.3). Only one ovary, usually the right, is typically developed in orectolobiform sharks (Penfold and Wyffels 2019; Rêgo et al. 2019). Initial results suggested that either the right or left ovary may develop in whale sharks, based on dissections of three immature animals (Nozu et al. 2015). The ovary produces oocytes (immature egg cells) through a process named oogenesis. In juvenile sharks, these oocytes are small with little or no yolk; the maximum size observed by Nozu et al. (2015) in three immature whale sharks (up to 7.6 m TL) was

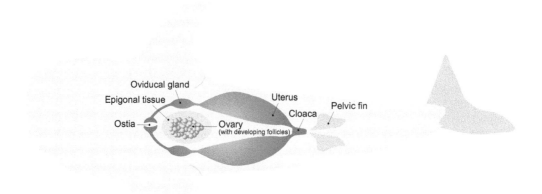

Figure 2.3 Diagrammatic representation of the adult female whale shark reproductive system. Not to scale. Design by Madeleine Pierce.

Figure 2.4 External ovary appearance in an 8.7 m TL maturing female whale shark. (Photo: Alistair Dove/ Georgia Aquarium.)

<1.8 mm. By contrast, oocytes in an 8.7 m, 6.2 t maturing female were around 8 mm in diameter in an ovary that weighed 12.1 kg (A. Dove unpubl. data, and see Figure 2.4).

Developing oocytes are enveloped by follicle cell layers, comprising inner granulosa cells and outer theca cells. No detailed description of the ovarian maturation process or mature ovaries is available from whale sharks. Based on typical shark development, as the oocyte maturation process proceeds, these cells start to synthesize and secrete steroid hormones, the oocytes enlarge, and yolk formation begins in a process called vitellogenesis. The largest of these oocytes are released (ovulation), and this "free" ovulated oocyte is now referred to as an ovum. The ovum passes through the ostia into one of the oviducts and then into the oviducal gland, where it is fertilized if the shark has mated. The oviducal gland is a discus-shaped organ around 15 cm in diameter, with a finely, transversely ridged and presumably secretory epithelial lining to the lumen. This then produces a protective egg capsule for the ovum, which passes into the uterus with the now-developing embryo within. The paired uteri are highly developed to support egg retention and embryonic growth in viviparous sharks. In the 8.7 m "maturing" female, no egg capsules or embryos were evident in the lumen of the oviducal glands or uteri, respectively, and the uteri were fairly simple tubes leading anteriorly from the cloaca, with little sign of any significant gestational lining at that stage.

Although knowledge about whale shark reproduction is obviously limited, their reproductive mode appears to be most similar to that of the nurse shark *G. cirratum* (Castro 2000, 2013; Awruch 2015). Castro (2013) noted that the two species share a variation of "normal" yolk-sac viviparity, which he referred to as serial yolk-sac viviparity. Both species produce relatively large litters. Because oocytes are usually ovulated in pairs, at discrete intervals, ovulation is a prolonged process that lasts at least 2–3 weeks in nurse sharks and may last far longer in whale sharks, as they have a much larger litter. This could explain the large variation in embryonic development observed among the fetal pups in the "megamamma". The first pairs of oocytes to be ovulated and fertilized may plausibly be more developmentally advanced by several weeks compared to the last pairs to be fertilized. In nurse sharks too, females in mid-gestation can, at the same time, be carrying some developing pups that are still inside their egg cases, while more advanced pups have already absorbed their yolk-sac and are ready for birth (Castro 2000, 2013). Birthing itself in nurse sharks takes place over at least several days (Castro 2000).

Anatomical changes in the reproductive tract as a shark grows can be used to determine the onset of maturity in individual sharks. In females, the diameter and presence of visible yolk in the largest ovarian follicles, the developmental state of the oviducal gland, and the width and structure of the uterus are common characters that are used to determine maturity (Walker 2020). Observations of developing offspring in a female whale shark are, as discussed earlier, limited to a single individual, and few other data on the reproductive anatomy of adult female whale sharks are available. Nozu et al. (2015) performed dissections and histological and endocrinological studies on three immature female whale sharks of 4.03, 6.65, and 7.61 m TL from Japan. Elsewhere, females measuring 8.2 m (Rohner et al. 2015), 8.7 m (Pai et al. 1983; A. Dove pers. obs.), and 8.8 m TL (Beckley et al. 1997) were also immature. The smallest female whale shark that was definitely mature is that 10.6 m "megamamma" from Taiwan. Hsu et al. (2014b) reported that two mature females of 9.6 m and 11.9 m TL had also been caught off Taiwan. We assume, therefore, that maturity in female whale sharks occurs at 9–10 m TL.

The size, and specifically the age at which whale sharks produce their first litter, is a key input for demographic models, so refining these parameters is an important goal for researchers. A recent and particularly exciting technological advance is the use of waterproofed ultrasound units that can be used by divers and not solely at the surface (Tomita et al. 2018; Murakumo et al. 2020). The probe used for ultrasound imaging is placed on and moved along the abdominal region between the pectoral and pelvic fins on either side of the shark's body to visualize the ovary and reproductive tract (Figure 2.5). The Okinawa Churaumi Aquarium in Japan holds a captive female shark that measures about 8.0 m TL at the time of writing. Two of us (RM and KM) have been able to repeatedly examine this female and can precisely locate and identify reproductive organs including the thin, undeveloped uterus and small undeveloped ovary using ultrasonography (Matsumoto et al. In revision). As this female matures, changes to these structures can be tracked and documented.

The ability to examine reproductive anatomy in a free-swimming whale shark provides a clear opportunity to gather data on maturity in wild sharks. Provisional field data from female whale sharks are available from Darwin Island in the Galapagos Archipelago (Matsumoto et al. In revision). The physical size of these large (>10 m TL) whale sharks presents a technical challenge for the current ultrasound equipment, as the dermal tissue on the sharks' flanks is so thick that it is often difficult to obtain clear images of their internal structures. However, the presence of ovarian follicles was documented in two female sharks estimated to be 11–12 m TL. Eighteen follicles could be distinguished in one shark, ranging from 29 to 84 mm, and a single follicle (76 mm diameter)

Figure 2.5   Rui Matsumoto using an underwater ultrasound unit on a female whale shark at St. Helena Island.
(Photo: Simon Pierce.)

was observed in the other individual. Pre-ovulatory follicle size is a useful indicator of maturity, but this character is ambiguous without specifying the follicle diameter to be used as a diagnostic (Walker 2020). During dissections of the school shark *Galeorhinus galeus*, where the presence of yolk can be directly observed, 0–3 mm follicles all appeared white and were, therefore, all categorized as pre-vitellogenic (immature). Vitellogenic (mature) oocytes were yellowish due to the presence of yolk and all measured >3 mm diameter, providing a threshold of 3 mm for follicle size in mature sharks (Walker 2005). Such detailed examination has not yet been possible for whale sharks, although the thousands of small ovarian follicles in the "megamamma", as mentioned above, all measured <10 mm in diameter (Joung et al. 1996).

As an interim exercise, before the captive female reaches maturity, we can provisionally estimate the maximum follicle diameter (MFD) for whale sharks by using verified results from other, better-known species. We can then assess if these two sharks from the Galapagos were mature, based on the follicle measurements obtained from ultrasound. We used MFD results from studies of other shark species whose young develop solely from yolk input, using the mean value or median if they provided a size range for MFD (Jones and Geen 1977; Tanaka et al. 1990; Peres 1991; Ebert 1996; Watson and Smale 1998; Capapé et al. 2005; Huveneers et al. 2007; Porcu et al. 2014). We found a strong positive correlation between MFD and the birth size (BS) of pups in these species (Figure 2.6). To extrapolate these results to whale sharks, we used this regression equation (and least-square methods) to build a 95% prediction interval for MFD based on their known size at birth, for which we took a median value of 61 cm TL. The mean MFD in mature whale sharks is provisionally estimated to be 102 mm in diameter.

The MFDs obtained from both whale sharks in the Galapagos (84 and 76 mm diameter, respectively) fell within the 95% prediction interval (Figure 2.6), indicating that they were mature. These results do not themselves refine our current estimate of size at maturity, as both of these sharks were estimated to be slightly larger than the 10.6 m TL "megamamma", but it shows that continued use of ultrasound in large >8 m sharks can aid in the data collection necessary to answer this question.

A useful independent verification tool for this method also now exists. Thanks to the recent development of in-water blood sampling techniques for free-swimming whale sharks (Figure 2.7), circulating reproductive hormone levels can now be determined from their plasma (Matsumoto et al. In revision). When baseline hormone measures for mature sharks are available (from further field sampling or the soon-to-be maturing captive female shark), these methods will substantially

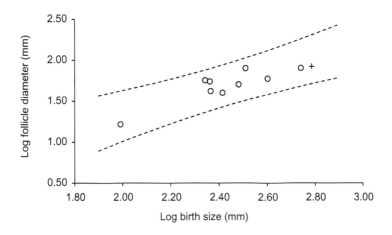

Figure 2.6 Relationship between maximum follicle diameter (MFD) and birth size (BS) based on nine yolk-sac viviparous shark species (open circles). A cross shows the estimated relationship for whale sharks (102 mm MFD; see text). Dotted lines show 95% prediction intervals.

Figure 2.7 Kiyomi Murakumo taking a blood sample from the base of the pelvic fins in a free-swimming whale shark at St. Helena Island. (Photo: Simon Pierce.)

expand the possibilities for minimally invasive data collection on whale shark reproduction and maturity assessment in the wild.

The "megamamma" shark was not explicitly reported to be visibly pregnant upon capture (Joung et al. 1996; Schmidt et al. 2010), although it is possible this was considered too obvious to mention! When information from dissections is not available, the size at maturity in wild female elasmobranchs has been assigned based on visible pregnancy, as has been done in reef manta rays *Mobula alfredi* (Marshall and Bennett 2010; Deakos 2012; Stevens et al. 2018) and subsequently verified by applying the same ultrasound techniques to this species (Murakumo et al. 2020). Similarly, in the absence of other data, we have routinely made inferences on pregnancy in whale sharks based on the distended appearance of the abdomen immediately posterior to the pelvic fins. This is particularly evident in large female whale sharks (Figure 2.8). However, repeated ultrasound examination of this region has revealed only imagery of dermal tissue and underlying musculature in large female whale sharks in the Galapagos Islands (Matsumoto et al. In revision). In the absence of supporting

Figure 2.8   A large female whale shark showing the characteristic "bump" posterior to the pelvic fins, which appears to be a sexually dimorphic feature. (Photo: Simon Pierce.)

evidence and the development of this appearance in large but known immature female whale sharks, i.e. the captive 8.0 m TL individual in Japan and an 8.7 m TL shark from the Georgia Aquarium, it appears that this "bump" may be a sexually dimorphic feature that is peculiar to female whale sharks, and hence an unreliable indicator of pregnancy.

2.3.2 Male Reproductive Anatomy and Maturity

Reproductive adaptations in male sharks revolve around producing sperm and transferring them to the female, as described in detail by Conrath and Musick (2012) and Walker (2020). Limited specific information is available on the internal organization of male whale sharks, but they are presumed to be generally similar to other sharks (Figure 2.9). The paired testes function in spermatogenesis (production of sperm) and steroidogenesis (secretion of steroid hormones). Their morphology and structure vary among species. Mature sperm travels through the efferent ductules and into the epididymis, which is continuous with the ductus deferens and the seminal vesicle. Sperm form into clumps in the latter two structures, where they are stored in readiness for mating.

Adult males have paired external "claspers" – tube-like extensions from each pelvic fin – that function as copulatory organs. At maturity, the claspers are rigid, calcified structures that can articulate freely at the base. Siphon sacs, which are muscular bladders under the skin, hold seawater to flush sperm from the longitudinal clasper groove into the female's oviduct. The clasper tip is armed with hooks or claws in many species and expands when inserted into the female's cloaca during mating. This allows the male to maintain physical contact while delivering sperm directly into the female's reproductive tract.

The claspers are visible in free-swimming male sharks. As their mature, calcified condition is required for successful mating, the external appearance of the claspers is the most common way to determine the maturity status of individual male sharks, including whale sharks (Norman and Stevens 2007; Awruch 2015; Rohner et al. 2015). Immature male whale sharks have relatively small, thin claspers. Mature males have thick claspers that extend well past the pelvic fins, with the tip often having a "cauliflowered" appearance (Figure 2.10; Norman and Stevens 2007). Rohner et al. (2015) used calibrated parallel-mounted lasers in the field to measure male sharks and their clasper length to quantify the size at maturity off eastern Africa. Eight of the measured sharks were mature, ranging from 8.2 to 10.3 m TL, with inner clasper lengths from 75 to 106 cm. The largest immature male was 9.3 m TL, with inner clasper lengths in immature sharks ranging from 26 to 84 cm. Within this population, the mean length of maturity (i.e. for 50% of male sharks) was estimated to be 9.2 m TL.

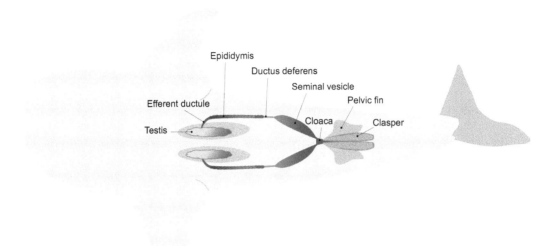

Figure 2.9 Diagrammatic representation of the adult male whale shark reproductive system. Not to scale. Design by Madeleine Pierce.

Figure 2.10 A mature male whale shark with calcified claspers. (Photo: Simon Pierce.)

The relative ease of assessing clasper length and appearance has allowed the collection of a large quantity of population-level data on male maturity, and the interesting observation that the length at maturity differs among locations. These eastern African male sharks have the largest mean length at maturity documented to date. The 50% maturity metric for sharks off Qatar was 7.3 m TL (Robinson et al. 2016), 8.1 m TL off Western Australia (Norman and Stevens 2007), 6.8 m TL at Donsol in the Philippines (McCoy et al. 2018), and 7.0 m TL in the Mexican Atlantic (Ramírez-Macías et al. 2012). Although some variance is probably due to differences in the techniques employed, such as visual estimates vs. laser photogrammetry (Rohner et al. 2011; Perry et al. 2018), along with bias in visual estimates among observers, our own personal observations across sites make us confident that real differences exist. Such variation is not uncommon among shark species. Larger sizes at maturity (and maximum size) have been tentatively been associated with higher latitudes and, correspondingly, lower mean water temperatures (Taylor et al. 2016). This ecological phenomenon, when observed in a single species, is known as James's rule (Blackburn et al. 1999). Whether this is the driver of variation in whale sharks is unclear, but it is interesting that male whale sharks at similar latitudes, for instance on the Atlantic vs. Pacific coast of Mexico, mature at different sizes (Ramírez-Macías et al. 2012). James's rule has typically been associated with endothermic species, in which larger body size can improve heat retention in cooler temperatures. A faster growth rate in warmer water is one potential explanation for the same pattern when seen in ectotherms, like most sharks and rays (Neer and Thompson 2005).

A male whale shark named Jinta has been held in the Okinawa Churaumi Aquarium in Japan since 1995, providing a unique opportunity to observe the maturation process in an individual shark (Matsumoto et al. 2019). Jinta grew from 4.6 m in 1995 to 8.7 m TL in 2018. Clasper elongation took place over 11 months, from August 2011, when Jinta was 8.5 m TL. During this period, the shape of the claspers changed, the internal cartilaginous structure became visible, and each developed the cauliflower-like tip. Jinta began to exhibit a new behavior in April 2012, inverting his body and crossing the claspers over the midline (Figure 2.11), which has been noted as a mating display in other shark species (Pratt and Carrier 2001). From August 2012, a white milky fluid was released from the claspers when this behavior was displayed. Similar behavior has been observed in YuShan, an 8.1 m TL maturing male whale shark in the Georgia Aquarium (Chapter 9). The frequency of Jinta's mating displays, including both clasper rotation and actively chasing the female whale shark held in the same aquarium, has subsequently been increasing every year. Sperm released from the rotated clasper were successfully collected by a diver in 2016 (Figure 2.12). Individual sperm consisted of a head, including an acrosome, a midpiece, and a flagellum, similar to other sharks (Tanaka 1995). The head of the sperm was moderately helical, with a comparatively shorter midpiece and a long, slender flagellum. Sperm survived for 3 days in a bottle of seawater after collection. The steroid hormone testosterone is likely to be involved in several reproductive functions including mating behavior (Rasmussen and Murru 1992; Rasmussen and Gruber 1993), spermiogenesis (Tricas et al. 2000; Henningsen et al. 2008), and clasper elongation (Heupel et al. 1999; Awruch et al. 2008). Jinta's testosterone level was low, albeit with some evidence for seasonal variation, prior to maturity. Testosterone level increased during clasper elongation, and remained high, with no significant relationship to seasonal water temperature, once the shark was mature (Matsumoto et al. 2019).

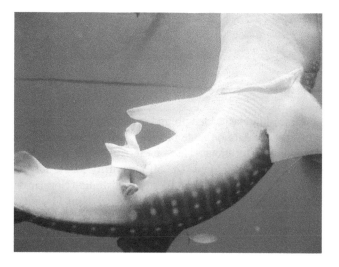

Figure 2.11   Clasper flexion and body inversion in "Jinta", a mature male whale shark in the Okinawa Churaumi Aquarium. (Photo: Kiyomi Murakumo.)

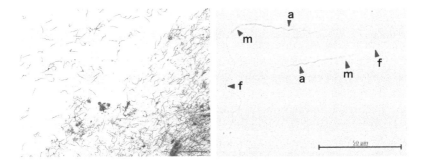

Figure 2.12   Microscopic morphology of whale shark sperm collected from a mature male in Okinawa Churaumi Aquarium. Left: Hemacolor rapid staining of sperm smear. Right: sperm structure; acrosome (a), midpiece (m), and flagellum (f). (Photos: Kiyomi Murakumo.)

2.4 REPRODUCTIVE CYCLE, SEASONALITY, AND BEHAVIOR

2.4.1 Female Reproductive Cycle and Seasonality

Female viviparous sharks all follow the same sequence of reproductive development: (a) ovulation, (b) fertilization, (c) embryonic development, and (d) the birth of free-swimming pups (Awruch 2015). The duration of each step varies among species. The first stage, known as the ovarian cycle, is the period during which follicles enlarge through vitellogenesis and then ovulate. Fertilization occurs shortly after ovulation, as the ovum passes through the oviduct and oviducal gland. The oviducal gland can, in many sharks, retain sperm to fertilize multiple ova over an extended period of time, up to a documented 45 months in the brown-banded bamboo shark *Chiloscyllium punctatum* (Bernal et al. 2015). This is hypothesized to have been the case in the "megamamma" shark from Taiwan after a genetic paternity analysis on a subset of the pups found that they had a single male parent (Schmidt et al. 2010). After the first reproductive cycle of a mature female shark, a new ovarian cycle can – in some species – proceed at the same time as embryonic development from the previous cycle, minimizing the time between litters. This allows these shark species to produce annual or even multiple litters per year (Walker 2020). However, in the species that produce large (>40 mm) follicles, which includes whale sharks, vitellogenesis typically occurs over 2 or 3 years, or even longer (Walker 2020). The length of the ovarian cycle, then, is the primary "bottleneck" that can extend the overall duration of the reproductive interval (the time between litters), which can be 3 years or more in some species (Walker 2020). Some sharks also introduce (e) a resting period, allowing time for the liver to sequester nutrients in preparation for the next cycle of vitellogenesis (Conrath and Musick 2012). This can result in female sharks in the same population having different reproductive intervals due to the variable inclusion of this 1- or 2-year resting period (Bansemer and Bennett 2009; Driggers and Hoffmayer 2009).

To discuss the reproductive cycle of female whale sharks we have to, once again, rely heavily on observations of the "megamamma" from the study by Joung et al. (1996). If the ovarian cycle was a year or less in that shark, we would have expected obvious development of yolked ovarian follicles; however, the authors reported that "thousands of small ova, each less than 1 cm in diameter, were found in the ovary". Follicles measured by ultrasound in the wild females from the Galapagos were far larger than this, and no uterine contents were observed in these sharks (Matsumoto et al. In revision), so it seems clear that follicle development and embryonic development were not concurrent in these sharks, and thus, a biennial or longer reproductive interval is most likely, at least in Pacific whale sharks.

Different shark species can have highly synchronized seasonal reproductive cycles, mate continuously through the year, or display some combination of both (Conrath and Musick 2012). Diapause can also occur, wherein embryonic development is delayed to regulate the time of parturition (Walker 2020). Nurse sharks have a biennial reproductive cycle off Florida in the USA (Castro 2000; Pratt and Carrier 2001), although individual females can introduce a resting stage and may hence reproduce less frequently (Pratt et al. 2018). The lack of data available for whale sharks makes it rather fruitless to speculate much further at this stage, although we can add a few notes. Hsu et al. (2014a) observed the first neonatal whale shark in Taiwan waters (78 cm TL) in October 2013, close to where the near-term "megamamma" pregnant female was caught in July 1995. Given the growth rate of neonates (discussed below), they posited that parturition could occur seasonally in late summer to early autumn in that area (Hsu et al. 2014a). However, three even smaller neonates were found in nearby Donsol (Philippines) earlier in the year, in March 2009 (Aca and Schmidt 2011) and again recently in March 2020 (Miranda et al. 2020). The ecology of whale shark pups is discussed further in Chapter 7. In the absence of other mature females becoming available for dissection, knowledge of wild whale shark reproduction is most likely to advance through the use of the new techniques such as ultrasonography and blood sampling of free-swimming sharks discussed above.

2.4.2 Male Reproductive Cycle and Seasonality

Sperm production in male sharks can be continuous, with little or no seasonal change in testicular development or spermatogenesis, or seasonal, often with a defined peak associated with a specific mating season (Awruch 2015). In evolutionary terms, male sharks will presumably benefit by mating as frequently as possible, but there is also an energetic cost to maintaining reproductive readiness year round. If female sharks are themselves on a seasonal cycle, it may be more efficient to match them. Most of the elasmobranchs studied to date seem to take this approach (Awruch 2015; Walker 2020), although sperm storage can render a synchronous cycle unnecessary (Awruch 2015). In some sharks, such as the spotted catshark *Scyliorhinus canicula* (Garnier et al. 1999) and the epaulette shark *Hemiscyllium ocellatum* (Heupel et al. 1999), seasonal variation in sperm production is associated with observable changes in clasper shape through the year, with swollen claspers corresponding with the mating season (Awruch 2015).

Presumed courtship and mating behaviors have been observed at St. Helena Island, in the equatorial mid-Atlantic, between January and March (Perry et al. 2020). A mature male was observed in what was apparently a misdirected mating display towards the hull of a boat(!) at the Archipelago of São Pedro and São Paulo, also in the equatorial mid-Atlantic, in April 2010 (Macena and Hazin 2016). This individual also had abraded claspers, potentially an indication of recent mating activity (Pratt and Carrier 2001; Macena and Hazin 2016). Female whale sharks with presumed bite marks on their pectoral fins have also been seen at both of these locations (Macena and Hazin 2016; Perry et al. 2020). Similar marks are commonly used as an indicator of mating in other species, such as nurse sharks, as they are produced by a male grasping the female's pectoral fin with his teeth to facilitate coupling (Pratt and Carrier 2001). Jinta, the male shark in the Okinawa Churaumi Aquarium, often chases the large (immature) female with which he shares the exhibit, particularly during the summer months from May to September when water temperatures range between 23°C and 29°C (R. Matsumoto pers. obs.). Reproductive behaviors, such as those shown in Figure 2.11, are most frequently observed in June (early summer, ca. 26°C) every year. However, such behaviors are also occasionally observed in winter (R. Matsumoto pers. obs.). Jinta was first observed biting the pectoral fin tip of this female in February (mid-winter, 21.2°C) 2018. A second observation of fin-biting approach was recorded in May (early summer, 23.4°C) 2020, when Jinta bit the broad area on the female's pectoral tip (Figure 2.13), resulting in characteristic abrasions that may be useful as a diagnostic for wild mating behavior in this species (Figure 2.14). Mating observations are strongly seasonal in Atlantic nurse sharks (Castro 2000; Pratt and Carrier 2001) which are the

Figure 2.13   Attempted mating behavior by a mature male whale shark (above, and left) towards a large immature female shark in the Okinawa Churaumi Aquarium. (Photos: Kiyomi Murakumo.)

Figure 2.14   Pectoral fin abrasions on a large immature female whale shark in the Okinawa Churaumi Aquarium caused by a mature male whale shark biting the fin during attempted mating. Left panel shows the fin, with inset; right panel shows the bite marks on the fin tip. (Photos: Kiyomi Murakumo.)

related species most similar to whale sharks in their reproduction. Mature male nurse sharks typically visit a specific mating location each year, whereas mature females return to the site in alternate years (Pratt and Carrier 2001), or less frequently (Pratt et al. 2018), in keeping with their biennial or longer reproductive cycle.

This question mark around reproductive seasonality in male whale sharks may be resolved by collecting field data on testosterone levels in these sharks or by using photogrammetry or a similar technique (e.g. Rohner et al. 2015) to assess seasonal variation in mature clasper morphology, as has been documented in cat and epaulette sharks. However, a practical caveat is that mature males may be seasonal visitors to these equatorial mid-Atlantic sites for other reasons, such as food availability, with the potential for mating opportunities being an occasional bonus (Macena and Hazin 2016; Perry et al. 2020). Sampling for such work may then need to take place in multiple locations to test for seasonal variation.

2.5 SIZE AT BIRTH AND EARLY GROWTH

Several of the larger fetal pups from the "megamamma" were still alive at the time of dissection and were placed into containers with seawater (Joung et al. 1996). One 60 cm (1 kg) pup was subsequently placed into an aquarium, where it survived for several months. This pup started to feed 17 days after its premature "birth" and grew rapidly to 139 cm and 20.4 kg after 120 days (Chang et al. 1997). Unfortunately, it died from septicemia (blood poisoning) incurred from a scratch. A different 60 cm pup was successfully raised in another aquarium for over 3 years, growing to 3.7 m TL over 1,157 days (Nishida 2001, cited in Rowat and Brooks 2012). These two pups provide the best information currently available on early growth rates in whale sharks. However, their maintenance in a controlled environment means that these results cannot necessarily be extrapolated to the wild.

Few neonate whale sharks have been recorded in the wild across the world. Rowat and Brooks (2012) summarized the data available at that time, identifying just 19 young whale sharks measuring less than 1.5 m TL. The smallest wild pup documented to date was 46 cm TL, found in the Philippines (Aca and Schmidt 2011), which still had a clear vitelline scar from prior yolk-sac attachment (indicating recent birth). A slightly larger pup (64 cm TL) captured near the same locale a few weeks later had a nearly healed scar. Wolfson (1983) documented faint vitelline scars on pups of 55, 62, and 63 cm TL. The scar on a 78 cm TL pup from Taiwan was also nearly healed (Hsu et al. 2014a). This character, therefore, appears to be a reliable indicator that a given pup is a young-of-the-year, i.e. is weeks to months of age.

Some of these free-swimming pups were smaller than the largest fetal pups from the "megamamma" shark, and one presumably free-swimming 94 cm TL pup was caught in a gillnet in India with a significant (300 g) external yolk-sac still attached (Manojkumar 2003). This suggests that natural size at birth may be somewhat more variable than found in the "megamamma" (Rowat and Brooks 2012). Regardless, whale sharks have the smallest pups in relative terms, at only ~5% of maternal length, across shark species. Most species have a pup length that averages 27% of the mother (Cortés 2000).

2.6 ESTIMATING AGE AND GROWTH

2.6.1 Vertebral Aging

Assigning the age of individual sharks at different lengths and life stages is one of the most important foundations of demographic research. The general biology underlying this technique has been summarized by Natanson et al. (2018). Shark skeletons are made of cartilage, and the hard cartilaginous centra of their vertebral column are the most commonly used structure in aging studies. Unlike bone in humans, cartilage cannot be resorbed to free up extra calcium for the body and instead continually accretes as the shark grows. Seasonal cycles of growth, due to temperature changes or feeding opportunities, produce mineralized band pairs that are visible in sections of the centra. It is generally assumed that the growth of sharks, as ectotherms, is related to water temperature, with faster growth in warmer water (creating a light band) and slower growth in cooler water (resulting in a denser, darker band). Although there has been some dispute about the underlying physiological mechanisms, multiple species of sharks have been aged by counting these band pairs, with a light band and dark band forming one annulus (Natanson et al. 2018).

Unfortunately, age determination "…continues to be one of the most challenging and nuanced aspects of life-history investigation in elasmobranchs" (Natanson et al. 2018). Technical considerations include the possibility of different band pair counts in different centra from the same shark, varying results from different visualization hardware, and ensuring standardization among the people doing the counting (Natanson et al. 2018). There are also biological differences in band pair formation among species. Whereas some sharks have relatively clear seasonal band pairs, other sharks including the basking shark *Cetorhinus maximus* (Natanson et al. 2008) and multiple wobbegong sharks *Orectolobus* spp. (Chidlow et al. 2007; Huveneers et al. 2013) form band pairs in conjunction with periods of growth, i.e. growing faster (producing a light band) following enhanced feeding opportunities, followed by the formation of a darker band when growth slows due to reduced feeding. Band pair formation is thus associated with prey availability, which is not necessarily seasonal. That means these species cannot be directly aged from their vertebra. There is also substantial evidence that band pairs have been systematically and substantially undercounted in many long-lived shark species, as their band pairs either cease to form or become so thin that they are unresolvable in larger individuals (Harry 2018; Natanson et al. 2018). This inflates estimates of growth rate and underestimates potential longevity.

To counter these issues, multiple methods have been developed to test and validate the results from vertebral aging studies (Natanson et al. 2018). Ong et al. (2020) used one of the current standards, bomb radiocarbon dating, to test the hypothesis of annual band pair deposition in whale shark centra from the Indo-Pacific. The bomb radiocarbon technique utilizes the rapid increase in atmospheric levels of the carbon 14 isotope that was created by atmospheric testing of thermonuclear bombs in the 1950s–1960s. This isotope saturated the atmosphere and then quickly diffused into the oceans to be uptaken by all marine organisms living at the time. The authors obtained vertebrae from a 10 m TL female shark from Pakistan and a 9.9 m TL male from Taiwan (with 50 and 35 band pairs, respectively), which were compared to reference chronologies of C14 from teleosts and coral species. These two individuals were included with vertebral samples from 18 other sharks from Taiwan ranging from 3.0 to 5.96 m TL. The width of annuli decreased with age, narrowing substantially in the oldest sharks (Figure 2.15), with counts of band pairs from the 20 sharks ranging from 15 to 50. Comparison with the reference chronologies indicated an annual formation of band pairs in the two large sharks.

These data are useful for informing the debate on the periodicity of band pair deposition in whale sharks. Previous work had hypothesized biannual formation of band pairs in Taiwanese whale sharks (Hsu et al. 2014b), but methodological updates and reanalyses of a subset of these samples, included in the Ong et al. (2020) study, amended these results to annual band pair formation in this population. The question had also been raised whether the whale shark, as a highly mobile animal, is necessarily subject to annual variation in temperature and growth, as opposed to the aperiodic growth based on feeding opportunities posited for its ecological analog, the planktivorous basking shark (Natanson et al. 2008). Growth parameters can vary across the range of broadly distributed sharks, so additional validation studies will be useful for certainty (Natanson et al. 2018). However, assuming that annual band formation is the case, at least for part of the

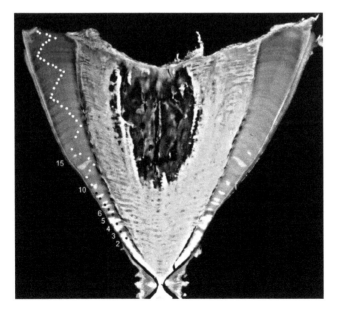

Figure 2.15 Vertebral centra showing band pairs in a ~50-year-old whale shark. (Photo: Steven Campana, with annotations by Joyce Ong. Published in Ong et al. (2020): https://doi.org/10.3389/fmars. 2020.00188.)

whale shark's growth, the Ong et al. (2020) results help to verify band pair counts from southern Africa provided by Wintner (2000) and Rohner et al. (2015). Both studies used X-radiography of vertebra from stranded sharks ($n=15$ and $n=2$, respectively) to visualize band pairs in specimens from 5.4 to 9.2 m TL, obtaining band pair counts up to 31 from an 8.9 m TL male shark. Although the practical difficulties of collecting vertebra in the field meant there were some differences in the location of extracted vertebra among these 17 individuals, which can result in differing band pair counts (Natanson et al. 2018), we can tentatively include these results here with those from Ong et al. (2020) to estimate the growth parameters for whale sharks within the Indo-Pacific subpopulation (Pierce and Norman 2016).

Using these combined length-at-age data, we can improve our estimates of whale shark growth parameters by using Bayesian von Bertalanffy growth functions (VBGF) with informative priors. The use of informative priors, i.e. integrating our existing knowledge of whale shark biology into the model, is particularly helpful for situations like this: where few data points are available, and these do not span the species' potential age and length range. We used a similar modeling approach to that used by Pardo et al. (2016) to evaluate the productivity of the threatened spinetail devil ray *Mobula mobular*. In this case, we have only 33 length-at-age records in total, and these are predominantly from medium to large juvenile male sharks. Informative priors help to constrain the growth parameter estimates based on the life-history information we do have, although the priors are not necessarily exact equivalents. Specifically, here we used a prior for the L_0 parameter (the y-intercept of the VBGF) based on known size at birth; here set at 60 cm TL. We also set the asymptotic size parameter L_{inf} based on the equation by Froese and Binohlan (2000) using a maximum length estimate of 18.8 m TL (McClain et al. 2015), as well as a more conservative estimate of 15 m TL (to approximate a more "typical" maximum length within whale shark populations; see Section 2.6.2) to assess whether a change in this prior would affect growth parameter estimates.

Our estimates of the growth parameter k using both priors hover around 0.02 year^{-1}, with a median posterior k estimate of 0.017 in the model based on an 18.8 m TL max size, and 0.023 year^{-1} in the model based on a 15 m TL max size (Figure 2.16 and Table 2.1). These values are similar to those obtained by dedicated growth studies for the species (Meekan et al. 2020), which indicate that whale sharks grow slower than most other chondrichthyan fishes. However, the main caveat of all growth studies up to date is the lack of data for the smallest (<3 m TL) and largest (>10 m TL) individuals; there is only a single data point for individuals estimated to be >40 years-old. In the absence of more data, even these combined growth parameter estimates should be taken with a grain of salt. Furthermore, we anticipate that future growth studies will confirm that female whale sharks grow slower than males (Meekan et al. 2020).

2.6.2 Field-Based Growth Estimates

Growth models for the whale shark can also be built using empirical data, i.e. measured changes in length between known time intervals in differently sized sharks. The benefit of this approach is that larger sample sizes can be obtained; the drawback is that it is challenging to obtain accurate measurements from free-swimming whale sharks and then to repeat this exercise at a later time for comparative purposes. Our methods for measuring wild whale sharks have evolved through the years, from using tape measures or knotted ropes and swimming along the sharks, comparisons with boats or snorkelers, or extrapolations based on dorsal fin height (Meekan et al. 2006; Rowat and Brooks 2012; Perry et al. 2018), moving to laser photogrammetry (Figure 2.17; Rohner et al. 2011; Jeffreys et al. 2013) and, increasingly, stereo-videography (Sequeira et al. 2016; Meekan et al. 2020). This last approach seems to generate the most precise measurements available to date, although analysis is more complex. This precision does provide substantial benefit though, as the species' slow growth rate can confound less precise techniques, even over periods of several years. As an example, some of us (Rohner et al. 2015) even found that 6/13 individual sharks apparently

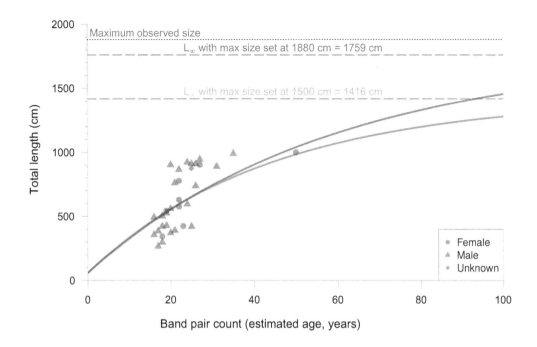

Figure 2.16 Estimated von Bertalanffy growth functions for the whale shark based on informative priors. The two scenarios gray and red lines represent estimated maximum lengths of 18.8 and 15 m, respectively. Size at birth was set at 60 cm.

Table 2.1 Summary of Posterior Estimates of von Bertalanffy Growth Parameters for Whale Sharks Fitted Using Two Bayesian Models with Differing Priors

Model	L_{inf}	K	L_0	σ^2
Max size = 1880 cm	1758 cm (1532–1979]	0.017 year⁻¹ [0.014–0.021]	59 cm [40–78]	0.311 [0.241–0.409]
Max size = 1500 cm	1420 cm [1205–1635]	0.023 year⁻¹ [0.018-0.03]	59 cm [39–78]	0.316 [0.246–0.408]

Median estimates are shown, while the numbers in brackets are the 95% credible intervals. σ^2 is the estimated error term in the model.

shrank over periods exceeding 340 days using laser photogrammetry which, although not necessarily impossible (Davenport and Stevens 1988), is definitely unlikely.

Rather than delving too deeply into the actual methodologies for measuring sharks here (instead, see the citations above), we will focus on the results. Whale sharks grow fast when they are young, as discussed in Section 2.5. By the time they reach 3–4 m TL and start to frequent constellation areas, where most research is conducted, this initial growth spurt seems to have slowed considerably. This is a normal pattern in fishes, whose growth (in either length or body mass) through time can generally be described by a curve, flattening out (= reaching an asymptote) as they approach maximum size. The VBGF, introduced above, is the most common growth curve used in fisheries biology (Kacev et al. 2017), and comparing VBGF predictions from either age- (Section 2.6.1) or growth-based models with field data is a useful way to verify their general accuracy.

Three field studies, from eastern Africa (Rohner et al. 2015), the Maldives (Perry et al. 2018), and Western Australia (Meekan et al. 2020), have followed this approach for whale sharks. These results were summarized by Meekan et al. (2020), who found that existing VBGF curves for the species, based largely on juvenile growth rates, consistently overestimate the growth of sharks as they

Figure 2.17   Chris Rohner using a parallel laser photogrammetry system to measure a whale shark. (Photo: Simon Pierce.)

approach maturity. This had the analytical byproduct of overestimating the VBFG-derived asymptotic size of whale sharks, which has usually been reported as 18–22 m TL (Hsu et al. 2014b; Perry et al. 2018; Meekan et al. 2020; Ong et al. 2020). Although this corresponds quite well with the largest whale sharks that have ever been recorded from fisheries catches, which measured 18–20 m TL (McClain et al. 2015), the revised asymptotic size based on the Australian field data was 8.5–8.9 m TL for males and 14.55 m TL for females (Meekan et al. 2020). It is important to note that asymptotic length is the average length at maximum size for individual sharks; clearly, on a population level, some individuals will reach far larger sizes than others. These revised asymptotic lengths fit well with other field data, with population-level studies of size-frequency in whale sharks rarely recording males exceeding 10 m TL or females exceeding 15 m TL (Rohner et al. 2015; Norman et al. 2017). Empirical data on female shark growth are relatively lacking, with less than 20 individuals included across these three studies due to the male-dominated population at each site (Rohner et al. 2015; Perry et al. 2018; Meekan et al. 2020). It appears, though, that female whale sharks grow substantially slower to their larger asymptotic size (Meekan et al. 2020), which has significant implications for both their general ecology (see Chapter 7) and their age at maturity (see below).

This suggests that whale sharks have determinate growth, at least for practical purposes; they do not necessarily continue measurable growth throughout their life. This is supported by results from field studies, such as "Stumpy" the (male) whale shark in Western Australia, whose length has remained around 7.4 m TL over two decades (Norman and Morgan 2016), and Jinta in the Okinawa Churaumi aquarium, whose growth slowed to 0.6–5.6 cm per year after reaching >8 m TL, in contrast to a growth of ~27 cm per year from 4.6 to 8 m TL (Matsumoto et al. 2019). This complicates the calculation of their potential longevity, estimated at 129 years for males from Maldives data (Perry et al. 2018), as it means their age is not necessarily related to their size once they attain close to their asymptomatic length. Additional use of radiocarbon dating, possibly extending this to eye lens studies as conducted for Greenland sharks *Somniosus microcephalus* (Nielsen et al. 2016; Natanson et al. 2018; Edwards et al. 2019), may be needed to resolve the currently open question of the whale shark's lifespan.

2.7 BODY CONDITION AND HEALTH ASSESSMENT

The accessibility of effective techniques such as laser photogrammetry (Rohner et al. 2011; Jeffreys et al. 2013) and, preferably in many situations, stereo-videography (Delacy et al. 2017;

Meekan et al. 2020), makes the collection of accurate length measurements for individual whale sharks a routine exercise at this point. That raises the possibility of collecting other measurements of the shark's body, not just TL and its proxies. In particular, metrics that correlate with girth have some potential to expand current ecological and biological studies. Girth measurements are commonly used to quantify "body condition" in marine mammals to evaluate individual-level health (Castrillon and Bengtson Nash 2020), which is also possible in sharks (Gallagher et al. 2014). As a large species that is readily accessible for in-water studies, it is worth briefly exploring whether such techniques may also be applicable to whale sharks.

Whale sharks, despite the name, are obviously quite different from the cetaceans that are a common focus for these techniques (Castrillon and Bengtson Nash 2020). Notably, for this methodological comparison, cetaceans store most of their high-energy lipid in their dermal blubber (Castrillon and Bengtson Nash 2020), whereas sharks primarily store energy as lipids in their large livers (Del Raye et al. 2013; Gallagher et al. 2014; Corsso et al. 2018). Energy storage and use in white sharks *Carcharodon carcharias* has been examined through dynamic measurement of body density through glide rate via satellite-linked tags (Del Raye et al. 2013) – which has a significant advantage, in that it measures total body energy stores (Castrillon and Bengtson Nash 2020) – but is outside the scope of this chapter. Similarly, large-scale collection of liver mass relative to body mass (hepatosomatic index) is demonstrably informative (Hussey et al. 2009; Corsso et al. 2018), but neither of these masses can be obtained from live, unrestrained whale sharks. Here, we focus instead on techniques that quantify the shark's basic body shape (morphometry) in conjunction with length estimation, as these have the potential to be a practical and cost-effective way to collect population-level data through time and build directly on existing methods.

Studies on live sharks have to date focused on measuring restrained individuals (Gallagher et al. 2014; Irschick and Hammerschlag 2014) rather than in-water measurements of free-swimming animals. However, the span condition analysis (SCA) metric proposed by Irschick and Hammerschlag (2014) may be modifiable for use in whale sharks. As it is challenging to measure the full girth accurately in the field, Irschick and Hammerschlag (2014) assessed shark condition by measuring across the flanks and back of the shark, in front of the pectoral and either side of the first dorsal fin, and taking the circumference of the caudal keel (base of the tail). These four measures were then added together and standardized using the pre-caudal length of the shark to obtain an index of condition (i.e. SCA). This appears to work well for sharks in which the body shape does not naturally change with growth. They found that the use of SCA was provisionally viable in the tiger *Galeocerdo cuvier*, nurse, and blacktip *Carcharhinus limbatus* sharks, but not the bull shark *C. leucas* which showed signs of changing body shape with increasing size. The use of SCA was then tested further on tiger sharks by also examining their blood chemistry. Plasma triglyceride concentrations had a significant positive relationship with condition values, meaning that fatter sharks have larger energy reserves than thin ones (Gallagher et al. 2014).

The usefulness of SCA or similar measures of condition in whale sharks has yet to be evaluated. Some common refrains from both shark and marine mammal studies are the need to compare condition between ecologically relevant groups, such as a common size range and maturity stage, including sex and reproductive phase (Hussey et al. 2009; Castrillon and Bengtson Nash 2020). Measuring free-swimming sharks is also a different exercise, as circumference could not be measured accurately, and the use of two-dimensional images from still photography (rather than a full over-the-back measurement with a tape measure as for SCA) clearly requires some changes. It may be possible to measure individual whale sharks from both above the shark (i.e. upper image, Figure 2.18) or from the side (i.e. lower image, Figure 2.18), with these two measures used either separately or combined. Body flexion during swimming can reduce the precision of laser photogrammetry (Rohner et al. 2015), but the use of stereo-videography may be feasible. Two-dimensional girth metrics are frequently collected in cetacean studies using photogrammetric techniques (Castrillon and Bengtson Nash 2020), similar to those used for whale shark growth

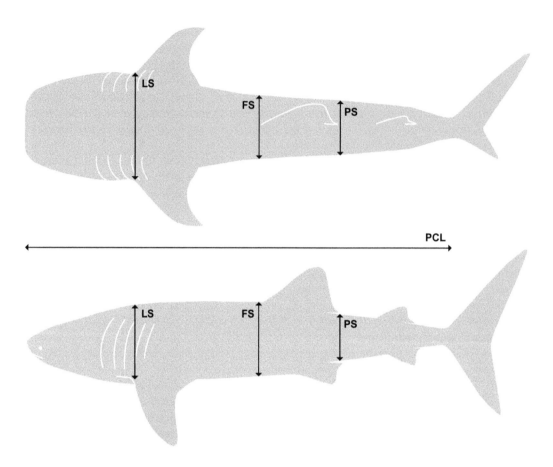

Figure 2.18 A potential modification of the span condition analysis metric for shark body condition (created by Irschick and Hammerschlag 2014) applied to a generic whale shark body outline. LS=lateral span; FS=frontal span; PS=proximal span; PCL=pre-caudal length. Design by Madeleine Pierce.

studies, and this literature is worth exploring for ideas. Poor condition, as assessed through reduced weight, has been associated with higher natural mortality in scalloped hammerhead shark *Sphyrna lewini* pups (Duncan and Holland 2006) and Brazilian sharpnose sharks *Rhizoprionodon lalandii* decreased in condition during an El Niño climate event (Corsso et al. 2018). The potentially wide scope of morphometric analyses makes this an interesting topic for further research.

Recent use of in-water blood sampling techniques for whale sharks (i.e. Matsumoto et al. In revision) may allow for the future study of condition directly, as well as individual whale shark health derived from biomarkers, such as metabolites (Dove et al. 2010, 2012; Wyatt et al. 2019), biochemical analyses of diet using stable isotopes and fatty acids (Chapter 8; Wyatt et al. 2019), or changes in hormone levels relating to the reproductive cycle.

2.8 POPULATION PRODUCTIVITY

Whale sharks are huge, long-lived, and slow-growing. These innate characteristics are part of the reason why the species is so iconic and under natural circumstances help to maintain a stable population in the face of changing conditions. Unfortunately, this "slow" life-history strategy also leaves the whale shark highly susceptible to decline when humans introduce unnatural threats such as fishing, shipping, and pollution (see Chapter 11). The additional stress and mortality from these pressures can outpace the sharks' natural ability to replace themselves through breeding.

We can use the preceding information on whale shark reproduction, age, and growth to esti-mate their biological productivity, known as the species' "maximum intrinsic rate of population increase", or r_{max}. This metric provides the theoretical maximum growth rate of depleted popula-tions in the absence of density-dependent regulation (i.e. unconstrained recovery). In conservation and management terms, this provides us with an indication of the biological limits on the whale shark's ability to recover from their current endangered state (Pierce and Norman 2016).

Two previous papers have created provisional demographic models for whale sharks. There was a detailed species-specific assessment by Bradshaw et al. (2007), and the species was included in a broader comparison among elasmobranchs by García et al. (2008). We can use the new biological data summarized above, using a modified Euler–Lotka equation that does not require age-specific life-history parameters, to update prior estimates for r_{max} for whale sharks.

2.8.1 Estimating Productivity

Here, we follow the approach used by Pardo et al. (2016) to estimate r_{max} for the spinetail devil ray by using an unstructured derivation of the Euler–Lotka model. The model is based on the prin-ciple that a breeding female whale shark, in this case, only has to produce one mature female in her lifetime to replace herself, and thereby ensure a stable population. The input variables required by this model are (a) an estimate for annual reproductive output, (b) age at first reproduction, (c) survival to maturity, and (d) natural mortality.

The model is formulated as shown here:

$$l_{\alpha_{mat}} b = e^{r_{max}} \; \alpha_{mat} - e^{-M} \left(e^{r_{max}} \right)^{\alpha_{mat} - 1}$$

where $l_{\alpha_{mat}}$ is the proportion of pups that survive to maturity, b is the annual reproductive output, α_{mat} is the age at maturity, and M is the instantaneous natural mortality rate. As the name indicates, the maximum intrinsic rate of population increase is the *maximum* theoretical population growth rate in the *absence of density dependence* and is equivalent to the fishing mortality that will drive a species to extinction F_{ext} (Dulvy et al. 2004). Annual reproductive rate b is estimated based on litter size l, the proportion of females in each litter p_f (i.e. the sex ratio), and the breeding interval i: $b = l * p_f / i$. The derivation of these input parameters is detailed below.

2.8.1.1 Annual Reproductive Output

The annual reproductive output of female whale sharks is calculated as half the total litter size (i.e. considering only female pups), divided by the breeding interval. We only have the one well-documented whale shark litter to consider here, the "megamamma" specimen from Taiwan (Joung et al. 1996). In that paper, litter size is cited variably as "about 300", "more than 300", and 307, whereas a follow-up study cites the litter size as 304 (Schmidt et al. 2010). For this model, then, we assume that ~300 pups is normal for whale sharks. Although litter size in some species does cor-relate with the mother's length, and the 10.6 m TL "megamamma" is not, in context, a particularly huge whale shark, no clear relationship has been found between maternal size and litter size in the related nurse shark (Castro 2000), or two unrelated species with potentially large litters, blue sharks *Prionace glauca*, or tiger sharks (Whitney and Crow 2007).

The sex ratio of pups from the "megamamma" was approximately 1:1 (123 F: 114 M, which was not significantly different from equal, in the available pups). We are not aware of data from other shark species that show a significant variation from 1:1 in sex ratio at birth, so it is reasonable to assume that female whale shark pups will comprise half of a given litter (i.e. p_f=0.5). Bradshaw et al. (2007) followed this approach, using an invariant value of 150 female pups in their models. To capture our uncertainty in these variables, here we assume random variation from 125 to 175 (female) pups within each litter, with 150 as the mean value, and uniform distribution.

Next, we can estimate the sharks' breeding interval, or reproductive periodicity. Unfortunately, we have limited field data from whale sharks to help us estimate this parameter (see Section 2.4.1 for a detailed discussion). Bradshaw et al. (2007) modeled two scenarios for breeding intervals, considering both annual and biennial cycles, while noting that an assumption of annual breeding was rather aggressive in light of the uncertainty involved. Gestation is 5–6 months in *G. cirratum* (Castro 2000), and biennial reproduction in that species has been confirmed by both field (Pratt and Carrier 2001) and dissection studies (Castro 2000, 2009). Biennial seems to be a reasonable "default" position for large sharks (Castro 2009), although some (blue sharks, for instance) have an annual cycle, and others (e.g. tiger sharks in Hawaii; Whitney and Crow 2007) appear to be triennial.

As whale sharks are yolk-sac viviparous, with a fixed maternal input, they are unlikely to have an unusually long gestation period relative to other species. Given the limited available data, though, we have included two different scenarios in our modeling. First, we assume a biennial breeding interval, with a range of 2–3 years to capture some probable natural variation. Second, we include a more conservative scenario, with a 4–5 years breeding interval, to factor in the possibility of a longer reproductive cycle or incorporation of resting periods.

A real-world example of natural variability in elasmobranchs comes from the reef manta ray *M. alfredi*, which can give birth annually (Marshall and Bennett 2010), but interbirth intervals of 2–7 years have been noted from field studies of the species (Stewart et al. 2018).

2.8.1.2 Age at First Reproduction

Bradshaw et al. (2007) used different scenarios for this, considering either a ~12 or ~24-year non-reproductive state. Data remain sparse on this particular topic. Field-based growth work from the Maldives estimated male maturity to be ~25 years (Perry et al. 2018), which is cautiously supported by additional male growth data from Western Australia (Meekan et al. 2020) and vertebral band-counts (Ong et al. 2020). "Jinta", a captive male whale shark in Okinawa, reached maturity at over 25 years (Matsumoto et al. 2019). As above, additional field-based growth work by Meekan et al. (2020) from Western Australia suggests that female whale sharks have a slower growth rate, which is common across shark species (i.e. Cortés 2000). Age at first reproduction probably follows the age at first maturity by ~2 years, assuming biennial reproduction as discussed above. We include two scenarios here, acknowledging the limited data available for whale shark growth in general, and particularly for females, which may reach maturity later than males. Thus, we modeled two scenarios, both with normal distributions: first, mean age of first reproduction of 30 years (S.D.=2) and second with a mean age of 40 years (S.D.=2).

2.8.1.3 Survivorship

Survival is one of the most difficult life-history parameters to estimate, and there are two different survival parameters required by this model: the proportion of whale sharks that survive to maturity ($l_{\alpha mat}$) and adult instantaneous natural mortality (M). Direct estimates of these are not available, so we need to parameterize these using a number of different sources.

Survival to maturity is perhaps the hardest variable to estimate for this model. We set a lower bound for this parameter based on data from juvenile survival from Tanzania (CA Rohner unpubl. data), where apparent survival was estimated as 0.71 per year. Accounting for around half of that estimated "mortality" is in fact assumed to be emigration (~0.15 per year); i.e., these sharks are leaving the site, not dying. To acknowledge this, we used an annual survival proportion of 0.85 over a 40-year maturity range, resulting in the lower bound of survival to maturity being 0.002 year^{-1}. As an upper bound, we assumed that the proportion of juveniles surviving to maturity is 0.215, based on a yearly survival of 0.95 over a 30-year period (i.e. 0.95^{30}). We used a uniform distribution to draw values from this range and, to ground our values, we also used the Lorenzen (1996) estimator of natural mortality, which provides weight-based mortality estimates. As we have created a growth curve (above), we can estimate the length for each age up to maturity and then convert length to weight using the published length–weight conversion equation from whale shark fishery data in Taiwan (Hsu et al. 2012). This results in estimates of survival to maturity between 0.082 (α_{mat}=35 years) and 0.062 (α_{mat}=45 years), which are well within our upper and lower bounds.

The instantaneous natural mortality of adult female whale sharks is expected to be very low, given their large size and presumed long lifespan. Furthermore, at their current low densities (due to documented population declines) and hence maximum rate of population increase, mortality should be lower than if their numbers were approaching environmental carrying capacity. The Hoenig (1983) method for estimating M used by Ong et al. (2020) results in an estimate of 0.09 year^{-1}, based on combined-sex data and maximum age of 50. This is equivalent to an adult annual survival probability of 0.91, which we set as the highest bound for natural mortality, with a lower bound of 0.02 year^{-1}, equivalent to annual survival of 0.98. We draw from a uniform distribution between these bounds to account for uncertainty in adult natural mortality.

Based on the rationale outlined above, we set distributions for all input parameters as shown in Table 2.2. To account for uncertainty in estimates, we used Monte Carlo simulation where we iteratively estimated r_{max} using the equation above by drawing parameter values from these distributions 50,000 times, thus obtaining a distribution of r_{max} values for each scenario.

2.8.2 Productivity Results

Median r_{max} estimates for all four scenarios ranged between 0.083 year^{-1} and 0.122 year^{-1}, with their distributions overlapping considerably (Figure 2.19). When compared to other chondrichthyans, whale sharks have among the lowest r_{max} values in the taxon (Figure 2.20). Interestingly, our r_{max} estimates for the whale shark were considerably lower than the estimate from Pardo et al (2016; light triangle in Figure 2.20); this is likely due to the different parameterization of survival to maturity, which is better grounded in recent data in our current analysis and results in a much lower estimate range of survival to maturity, and hence lower r_{max} estimates as well.

2.9 CONCLUSIONS

Whale sharks are well known for hogging many superlatives among sharks. They are the largest species, have the biggest litter (although we only have one example), and are one of the few circumglobal species. Nevertheless, a lot of questions about their broader life history remain unanswered. It surprises many that, as of 2020, we still have only a single pregnant female whale shark from which

Table 2.2 Input Parameters and Distributions (With Upper and Lower Bounds for Uniform Distributions; Mean and Standard Deviation in Normal Distributions) Used to Model Whale Shark Productivity, r_{max}

Parameter	Distribution	Alternate Scenario
Litter size (*l*)	Uniform (250, 350)	
Breeding interval (*i*)	Uniform (2, 3)	Uniform (4, 5)
Age at first reproduction (α_{mat})	Normal (30; 2 S.D.)	Normal (40; 2 S.D.)
Instantaneous natural mortality (*M*)	Uniform (0.02, 0.09)	
Survival to maturity ($l_{\alpha mat}$)	Uniform (0.002, 0.215)	

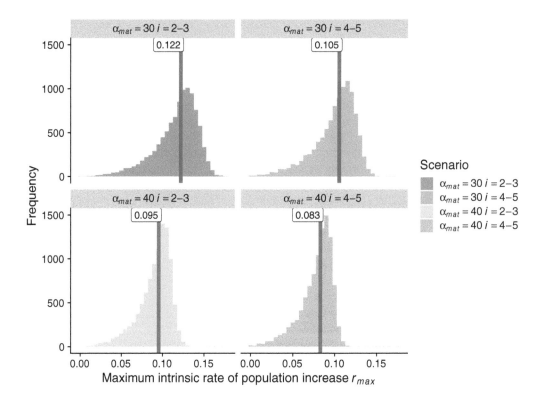

Figure 2.19   Estimated distributions of the maximum intrinsic rate of population increase (r_{max}) for whale sharks for four life-history scenarios with different assumptions of age at maturity (α_{mat}) and breeding interval (i).

to extrapolate fundamental aspects of the species' biology. That said, much has been learned over the past few years. Adaptation of technologies like ultrasound for underwater use provides a way to non-invasively assess maturity (Matsumoto et al. In revision) and potentially the reproductive cycle of female whale sharks. Hopefully, we can examine more "megamammas" using this method. The reproductive ecology of adult whale sharks is likely to have a major influence on the movements of both sexes, particularly if they use specific areas for mating and birthing (see Chapter 7 for more discussion), so new innovation in this research area could be hugely informative.

Whale sharks seem to have evolved a strategy of producing a large number of relatively tiny pups, which doubtless have a high mortality rate in the first few years, but the few that survive this fraught period probably have a good chance of surviving to adulthood. As we have explored in the preceding sections, this particular survivorship parameter is one of the most uncertain. The situation is made even more complex in whale sharks than most other sharks, as many individual whale sharks probably move offshore as they mature, where it is challenging to find them. That makes it difficult for researchers to distinguish mortality from a simple change of address. All scenarios, however, point to whale sharks having a slow life history, with low population productivity; i.e., a depleted population will take a long time to rebound, even under ideal circumstances. This, of course, has significant implications for the species' conservation status (Chapter 12). Even low additive mortality from human pressures could suppress their recovery.

Our lack of knowledge of the adult lives of whale sharks presents a fascinating mystery. It appears likely that the species has determinate growth (Meekan et al. 2020), in that individuals cease to grow once they reach their own "maximum" size. That makes it challenging to estimate the potential longevity of large whale sharks, as their age may not correspond to their length (Harry 2018).

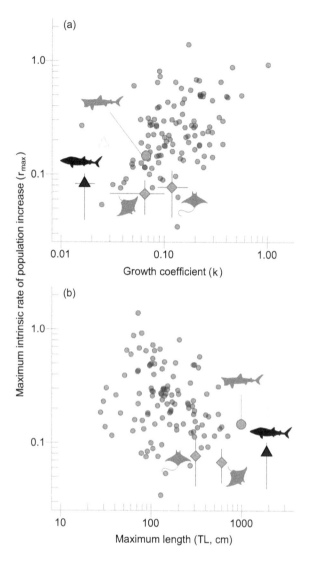

Figure 2.20 Comparison of maximum intrinsic rate of population increase (r_{max}) for 11 elasmobranch species arranged by (a) growth coefficient k and (b) maximum size based on data from Pardo et al. (2016). The whale shark is shown by the black triangle with the r_{max} estimate from Pardo et al. (2016) denoted by the light gray triangle; the spinetail devil ray (left) and "manta rays" (*M. alfredi* and *M. birostris*; right) are represented by the gray diamonds and the basking shark with the gray circle. Sources and attributions for the silhouettes used are included in the acknowledgments.

More samples from large (and small) sharks, if these become available from fisheries or strandings, will also be hugely useful for further validation of these initial results. While we now have a validated age for a 50-year-old 10 m female shark (Ong et al. 2020), we know that at least some individuals grow much larger; their ultimate lifespan remains an open – and intriguing – question. More immediately, the possibility that female whale sharks may grow significantly slower than males is a high priority for further research, as they may be even more susceptible to human-induced decline than our current analyses would suggest.

ACKNOWLEDGMENTS

We thank our respective institutions, collaborators, and funders for supporting the work underpinning this chapter. Figure 2.20 includes several images that are used with attribution and thanks to the creators. The devil ray silhouette is a public domain image from http://phylopic.org/image/914af720-568d-47ef-ada8-ed5a10660266/. The manta ray icon was created by http://www.freepik.comFreepik from http://www.flaticon.comFlaticon and sourced from http://www.flaticon.com/free-icon/manta-ray-shape_47440. The whale shark image was created by Scarlet23 (and vectorized by T. Michael Keesey) and is available at http://phylopic.org/image/ef0b81ab-a39e-4780-a572-e2849a23226b/. The basking shark image was designed by http://alphateck.com/NickBotner and sourced from http://alphateck.com/free-vector-file-20-shark-silhouettes/.

REFERENCES

Aca, E.Q., and J. V. Schmidt. 2011. Revised size limit for viability in the wild: Neonatal and young-of-the-year whale sharks identified in the Philippines. *Asia Life Sciences* 20:361–68.

Akhilesh, K. V., C. P. R. Shanis, W. T. White, et al. 2013. Landings of whale sharks *Rhincodon typus* Smith, 1828 in Indian Waters since protection in 2001 through the Indian Wildlife (Protection) Act, 1972. *Environmental Biology of Fishes* 96:713–22.

Awruch, C. A. 2015. Reproduction Strategies. In *Fish Physiology*, eds. R. E. Shadwick, A. P. Farrell, and C. J. Brauner, 255–310. Cambridge: Academic Press.

Awruch, C. A., N. W. Pankhurst, S. D. Frusher, and J. D. Stevens. 2008. Endocrine and morphological correlates of reproduction in the draughtboard shark *Cephaloscyllium laticeps* (Elasmobranchii: Scyliorhinidae). *Journal of Experimental Zoology. Part A, Ecological Genetics and Physiology* 309:184–97.

Bansemer, C. S., and M. B. Bennett. 2009. Reproductive periodicity, localised movements and behavioural segregation of pregnant *Carcharias taurus* at Wolf Rock, Southeast Queensland, Australia. *Marine Ecology Progress Series* 374:215–27.

Baughman, J. L. 1955. The oviparity of the whale shark, *Rhineodon typus*, with records of this and other fishes in Texas waters. *Copeia* 1955:54–55.

Beckley, L. E., G. Cliff, M. J. Smale, and L. J. V. Compagno. 1997. Recent strandings and sightings of whale sharks in South Africa. *Environmental Biology of Fishes* 50:343–48.

Bernal, M. A., N. L. Sinai, C. Rocha, M. R. Gaither, F. Dunker, and L. A. Rocha. 2015. Long-term sperm storage in the brownbanded bamboo shark *Chiloscyllium punctatum*. *Journal of Fish Biology* 86:1171–76.

Blackburn, T. M., K. J. Gaston, and N. Loder. 1999. Geographic gradients in body size: A clarification of Bergmann's rule. *Diversity and Distributions* 5:165–74.

Bradshaw, C. J. A., H. F. Mollet, and M. G. Meekan. 2007. Inferring population trends for the world's largest fish from mark-recapture estimates of survival. *The Journal of Animal Ecology* 76:480–89.

Capapé, C., Y. Diatta, A. A. Seck, O. Guélorget, J. B. Souissi, and J. Zaouali. 2005. Reproduction of the sawback angelshark *Squatina aculeata* (Chondrichthyes: Squatinidae) off Senegal and Tunisia. *Cybium* 29:147–57.

Carrillo-Briceño, J. D., J. A. Villafaña, C. De Gracia, F. F. Flores-Alcívar, R. Kindlimann, and J. Abella. 2020. Diversity and paleoenvironmental implications of an elasmobranch assemblage from the Oligocene–Miocene boundary of Ecuador. *PeerJ* 8:e9051.

Castrillon, J., and S. Bengtson Nash. 2020. Evaluating cetacean body condition; a review of traditional approaches and new developments. *Ecology and Evolution* 10:6144–62.

Castro, J. I. 2000. The biology of the nurse shark, *Ginglymostoma cirratum*, off the Florida east coast and the Bahama Islands. *Environmental Biology of Fishes* 58:1–22.

Castro, J. I. 2009. Observations on the reproductive cycles of some viviparous North American sharks. *Aqua, International Journal of Ichthyology* 15:205–22.

Castro, J. I. 2013. A primer on shark reproduction for aquarists. In *Reproduction of Marine Life, Birth of New Life! Investigating the Mysteries of Reproduction*, ed. E. K. Sato, 52–69. Motobu: Okinawa Churashima Foundation.

Chang, W.-B., M.-Y. Leu, and L.-S. Fang. 1997. Embryos of the whale shark, *Rhincodon typus*: Early growth and size distribution. *Copeia* 1997:444–46.

Chapman, D. D., B. Firchau, and M. S. Shivji. 2008. Parthenogenesis in a large-bodied requiem shark, the blacktip *Carcharhinus limbatus. Journal of Fish Biology* 73:1473–77.

Chapman, D. D., M. S. Shivji, E. Louis, J. Sommer, H. Fletcher, and P. A. Prodöhl. 2007. Virgin birth in a hammerhead shark. *Biology Letters* 3:425–27.

Chidlow, J. A., C. A. Simpfendorfer, and G. R. Russ. 2007. Variable growth band deposition leads to age and growth uncertainty in the western wobbegong shark, *Orectolobus hutchinsi. Marine and Freshwater Research* 58:856–65.

Cicimurri, D. J., and J. L. Knight. 2009. Late Oligocene sharks and rays from the Chandler Bridge Formation, Dorchester County, South Carolina, USA. *Acta Palaeontologica Polonica* 54:627–47.

Conrath, C. L., and J. A. Musick. 2012. Reproductive biology of elasmobranchs. In *Biology of Sharks and Their Relatives, Second Edition*, eds. J. C. Carrier, J. A. Musick, and M. R. Heithaus, 291–311. Boca Raton, FL: CRC Press.

Corsso, J. T., O. B. F. Gadig, R. R. P. Barreto, and F. S. Motta. 2018. Condition analysis of the Brazilian sharpnose shark *Rhizoprionodon lalandii*: Evidence of maternal investment for initial post-natal life. *Journal of Fish Biology* 93:1038–45.

Cortés, E. 2000. Life history patterns and correlations in sharks. *Reviews in Fisheries Science* 8:299–344.

Davenport, S., and J. D. Stevens. 1988. Age and growth of two commercially important sharks (*Carcharhinus tilstoni* and *C. sorrah*) from northern Australia. *Marine and Freshwater Research* 39:417–33.

Deakos, M. H. 2012. The reproductive ecology of resident manta rays (*Manta alfredi*) off Maui, Hawaii, with an emphasis on body size. *Environmental Biology of Fishes* 94:443–56.

Delacy, C. R., A. Olsen, L. A. Howey, D. D. Chapman, E. J. Brooks, and M. E. Bond. 2017. Affordable and accurate stereo-video system for measuring dimensions underwater: A case study using oceanic whitetip sharks *Carcharhinus longimanus. Marine Ecology Progress Series* 574:75–84.

Del Raye, G., S. J. Jorgensen, K. Krumhansl, J. M. Ezcurra, and B. A. Block. 2013. Travelling light: White sharks (*Carcharodon carcharias*) rely on body lipid stores to power ocean-basin scale migration. *Proceedings of the Royal Society B: Biological Sciences* 280:20130836.

Dove, A. D. M., J. Arnold, and T. M. Clauss. 2010. Blood cells and serum chemistry in the world's largest fish: The whale shark *Rhincodon typus. Aquatic Biology* 9:177–83.

Dove, A. D. M., J. Leisen, M. Zhou, J. J. Byrne, K. Lim-Hing, H. D. Webb, L. Gelbaum, M. R. Viant, J. Kubanek, and F. M. Fernández. 2012. Biomarkers of whale shark health: A metabolomic approach. *PloS One* 7:e49379.

Driggers, W. B., and E. R. Hoffmayer. 2009. Variability in the reproductive cycle of finetooth sharks, *Carcharhinus isodon*, in the northern Gulf of Mexico. *Copeia* 2009:390–3.

Dudgeon, C. L., L. Coulton, R. Bone, J. R. Ovenden, and S. Thomas. 2017. Switch from sexual to parthenogenetic reproduction in a zebra shark. *Scientific Reports* 7:40537.

Dulvy, N. K., J. R. Ellis, N. B. Goodwin, A. Grant, J. D. Reynolds, and S. Jennings. 2004. Methods of assessing extinction risk in marine fishes. *Fish and Fisheries* 5:255–76.

Duncan, K. M., and K. N. Holland. 2006. Habitat use, growth rates and dispersal patterns of juvenile scalloped hammerhead sharks *Sphyrna lewini* in a nursery habitat. *Marine Ecology Progress Series* 312:211–21.

Ebert, D. A. 1996. Biology of the sevengill shark *Notorynchus cepedianus* (Peron, 1807) in the temperate coastal waters of southern Africa. *South African Journal of Marine Science* 17:93–103.

Ebert, D. A., S. L. Fowler, and L. J. V. Compagno. 2013. *Sharks of the World: A Fully Illustrated Guide.* Plymouth: Wild Nature Press.

Edwards, J. E., E. Hiltz, F. Broell, et al. 2019. Advancing research for the management of long-lived species: A case study on the Greenland shark. *Frontiers in Marine Science* 6:87.

Froese, R., and C. Binohlan. 2000. Empirical relationships to estimate asymptotic length, length at first maturity and length at maximum yield per recruit in fishes, with a simple method to evaluate length frequency data. *Journal of Fish Biology* 56:758–73.

Gallagher, A. J., D. N. Wagner, D. J. Irschick, and N. Hammerschlag. 2014. Body condition predicts energy stores in apex predatory sharks. *Conservation Physiology* 2:cou022.

García, V. B., L. O. Lucifora, and R. A. Myers. 2008. The importance of habitat and life history to extinction risk in sharks, skates, rays and chimaeras. *Proceedings of the Royal Society B: Biological Sciences* 275:83–89.

Garnier, D. H., P. Sourdaine, and B. Jégou. 1999. Seasonal variations in sex steroids and male sexual charac-teristics in *Scyliorhinus canicula*. *General and Comparative Endocrinology* 116:281–90.

Gudger, E. W. 1933. The whale shark in the waters around Ceylon. *Nature* 131:165.

Harry, A. V. 2018. Evidence for systemic age underestimation in shark and ray ageing studies. *Fish and Fisheries* 19:185–200.

Henningsen, A. D., F. L. Murru, L. E. L. Rasmussen, B. R. Whitaker, and G. C. Violetta. 2008. Serum levels of reproductive steroid hormones in captive sand tiger sharks, *Carcharias taurus* (Rafinesque), and com-ments on their relation to sexual conflicts. *Fish Physiology and Biochemistry* 34:437.

Heupel, M. R., J. M. Whittier, and M. B. Bennett. 1999. Plasma steroid hormone profiles and reproductive biol-ogy of the epaulette shark, *Hemiscyllium ocellatum*. *The Journal of Experimental Zoology* 284:586–94.

Hoenig, J. M. 1983. Empirical use of longevity data to estimate mortality rates. *Fishery Bulletin* 82:898–903.

Hsu, H. H., S. J. Joung, and K. M. Liu. 2012. Fisheries, management and conservation of the whale shark *Rhincodon typus* in Taiwan. *Journal of Fish Biology* 80:1595–1607.

Hsu, H. H., C. Y. Lin, and S. J. Joung. 2014a. The first record, tagging and release of a neonatal whale shark *Rhincodon typus* in Taiwan. *Journal of Fish Biology* 85:1753–56.

Hsu, H. H., S. J. Joung, R. E. Hueter, and K. M. Liu. 2014b. Age and growth of the whale shark (*Rhincodon typus*) in the north-western Pacific. *Marine and Freshwater Research* 65:1145–54.

Hussey, N. E., D. T. Cocks, S. F. J. Dudley, I. D. McCarthy, and S. P. Wintner. 2009. The condition conundrum: Application of multiple condition indices to the dusky shark *Carcharhinus obscurus*. *Marine Ecology Progress Series* 380:199–212.

Huveneers, C., J. Stead, M. B. Bennett, K. A. Lee, and R. G. Harcourt. 2013. Age and growth determination of three sympatric wobbegong sharks: How reliable is growth band periodicity in Orectolobidae? *Fisheries Research* 147:413–25.

Huveneers, C., T. I. Walker, N. M. Otway, and R. G. Harcourt. 2007. Reproductive synchrony of three sym-patric species of wobbegong shark (genus *Orectolobus*) in New South Wales, Australia: Reproductive parameter estimates necessary for population modelling. *Marine and Freshwater Research* 58:765–77.

Irschick, D. J., and N. Hammerschlag. 2014. A new metric for measuring condition in large predatory sharks. *Journal of Fish Biology* 85:917–26.

Jacoby, D. M. P., D. P. Croft, and D. W. Sims. 2012. Social behaviour in sharks and rays: Analysis, patterns and implications for conservation. *Fish and Fisheries* 13:399–417.

Jeffreys, G. L., D. Rowat, H. Marshall, and K. Brooks. 2013. The development of robust morphometric indices from accurate and precise measurements of free-swimming whale sharks using laser photogrammetry. *Journal of the Marine Biological Association of the United Kingdom* 93:309–20.

Jones, B. C., and G. H. Geen. 1977. Reproduction and embryonic development of spiny dogfish (*Squalus acan-thias*) in the Strait of Georgia, British Columbia. *Journal of the Fisheries Research Board of Canada* 34:1286–92.

Joung, S.-J., C.-T. Chen, E. Clark, S. Uchida, and W. Y. P. Huang. 1996. The whale shark, *Rhincodon typus*, is a livebearer: 300 embryos found in one 'megamamma' supreme. *Environmental Biology of Fishes* 46:219–23.

Kacev, D., T. J. Sippel, M. J. Kinney, S. A. Pardo, and C. G. Mull. 2017. An introduction to modelling abun-dance and life history parameters in shark populations. *Advances in Marine Biology* 78:45–87.

Krishna-Pillai, S. 1998. On a whale shark *Rhinodon typus* found accompanied by its young. *Marine Fisheries Information Service Technical and Extension Series* 152:15.

Lorenzen, K. 1996. The relationship between body weight and natural mortality in juvenile and adult fish: A comparison of natural ecosystems and aquaculture. *Journal of Fish Biology* 49:627–42.

Macena, B. C. L., and F. H. V. Hazin. 2016. Whale shark (*Rhincodon typus*) seasonal occurrence, abundance and demographic structure in the mid-equatorial Atlantic Ocean. *PloS One* 11:e0164440.

Manojkumar, P. P. 2003. An account on the smallest whale shark, *Rhincodon typus* (Smith 1828) landed at Calicut. *Marine Fisheries Information Service Technical and Extension Series* 176:9–10.

Marshall, A. D., and M. B. Bennett. 2010. Reproductive ecology of the reef manta ray *Manta alfredi* in south-ern Mozambique. *Journal of Fish Biology* 77:169–90.

Matsumoto, R., K. Murakumo, R. Nozu, et al. In revision. Underwater ultrasonography and blood sampling enable the first observations of reproductive biology in free-swimming whale sharks. *Endangered Species Research*.

Matsumoto, R., Y. Matsumoto, K. Ueda, M. Suzuki, K. Asahina, and K. Sato. 2019. Sexual maturation in a male whale shark (*Rhincodon typus*) based on observations made over 20 years of captivity. *Fishery Bulletin* 117:78–86.

McClain, C. R., M. A. Balk, M. C. Benfield, et al. 2015. Sizing ocean giants: Patterns of intraspecific size variation in marine megafauna. *PeerJ* 2015:e715.

McCoy, E., R. Burce, D. David, et al. 2018. Long-term photo-identification reveals the population dynamics and strong site fidelity of adult whale sharks to the coastal waters of Donsol, Philippines. *Frontiers in Marine Science* 5:271.

Meekan, M. G., C. J. A. Bradshaw, M. Press, et al. 2006. Population size and structure of whale sharks (*Rhincodon typus*) at Ningaloo Reef, Western Australia. *Marine Ecology Progress Series* 319:275–85.

Meekan, M. G., B. M. Taylor, E. Lester, et al. 2020. Asymptotic growth of whale sharks suggests sex-specific life-history strategies. *Frontiers in Marine Science* 7: 774.

Miranda, J. A., N. Yates, A. Agustines, et al. 2020. Donsol: An important reproductive habitat for the world's largest fish *Rhincodon typus*?" *Journal of Fish Biology*. doi:10.1111/jfb.14610.

Murakumo, K., R. Matsumoto, T. Tomita, Y. Matsumoto, and K. Ueda. 2020. The power of ultrasound: Observation of nearly the entire gestation and embryonic developmental process of captive reef manta rays (*Mobula alfredi*). *Fishery Bulletin* 118: 1–8.

Musick, J. A., J. K. Ellis, and W. C. Hamlett. 2005. Reproductive evolution of chondrichthyans. In *Reproductive Biology and Phylogeny of Chondrichthyes: Sharks, Batoids and Chimaeras*, ed. W. C. Hamlett, 45–71. Enfield: Science Publishers.

Nakano, H., and J. D. Stevens. 2008. The biology and ecology of the blue shark, *Prionace glauca*. In *Sharks of the Open Ocean: Biology, Fisheries and Cconservation*, eds. M. D. Camhi, E. K. Pikitch, and E. A. Babcock, 140–51. Oxford: Blackwell Scientific Publications.

Natanson, L. J., A. H. Andrews, M. S. Passerotti, and S. P. Wintner. 2018. History and mystery of age and growth studies in elasmobranchs. In *Shark Research: Emerging Technologies and Applications for the Field and Laboratory*, eds. J. C. Carrier, M. R. Heithaus, C. A. Simpfendorfer, 177–99. Boca Raton, FL: CRC Press.

Natanson, L. J., S. P. Wintner, F. Johansson, et al. 2008. Ontogenetic vertebral growth patterns in the basking shark *Cetorhinus maximus*. *Marine Ecology Progress Series* 361:267–78.

Naylor, G. J. P., J. N. Caira, K. Jensen, K. A. M. Rosana, N. Straube, and C. Lakner. 2012. Elasmobranch phylogeny: A mitochondrial estimate based on 595 species. In *Biology of Sharks and Their Relatives*, Second edition, eds. J. C. Carrier, J. A. Musick, and M. R. Heithaus, 31–56. Boca Raton, FL: CRC Press.

Neer, J. A., and B. A. Thompson. 2005. Life history of the cownose ray, *Rhinoptera bonasus*, in the northern Gulf of Mexico, with comments on geographic variability in life history traits. *Environmental Biology of Fishes* 73:321–31.

Nielsen, J., R. B. Hedeholm, J. Heinemeier, et al. 2016. Eye lens radiocarbon reveals centuries of longevity in the Greenland shark (*Somniosus microcephalus*). *Science* 353:702–4.

Nishida, K. 2001. Whale shark—the world's largest fish. In *Fishes of the Kuroshio Current, Japan*, eds. T. Kakabo, Y. Machida, K. Yamaoka, and K. Nishida, 20–35. Osaka: Osaka Aquarium Kaiyukan.

Norman, B. M., J. A. Holmberg, Z. Arzoumanian, et al. 2017. Undersea constellations: The global biology of an endangered marine megavertebrate further informed through citizen science. *Bioscience* 67:1029–43.

Norman, B. M., and D. L. Morgan. 2016. The return of 'Stumpy' the whale shark: Two decades and counting. *Frontiers in Ecology and the Environment* 14:449–50.

Norman, B. M., and J. D. Stevens. 2007. Size and maturity status of the whale shark (*Rhincodon typus*) at Ningaloo Reef in Western Australia. *Fisheries Research* 84:81–86.

Nozu, R., K. Murakumo, R. Matsumoto, M. Nakamura, K. Ueda, and K. Sato. 2015. Gonadal morphology, histology, and endocrinological characteristics of immature female whale sharks, *Rhincodon typus*. *Zoological Science* 32:455–58.

Ong, J. J. L., M. G. Meekan, H. H. Hsu, L. P. Fanning, and S. E. Campana. 2020. Annual bands in vertebrae validated by bomb radiocarbon assays provide estimates of age and growth of whale sharks. *Frontiers in Marine Science* 7: 1–7.

Pai, M. V., G. Nandakumar, and K. Y. Telang. 1983. On a whale shark *Rhineodon typus* Smith landed at Karwar, Karnataka. *Indian Journal of Fisheries* 30:157–60.

Pardo, S. A., H. K. Kindsvater, E. Cuevas-Zimbrón, O. Sosa-Nishizaki, J. C. Pérez-Jiménez, and N. K. Dulvy. 2016. Growth, productivity, and relative extinction risk of a data-sparse devil ray. *Scientific Reports* 6:1–10.

Parsons, G. R., E. R. Hoffmayer, J. M. Hendon, and W. V. Bet-Sayad. 2008. A review of shark reproductive ecology: Life history and evolutionary implications. In *Fish Reproduction*, eds. M. A. Rocha, A. Arukwe, B. G. Kapoor, 435–69. Enfield: Science Publishers.

Penfold, L. M., and J. T. Wyffels. 2019. Reproductive science in sharks and rays. In *Reproductive Sciences in Animal Conservation*, eds. P. Comizzoli, J. Brown, and W. Holt, 465–88. Cham: Springer.

Peres, M. B. 1991. Sexual development, reproductive cycle and fecundity of the school shark *Galeorhinus galeus* of south Brazil. *Fishery Bulletin* 89:566–667.

Perry, C. T., E. Clingham, D. H. Webb, et al. 2020. St. Helena: An important reproductive habitat for whale sharks (*Rhincodon typus*) in the south central Atlantic. *Frontiers in Marine Science* 7:899.

Perry, C. T., J. Figueiredo, J. J. Vaudo, J. Hancock, R. Rees, and M. Shivji. 2018. Comparing length-measurement methods and estimating growth parameters of free-swimming whale sharks (*Rhincodon typus*) near the South Ari Atoll, Maldives. *Marine and Freshwater Research* 69:1487–95.

Pierce, S. J., and B. Norman. 2016. *Rhincodon typus*. The IUCN red list of threatened species: e.T19488A2365291.

Porcu, C., M. F. Marongiu, M. C. Follesa, et al. 2014. Reproductive aspects of the velvet belly lantern shark *Etmopterus spinax* (Chondrichthyes: Etmopteridae), from the central western Mediterranean Sea. Notes on gametogenesis and oviducal gland microstructure. *Mediterranean Marine Science* 15:313–26.

Pratt, H. L., and J. C. Carrier. 2001. A review of elasmobranch reproductive behavior with a case study on the nurse shark, *Ginglymostoma cirratum*. *Environmental Biology of Fishes* 60:157–88.

Pratt, H. L., T. C. Pratt, D. Morley, et al. 2018. Partial migration of the nurse shark, *Ginglymostoma cirratum* (Bonnaterre), from the Dry Tortugas Islands. *Environmental Biology of Fishes* 101:515–30.

Ramírez-Macías, D., M. Meekan, R. De La Parra-Venegas, F. Remolina-Suárez, M. Trigo-Mendoza, and R. Vázquez-Juárez. 2012. Patterns in composition, abundance and scarring of whale sharks *Rhincodon typus* near Holbox Island, Mexico. *Journal of Fish Biology* 80:1401–16.

Rasmussen, L. E. L., and S. H. Gruber. 1993. Serum concentrations of reproductively-related circulating steroid hormones in the free-ranging lemon shark, *Negaprion brevirostris*. In *The Reproduction and Development of Sharks, Skates, Rays and Ratfishes*, eds. L. S. Demski and J. P. Wourms, 167–74. Dordrecht: Springer Netherlands.

Rasmussen, L. E. L., and F. L. Murru. 1992. Long-term studies of serum concentrations of reproductively related steroid hormones in individual captive carcharhinids. *Marine and Freshwater Research* 43:273–81.

Rêgo, M. G., M. L. G. Araujo, M. E. G. Barros, et al. 2019. Morphological description of ovary and uterus of the nurse shark (*Ginglymostoma cirratum*) caught off at the Fortaleza Coast, northeast Brazil. *Brazilian Journal of Veterinary Research* 39:997–1004.

Robinson, D. P., M. Y. Jaidah, S. Bach, et al. 2016. Population structure, abundance and movement of whale sharks in the Arabian Gulf and the Gulf of Oman. *PloS One* 11:e0158593.

Robinson, D. P., W. Baverstock, A. Al-Jaru, K. Hyland, and K. A. Khazanehdari. 2011. Annually recurring parthenogenesis in a zebra shark *Stegostoma fasciatum*. *Journal of Fish Biology* 79:1376–82.

Rohner, C. A., A. J. Richardson, A. D. Marshall, S. J. Weeks, and S. J. Pierce. 2011. How large Is the world's largest fish? Measuring whale sharks *Rhincodon typus* with laser photogrammetry. *Journal of Fish Biology* 78:378–85.

Rohner, C. A., A. J. Richardson, C. E. M. Prebble, et al. 2015. Laser photogrammetry improves size and demographic estimates for whale sharks. *PeerJ* 3:e886.

Rowat, D., and K. S. Brooks. 2012. A review of the biology, fisheries and conservation of the whale shark *Rhincodon typus*. *Journal of Fish Biology* 80:1019–56.

Schmidt, J. V., C. C. Chen, S. I. Sheikh, M. G. Meekan, B. M. Norman, and S. J. Joung. 2010. Paternity analysis in a litter of whale shark embryos. *Endangered Species Research* 12:117–24.

Sequeira, A. M. M., M. Thums, K. Brooks, and M. G. Meekan. 2016. Error and bias in size estimates of whale sharks: Implications for understanding demography. *Royal Society Open Science* 3:150668.

Stevens, G. M. W., J. P. Hawkins, and C. M. Roberts. 2018. Courtship and mating behaviour of manta rays *Mobula alfredi* and *M. birostris* in the Maldives. *Journal of Fish Biology* 93:344–59.

Stewart, J. D., F. R. A. Jaine, A. J. Armstrong, et al. 2018. Research priorities to support effective manta and devil ray conservation. *Frontiers in Marine Science* 5:314.

Tanaka, S. 1995. Comparative morphology of the sperm in chondrichthyan fishes. In *Advances in Spermatozoal Phylogeny and Taxonomy*, eds. B. G. M. Jamieson, J. Ausio, J.-L. Justine, 313–20. Paris: Memoires du Museum National d'Histoire Naturelle.

Tanaka, S., Y. Shiobara, S. Hioki, et al. 1990. The reproductive biology of the frilled shark, *Chlamydoselachus anguineus*, from Suruga Bay, Japan. *Japanese Journal of Ichthyology* 37:273–91.

Taylor, G. 2019. *Whale Sharks: The Giants of Ningaloo Reef.* 2nd Revised Edition. Perth: Scott Print.

Taylor, S. M., A. V. Harry, and M. B. Bennett. 2016. Living on the edge: Latitudinal variations in the reproductive biology of two coastal species of sharks. *Journal of Fish Biology* 89:2399–418.

Teshima, K., Kamei, Y., Toda, M., Uchida, S. 1995. Reproductive mode of the tawny nurse shark taken from the Yaeyama Islands, Okinawa, Japan with comments on individuals lacking the second dorsal fin. *Bulletin of Seikai National Fisheries Research Institute* 73:1–12.

Tomita, T., K. Murakumo, K. Ueda, H. Ashida, and R. Furuyama. 2018. Locomotion Is not a privilege after birth: Ultrasound images of viviparous shark embryos swimming from one uterus to the other. *Ethology* 125:122–6.

Tricas, T. C., K. P. Maruska, and L. E. Rasmussen. 2000. Annual cycles of steroid hormone production, gonad development, and reproductive behavior in the Atlantic stingray. *General and Comparative Endocrinology* 118:209–25.

Walker, T. I. 2005. Reproduction in fisheries science. In *Reproductive Biology and Phylogeny of Chondrichthyes: Sharks, Batoids and Chimaeras*, ed. W. C. Hamlett, 81–127. Enfield: Science Publishers.

Walker, T. I. 2020. Reproduction of chondrichthyans. In *Reproduction in Aquatic Animals*, eds. M. Yoshida, J. F. Asturiano, 193–223. Singapore: Springer.

Watson, G., and M. J. Smale. 1998. Reproductive biology of shortnose spiny dogfish, *Squalus megalops*, from the Agulhas Bank, South Africa. *Marine and Freshwater Research* 49:695–703.

Watson, L., and M. Janse. 2017. Reproduction and husbandry of zebra sharks, *Stegostoma fasciatum*, in Aquaria. In *The Elasmobranch Husbandry Manual II: Recent Advances in the Care of Sharks, Rays and Their Relatives*, eds. M. Smith, D. Warmolts, D. Thoney, R. Hueter, M. Murray, and J. Excurra, 421–32. Columbus: Ohio Biological Survey.

Whitney, N. M., and G. L. Crow. 2007. Reproductive biology of the tiger shark (*Galeocerdo cuvier*) in Hawaii. *Marine Biology* 151:63–70.

Wintner, S. P. 2000. Preliminary study of vertebral growth rings in the whale shark, *Rhincodon typus*, from the east coast of South Africa. *Environmental Biology of Fishes* 59:441–51.

Wolfson, F. H. 1983. Records of seven juveniles of the whale shark, *Rhiniodon*. *Journal of Fish Biology* 22:647–55.

Wyatt, A. S. J., R. Matsumoto, Y. Chikaraishi, et al. 2019. Enhancing insights into foraging specialization in the world's largest fish using a multi-tissue, multi-isotope approach. *Ecological Monographs* 89:e01339.

CHAPTER 3

Whale Shark Sensory Biology and Neuroanatomy

Kara E. Yopak and Emily E. Peele
University of North Carolina Wilmington

CONTENTS

3.1 INTRODUCTION

Although many aspects of the biology of whale sharks are poorly known, even less information is available on their sensory biology and neuroanatomy. The whale shark possesses the same battery of highly developed sensory systems shared across all cartilaginous fishes (see Collin et al. 2016 for review), including hearing, chemoreception, vision, mechanoreception, and electroreception, yet the relative importance of these different systems, and how these sensory cues are processed in the brain, remains an underexplored area of research.

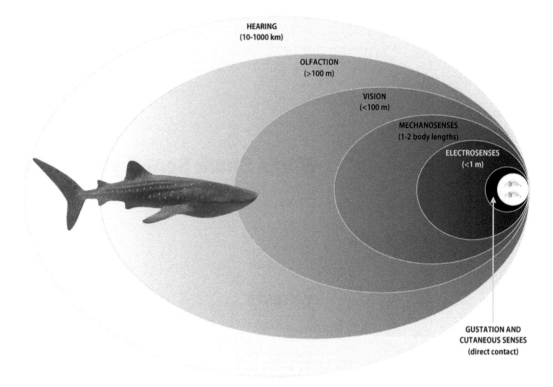

Figure 3.1 The detection distance over which each sensory modality is primarily used. Auditory cues travel the greatest distance in water and can be detected by sharks from ten to hundreds of kilometers away. Olfactory cues can be detected by sharks from over 100 m away. The ability to detect visual cues is used at distances up to 100 m, and the mechanosensory lateral line is only used from 1 to 2 body lengths away from the shark. The electrosensory system is used for close-range detection of electrical stimuli up to 30 cm, where gustation is only used upon contact with the source stimulus.

3.2 SENSORY SYSTEMS

3.2.1 Introduction

Sharks and their relatives possess a suite of highly specialized senses to aid in their ability to avoid predators, find prey, and navigate through the aquatic environment. As each sense is effective over different distances (Figure 3.1), sharks are capable of using multiple sensory systems in unison and switching between them as they approach a stimulus (Gardiner et al. 2012). Anatomical, physiological, and molecular differences suggest the relative importance of each sense may vary between species (Collin et al. 2016), but there is still much to learn about the sensory capabilities of sharks and rays. Even less is known about the sensory system of whale sharks in particular, because they are difficult to study in both field and captive environments. Predictions for sensory capabilities can be made for the whale shark, based on characteristics of other related species in the order Orectolobiformes and other planktivorous sharks, but further work is necessary to understand the sensitivity and acuity of each sensory system in the whale shark to better describe their influence on ecology and behavior.

3.2.2 Hearing

Sound in the aquatic environment is an important sensory cue for animals living in the ocean because it can travel distances from tens to hundreds of kilometers (Urick 1979). As sound travels such great distances in water, it is often one of the first stimuli used by animals to locate sources of prey. Sound is also used by aquatic animals to avoid potential predators, attract, or find mates, and may aid in long-distance navigation (Collin 2012).

The inner ears of sharks are similar to other vertebrates and are composed of three fluid-filled semicircular canals used to sense head movement and maintain equilibrium using otoconia, which are tiny crystal structures that help the animal sense its orientation in three-dimensional space (Fay and Popper 1980). The canals originate from the three otoconial organs (called the saccule, lagena, and utricle), each of which contains a sensory epithelium with hair cells and otoconial mass that, when it moves, triggers the hair cells to start a nerve impulse that allows sharks to perceive sound (Evangelista et al. 2010). Sharks also possess a structure in the inner ear, called the macula neglecta, which also contains sensory hair cells and aids in sound detection. Differences in the morphology of this structure across species may be related to foraging behavior (Corwin 1978).

Of the two components of sound, particle motion, and pressure, fishes are only capable of detecting the former (Fay 1984). This sound component is described by acceleration, velocity, and displacement and helps sharks display a higher sensitivity to hearing lower frequency sounds from 40 to 1,000 Hz (Casper and Mann 2009). This frequency of sound helps sharks detect noise generated by wounded prey, water movement, and human-produced (i.e. anthropogenic) noises, such as boat motors. The sensitivity range of whale sharks is unknown, although a study on the nurse shark *Ginglymostoma cirratum* indicated that their hearing was most sensitive between 300 and 600 Hz (Casper and Mann 2006). Although no research has been done on the hearing capabilities of the whale shark, previous research on their residency patterns near the island of Utila, in Honduras, indicated they were frequently sighted near tuna feeding locations (Chapter 4) where fish jump and splash, presumably making a significant amount of noise. As whale sharks consume tuna eggs and small bait fishes associated with tuna, they may have been able to audibly detect and navigate towards these locations and improve their chances of finding spawning animals (Fox et al. 2013).

Previous research on the variation of the inner ear in elasmobranchs has indicated that morphology varies across species, correlated with both feeding strategy and phylogeny, and therefore reflects the different perceptual requirements of each species. The inner ear of the spotted wobbegong shark *Orectolobus maculatus*, a relative of the whale shark, contains a large sacculus and its morphology is similar to other elasmobranch species that exhibit a comparable "ambush" hunting strategy, in which they sit and wait for prey to swim over them as they lie on the bottom (Evangelista et al. 2010). The whale shark possesses the largest inner ear in the animal kingdom, the size of which may allow for higher sensitivity to sounds (Muller 1999, Myrberg 2001). However, further research on the hearing sensitivity and morphology of the inner ear of whale sharks is needed to better understand their hearing capabilities.

3.2.3 Chemoreception

The chemosensory system of sharks is comprised of the olfactory (smell) and gustatory (taste) senses. The sense of smell is highly developed in elasmobranchs and is used to avoid predators, find prey, communicate with conspecifics, and locate potential mates, and for pelagic navigation (Collin 2012, Nosal et al. 2016). The ability for odorants to travel in water allows sharks to use this sense from long-range distances of over 100 m (Figure 3.1), but is dependent on the concentration of the odor, the movement of water, and the swimming speed of the shark itself (Collin et al. 2016). Taste is used to determine the palatability of prey and is sensed upon contact with a potential food source.

3.2.3.1 Olfaction (smell)

The ability for sharks to orient and navigate towards a smell is called chemotaxis. Directed swimming begins when an odor carried by water reaches the nasal openings or nares, which are

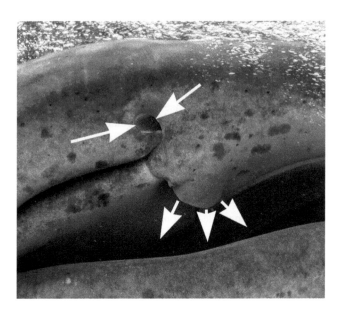

Figure 3.2   The nares, showing water flow pattern. The rosette of olfactory lamellae (rather like gills) are visible. (Source: Alistair D. M. Dove/Georgia Aquarium.)

analogous to nostrils. Sharks in the order Orectolobiformes, including the whale shark, also possess small extensions of skin off of the nares, called barbels. Although the exact function of these structures in sharks is still unknown, barbels in both the nurse shark and the Japanese wobbegong (*O. japonicus*) displayed mechanoreceptive functions (Nevatte et al. 2017). Once water flows into the nares, it passes over the olfactory epithelium and then out below the barbels and into the mouth (Figure 3.2). The olfactory epithelium is arranged into primary and secondary folds, or lamellae, collectively called the olfactory rosette. These increase the surface area and therefore detection capability for an odor (Collin 2012). There is a considerable amount of variation in the structure of the olfactory rosette in elasmobranchs. Previous research has indicated that the long-range odor detection requirements of pelagic species have led to an enlargement of these structures throughout evolution (Schluessel et al. 2008). Although not assessed in detail, the olfactory capsules that house the olfactory epithelium and bulbs have been qualitatively described as being moderately large and spherical in the whale shark (Dennison 1937). Therefore, it is reasonable to assume that the whale shark may have correspondingly large olfactory structures, which may indicate an enhanced sensitivity or olfactory acuity. Once an odor contacts the olfactory epithelium, it can be detected by olfactory receptor neurons which send information to the olfactory bulbs (OBs) in the brain (Yopak et al. 2015).

Whale sharks feed on a range of plankton species and small pelagic animals and can be found in aggregations in the presence of ample prey abundance (Martin 2007; see Chapter 8). One hypothesis for how whale sharks locate these areas of high productivity is the detection of dimethyl sulfide (DMS), which is produced by phytoplankton when preyed upon by zooplankton. DMS is used by seabirds (Nevitt et al. 1995) and possibly basking sharks (Sims and Quayle 1998) to locate areas of high plankton abundance. Dove (2015) investigated the olfactory capabilities of whale sharks in captivity and found that exposure to both DMS and a filtrate of homogenized krill both elicited a feeding response by the sharks. He further speculated that whale sharks may smell aerosolized DMS in the boundary layer of air just above the water, whenever they feed in the active surface suction feeding mode and the top jaw and nares are held clear of the water. These results indicate the importance of odor detection for whale sharks in locating sources of prey.

3.2.3.2 Gustation (taste)

The second chemosensory system, gustation, has received less study in elasmobranch fishes. Sharks possess taste buds covered with sensory receptor cells within small projections, or papillae, in the mouth and gill arches, similar to those found in bony fishes. They are also found in the highest concentrations near the front of the mouth, indicating that biting prey is important for determining palatability (Collin 2012). The spotted wobbegong exhibits the lowest recorded density of taste buds in elasmobranchs, at four taste buds per cm^2, compared to the eastern fiddler ray (*Trygonorrhina fasciata*), which has 159 taste buds per cm^2. However, the density of taste buds has not yet been related to higher sensitivity in gustatory perception (Atkinson 2011). No research has been done on taste bud density or capability in the whale shark, though a study on captive whale sharks indicated exposure to food odorants may have aided in eliciting a feeding response (Dove 2015).

3.2.4 Vision

The ability to detect light underwater is an important sense for marine animals to aid in locating prey, communicating with other members of their species, and avoiding potential predators, at distances of up to 100 m, depending on conditions. Units of light, or photons, have characteristic wavelengths that together form a spectrum of colors that can vary in intensity and composition with depth of water (Collin 2012). Due to the change in light intensity with depth and water clarity, the eyes of elasmobranchs have evolved to best suit the environment in which they live. Sharks inhabiting shallow and brightly lit waters have the largest eyes, which may improve photon detection sensitivity (Lisney and Collin 2007). Whale sharks, in comparison, seem to have small eyes relative to their body size, but only two measurements of an eyeball have been published and they disagree: one was recorded to be 30 mm in diameter, removed from a 12.18 m shark (Karbhari and Josekutty 1986), and the other 65 mm from a shark of unspecified length (but probably immature) (Tomita et al. 2020); unpublished data suggest that the latter figure is more likely to be correct (A. Dove, pers comm.). To put this into context, a 4.3 m white shark *Carcharodon carcharias* had an axial eye diameter of 37 mm, while a 2.78 m bigeye thresher *Alopias superciliosus* had an axial eye diameter of over 60 mm (Lisney and Collin 2007). The large relative eye size in these species may indicate a higher reliance on vision in prey capture (Lisney and Collin 2007). In contrast, the whale shark's relatively small eye size suggests that they may not rely as heavily on vision for feeding and navigation as they do on other sensory systems (Clark and Nelson 1997). This species does, however, possess unique coloration patterns, which may be recognizable by conspecifics, though these patterns likely act primarily as camouflage within their aquatic habitat (Wilson and Martin 2003).

The eyes of sharks are positioned laterally on either side of the head, giving them almost a 360° field of view. Sharks also can increase the visual field of view with sinusoidal swimming, where head movements while swimming can compensate for blind spots between the eyes and in front of the head (Collin et al. 2016). Most sharks possess an upper and lower eyelid to protect the majority of the eye from injury, and some species also have a mobile third eyelid (or nictitating membrane) that affords additional protection (Gruber 1977). By contrast, the whale shark does not appear to possess eyelids or nictitating membrane. Rather, the outer surface of their eyeballs is covered in around 2,000 denticles that have a "bramble-like" structure suggestive of a function in abrasion resistance, and this is a seemingly unique adaptation among known sharks (Tomita et al. 2020). Whale sharks also have the capacity to retract and rotate the eyeball most of the way back into the head, again as a protective mechanism (Tomita et al. 2020).

Detection of light begins with photons entering the eye through the cornea, where it then passes through the pupil. Pupil shape can vary in different species of elasmobranchs, from circular to slit or crescent shaped, all of which have different benefits in resolution and contrast ability. Whale sharks have a circular pupil, the shape of which is observed in species that do not need to cope with large fluctuations in light intensity (Hart et al. 2006). The combined small eye and circular pupil shape of this species may limit the number of photons reaching the retina, decreasing eye sensitivity (Lisney and Collin 2007).

After light enters the eye, it passes through the lens and is sampled by the back of the retina in cells called photoreceptors. These cells convert photons to nerve signals that are passed to the brain for higher processing. These photoreceptors are also distinguished from each other by the light conditions in which they operate, where rods function in low light conditions and cones operate in bright light conditions and are responsible for color vision. Most elasmobranch species possess both rods and cones, but the proportion of each cell type varies in different species based on the environment in which they live. Species that inhabit bright light conditions have a higher number of cones, whereas species that live in low light habitats tend to have a higher proportion of rods (Collin 2012). Both rods and cones also have an outer segment, which contains a protein, opsin, attached to chromophore, or visual pigment. The type of visual pigment found in different photoreceptors determines the wavelength of light it is most sensitive to and has led to the differentiation between one rod and four classes of cone photoreceptors throughout evolution (Collin 2010). Not all species possess all five photoreceptor types, but the ratio of activity between at least two types of cones allows for color vision. Most sharks, including two species of orectolobid wobbegongs, only have one rod and one cone type, meaning they are likely colorblind (Hart et al. 2011). If its retinal topography is similar to the closely related nurse shark (Gruber and Cohen 1978), whale shark may have a rod-dominated retina and display cone monochromacy, meaning they likely cannot discriminate colors or display high visual acuity. Although much more work is required to fully understand the visual system of the whale shark, it is unlikely this species relies heavily on vision for prey capture. In an analysis of the genome of the whale shark, Hara et al. (2018) determined that the predicted structure of opsins in whale sharks means they would have the most "blue-shifted" light sensitivity of any known shark, with a peak sensitivity at 478 nm. Increased blue sensitivity of this sort is correlated with greater maximum swimming depth and we know that the whale shark is the deepest diving fish (Tyminski et al. 2015), so this certainly makes sense.

3.2.5 Mechanoreception (touch)

The mechanosenses in fishes are used to detect water movements at close range (within 1–2 body lengths). Detection of water currents can help with navigation and orientation, while the perception of small currents generated by predators, prey, and conspecifics can aid in their location (Collin et al. 2016, Kalmijn 1989). The directed movement of an animal to align with water currents is known as rheotaxis, and it aids in the ability to migrate, detect chemical cues, and maintain their position in the water column (Peach 2001). The response of whale sharks to water currents has also been recorded in their migration to nutrient-rich areas of upwelling (Motta et al. 2010) and positive rheotaxis in the Indian Ocean near the Seychelles (Rowat and Gore 2007), although they will also travel actively and independently of surface currents (Sleeman et al. 2010).

The mechanosensory lateral line system is stimulated by differences in movement between an animal's own body and the surrounding water. This movement of water is detected by tiny sensory organs called neuromasts that found along the side of the body from the head to the tail. These neuromasts are covered with sensory hair cells and have an overlying gelatinous cap called a cupula. When there is change in water pressure, deflection in the overlying cupula moves the hair cells (much like the hair cells in the auditory system), stimulating the underlying neurons or nerve cells. These neurons send sensory information to the brain from the stimulated mechanoreceptors and signal the velocity, frequency, intensity, and location of the pressure change along the body (Bodznick and Northcutt 1980, Kalmijn 1989). This system is sensitive to low frequencies between 1 and 200 Hz, such as the displacement of water by wounded fishes, which is highly relevant to predatory sharks (Collin 2012) but of uncertain significance for filter feeding species like whale sharks.

The major types of neuromasts are differentiated from one another based on external morphology. The two main types of lateral line organs are canal neuromasts, contained within canals along the body, and superficial neuromasts, which are directly exposed to the water. Different morphologies and positions of these neuromasts aid in determining their function, but it is generally thought canal neuromasts serve to detect acceleration, whereas superficial neuromasts are better adapted to detecting velocity (Collin 2012).

Superficial neuromasts are found in higher abundance in pelagic shark species (Reif 1985), though the ecological function of this relationship is still not understood. Research on the lateral line in wobbegong sharks indicated a distribution of neuromasts similar to other elasmobranch species, though both canal and superficial neuromasts were concentrated dorsally. These could aid in detecting water flow above the head in these demersal species (Theiss et al. 2012). Further research on the distribution and location of neuromasts along the body and their relationship with the ecology of elasmobranchs across different habitats is needed to better understand neuromast use in these fishes (Gardiner et al. 2012). The lateral line system of the whale shark has not been studied. Future research into this sensory system may indicate whether the system is adapted to their epipelagic habitat, similar to other open ocean species, or specialized for their planktivorous feeding style, or both.

3.2.6 Electroreception

The electrosensory system in elasmobranchs serves to aid in the close-range (less than 1 m) detection of low-intensity (less than 5 nV/cm) electrical stimuli produced by other organisms or from anthropogenic sources. Detection of these electric fields assists in finding prey, detecting predators, aids in within-species communication, and allows these fishes to detect temperature gradients (Brown 2003, Kempster et al. 2012). Elasmobranchs may also use their electroreception for geomagnetic navigation, either by following currents *via* passive electro-orientation or by detecting their direction of movement based on the earth's magnetic field *via* active electro-orientation (Kalmijn 1982, Anderson et al. 2017).

The anatomy of the electrosensory system of elasmobranchs consists of ampullary receptors, also known as ampullae of Lorenzini, lying beneath the dermis and embedded in an ampullary canal. Small capsules at the base of the canal contain clusters of sensory receptor cells. The canals are filled with a hydrogel that promotes a voltage gradient on the interior length of the canal (Brown et al. 2004). As the sensory cells detect differences in voltage potential, they signal primary afferent neurons, which send this information to the brain in addition to the location along the body where the signal was received (Bodznick and Northcutt 1980, Montgomery et al. 2012).

The ampullae are clustered into discrete groups on the head of sharks. Their spacing and organization may indicate feeding strategy, phylogeny, or habitat (Kempster et al. 2012). The arrangement of electrosensory pores near the snout in the basking shark is thought to have resulted from the unique head morphology of this species and may assist in their ability to determine the presence of zooplankton throughout their daily diel vertical migration through the water column (Kempster and Collin 2011a). In the third planktivorous shark species, the megamouth shark *Megachasma pelagios*, the pores all face in the direction of forward movement to aid the detection of plankton as the sharks swim through the water column (Kempster and Collin 2011b). Orectolobiform sharks display a large range in pore abundance, from 297 pores in the ornate wobbegong *Orectolobus ornatus*, primarily located on the dorsal surface of the head, to 1180 pores in the nurse shark, with a greater percentage of pores located ventrally. The variation in abundance and positioning (dorsal or ventral) of electrosensory pores in these species may relate to differences in their feeding strategies or the environments in which they live (Kempster et al. 2012). Many species of oceanic pelagic elasmobranchs have a reduced number of pores compared to coastal pelagic species, thought to relate to their increased reliance on vision in clear water habitats (Kempster et al. 2012). The electrosensory system of the whale shark is yet to be described, though based on feeding strategy and phylogeny, we can predict that they will have relatively few pores, with those clustered near the head to aid in detecting plankton.

3.3 THE BRAIN AND ITS IMPLICATIONS FOR BEHAVIOR

3.3.1 Introduction

All jawed vertebrates, or gnathostomes, possess a common brain "blueprint" made up of the same six major structures. This blueprint first appeared evolutionarily in early cartilaginous fishes and consists of the forebrain (OB(s), telencephalon, and diencephalon), midbrain (mesencephalon), and hindbrain (cerebellum and medulla) (Striedter 2005). This same general structure is conserved across all more-derived vertebrates, including humans.

Various features of the brain of several other orectolobids have been described, including wobbegongs and nurse sharks (Ebbesson and Campbell 1973, Graeber et al. 1973, Northcutt 1978, Yopak et al. 2007), as have other large-bodied, planktivorous shark and ray species, such as basking sharks, megamouth sharks, and manta rays (Ari 2011, Ito et al. 1999, Kruska 1988). However, few studies have discussed features of the brain of the whale shark (Sato 1986). Only one study has quantified brain size and brain organization as compared to a wide range of other species (Yopak and Frank 2009). Given the rarity of the tissue, Yopak and Frank (2009) examined the hemisected brains of two juvenile (6–6.5 m) male whale shark specimens (Figure 3.3) and the brain of one preserved neonate specimen *in situ* using magnetic resonance imaging (MRI) (Figure 3.4). This technique

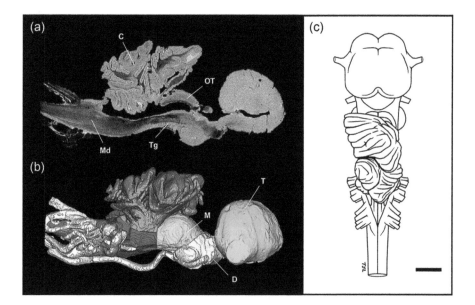

Figure 3.3 The brain and its major regions of the whale shark *Rhincodon typus*. (a) Sagittal slice of the brain from a juvenile specimen, acquired using magnetic resonance imaging (MRI). (b) Digital segmentation of the major structures of the brain and cranial nerves, performed using ITK-SNAP (Yushkevich et al. 2006). (c) An estimated dorsal drawing of the brain of *R. typus*, composed by T. Lisney. Brain structures identified as follows: T, telencephalon in green; D, diencephalon in yellow; M, mesencephalon in teal, divided into OT, optic tectum, and Tg, tegmentum; C, cerebellum in blue; Md, medulla oblongata in red. Scale bar corresponds to 1 cm. (Adapted with permission from Yopak, K. E., and L. R. Frank. 2009. *Brain, Behavior, and Evolution* 74:121–42.)

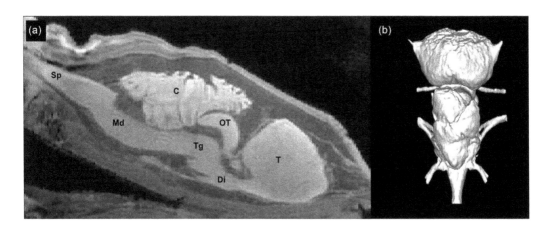

Figure 3.4 The brain of a neonate whale shark, (a) a sagittal slice of the brain *in situ*, acquired using MRI. (b) Digital segmentation of the total brain using ITK-SNAP (Yushkevich et al. 2006). Brain structures identified as follows: T=telencephalon; Di=diencephalon; mesencephalon, divided into OT=optic tectum and Tg=tegmentum; C=cerebellum; Md=medulla oblongata; Sp=spinal cord. (Adapted with permission from Yopak, K. E., and L. R. Frank. 2009. *Brain, Behavior, and Evolution* 74:121–42.)

allows for high-resolution images of soft tissue structures non-invasively and non-destructively (see Yopak et al. 2018 for details on this methodology in elasmobranchs).

3.3.2 Anatomy

The brain of all sharks is housed in a cartilaginous cranium (equivalent to a skull) within the head, oriented in a medial (on the mid-line of the body) position. The cranium in whale sharks is roughly oblong, capacious, and set well forward; it starts just behind the maxilla or top jaw. The brain lies at the posterior end of this space, roughly level with the eyes, suspended in a large volume of cerebrospinal fluid by a tough cobweb of connective tissue filaments that likely represents part of the meninges (brain lining) called arachnoid tissue in mammals (A. Dove pers. comm.). The function of the large space and volume of cerebrospinal fluid is unknown.

The front-most region of the brain is the paired OBs, positioned caudal to the olfactory epithelium and associated with processing odor cues (see Section 3.3.2., Meredith and Kajiura 2010, Schluessel et al. 2008). Olfactory bulbs can vary considerably in size and morphology across species (Yopak et al. 2015), ranging from elongated, sausage-shaped cylinders in hammerhead sharks and rays, to two ellipsoid-shaped "hemi-bulbs" in most other sharks and holocephalans (Dryer and Graziadei 1993). The OBs in whale shark have not yet been described. However, other orectolobids, including the tawny nurse shark and ornate wobbegong, show the distinct hemi-bulb morphology (Yopak et al. 2015).

Olfactory information is then sent from the OBs on to the second region of the forebrain, the telencephalon. In cartilaginous fishes, the telencephalon is a bulbous structure with two bilaterally symmetrical hemispheres, divided into a dorsally positioned pallium and a ventral subpallium. Although previously believed to be dominated solely by olfactory input (Ariëns Kappers 1906), it is now known that the telencephalon is involved in a range of high-order neural processing tasks, including multimodal sensory integration, complex behavioral control, and higher cognitive functions including learning, memory, and social intelligence (Hofmann and Northcutt 2012, Schluessel 2015, Guttridge and Schluessel 2018). This region can range from comprising only 20% of the brain (e.g. chimaeras *Callorhinchus* spp., and kitefin sharks *Dalatias* spp.), up to as much as 67% in the great hammerhead *Sphyrna mokarran* (Yopak et al. 2007).

The third region of the forebrain, the diencephalon, serves as a multisensory relay station to the telencephalon. It plays a central role in homeostatic behaviors, regulation of the production of hormones, and interfaces with the animal's endocrine system (see Smeets 1998 for review). This structure occupies 8% of the brain in juvenile whale sharks (Yopak and Frank 2009), compared to only 1.4% in shortfin mako shark *Isurus oxyrinchus*, although it comprises over 10% of the brain in some species (e.g. sand tiger shark *Carcharias taurus*) (Yopak 2012). However, as the diencephalon is associated with a number of reproductive behaviors, there may be changes in the size of this structure through the animal's lifespan (referred to as ontogenetic shifts, Section 3.3.3.3) that coincides with the onset of sexual maturity (e.g. Lisney et al. 2017). As noted, only juvenile whale sharks have been examined to date.

The mesencephalon, or midbrain, is divided into two major areas in cartilaginous fishes: a dorsally positioned tectum and a ventral tegmentum (Smeets 1998). The majority of research has focused on the optic tectum, which consists of paired lobes that form the roof of the mesencephalon (Figure 3.3). The optic tectum region is most readily associated with vision and visual processing because it receives direct input from the retina (see Collin et al. 2016, Yopak 2012 for review), although it is also involved in the processing of multisensory information. The size of the optic tectum varies considerably across cartilaginous fishes (Yopak and Lisney 2012) and can occupy up to as much as 18–19% of the brain in lamniform species, such as the shortfin mako and the crocodile shark *Pseudocarcharias kamoharai*, which may indicate an enhanced reliance on vision (Yopak 2012).

Cartilaginous fishes are believed to be the first group to have a confirmed true cerebellum (Montgomery and Bodznick 2010, Sugahara et al. 2016), which coincided in evolutionarily time with the appearance of jaws and paired fins in these fishes (Northcutt 2002). The full function of the cerebellum has been an area of considerable debate and speculation, although it is generally agreed that this structure plays a role in target tracking and is responsible for effective regulation of motor skills (Montgomery et al. 2012). Morphologically, the cerebellum varies substantially in size, convolution (also called "foliation"), and symmetry among species of cartilaginous fishes (Ari 2011, Lisney et al. 2008, Puzdrowski and Gruber 2009, Yopak et al. 2007).

Lastly, the medulla oblongata is the most caudal component of the hindbrain, acting as a relay station between the brain and spinal cord. It receives primary projections from the acoustic and vestibular system, electroreceptors, and mechanoreceptive lateral line and regulates a number of homeostatic functions (Smeets et al. 1983). In addition to variation in overall in the size and morphology of this brain region, some attention has been given to the central termination site for the electroreceptors (called the dorsal octavolateralis nucleus or DON) and the lateral line (medial octavolateralis nucleus or MON). These regions vary considerably in size, possibly corresponding with sensory specialization (Yopak and Montgomery 2008). Although the overall size of the medulla has been quantified in whale sharks (Yopak and Frank 2009), the size of the MON and DON have not. The only studies to explore variation in MON and DON in other orectolobids (e.g. *Orectolobus* spp., Kajiura et al. 2010) showed a relative reduction in the size of these structures in comparison with 36 other elasmobranch species.

3.3.3 Neuroecology: Understanding the Neurological Basis of Behavior

Whale sharks, along with the two other plankton-feeding shark species, the megamouth and basking shark, represent an enormous lifestyle divergence in chondrichthyan evolution compared to most other sharks. All are characterized by extremely large body sizes and unique filter feeding mechanisms that apparently evolved independently (see Chapter 1). In the field of comparative neuroscience, the study of species with unique behavioral and morphological specializations is often critical when teasing apart evolutionary trends in brain organization. Outliers tend to be the most informative when attempting to link form and function.

3.3.3.1 Encephalization

Brain size relative to an animal's body size (termed encephalization) varies considerably across species (Yopak 2012). Sharks and their relatives possess similar brain weight to body weight ratios to other vertebrates, including birds and mammals (Northcutt 1978). Although many orectolobids have small brains relative to their body size, in comparison with other sharks, some of the closest relatives to whale sharks, the ginglymostomatids (nurse sharks, e.g. *Nebrius* and *Ginglymostoma*), are more highly encephalized (Yopak et al. 2007). The whale shark shows a dramatic reduction in relative brain size, however, compared to similar large-bodied shark species, including the megamouth, basking, and great white sharks (Ito et al. 1999, Kruska 1988, Yopak and Frank 2009, Yopak et al. 2007). In the case of the planktivorous shark species, Striedter (2005) suggested that this character may not necessarily represent an example of an independent decrease in brain size, but rather be "gigantism," where we see an evolutionary increase in body size without a corresponding increase in brain size.

Larger brain sizes have been correlated with habitat, lifestyle, and cognitive capabilities in a range of shark species (Yopak 2012), although the link between encephalization and enhanced cognitive capabilities continues to be vigorously debated across vertebrates (Herculano-Houzel 2009, Kotrschal et al. 2013, Willemet 2013). Previous work has suggested that the low levels of encephalization seen in planktivorous predators may reflect the more passive, opportunistic, filter-feeding modality they employ (Nilsson et al. 2000), which could be less cognitively demanding in terms of sensory and/or motor requirements in comparison with more agile predatory species hunting active prey (Yopak et al. 2007). However, diet alone cannot fully account for encephalization, as the planktivorous mobulid rays are among the most highly encephalized group in the elasmobranch clade (Ari 2011, Fossi 1984).

Encephalization has also been linked to the degree of maternal investment during development. Matrotrophic shark and rays (where the developing embryo receives additional provisioning from the mother, such as via a placental connection) have brains that are 20%–70% larger than lecithotrophic species that do not provide investment beyond the yolk-sac (Mull et al. 2011). Whale sharks are currently described as a lecithotrophic livebearer, which is also called yolk-sac viviparity (see Chapter 2). This describes a reproductive mechanism by which, although born free-swimming, embryos are nourished via yolk sacs with no direct transfer of nutrients from mother to pup. This low degree of maternal investment may provide some insights into the small relative brain size of the whale shark. However, much more work is required in this field, as most of our current understanding of whale shark reproduction stems from the physical examination of a single pregnant female with 304 pups in various stages of development (Joung et al. 1996).

3.3.3.2 Brain Organization

Patterns of brain organization, or the relative development of the major brain regions, have long been considered a useful predictor for functional specialization. According to the principle of proper mass (Jerison 1973), the relative size of a brain region should reflect, to some extent, the importance of the neural function of that brain region. Although the exact mechanisms through which the brain evolves is unresolved (see Barton and Harvey 2000, Finlay and Darlington 1995, Yopak et al. 2010), evidence has shown that that a combination of phylogenetic, adaptive, and life history characteristics correlates with the variation in brain organization within cartilaginous fishes (Yopak 2012). A major finding in shark neuroanatomy is that species that live in similar environments tend to have similar patterns of brain organization. For example, reef-associated species have relatively large, well-developed telencephalons, relatively large optic tectums, large highly foliated cerebellums, and relatively small OBs (Yopak and Lisney 2012, Yopak et al. 2015, 2007). This potentially reflects the visual and spatial demands of well-lit, spatially complex reef habitats. In contrast, the brains of deep-sea sharks are suggestive of lower activity levels and a specialization of non-visual senses. They are characterized by relatively small telencephalons and optic tectums, small, smooth cerebellums, relatively enlarged olfactory bulbs, and enlargement of the regions of the brain that receive input from the animal's electroreceptors (DON) and lateral line (MON) (Yopak and Lisney 2012, Yopak et al. 2015, Yopak and Montgomery 2008). These results suggest that, although nearly all cartilaginous fishes possess the same well-developed sensory systems (see Section 3.3.2.), the relative importance of these systems varies between species (Collin et al. 2016).

Patterns of brain organization in the whale shark have been examined in comparison with a wide range of other shark species (Yopak and Frank 2009). Although a member of the Orectolobiformes, whale sharks share more neural characteristics with lamniform sharks, such as thresher (Alopias spp.) and basking sharks, than with their bottom-dwelling orectolobid relatives. Shared features include a relatively average to small telencephalon that occupies 42% of the brain and a large, highly foliated cerebellum that comprises nearly 30% (Figure 3.3). Given its association with higher cognitive functions such as learning and memory (Schluessel 2015; Guttridge and Schluessel 2018), variation in the size of the telencephalon has been correlated with navigating in spatially complex habitats, increased sociality, and strategic hunting (Yopak 2012). Demski and Northcutt (1996) noted that, in particular, protrusion of the dorsal pallium of the telencephalon is seen in species with more complex social behaviors, such as dominance hierarchies and other forms of "social intelligence." Correspondingly, the largest telencephalons with the most well-developed dorsal pallia are found in members of the families Carcharhinidae, Sphyrnidae (Yopak et al. 2007), Dasyatidae, and Myliobatidae (Ari 2011, Lisney et al. 2008). Most of these species live in reef-associated habitats and often display complex social structure relative to other sharks. Although the overall telencephalon was average in size in whale sharks, compared to other species, the dorsal pallium was moderately well-developed, certainly more so than in the basking shark (Sato 1986). Although not generally regarded as social species, whale sharks aggregate predictably in large numbers in some areas throughout their distribution, with noted segregation by sex and size (see Chapter 7).

An increase in the size of the mesencephalon (or midbrain) has previously been linked to an increased reliance on vision; the largest midbrains (in particular, the optic tectum, Section 3.3.3.2) have been documented in pelagic coastal/oceanic and oceanic sharks such as shortfin mako, white, and thresher sharks (Yopak and Lisney 2012). These species also have the absolutely and/or relatively largest eyes among cartilaginous fishes and are assumed to rely heavily on vision (Lisney and Collin 2007). Surprisingly, the mesencephalon of the whale shark was relatively small; despite living in well-lit, open ocean habitats, it comprises just 6.5% of the brain, which is similar to other orectolobiforms. Given the retinal topography of its close relatives, it is unlikely that this species has high visual acuity beyond close range ~ 3–5 m (Martin 2007), which may be reflected in its relatively small mesencephalon, although the optic tectum alone has yet to be described.

The most notable characteristic of the whale shark brain is an extremely large, highly foliated cerebellum (Figure 3.3), which differs from the pattern seen in most orectolobids (Yopak et al. 2007). Although many other species that dwell in pelagic or coastal pelagic habitats also share this feature, the whale shark has one of the largest cerebellums documented to date (Yopak and Frank 2009). Of the other shark species currently described, only the basking shark (28%), bigeye (Alopias superciliosus; 32%), and common thresher (A. vulpinus; 31%) sharks have a cerebellum that occupies such a large proportion of the brain (see Section 3.3.3.2, and Yopak 2012). Although the function of the cerebellum in sharks is as yet unresolved, variation in size and foliation of this structure likely reflects variation in cerebellar-dependent function and behaviors (Yopak et al. 2017), and may allow for greater control of swimming movement in the three-dimensional world of the open ocean. There is a clear trend of higher levels of cerebellar foliation occurring in large-bodied species that occupy oceanic and/or spatially complex habitats (Montgomery et al. 2012). In contrast, low levels of foliation are observed in benthic and bathyal benthopelagic species, including wobbegongs, swellsharks

(Yopak et al. 2007), and longnose chimaeras (Yopak and Montgomery 2008), which is potentially related to their lower activity levels and close association with the substrate. Prior to the description of the whale shark brain, large and highly foliated cerebellums had only been documented in active, agile species (Lisney et al. 2008, Yopak et al. 2007). However, the high levels of foliation seen in the whale shark supported the notion that this characteristic might reflect a level of proprioception or the ability of an organism to detect motion, position, and equilibrium of its body, to a greater extent (Yopak and Frank 2009). Indeed, the whale shark has been known to make vertical migrations, sometimes down to depths of >1,900 m (Tyminski et al. 2015), travel extensive distances (see Chapter 6), and can suspend vertically in the water column during feeding (e.g. Motta 2004).

Given all of the evidence documented thus far, it appears that a combination of phylogenetic and adaptive factors has influenced the evolution of the brain of the whale shark. Although it shares some neural features with closely related species, analyses by Yopak and Frank (2009) grouped the brains of the whale shark and basking shark as being more statistically similar to each other than to any other shark species, providing evidence for convergent evolution between these taxonomically divergent species that have evolved analogous lifestyles.

When dealing with rare specimens, it is often the case that tissue may be incomplete, damaged, or unable to be processed using standard techniques. To date, our knowledge of the whale shark brain comes from few specimens, obtained from aquarium mortalities. There are key features of the brain that have yet to be described, such as the olfactory bulbs and optic tectum. It is also important to note again that the specimens described by Yopak and Frank (2009) were both juvenile males. As it is currently unknown the degree to which the brain exhibits sexually dimorphic characteristics (Ari 2011, Kempster et al. 2014), it is uncertain whether these results are representative of both sexes. Our understanding of the whale shark brain will only improve along with the collection of additional specimens and the description of additional species to contextualize these data.

3.3.3.3 Ontogenetic Shifts: Does Brain Organization Change with Growth?

All fishes, as well as some amphibians and reptiles, experience indeterminate growth (Sebens 1987). Similarly, their brains and other organs continue to grow throughout their lifespan, unlike other vertebrates, including humans, which produce few new neurons as adults (Gage 2002). The life histories of most sharks are characterized by ontogenetic shifts in habitat and diet, as well as behavioral shifts associated with sexual maturity (e.g. Grubbs 2010). Limited studies have shown that there are changes in the relative size of major brain regions as an animal ages (Lisney et al. 2007, Lisney et al. 2017), which might suggest a corresponding functional shift in the relative importance of different sensory systems throughout life. For example, in six species of shark and one species of ray, Lisney et al. (2007) demonstrated an ontogenetic shift from a relatively large optic tectum in juveniles to relatively large olfactory bulbs in adults of the same species, which may reflect a shift in the importance of vision versus olfaction as these species mature.

Similarly, there are ontogenetic shifts in the brain of the whale shark. Neonate whale sharks have a larger brain size relative to their body mass, with the brain nearly filling the endocranial space (Figure 3.3). This dramatically differs from the 6 m juvenile (and likely adult) condition, where Yopak and Frank (2009) noted a vast discrepancy between endocranial space and brain tissue. This is similar to the basking shark, in which the brain occupies only 1/16th of the cranial cavity (Kruska 1988), the rest being occupied by cerebrospinal fluid and arachnoid tissue. Furthermore, foliation of the cerebellum exists in early development (Figure 3.3), although the depth of folds did not appear to be as numerous or as pronounced, indicating ontogenetic shifts in cerebellar foliation (Yopak and Frank 2009). Similarly, Lisney et al. (2017) showed an increase in the degree of cerebellar foliation in the blue spotted stingray (*Neotrygon trigonoides*, originally published as *N. kuhlii*) from juveniles to sexually mature adults, which may reflect improved locomotor performance or the expansion of activity space. Resolution was not high enough to provide detailed insights into ontogenetic shifts in brain organization in whale sharks beyond these gross characteristics, so additional differences may be likely given the extreme changes in body size, late age of maturity, and highly migratory nature of these animals. Indeed, a better understanding of the whale shark brain throughout life may be able to provide critical insights into their behavior.

3.4 SUMMARY AND FUTURE DIRECTIONS

Our current understanding of the sensory biology and neuroanatomy of whale sharks is limited. More work is required in this area. Although based on a handful of opportunistic specimens and/or inferences made from close relatives, the current data show that the nervous system of the whale shark reflects both its primary habitat and the unique morphological and behavioral specializations exhibited by this species. Their large size and pelagic lifestyle make whale shark behavior difficult to study in the wild, and few aquaria worldwide are equipped to house the world's largest fish (see Chapter 9). However, a better understanding of how the whale shark senses its environment and how these signals are processed in the brain will enable a much greater understanding of the feeding behavior, movement patters, and sociality of this unique and mysterious fish.

ACKNOWLEDGMENTS

The authors would like to thank the Georgia Aquarium, particularly Dr. Bruce Carlson, Dr. Tonya Clauss, and Tim Binder, who originally provided the two juvenile whale shark brain samples described in this chapter, published in Yopak and Frank (2009), and the Marine Vertebrates Collection at SIO, which allowed for the neonate specimen to be imaged for the original publication. Special thanks to T. Lisney for his incredible brain illustration. KEY acknowledges funds from the University of North Carolina Wilmington during the writing of this chapter.

REFERENCES

Anderson, J. M., Clegg, T. M., Veras, L. V. M. V. Q., and K. M. Holland. 2017. Insight into shark magnetic field perception from empirical observations. *Scientific Reports.* 7:11042.

Ari, C. 2011. Encephalization and brain organization of mobulid rays (myliobatiformes, elasmobranchii) with ecological perspectives. *The Open Anatomy Journal* 3:1–13.

Ariëns Kappers, C. U. 1906. The structure of the teleostean and selachian brain. *Journal of Comparative Neurology and Psychology* 16:1–109.

Atkinson, C. 2011. The gustatory system of elasmobranchs: Morphology, distribution and development of oral papillae and oral denticles. *Biology Open* 5:1759–69.

Barton, R. A., and P. H. Harvey. 2000. Mosaic evolution of brain structure in mammals. *Nature* 405:1055–58.

Bodznick, D., and R. G. Northcutt. 1980. Segregation of electro-and mechanoreceptive inputs to the elasmobranch medulla. *Brain Research* 195:313–21.

Brown, B. R. 2003. Neurophysiology: Sensing temperature without ion channels. *Nature* 421:495.

Brown, B. R., Hughes, M. E., and C. Russo. 2004. Thermoelectricity in natural and synthetic hydrogels. *Physical Review E* 70:031917.

Casper, B. M., and D. A. Mann. 2006. Evoked potential audiograms of the nurse shark (*Ginglymostoma cirratum*) and the yellow stingray (*Urobatis jamaicensis*). *Environmental Biology of Fishes* 76:101–08.

Casper, B. M., and D. A. Mann. 2009. Field hearing measurements of the Atlantic sharpnose shark *Rhizoprionodon terraenovae*. *Journal of Fish Biology* 75:2768–76.

Clark, E., and D. R. Nelson. 1997. Young whale sharks, *Rhincodon typus*, feeding on a copepod bloom near La Paz, Mexico. *Environmental Biology of Fishes* 50:63–73.

Collin, S. P. 2010. Evolution and ecology of retinal photoreception in early vertebrates. *Brain Behavior and Evolution* 75:174–85.

Collin, S. P. 2012. The neuroecology of cartilaginous fishes: sensory strategies for survival. *Brain, Behavior and Evolution* 80:80–96.

Collin, S. P., Kempster, R., and K. Yopak. 2016. Sensing the environment. In *Fish Physiology: Physiology of Elasmobranch Fishes*. Eds. R. E. Shadwick, A. P. Farrell and C. J. Brauner. New York: Elsevier. pp. 19–99.

Corwin, J. T. 1978. The relation of inner ear structure to the feeding behavior in sharks and rays. *Scanning Electron Microscopy* 2:1105–12.

Demski, L. S., and R. G. Northcutt. 1996. The brain and cranial nerves of the white shark: an evolutionary perspective. In *Great White Sharks: The Biology of Carcharodon Carcharias*. Eds. A. P. Klimley and D. Ainley. San Diego, CA: Academic Press, pp. 121–30.

Dennison, R. H. 1937. Anatomy of the head and pelvic fin of the whale shark, *Rhincodon. Bulletin of the American Museum of Natural History New York* 73:477S–515S.

Dove, A. D. M. 2015. Foraging and ingestive behaviors of whale sharks, *Rhincodon typus*, in response to chemical stimulus cues. *The Biological Bulletin* 228:65–74.

Dryer, L., and P. P. C. Graziadei. 1993. A pilot study on morphological compartmentalization and heterogeneity in the elasmobranch olfactory bulb. *Anatomy and Embryology* 190:41–51.

Ebbesson, S. O. E., and C. B. G. Campbell. 1973. On the organization of cerebellar afference pathways in the nurse shark (*Ginglymostoma ciratum*). *Journal of Comparative Neurology* 152:233–54.

Evangelista, C., M. Mills, U. E. Siebeck, and S. P. Collin. 2010. A comparison of the external morphology of the membranous inner ear in elasmobranchs. *Journal of Morphology* 271:483–95.

Fay, R. R. 1984. The goldfish ear codes the axis of acoustic particle motion in three dimensions. *Science* 225:951–54.

Fay, R. R., and A. N. Popper. 1980. Structure and function in teleost auditory systems. In *Comparative Studies of Hearing in Vertebrates*. Munich: Springer, pp. 3–42.

Finlay, B. L., and R. B. Darlington. 1995. Linked regularities in the development and evolution of mammalian brains. *Science* 268:1578–84.

Fossi, A. 1984. Quelques aspects de la biologie des raies du genre mobula du golfe de californie. Memoire de fin d'etudes. MSc Thesis. Cherbourg, France. Institut national des Techniques de la Mer.

Fox, S., I. Foisy, R. De La Parra Venegas, B. Galván Pastoriza, R. Graham, E. Hoffmayer, J. Holmberg, and S. Pierce. 2013. Population structure and residency of whale sharks *Rhincodon typus* at Utila Bay islands, Honduras. *Journal of Fish Biology* 83:574–87.

Gage, F. 2002. Neurogenesis in the adult brain. *Journal of Neuroscience* 22:612–13.

Gardiner, J. M., Hueter, R. E., Maruska, K. P., et al. 2012. Sensory physiology and behavior of elasmobranchs. In *Biology of Sharks and Their Relatives*, 2nd edition Eds. J.C. Carrier, J.A. Musick and M.R. Heithaus. New York: CRC Press.

Graeber, R. C., Ebbesson, S. O. E., and J. A. Jane. 1973. Visual discrimination in sharks without optic tectum. *Science* 180:413–15.

Grubbs, R. 2010. Ontogenetic shifts in movements and habitat use. In *Biology of Sharks and Their Relatives*, 2nd edition, Eds. Carrier, Musick and Heithaus. New York: CRC Press, pp. 319–50.

Gruber, S. H. 1977. The visual system of sharks: Adaptations and capability. *American Zoologist* 17:453–69.

Gruber, S. H., and J. L. Cohen. 1978. Visual system of the elasmobranchs: State of the art 1960–1975. In *Sensory Biology of Sharks, Skates, and Rays*, Eds Hodgson and Mattewson. Arlington, VA: Office of Naval Research, pp. 11–116.

Guttridge, T., and V. Schluessel. 2018. Sharks - Elasmobranch cognition. In *Field and Laboratory Methods in Animal Cognition: A Comparative Guide*. Eds. Bueno-Guerra and Amici. New York: Cambridge University Press, pp. 354–80.

Hara, Y., Yamaguchi, K., Onimaru, K. et al. 2018. Shark genomes provide insights into elasmobranch evolution and the origin of vertebrates. *Nature Ecology and Evolution* 2:1761–1771.

Hart, N. S., Lisney, T. J., and S. P. Collin. 2006. Visual communication in elasmobranchs. *Communication in Fishes* 2:337–92.

Hart, N. S., Theiss, S. M., Harahush, B. K., and S. P. Collin. 2011. Microspectrophotometric evidence for cone monochromacy in sharks. *Naturwissenschaften* 98:193–201.

Herculano-Houzel, S. 2009. The human brain in numbers: a linearly scaled-up primate brain. *Frontiers in Human Neuroscience* 3:31.

Hofmann, M. H., and R. G. Northcutt. 2012. Forebrain organization in elasmobranchs. *Brain Behavior and Evolution* 80:142–51.

Ito, H., Yoshimoto, M., and H. Somiya. 1999. External brain form and cranial nerves of the megamouth shark, *Megachasma pelagios*. *Copeia* 1999:210–13.

Jerison, H. J. 1973. *Evolution of the Brain and Intelligence*. New York: Academic Press.

Joung, S. J., Chen, C. T., Clark, E., et al. 1996. The whale shark, *Rhincodon typus*, is a livebearer: 300 embryos found in one 'megamamma' supreme. *Environmental Biology of Fishes* 46:219–23.

Kajiura, S. M., Cornett, A. D., and K. E. Yopak. 2010. Sensory adaptations to the environment: Electroreceptors as a case study. In *Sharks and Their Relatives II: Biodiversity, Adaptive Physiology, and Conservation*. Eds Carrier, Musick and Heithaus. New York: CRC Press.

Kalmijn, A. J. 1982. Electric and magnetic field detection in elasmobranch fishes. *Science* 218:916–18.

Kalmijn, A. J. 1989. Functional evolution of lateral line and inner ear sensory systems. In *The Mechanosensory Lateral Line*. Munich: Springer. pp. 187–215.

Karbhari, J. P., and C. J. Josekutty. 1986. On the largest whale shark *Rhincodon typus* (Smith) landed alive at Cuffe Parade, Bombay. *Marine Fisheries Information Service, Technical and Extension Series* 66:31–5.

Kempster, R. M., and S. P. Collin. 2011a. Electrosensory pore distribution and feeding in the basking shark *Cetorhinus maximus* (Lamniformes: Cetorhinidae). *Aquatic Biology* 12:33–36.

Kempster, R. M., and S. P. Collin. 2011b. Electrosensory pore distribution and feeding in the megamouth shark *Megachasma pelagios* (Lamniformes: Megachasmidae). *Aquatic Biology* 11:225–28.

Kempster, R. M., Garza-Gisholt, E., Egeberg, C., et al. 2014. Sexual dimorphism of the electrosensory system: A quantitative analysis of nerve axons in the dorsal anterior lateral line nerve of the blue-spotted fantail stingray (*Taeniura lymma*). *Brain, Behavior and Evolution* 81:226–35.

Kempster, R. M., McCarthy, I. D., and S. P. Collin. 2012. Phylogenetic and ecological factors influencing the number and distribution of electroreceptors in elasmobranchs. *Journal of Fish Biology* 80:2055–88.

Kotrschal, A., Rogell, B., Bundsen, A., et al. 2013. Artificial selection on relative brain size in the guppy reveals costs and benefits of evolving a larger brain. *Current Biology* 23:168–71.

Kruska, D. C. T. 1988. The brain of the basking shark (*Cetorhinus maximus*). *Brain, Behavior and Evolution* 32:353–63.

Lisney, T. J., Bennett, M. B., and S. P. Collin. 2007. Volumetric analysis of sensory brain areas indicates ontogenetic shifts in the relative importance of sensory systems in elasmobranchs. *Raffles Bulletin of Zoology* 14:7–15.

Lisney, T. J., and S. P. Collin. 2007. Relative eye size in elasmobranchs. *Brain, Behavior, and Evolution* 69:266–79.

Lisney, T. J., Yopak, K. E., Camilieri-Asch, V., and S. P. Collin. 2017. Ontogenetic shifts in brain organization in the bluespotted stingray *Neotrygon kuhlii* (Chondrichthyes: Dasyatidae). *Brain, Behavior and Evolution* 89:68–83.

Lisney, T. J., Yopak, K. E., Montgomery, J. C., and S. P. Collin. 2008. Variation in brain organization and cerebellar foliation in chondrichthyans: Batoids. *Brain, Behavior and Evolution* 72:262–82.

Martin, R. A. 2007. A review of behavioural ecology of whale sharks (*Rhincodon typus*). *Fisheries Research* 84:10–16.

Meredith, T. L., and S. M. Kajiura. 2010. Olfactory morphology and physiology of elasmobranchs. *Journal of Experimental Biology* 213:3449–56.

Montgomery, J. C., and D. Bodznick. 2010. Functional origins of the vertebrate cerebellum from a sensory processing antecedent. *Current Zoology* 56:277–84.

Montgomery, J. C., Bodznick, D., and K. E. Yopak. 2012. The cerebellum and cerebellar-like structures of cartilaginous fishes. *Brain, Behavior and Evolution* 80:152–65.

Motta, P. J. 2004. Prey capture behavior and feeding mechanisms of elasmobranchs. In *Biology of Sharks and Their Relatives*. Eds. Carrier, Musick and Heithaus. London: CRC Press.

Motta, P. J., Maslanka, M., Hueter, R. E., et al. 2010. Feeding anatomy, filter-feeding rate, and diet of whale sharks *Rhincodon typus* during surface ram filter feeding off the Yucatan peninsula, Mexico. *Zoology* 113:199–212.

Mull, C., Yopak, K. E. and N. Dulvy. 2011. Does more maternal investment mean a larger brain? Evolutionary relationship between reproductive mode and brain size in chondrichthyans. *Marine and Freshwater Research* 62:567–75.

Muller, M. 1999. Size limitations in semicircular duct systems. *Journal of Theoretical Biology* 198:405–37.

Myrberg, A. A. 2001. The acoustical biology of elasmobranchs. In *The Behavior and Sensory Biology of Elasmobranch Fishes: An Anthology in Memory of Donald Richard Nelson*. Munich: Springer. pp. 31–46.

Nevatte, R. J., Williamson, J. E., Vella, N. G. F., et al. 2017. Morphometry and microanatomy of the barbels of the common sawshark *Pristiophorus cirratus* (Pristiophoridae): Implications for pristiophorid behaviour. *Journal of Fish Biology* 90:1906–25.

Nevitt, G. A., Veit, R. R., and P. Kareiva. 1995. Dimethyl sulphide as a foraging cue for antarctic procellariiform seabirds. *Nature* 376:680.

Nilsson, G. E., Routley, M. H., and G. M. C. Renshaw. 2000. Low mass-specific brain na+/k+-ATPase activity in elasmobranch compared to teleost fishes: Implications for the large brain size of elasmobranchs. *Proceedings of the Royal Society London B*. 267:1335–39.

Northcutt, R. G. 1978. Brain organization in the cartilaginous fishes. In *Sensory Biology of Sharks, Skates, and Rays*. Eds. Hodgson and Mathewson. Arlington, VA: Office of Naval Research, pp. 117–94.

Northcutt, R. G. 2002. Understanding vertebrate brain evolution. *Integrative and Comparative Biology* 42:743–56.

Nosal, A. P., Chao, Y., Farrara, J. D., et al. 2016. Olfaction contributes to pelagic navigation in a coastal shark. *PLoS One* 11:e0143758.

Peach, M. 2001. The dorso-lateral pit organs of the Port Jackson shark contribute sensory information for rheotaxis. *Journal of Fish Biology* 59:696–704.

Puzdrowski, R. L., and S. Gruber. 2009. Morphologic features of the cerebellum of the Atlantic stingray, and their possible evolutionary significance. *Integrative Zoology* 4:110–22.

Reif, W. E. 1985. Squamation and ecology of sharks. *Senckenbergische Naturforschende Gesellschaft*.

Rowat, D., and M. Gore. 2007. Regional scale horizontal and local scale vertical movements of whale sharks in the Indian Ocean off Seychelles. *Fisheries Research* 84:32–40.

Sato, Y. 1986. Brain patterns of the whale and basking sharks, *Rhincodon typus* and *Cetorhinus maximus* in relation to systematics. *Zoological Science* 3:1115.

Schluessel, V. 2015. Who would have thought that 'jaws' also has brains? Cognitive functions in elasmobranchs. *Animal Cognition* 18:19–37.

Schluessel, V., Bennett, M. B., Bleckmann, H., et al. 2008. Morphometric and ultrastructural comparison of the olfactory system in elasmobranchs: the significance of structure-function relationships based on phylogeny and ecology. *Journal of Morphology* 269:1365–86.

Sebens, K. P. 1987. The ecology of indeterminate growth in animals. *Annual Review of Ecology and Systematics* 18:371–407.

Sims, D. W., and V. A. Quayle. 1998. Selective foraging behaviour of basking sharks on zooplankton in a small-scale front. *Nature* 393:460.

Sleeman, J., Meekan, M., Fitzpatrick, B., et al. 2010. Oceanographic and atmospheric phenomena influence the abundance of whale sharks at Ningaloo reef, Western Australia. *Journal of Experimental Marine Biology and Ecology* 382:77–81.

Smeets, W. J. A. J. 1998. Cartilaginous fishes. In *The Central Nervous System of Vertebrates*. Eds. Nieuwenhuys, Ten Donkelaar and Nicholson. Berlin: Springer-Verlag, pp. 551–654.

Smeets, W. J. A. J., Nieuwenhuys, R., and B. L. Roberts. 1983. *The Central Nervous System of Cartilaginous Fishes: Structural and Functional Correlations*. New York: Springer-Verlag.

Striedter, G. F. 2005. *Principles of Brain Evolution*. Sunderland: Sinauer Associates, Inc.

Sugahara, F., Pascual-Anaya, J., Oisi, Y., et al. 2016. Evidence from cyclostomes for complex regionalization of the ancestral vertebrate brain. *Nature* 531:97–100.

Theiss, S. M., Collin, S. P., and N. S. Hart. 2012. The mechanosensory lateral line system in two species of wobbegong shark (Orectolobidae). *Zoomorphology* 131:339–48.

Tomita, T., Murakumo, K., Komoto, S., et al. 2020. Armored eyes of the whale shark. *PLOS ONE*. 15(6): e0235342.

Tyminski, J.P., de la Parra-Venegas, R., González-Cano, J., and R. E. Hueter. 2015. Vertical movements and patterns in diving behavior of whale sharks as revealed by pop-up satellite tags in the Eastern Gulf of Mexico. *PLoS ONE* 10(11):e0142156.

Urick, R. J. 1979. *Sound propagation in the sea*. Catholic University of America. Washington DC: Dept. of Mechanical Engineering.

Willemet, R. 2013. Reconsidering the evolution of brain, cognition, and behavior in birds and mammals. *Frontiers in Psychology* 4:396.

Wilson, S. G., and R. A. Martin. 2003. Body markings of the whale shark: Vestigial or functional? *Western Australian Naturalist* 24:115–17.

Yopak, K. E. 2012. Neuroecology in cartilaginous fishes: The functional implications of brain scaling. *Journal of Fish Biology* 80:1968–2023.

Yopak, K. E., Carrier, J. C., and A. P. Summers. 2018. Imaging technologies in the field and laboratory. In *Shark research: Emerging Technologies and Applications for the Field and Laboratory*. Eds. Carrier, Heithaus and Simpfendorfer. New York: CRC Press, pp. 157–76.

Yopak, K. E., and L. R. Frank. 2009. Brain size and brain organization of the whale shark, *Rhincodon typus*, using magnetic resonance imaging. *Brain, Behavior, and Evolution* 74:121–42.

Yopak, K. E., and T. J. Lisney. 2012. Allometric scaling of the optic tectum in cartilaginous fishes. *Brain, Behavior and Evolution* 80:108–26.

Yopak, K. E., and J. C. Montgomery. 2008. Brain organization and specialization in deep-sea chondrichthyans. *Brain, Behavior, and Evolution* 71:287–304.

Yopak, K. E., Lisney, T. J., and S. P. Collin. 2015. Not all sharks are "swimming noses": Variation in olfactory bulb size in cartilaginous fishes. *Brain Structure and Function* 220:1127–11143.

Yopak, K. E., Lisney, T. J., Collin, S. P., and J. C. Montgomery. 2007. Variation in brain organization and cerebellar foliation in chondrichthyans: Sharks and holocephalans. *Brain, Behavior, and Evolution* 69:280–300.

Yopak, K. E., Lisney, T. J., Darlington, R. B., et al. 2010. A conserved pattern of brain scaling from sharks to primates. *Proceedings of the National Academy of Sciences* 107:12946–51.

Yopak, K. E., Pakan, J., and D. Wylie. 2017. The cerebellum of non-mammalian vertebrates. In *Evolution of Nervous Systems*, 2nd edition. Ed. Kaas. New York: Elsevier, pp. 373–85.

Yushkevich, P. A., Piven, J., Cody Hazlett, H., et al. 2006. User-guided 3D active contour segmentation of anatomical structures: Significantly improved efficiency and reliability. *Neuroimage* 31:1116–28.

CHAPTER 4

Parasites and Other Associates of Whale Sharks

Alistair D. M. Dove
Georgia Aquarium

David P. Robinson
Sundive Research

CONTENTS

4.1 INTRODUCTION

An apparently universal truth about whale sharks is that they are never alone. To even the casual observer, every whale shark seems to be accompanied by cadre of hangers-on, from fish escorts riding their bow wave, to more tenacious remoras, to smaller and more cryptic animals like parasitic copepods, which can be observed on the skin at close range with the naked eye. There is, however, very little peer-reviewed scientific literature describing these symbiotic relationships. Early records of whale shark encounters reported a variety of animals associated with them, including a range of invertebrate commensals and parasites, and several species of fish, especially tuna (Springer 1957, Iwasaki 1970, Rao et al. 1986, Silas 1989, Matsunaga et al. 2003, Rowat & Brooks 2012).

In this chapter, we review what is known about the symbionts of whale sharks. We do this in ascending order of body complexity, which is a largely arbitrary choice that in no way reflects the significance of the relationships described. All such symbiotic relationships fall on a spectrum according to the relative benefits to the two organisms participating in the relationship, but the three most common types are *mutualism*, where both participants benefit; *commensalism*, where one participant benefits and the other is unaffected; and *parasitism*, where one participant benefits at the expense (however small) of the other (Lafferty et al. 2008). In the case of whale sharks and other megafauna, there are a range of other looser relationships which we here term "associates", which is meant to describe species that have some sort of other facultative relationships that do not fit the more typical ecological categories of parasite, commensal, mutualist, predator or prey.

There is a well-established positive relationship between both parasite diversity and prevalence and the body size of the host animal (Lafferty & Kuris 2002) so, as the largest of all fishes, we should expect whale sharks to host a diverse array of parasites. While there are several species summarized below, there are still relatively few in comparison with other large-bodied marine species. The reason for the lack of published literature on the subject is almost certainly under-sampling; few whale sharks have ever been necropsied in the fashion necessary to gather rigorous systematic information about parasitic symbionts or infectious diseases. For that reason, the list below should not be considered comprehensive, but rather a first foray into what we expect will prove a much longer list at some future date.

4.2 MICROBIAL SYMBIONTS

Microbial symbionts include the viruses, bacteria and fungi. These organisms frequently have the potential to cause serious diseases, so they are often grouped together as microbial pathogens or just pathogens. There are essentially no published studies of the microbial symbionts of whale sharks. Given the explosive growth of the fields of genomics and microbiomics, we expect that this situation will change in the very near future, but at the time of publication it remains a gaping hole in our understanding of the biology of this species.

4.2.1 Viruses

There are no peer-reviewed publications about viral infections of whale sharks. Indeed, viruses are poorly known among sharks in general, although almost certainly present (Garner 2013). This reflects a lack of available primary or perpetual cell culture lines for sharks, which are an important tool used for diagnosing viral infections. In other animal groups, cell cultures are used to detect viral infections in fluid samples (usually blood plasma or serum) through the tendency of viruses to modify the appearance of cells in a monoculture through cytopathic effects (La Patra 2003). Recent advances in DNA/RNA technology will certainly help scientists explore the diversity of viruses in sharks because they do not rely on a need for cell lines. Rather, they detect directly the presence of virus DNA/RNA sequence among the shark DNA/RNA extracted from tissue samples, implying current or recent past infection (Kiselev et al. 2020).

4.2.2 Bacteria

As with the viruses, there are no peer-reviewed publications about bacterial infections of whale sharks. It is known that some sharks harbor bacterial populations in the bloodstream and other tissues of a significant portion of individuals, without displaying any overt signs of disease (Grimes et al. 1993; Mylniczenko et al. 2007; Tao et al. 2014), even though this would be considered septicemia in a mammal. The nature of the relationships of these bacteria to the host shark is not known for certain, but it is possible that they form a sort of mutualism, because many of the bacteria so far recovered have tested positive for presence of the urease enzyme system (Wood et al. 2019). This finding suggests that the bacteria may be a part of the way the shark regulates the chemical urea, which they naturally carry in their blood to aid with osmotic (water and salt) balance. Our knowledge of bacterial symbionts of whale sharks, be they pathogens or mutualists, is likely to increase dramatically as DNA-based tools are brought to bear on samples from whale sharks.

4.2.3 Fungi

Like the viruses and bacteria, there are no peer-reviewed publications about fungal infections of whale sharks. Fungal infections in sharks are generally thought to be uncommon (Garner 2013), but a few recent publications have characterized pathogenic fungi infecting other shark species and causing skin lesions and neurological signs like loss of equilibrium, uncoordinated swimming, and ultimately death (Marancik et al. 2011). Microsporans are a divergent group of fungus-like parasitic organisms that are quite common in bony fish and many arthropod groups, but none of these have been described from whale sharks either.

4.2.4 Protistan parasites

Protists are not a natural taxonomic group, but rather an extraordinarily diverse assortment of unrelated animal-like taxa that are generally treated together because they share the common feature of having a body that consists of a single cell. We follow this same convention here. Many protist groups include large numbers of parasitic species, especially among the amoebae, flagellates and ciliates, but none has so far been described from whale sharks.

4.3 METAZOAN PARASITES

Like the protists, metazoan parasites are not a natural taxonomic group, but a catch-all term applied to parasites with bodies consisting of more than a single cell. Among metazoans, the parasitic lifestyle has evolved independently on dozens of occasions in different lineages (Weinstein & Kuris 2016). Classical parasitologists recognize several major groups of obligate metazoan parasites, namely, the myxozoans (highly derived cnidarians); the roundworms or nematodes; the flatworms or platyhelminths (which includes the tapeworms, monogeneans, true digenean trematodes and aspidogastreans); the spiny-headed worms or acanthocephalans; the parasitic arthropods such as some copepods, isopods and amphipods; and the annelid leeches (Benz et al. 2001). Given the high diversity of parasites known from other shark species that have been better surveyed (Benz et al. 2001, Dove et al. 2017), we should expect whale sharks to host a diverse array of tapeworms, monogenean flatworms and parasitic copepods. Once again, however, we are hampered by under-sampling, but in this case, there are at least some published data, which we describe below and summarize in Table 4.1.

Table 4.1 Host–Parasite List for *Whale Sharks*

Higher Taxon	Family	Species	Locality	Reference
Copepoda	Pandaridae	*Pandarus rhincodonicus*	Western Australia	Norman et al. (2000)
Copepoda	Pandaridae	*Prosaetes rhinodontis*	Okinawa, Japan	Tang et al. (2010)
Copepoda	Caligidae	*Lepeophtheirus acutus*	Okinawa, Japan	Tang et al. (2013)
Isopoda	Gnathiidae	*Gnathia trimaculata*	Okinawa, Japan	Ota et al. (2012)
Cestoda	Caryophyllidae	*Wenyonia rhincodonti*[a]	Southern India	Malhotra et al. (2011)
Digenea	Syncoeliidae	*Paronatrema boholana*	Bohol, Philippines	Eduardo (2010)
Hirudinea	Piscicolidae	*Branchellion torpedinis*	Georgia Aquarium	Dove et al. (2017)

[a] May not be a valid record, see text.

4.3.1 Copepoda

Of all the parasite groups, we know the most about parasitic copepods on whale sharks because they are strict ectoparasites, are frequently observed on whale sharks in locations all over the world, and can be sampled non-lethally by picking or scraping them off the skin of living animals. These parasites are large relative to free-living copepods, but still quite small in absolute terms, with body sizes in the 4–10 mm range. At least three species have been described (Figure 4.1). All are in the caligoid group that includes many parasitic copepods of bony fishes, including the famous pathogenic sea louse of salmonids, *Lepeophtheirus salmonis*. Interestingly, parasitic copepods have never been noted from any encounter with whale sharks within the Arabian Gulf; it is possible that the increased temperature and salinity during the summer make the Gulf an unsuitable habitat (Robinson 2016), or it may result from some other unrecognized cause.

4.3.1.1 Pandarus rhincodonicus *Norman, Newbound and Knott, 2000*

This species has a flattened disc-like forebody roughly 0.5 cm wide, and it infects the skin and fins of whale sharks, which are the sole known host species to date. It was originally described from whale sharks on the coast of Western Australia (Norman et al. 2000) and is the most widespread and prominent of the ectoparasites seen on whale sharks. It can move actively across the surface of the host, but prefers to remain sedentary and can attract a certain amount of marine growth on the carapace. It is distinctive in having a prominently pigmented dorsal surface. The pigment is dark brown to black and forms an irregular bilaterally symmetrical pattern rather like a Rorschach blot (Figure 4.2a); patches of copepods may look like a rusty or brown smudge on the skin, and infections are commonly seen on the edges of the jaws and on many of the fins, especially the upper lobe of the caudal. Many of the larger individuals observed are mature females, which can be identified by the presence of two thread-like egg strands that trail from the posterior end of the organism. Unlike most free-living copepods, in which spherical eggs are carried in dense sacs, the eggs of *P. rhincodonicus* form serial strings of disc-shaped eggs, rather like a stack of casino chips.

This parasite does not seem to cause much in the way of tissue damage, but excess mucus production and localized epithelial erosions can occasionally be noted (A. Dove pers. obs.). Interestingly, *P. rhincodonicus* occurs in *very* high densities on whale sharks seen at the Tubbataha Reefs in the Philippines (Figure 4.2b), but has not been observed in such quantities on sharks seen elsewhere in Philippine waters, even relatively close by (Simon J. Pierce, pers. comm.). Why this parasite should be so abundant and prevalent at Tubbataha, which is a healthy coral reef marine protected area, and not at nearby sites, is unknown.

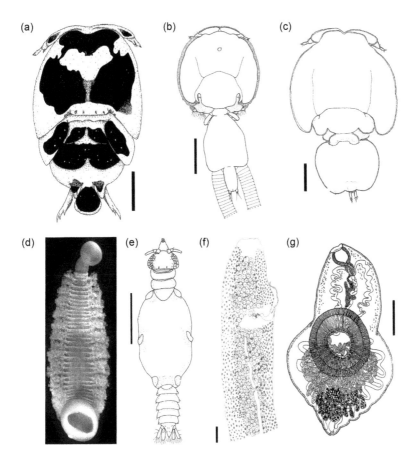

Figure 4.1 Parasites of whale sharks *Rhincodon typus*. (a) *Pandarus rhincodonicus* Norman, Newbound & Knott 2000; scale bar 1.6 mm © Kluwer Academic Publishing, reproduced with permission. (b) *Lepeophtheirus acutus* Heegaard 1943; scale bar 1.0 mm © Taylor & Francis, reproduced with permission. (c) *Prosaetes rhinodontis* (Wright 1876); scale bar 1.0 mm © Magnolia Press, reproduced with permission. (d) *Branchellion torpedinis* Savigny, 1822, length approx. 3 cm. (e) *Gnathia trimaculata* Coetzee et al. (2009), third-stage praniza larva; scale bar 1.0 mm © Kluwer Academic Publishing, reproduced with permission. (f) *Wenyonia rhincodonti* Malhotra et al. 2011; scale bar 0.1 mm © Springer Nature BV, reproduced with permission. (g) *Paronatrema boholana* Eduardo (2010); scale bar 1.0 mm.

4.3.1.2 Prosaetes rhinodontis *(Wright, 1876)*

Originally described in 1876 from specimens collected from whale sharks in the Seychelles, this species was re-described in 2010 from mature females collected from whale sharks held in net pens for the Okinawa Churaumi Aquarium in Japan (Tang et al. 2010). It was found only on the unique micro-habitat of the surface of the filter pads inside the mouth, but it nonetheless appears to be quite mobile. Unlike *P. rhincodonicus*, it lacks pigment in the carapace, but developing white egg strands are clearly and distinctively visible through the translucent body in the region of the inflated gonad segment behind the main carapace of females. *Dysgamus rhinodontis* and *D. atlanticus* collected from the Caribbean, and *Echthrogaleus pectinatus* collected from the Indian Ocean at Sri Lanka, are synonyms (Tang et al. 2010). Tang et al. (2012) subsequently described the male of the species, which is difficult and not always done with parasitic copepods because they are often small and dissimilar to the females. In this case, characterization of the male resolved the proper placement of the genus in the family Pandaridae.

Figure 4.2 (a) *Pandarus rhincodonicus* copepods on the first dorsal fin of a whale shark in Indonesia. (Picture: Alistair D. M. Dove.) (b) Whale shark from Tubbataha, Philippines, heavily infected with *P. rhincodonicus*. (Picture: Simon J. Pierce.)

4.3.1.3 Lepeophtheirus acutus *Heegaard, 1943*

This species is a member of the genus that includes the salmon louse *L. salmonis*, a known pest in salmonid aquaculture (Costello 2006). Tang et al. (2013) re-described *L. acutus* based on material from whale sharks collected in Okinawa, Japan, but unlike *Pandarus rhincodonicus* and *Pr. rhinodontis*, which both only infect whale sharks, *L. acutus* infects a range of elasmobranchs including the blue spotted ray *Taeniura lymma* and reef manta *Mobula alfredi* in the wild, and zebra sharks *Stegostoma tigrinum* (formerly *S. fasciatum*), white-tipped reef sharks *Triaenodon obesus*, pelagic stingrays *Pteroplatytrygon violacea*, and grey reef sharks *Carcharhinus amblyrhynchos* in various aquariums in Europe (Tang et al. 2013).

4.3.2 Isopoda

Like the copepods, the isopods are another group of crustaceans that contains many parasitic members but, of these, only one has been described from whale sharks.

4.3.2.1 Gnathia trimaculata *Coetzee, Smit, Grutter and Davies, 2009*

Originally described from blacktip reef sharks *Carcharhinus melanopterus* from the Great Barrier Reef in Australia, this gnathiid isopod was recorded from net-penned whale sharks and also from 24 other species of elasmobranchs collected in Japan by Ota et al. (2012) (Figure 4.1). These included other carpet sharks (Orectolobiformes), requiem sharks (Carcharhinidae), hammerhead sharks (Sphyrnidae), and a range of ray species, particularly among the Myliobatidae (eagle rays & devil rays), so clearly this species has quite liberal host specificity. Like other members of its family, *G. trimaculata* has an interesting life cycle that parallels that of terrestrial ticks in that the organism needs to leave the host periodically in order to molt its exoskeleton. Adults are free living, and mating takes place in refugia within coral reefs on the bottom, after which the first- and second-stage praniza larvae attach to and feed off the blood of teleost (bony fish) hosts, molting in between (Tanaka 2007). It is only the third molt instar that infects elasmobranch hosts, after which the terminal molt adult adopts its free-living habit. There are no data about pathogenesis of *G. trimaculata* infecting whale sharks; the organism does feed on blood, but the enormous difference in relative body size indicates that infection loads would need to be extremely high to induce a hemorrhagic anemia.

4.3.3 Platyhelminthes

Four related groups make up the neodermatans or parasitic flatworms: the true flukes (Digenea and Aspidogastrea), tapeworms (Cestoda) and monogeneans (Monogenea or Monogenoidea). Tapeworms and monogeneans are particularly diverse among elasmobranchs, whereas the true flukes are much more common and diverse in the gastrointestinal tract of ray-finned fishes (Poulin & Morand 2004). Just one tapeworm and one aberrant digenean have been described from whale sharks to date (Figure 4.1), although it seems certain that many other species await description.

4.3.3.1 Wenyonia rhincodonti *Malhotra, Jaiswal, Singh, Capoor and Malhotra, 2011*

This species is ostensibly a member of the tapeworm order Caryophyllidea, which is thought to be a basal group among tapeworms, based on the lack of body pseudo-segmentation (proglottids) and the very simple structure of the anterior attachment organ or scolex. Like most tapeworms, caryophyllideans live in the lumen of the intestine of their host. However, the original taxonomic description of *W. rhincodonti* by Malhotra et al. (2011) is poor, and there is good reason to suspect that this is not a valid taxon. *All* other *Wenyonia* species are known only from the intestines of African freshwater catfishes (!) and are moderately sized tapeworms with a simple but well-developed scolex (Schaeffner et al. 2011). *Wenyonia rhincodonti*, on the other hand, is <2mm in length and appears to have no scolex at all, based on the figure in the original description. In every respect, the figure closely resembles not a caryophyllidean, but a single proglottid (segment) shed from the strobila (the "tape") of a tetraphyllidean tapeworm.

In addition to confusion over the taxonomic description of this worm, there is also confusion about the host. It was described from 22 of 189 *Rhincodon typus* examined in Goa, India, and to the best of our knowledge, there has never been such a large sample of whale sharks dissected for science. In the description, *Rhincodon typus* is sometimes spelled as *Rhyncodon typus*, and they are described once as "deep sea fishes" and another time as "sharks", but there is no other discussion of the host or the site within the host.

There seems a good possibility that Malhotra et al. (2011) either misidentified the host (perhaps they meant the guitarfish *Rhinobatos typus*?), mischaracterized the parasite, or most likely both. Tapeworms in sharks often shed individual proglottids from the end of their "tape" or strobila, and in this case, it seems very likely that the authors have misidentified one of these free proglottids from a tetraphyllidean tapeworm as a whole, monozoic animal. This confusion is unlikely to be resolved until the spiral valve of a recently dead whale shark can be dissected using appropriate parasitological techniques, but in the meantime, it seems most wise to consider this species dubious at best or what taxonomists call *nomen dubium*.

4.3.3.2 Paronatrema boholana *Eduardo, 2010*

Most digeneans, or true flukes, inhabit the gastrointestinal tract or one of a number of other internal sites within the body of the host. The family Syncoeliidae includes digeneans that are unusual among the rest of the class in being truly ectoparasitic. Eduardo (2010) described *P. boholana* from

Bohol island in the Philippines, where it was found on the skin and gills of two dead whale sharks. These are small worms 2.5–3.5 mm in length, with a body divided into a tubular forebody and a flattened hindbody, separated by a well-developed muscular sucker on the ventral surface, used for attachment to the host. Two other species are published: *P. vaginicola* from *Squalus* dogfishes and *P. mantae* from manta rays (*Mobula* spp.), and we are aware of at least one other syncoeliid on the gills of tiger sharks *Galeocerdo cuvieri* (T. H. Cribb pers. comm.)

4.3.4 Hirudinea

The leech *Branchellion torpedinis* has very occasionally been observed on the skin of whale sharks at Georgia Aquarium (Dove et al. 2017). This species has not been observed on whale sharks in natural settings and most likely represents an opportunistic host switch within the aquarium from the more typical hosts for this large and distinctive ectoparasite, which are bottom-dwelling sharks and rays. Among these other hosts, the cow-nosed ray *Rhinoptera bonasus* seems especially susceptible. This parasite is not a jawed leech of the sort familiar to most hikers in the tropics, but part of a strictly aquatic family (Piscicolidae) that uses an eversible muscular and glandular pharynx to dissolve and then ingest skin, mucus and even blood from aquatic hosts (Tessler et al. 2018). It is quite pathogenic to smaller rays and can lead to a profound hemorrhagic anemia in them (Marancik et al. 2012), especially when it occurs in large numbers. The only pathogenesis in the whale shark appeared to be a slight thickening of the epithelium and extra mucus production in the area around the attachment site. Most leeches glue onto hard surfaces cocoons containing one to a few embryos, but in an apparent case of convergent evolution, *B. torpedinis* behaves much more like a monogenean flatworm, broadcasting cocoons into the water column; each of these contains a single embryo.

4.4 ASSOCIATES

In addition to the more orthodox parasites described above, where the nature of the relationship is fairly clear, there exists another suite of species that have much more nebulous relationships with whale sharks. These are primarily bony fishes of a number of different types, which are neither predators nor prey of whale sharks. Rather, they associate with the shark as a source of food to obtain shelter from predators or to reduce the energetic cost of movement (Rowat & Brooks 2012) by either drafting behind the shark or riding its leading pressure wave, much as dolphins ride the bow waves of ships.

Whale sharks can and do move very large distances (see Chapter 6), and changing environmental conditions along the journey may play a role in the diversity of species that are able to associate with them (Eckert & Stewart 2001, Eckert et al. 2002, Wilson et al. 2005, Rowat et al. 2006, Hsu et al. 2007, Rowat & Gore 2007, Wilson et al. 2007, Ketchum et al. 2012, Fox et al. 2013, Hueter et al. 2013, Acuña-Marrero et al. 2014, Berumen et al. 2014). Some of the more common fish associates of whale sharks are summarized in Table 4.2, and selected species of particular interest are discussed in more detail below, grouped by their dominant behavior. None of these behaviors are mutually exclusive and some fishes almost certainly engage in two or more, depending on conditions or at different times. Such dynamic and transient relationships are characteristic of more facultative symbioses.

4.4.1 Surfers and Drafters

For these species, the main function of their relationship with whale sharks appears to be efficient travel. By either riding the bow wave or drafting on or behind fins, the smaller fish gets a hydrodynamic boost, which reduces their energetic cost of transport (Figure 4.3).

Table 4.2 Some Fish Associates of Whale Sharks

Family	Species	Common Names	Association
Carangidae	Gnathanodon speciosus	Golden trevally	Bow rider (juveniles)
Carangidae	Naucrates ductor	Pilotfish	Bow rider
Carangidae	Elagatis bipinnulata	Rainbow runner	Preying on smaller fishes that shelter around the shark
Carangidae	Atule mate	Yellowtail scad	Shelter
Carangidae	Caranx lugubris	Black jack	Drafting. Preying on smaller fishes that shelter around the shark
Carangidae	Caranx ignobilis	Giant trevally	Drafting (Georgia Aquarium)
Carcharhinidae	Carcharhinus falciformis	Silky shark	Parasite removal (scratching on whale shark)
Carcharhinidae	Carcharhinus limbatus	Blacktip shark	Indirectly associated
Echeneidae	Echeneis naucrates	Live sharksucker or striped suckerfish	Transport, protection, coprophagy
Echeneidae	Remora remora	Shark sucker or remora	Transport, protection, coprophagy
Echeneidae	Remora albescens	White suckerfish	Transport, protection, coprophagy
Echeneidae	Remora osteochir	Marlin sucker	Transport, protection, coprophagy
Engraulidae	Engraulis spp.	Anchovy	Shelter
Mobulidae	Mobula alfredi	Reef manta ray	Indirectly associated
Rachycentridae	Rachycentron canadum	Cobia	Juveniles may mimic remoras
Scombridae	Rastrelliger kanagurta	Indian mackerel	Plankton feeder and seen feeding alongside and in front of juvenile whale sharks.
Scombridae	Euthynnus affinis, E. aleterattus	Mackerel tuna, Little tunny	Whale sharks are known to eat their eggs.
Stromatidae	Pampus argenteus	Silver pomfret	From literature (Tubb 1948)

Juvenile golden trevally *Gnathanodon speciosus* (Bleeker 1851) are commonly found associated with whale sharks around the world and they are frequently seen in large numbers at the Gulf of Tadjoura in Djibouti and at Mafia Island in Tanzania (S. J. Pierce pers. comm.). They are usually seen swimming immediately in front of the mouth of the whale shark, so this association appears to benefit the trevally in the form of energy saving through the practice of bow riding. It is possible that the trevally also receive some protection from predators, but the relationship seems unlikely to have any measurable impact on the whale shark and so may be considered a form of commensalism (Figure 4.3).

The pilotfish *Naucrates ductor* (Rafinesque 1810) associates commonly with whale sharks, as well as many other marine megafaunal species, and is most often seen bow riding (Figure 4.4). Their behavior is similar to that of juvenile golden trevally, to which they are related (Carangidae), but in the pilotfish, this behavior continues throughout the adult stage. This fact suggests that on the spectrum of relationships from obligate to facultative, the relationship between pilotfish and their hosts is more towards the obligate end. Pilotfish are predatory, epipelagic, oceanic fish, and their relationship with whale sharks is considered commensal (Fuller & Parsons 2019). Largely because the pilotfish is oceanic, the presence of *N. ductor* on whale sharks at coastal constellations has been speculated to indicate a recent arrival of sharks moving in from oceanic sites (David Rowat pers. comm.)

In addition to golden trevally and pilotfish, some other carangids (trevallies and jacks) have been observed swimming just above or just below the pectoral fins of whale sharks and gaining some boost to their cruising behavior. Black jacks *Caranx lugubris*, are often seen engaging in this behavior around the giant female whale sharks seen at Darwin Island in the Galapagos, often

Figure 4.3 Juvenile golden trevally, *Gnathanodon speciosus* bow riding on a whale shark. (Picture credit: David P. Robinson.)

Figure 4.4 Pilotfish, *N. ductor* bow riding a whale shark. (Picture credit: David P. Robinson.)

in courting pairs where the female is lightly colored, while the male is very dark. Giant trevally, *C. ignobilis* show similar behavior in association with the subadult whale sharks with which they share the Ocean Voyager habitat at Georgia Aquarium (A. Dove pers. obs.). Almaco jack, *Seriola rivoliana* (Carangidae: Naucratinae) have been observed drafting behind the first dorsal fin of whale sharks in St. Helena, South Atlantic (A. Dove unpublished data).

4.4.2 Scratchers

A number of different species have been observed scratching – sometimes referred to as "flashing" – on the skin of whale sharks. Whale sharks do not seem to be disturbed by this behavior, even when it comes from sizeable sharks. The most common place this behavior occurs is on the top of the body anterior to, and sometimes just behind, the first dorsal fin, but some animals also flash off the upper lobe of the caudal fin. Why animals flash is not clear; it may be to remove ectoparasites or dead skin, or perhaps just to scratch an itch.

The silky shark, *C. falciformis* (Müller & Henle 1839) has frequently been observed rubbing itself on whale sharks at Darwin Island in the Galapagos islands of Ecuador. These interactions may be beneficial to the silky shark because the behavior dislodges external parasites; silky sharks at that site are frequently infected with visible parasitic caligoid copepods on the skin and fins (A. Dove pers. obs.).

Among bony fishes, yellowfin tuna *Thunnus albacares*, kingston (*Decapterus maccarellus*), stonebrass (*D. muroadsi*), and pompano (*Trachinotus ovatus*) have all been observed engaging in scratching behavior on whale sharks in the waters surrounding St. Helena in the South Atlantic (A. Dove, unpublished data). The tuna draft behind the shark, occasionally flashing off the upper caudal lobe, whereas the *Decapterus* spp. and *T. ovatus* school around, and flash off, the dorsal surface of the head.

4.4.3 Suckers

Among the fish associates commonly found with whale sharks, the remoras or suckerfish deserve special discussion. Rather than drafting or bow riding their hosts, these extraordinary animals temporarily attach themselves to a wide range of marine megafauna using a modified dorsal fin that has migrated anteriorly to a position immediately on top of the head. This organ takes the form of a loculated oval disc with many transverse ridges and a single medial ridge connecting them all (Britz & Johnson 2012). The attachment afforded by this organ is very tenacious in the posterior direction; pulling on the remora's tail causes hinged spiny ridges within the disc to bite down into the skin of the host like tent poles. The attachment force of these ridges, which are derived from the ancestral dorsal fin spines, is enhanced by a flexible flap that forms a marginal valve seal around the edge of the disc, such that attachment is achieved through both friction and suction. Release occurs near-instantaneously by swimming forwards slightly and breaking the valve seal using a muscle at the front of the valve. The purpose of attachment to the host appears to be extremely efficient transport; indeed, using this adaptation, remoras probably spend little or no energy to move vast distances in the open ocean.

The four species of remora commonly found on whale sharks are also seen on turtles, dolphins, whales, manta rays and other sharks: *Remora*, *R. albescens*, *R. osteochir*, and *Echeneis naucrates*. While the live sharksucker or striped suckerfish, *E. naucrates* is probably the most well known of all whale shark associates (Figure 4.5); there are no remora species that are restricted exclusively to whale sharks. Despite the very specific adaptation for temporary attachment, remoras are strong independent swimmers, and the shark sucker in particular is often seen swimming clear of its host, foraging for food. Small remoras have frequently been observed inside the mouth of whale sharks at a number of localities, and the tails of tiny remoras can frequently be seen protruding from the spiracle of whale sharks.

The relatively more intimate association of remoras with whale sharks has led to speculation about the nature of their relationship. There is some evidence that remoras eat parasites off the skin of the whale shark and are therefore engaged in a form of mutualism (Cressey and Lachner 1970, Mucientes et al. 2009). In many years of observation at different locations around the world, we have never observed this directly. What we have observed, however, is coprophagy (eating feces) on a grand scale. When all of the remoras move from the head and fins to sites near the cloaca of the shark, it is a good indication that a defecation event is imminent, although how the remoras detect this is unknown.

It is hard to quantify the impact of remoras on whale sharks. At times their abundance can be so high that some negative effect on the host seems inevitable, at least in terms of the additional hydrodynamic drag that the shark must endure. But in general, they seem to tolerate their passengers with the usual indifference that they show to most external stimuli. Attachment of the remora's dorsal disc can leave lesions that persist for at least short periods of time (Figure 4.6), but whether these lesions ever form portals of entry for secondary infections has not been evaluated.

Figure 4.5 Live sharksuckers, *Echeneis naucrates* gathered beneath a whale shark. (Picture credit: Simon J. Pierce.)

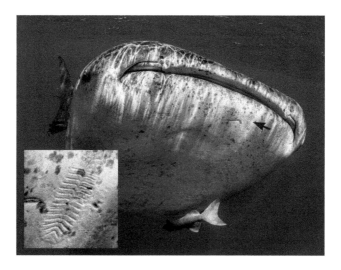

Figure 4.6 A mark left by recent attachment of a common shark sucker or remora, *Remora* on a whale shark in St. Helena. (Picture credit: Simon J. Pierce.)

4.4.4 Shelterers

A wide range of fish species can be observed schooling or shoaling around the large bodies of whale sharks. In many of these cases, the behavior is transient or fleeting, which suggests that this relationship is facultative and temporary at best. This behavior may be an example of simple thigmotaxis, where smaller animals are attracted to large objects (Vassilopoulou et al. 2004), or they may use close proximity to the large bodies of whale sharks as some form of protection from predators. Most often it involves smaller fishes such as anchovies (Engraulidae), scads (Carangidae: *Decapterus* spp.) and smaller mackerels like the Indian mackerel *Rastrelliger kanagurta* (Table 4.2) (Figure 4.7); the silver pomfret *Pampus argenteus* Euphrasen, 1788 (Stromatiidae) was also

Figure 4.7 A dense school of Indian mackerel, *Rastrelliger kanagurta* sheltering under a whale shark. Note also that there are golden trevally, *Gnathanodon speciosus* bow riding, and a live sharksucker, *Echeneis naucrates* near the right pectoral fin, so this shark is a mobile island of biodiversity, not even counting the parasites almost certainly present in and on the shark. (Picture credit: Simon J. Pierce.)

observed to engage in this behavior off the coast of India (Tubb 1948). We have observed kingston *D. macarellus* breaking away from their school to shelter on top of the head of whale sharks, from where rainbow runner (*Elagatis bipinnulata*) and almaco jack (*Seriola rivoliana*) dart in to try and pick them off, but where they are at least safe from attack from below by yellowfin tuna (*Thunnus albacares*) (A. Dove unpublished data).

4.4.5 It's Complicated

Some regular associates of whale sharks have relationships that do not appear to fit neatly into any of the informal categories above. Principal among these is the special and complex relationship between whale sharks and various species of tuna. This sometimes involves predation of tuna eggs by whale sharks (Heyman et al. 2001, Duffy 2002), which has been recorded at several locations around the world, including for the little tunny *Euthynnus alletteratus* (Rafinesque 1810) off the Yucatan Peninsula in Mexico (de la Parra Venegas et al. 2011), blackfin tuna *Thunnus atlanticus* (Lesson 1831) and skipjack tuna *Katsuwonus pelamis* (Linnaeus 1758) in the U.S. Gulf of Mexico (Hoffmayer et al. 2007), skipjack tuna in St. Helena (Clingham et al. 2016, Perry et al. 2020) and mackerel tuna *Euthynnus affinis* (Cantor 1849) in Qatari waters of the Arabian Gulf (Robinson et al. 2013) (Figure 4.8). Most often the tuna and the whale sharks are not directly observed together and the identity of the tuna is sometimes determined by DNA sequencing of the fish eggs on which the sharks are feeding. Other times, whale sharks are observed actively pursuing schools of tuna, as if waiting for them to stop and spawn.

While the preceding examples describe situations where whale sharks prey on tuna at the egg stage, sometimes tuna appear to benefit more from their relationship by preying on fish that aggregate around whale sharks, as described in Section 4.4. For example, at Nosy Be in Madagascar, whale sharks are almost exclusively associated with mackerel tuna, *E. affinis,* even though there appears to be no spawning of this fish taking place at that location. Instead, small baitfish tend to school around the heads of whale sharks and the tuna appear to associate with the sharks in order to feed on these baitfish. The tuna drive a bait ball to the surface by attacking them from beneath, which causes the baitfish to seek shelter around the head of a whale shark. But it is false shelter,

Figure 4.8   A whale shark among a school of mackerel tuna *Euthynnus affinis*. Whale sharks feed on the eggs
of this species at a number of constellation sites around the world, which are presumably also
mass spawning sites for the tuna, although this has not been observed directly. (Picture credit:
Simon J. Pierce.)

and often the tuna and whale shark both feed on the baitfish simultaneously, with the tuna around
the edges and the whale shark in the center of the bait ball (D. Robinson, pers. obs.). This situation
may represent a form of cooperative feeding, so the relationship between whale sharks and tuna may
sometimes form a true mutualism.

In animal-borne camera footage from whale sharks in St. Helena (A. Dove, unpublished
data), we have observed yellowfin tuna *Thunnus albacares* following whale sharks for prolonged
periods, drafting just behind the caudal fin. This may reduce somewhat the energetic cost of
transport for the tuna, but it seems more likely that the sharks are waiting for baitfish to be
attracted to the whale shark, facilitating feeding and reducing the energetic cost of foraging in
open water.

Regardless of who is following whom, commercial tuna fishers often seek out whale sharks and
fish close to them due to the reliable co-occurrence of the two species. This can result in incidental –
and sometimes deliberate – capture of whale sharks (Romanov 2002, Dagorn et al. 2013, Hall 1998,
Rowat & Brooks 2012). More is said in Chapters 11 and 12 about bycatch in tuna fisheries as a
source of whale shark mortality.

The west Atlantic or Caribbean manta ray, *Mobula* sp. cf. *birostris*, is commonly found in
close proximity to whale sharks in Mexico. They have been observed feeding on the same tuna
spawn as the whale sharks, so this association seems most likely to be an indirect link reflecting a
common food source. However, researchers have noted a certain level of interaction between the
two species that could be a form of exploitative feeding behavior on the part of the ray, similar to
that observed in tuna (C. Rohner pers. comm.), but this phenomenon needs further observation
and study.

Cobia *Rachycentron canadum* (Linnaeus 1766) are associated with whale sharks around the
world and are sometimes misidentified as small sharks and even juvenile whale sharks (Rowat &
Brooks 2012). It appears that juvenile cobia can be mimics of the live sharksucker *Echeneis naucrates*, although what benefit they get from this mimicry is unclear (Figure 4.9). It is possible that
the cobia may exploit the size of the whale shark to get close to prey items such as baitfish that gather
around them in a form or exploitative feeding, although their relationship may also involve a certain
amount of drafting (see Section 4.6.1).

Figure 4.9 Juvenile cobia, *Rachycentron canadum* apparently mimicking live sharksuckers, *Echeneis naucrates*, on a whale shark in Indonesia. (Picture credit: Alistair D. M. Dove.)

4.5 CONCLUSIONS

The most obvious conclusion of this chapter with regard to parasites and pathogens of whale sharks is that they are drastically under-studied. This is an enormous gap in our understanding of the biology of whale sharks, and one that is difficult to remedy. Collecting parasites or tissue samples for these kinds of analyses usually requires access to freshly dead host specimens, which are few and far between, since conservation measures have been enacted globally for whale sharks and access at fishery landing sites has dropped dramatically. Deliberate sacrifice for parasitological study is unethical given its conservation status and the cultural values attached to whale sharks by the public. So, we must rely on stranded animals, which occur rarely, and even more so in the proximity of available suitably trained scientific staff to conduct these kinds of studies. DNA-based technologies, while indirect and certainly no substitute for careful taxonomic efforts, offer some hope for potential detection of parasites, bacteria, fungi and viruses by their DNA traces rather than collection of specimens. That may be the best we can hope for in the future studies of parasites and microbes in whale sharks, while we wait for occasional stranded animals to give insight into the parasite fauna of this species, which is likely to be considerable in both abundance and diversity.

Aside from the more obligate parasites, a number of different fishes associate facultatively with whale sharks for a range of reasons described in Section 4.4. The nature of these associations varies considerably in time and at different locations around the world. Some of them may be exploitative, others predatory, some indirectly associated by a common resource and some even potentially mutualistic. The discovery of new types of relationships is certainly a possibility, so careful documentation of associated fauna and their behavioral interactions with whale sharks is of considerable interest. Studying fleeting, facultative relationships in the vast expanse of the open ocean is logistically challenging, but these relationships nonetheless warrant dedicated study because some of them appear to be quite complex and nuanced.

ACKNOWLEDGMENTS

AD appreciates his training in parasitology from Tom Cribb, past help with copepods from Ash Bullard and the late George Benz and help with pathology from everyone who goes to the Eastern Fish Health Workshop every year. Thanks to Simon Pierce and David Robinson for figure permissions.

REFERENCES

Acuña-Marrero, D., Jiménez, J., Smith, F., et al. 2014. Whale shark (*Rhincodon typus*) seasonal presence, residence time and habitat use at Darwin Island, Galapagos Marine Reserve. *PloS One* 9(12):e115946. https://doi.org/10.1371/journal.pone.0115946.

Benz, G. W., Bullard, S. A., and A. D. M. Dove. 2001. Metazoan parasites of fishes: Synoptic information and portal to the literature for aquarists. *Proceedings of the American Zoo and Aquarium Association* AZA: Silver Spring, MD, pp. 1–15.

Berumen, M. L., Braun, C. D., Cochran, J. E., et al. 2014. Movement patterns of juvenile whale sharks tagged at an aggregation site in the Red Sea. *PloS One* 9(7):e103536. https://doi.org/10.1371/journal. pone.0103536.

Britz, R., and D. Johnson. 2012. Ontogeny and homology of the skeletal elements that form the sucking disc of remoras (Teleostei, Echeneoidei, Echeneidae). *Journal of Morphology* 273(12). https://doi. org/10.1002/jmor.20105.

Clingham, E., Webb, D. H., De la Parra-Venegas, R., et al. 2016. Further evidence of the importance of St. Helena as habitat for whale sharks. *Q Science Proceedings 4th International Whale Shark Conference*. Doha, Qatar, May 2016. doi:10.5339/qproc.2016.iwsc4.11.

Coetzee, M. L., Smit, N. J., Grutter, A. S., and A. J. Davies. 2009. *Gnathia trimaculata* n. sp. (Crustacea: Isopoda: Gnathiidae), an ectoparasite found parasitising requiem sharks from off Lizard Island, Great Barrier Reef, Australia. *Systematic Parasitology* 72:97–112.

Costello, M. 2006. Ecology of sea lice parasitic on farmed and wild fish. *Trends in Parasitology* 22(10):475–483.

Cressey, R. F., and E. A. Lachner. 1970. The Parasitic Copepod Diet and Life History of Diskfishes (Echeneidae). *Copeia* 1970(2): 310–318. doi:10.2307/1441652.

Dagorn, L., Holland, K. N., Restrepo, V., and G. Moreno. 2013. Is it good or bad to fish with FADs? What are the real impacts of the use of drifting FADs on pelagic marine ecosystems? *Fish and Fisheries* 14(3): 391–415.

De la Parra-Venegas, R., Hueter, R. E., Gonzalez-Cano, J., et al. 2011. An unprecedented aggregation of whale sharks, *Rhincodon typus*, in Mexican coastal waters of the Caribbean Sea. *PloS One* 6(4).

Dove, A. D. M., Clauss, T. M., Marancik, D. P. and A. C. Camus. 2017. Some emerging diseases of elasmobranchs in aquariums. In: (M. Smith ed.) *Elasmobranch Husbandry Manual II – Recent Advances in the Care of Sharks, Rays and their Relatives*. Columbus, OH: Ohio Biological Survey, pp. 263–276.

Duffy, C. 2002. Distribution, seasonality, lengths, and feeding behaviour of whale sharks (*Rhincodon typus*) observed in New Zealand waters. *New Zealand Journal of Marine and Freshwater Research* 36:565–570.

Eckert S. A., and B. S. Stewart. 2001. Telemetry and satellite tracking of whale sharks, *Rhincodon typus*, in the Sea of Cortez, Mexico, and the north Pacific Ocean. *Environmental Biology of Fishes* 60(1–3):299–308.

Eckert, S. A., Dolar, L. L., Kooyman, G. L., et al. 2002. Movements of whale sharks (*Rhincodon typus*) in South-east Asian waters as determined by satellite telemetry. *Journal of Zoology* 257(1):111–115.

Eduardo, S. L. 2010. A new species of parasitic fluke, *Paronatrema boholana* Eduardo (Trematoda: Syncoeliidae) from the whale shark, *Rhincodon typus* Smith 1828 (Orectolobiformes: Rhincodontidae) off Bohol Island, Philippines. *Asia Life Sciences* 19(2):315–323.

Fox, S., Foisy, I., de la Parra-Venegas, R., et al. 2013. Population structure and residency of whale sharks *Rhincodon typus* at Utila, Bay Islands, Honduras. *Journal of Fish Biology* 83(3):574–87.

Fuller, L. N., and G. R. Parsons. 2019. A note on associations observed between sharks and teleosts. *Southeastern Naturalist* 183:489–498.

Garner, M. M. 2013. A retrospective study of disease in elasmobranchs. *Veterinary Pathology* 50(3):377–389.

Grimes, D. J., Jacobs, D., Swartz, D. G., et al. 1993. Numerical taxonomy of gram-negative, oxidase positive rods from carcharhinid sharks. *International Journal of Systematic Bacteriology* 43(1):88–98.

Hall, M. A. 1998. An ecological view of the tuna-dolphin problem: Impacts and trade-offs. *Reviews in Fish Biology and Fisheries* 8(1):1–34.

Heyman, W., Graham, R., Kjerfve, B., and R. Johannes. 2001. Whale sharks *Rhincodon typus* aggregate to feed on fish spawn in Belize. *Marine Ecology Progress Series* 215:275–282.

Hoffmayer, E. R., Franks, J. S., Driggers, W. B., et al. 2007. Observations of a feeding aggregation of whale sharks, *Rhincodon typus*, in the north central Gulf of Mexico. *Gulf and Caribbean Research* 19(2):69.

Hsu, H.-H., Joung, S. J., Liao, Y. Y., and K. M. Liu. 2007. Satellite tracking of juvenile whale sharks, *Rhincodon typus*, in the Northwestern Pacific. *Fisheries Research* 84(1):25–31.

Hueter, R. E., Tyminski, J. P., and R. de la Parra-Venegas. 2013. Horizontal movements, migration patterns, and population structure of whale sharks in the Gulf of Mexico and northwestern Caribbean Sea. *PLoS One* 8(8). doi:10.1371/annotation/491b9b6c-7f77-4fb0-b336–572078aec830.

Iwasaki, Y. 1970. On the distribution and environment of the whale shark, *Rhincodon typus*, in skipjack fishing grounds in the western Pacific Ocean. *Journal of the College of Marine Science and Technology Tokai University* 4:37–51.

Ketchum, J. T., Galván-Magaña, F., and A. P. Klimley. 2012. Segregation and foraging ecology of whale sharks, *Rhincodon typus*, in the southwestern Gulf of California. *Environmental Biology of Fishes* 96(6):779–795.

Kiselev, D., Matsvay, A., Abramov, I., et al. 2020. Current trend in diagnostics of viral infections of unknown etiology. *Viruses* 12:211. doi:10.3390/v12020211.

Lafferty, K. D., Allesina, S., Arim, M., et al. 2008. Parasites in food webs: The ultimate missing links. *Ecology Letters* 11: 533–546.

Lafferty, K. D., and A. M. Kuris. 2002. Trophic strategies, animal diversity and body size. *Trends in Ecology and Evolution* 17(11):507–513.

La Patra, S. 2003. General procedures for virology. In: AFS-FHS (American Fisheries Society-Fish Health Section). *FHS blue book: Suggested procedures for the detection and identification of certain finfish and shellfish pathogens* 2016 edition. Accessible at: http://afsfhs.org/bluebook/bluebook-index.php.

Malhotra, A., Jaiswal, N., Singh, H. R., et al. 2011. Taxometric analysis of helminthes of marine fishes. 1. *Pedunculacetabulum spinatum* n.sp., from *Chorinemus mandetta* and *Wenyonia rhincodonti* n.sp., from *Rhincodon typus*. *Journal of Parasitic Diseases* 35(2):222–229.

Marancik, D. P., Berliner, A. L., Cavin, J. M., et al. 2011. Disseminated fungal infection in two species of captive sharks. *Journal of Zoo and Wildlife Medicine* 42(4):686–693.

Marancik, D. P., Dove, A. D. M., and A. C. Camus. 2012. Experimental infection of yellow stingrays *Urobatis jamaicensis* with the marine leech *Branchellion torpedinis*. *Diseases of Aquatic Organisms* 101(1):51–60.

Matsunaga, H., Nakano, H., Okamoto, H. and Z. Suzuki. 2003. Whale shark migration observed by pelagic tuna fishery near Japan. In *16th Meeting of the Standing Committee on Tuna and Billfish*, pp. 1–7.

Mucientes, G., Queiroz, N., Pierce, S. J., et al. 2009. Is host ectoparasite load related to echeneid fish presence? *Research Letters in Ecology*. doi:10.1155/2008/107576.

Mylniczenko, N. D., Harris, B., and R. E. Wilborn. 2007. Blood culture results from healthy captive and free-ranging Elasmobranchs. *Journal of Aquatic Animal Health* 19(3):159–167.

Norman, B. M., Newbound, D. R., and B. Knott. 2000. A new species of Pandaridae (Copepoda), from the whale shark *Rhincodon typus* (Smith). *Journal of Natural History* 34(3):355–366.

Ota, Y., Hoshino, O., Hirose, M., et al. 2012. Third-stage larva shifts host fish from teleost to elasmobranch in the temporary parasitic isopod, *Gnathia trimaculata* (Crustace: Gnathiidae). *Marine Biology* 159:2333–2347.

Perry, C. T., Clingham, E., Webb, D. H., et al. 2020. St. Helena: An important reproductive habitat for whale sharks (*Rhincodon typus*) in the South Central Atlantic. *Frontiers in Marine Science* 7:576343.

Poulin, R., and S. Morand. 2004. *Parasite Biodiversity*. Washington, DC: Smithsonian Books, 216 pp.

Rao, K. S., Authors, L., Pauly, D., and D. Zeller. 1986. On the capture of whale sharks off Dakshina Kannada coast. *Marine Fisheries Information Service: Technical and Extension Series* 66:22–29.

Robinson, D. P. 2016. The ecology of whale sharks in the Arabian Gulf and the Gulf of Oman. PhD Thesis, Heriot-Watt University, UK

Robinson, D. P., Jaidah, M. Y., Jabado, R. W., et al. 2013. Whale sharks, *Rhincodon typus*, aggregate around offshore platforms in Qatari waters of the Arabian Gulf to feed on fish spawn. *PloS One* 8(3):e58255.

Romanov, E. V. 2002. Bycatch in the tuna purse-seine fisheries of the western Indian Ocean. *Fishery Bulletin* 100(1):90–105.

Rowat, D., Meekan, M. G., Engelhardt, U., et al. 2006. Aggregations of juvenile whale sharks (*Rhincodon typus*) in the Gulf of Tadjoura, Djibouti. *Environmental Biology of Fishes* 80(4):465–472.

Rowat, D., and K. S. Brooks. 2012. A review of the biology, fisheries and conservation of the whale shark *Rhincodon typus*. *Journal of Fish Biology* 80(5):1019–1056.

Rowat, D., and M. Gore. 2007. Regional scale horizontal and local scale vertical movements of whale sharks in the Indian Ocean off Seychelles. *Fisheries Research* 84(1):32–40.

Schaeffner, B. C., Jirku, M., Mahmoud, Z. N., and T. Scholz. 2011. Revision of *Wenyonia* Woodland 1923 (Cestoda: Caryophyllidea) from catfishes (Siluriformes) in Africa. *Systematic Parasitology* 79:83–107.

Silas, E. G. 1989. The whale shark: Is the species endangered or vulnerable? *Biology Education* 6(2): 78–90.

Springer, S. 1957. Some observations of the behavior of schools of fishes in the Gulf of Mexico and adjacent waters. *Ecology* 166–171.

Tanaka, K. 2007. Life history of gnathiid isopods – current knowledge and future directions. *Plankton and Benthos Research* 2(1):1–11.

Tang, D., Yanagisawa, M., and K. Nagasawa. 2010. Redescription of *Prosaetes rhinodontis* (Wright 1876) (Crustacea: Copepoda: Siphonostomatoida), an enigmatic parasite of the whale shark, *Rhincodon typus* Smith (Elasmobranchii: Orectolobiformes: Rhincodontidae. *Zootaxa* 2493:1–15.

Tang, D., Benz, G. W., and K. Nagasawa. 2012. Description of the male of *Prosaetes rhinodontis* (Wright 1876) (Crustacea: Copepoda: Siphonostomatoida), with a proposal to synonymize Cecropidae Dana 1849 and Amasteridae Isawa 2008 with Pandaridae Milne Edwards 1840. *Zoosymposia* 8:7–19.

Tang, D., Venmathi Maran, B.A., Matsumoto, Y., and K. Nagasawa. 2013. Redescription of *Lepeophtheirus acutus* Heegard 1943 (Copepoda: Caligidae) parasitic on two elasmobranch hosts off Okinawa-jima Island, Japan. *Journal of Natural History* 47(5): 581–596.

Tao, Z., Bullard, S. A., and C. R. Arias. 2014. Diversity of bacteria cultured from the blood of lesser electric rays caught in the Northern Gulf of Mexico. *Journal of Aquatic Animal Health* 26:225–232.

Tessler, M., Marancik, D., Champagne, D., Dove, A. D. M., Camus, A. C., and M. E. Siddall. 2018. Marine leech anticoagulant diversity and evolution. *Journal of Parasitology* 104(3):210–220.

Tubb, J. A. 1948. Whale sharks and devil rays in North Borneo. *Copeia* 3:222.

Vassilopoulou, V., Siapatis, A., Christides, G., and P. Bekas. 2004. The biology and ecology of juvenile pilot-fish (*Naucrates ductor*) associated with Fish Aggregating Devices (FADs) in eastern Mediterranean waters. *Mediterranean Marine Science* 5(1):61–70.

Weinstein, S. B. and A. M. Kuris. 2016. Independent origins of parasitism in Animalia. *Biology Letters* 12(7):20160324. doi:10.1098/rsbl.2016.0324.

Wood, C. M., Liew, H. J., De Boeck, G., et al. 2019. Nitrogen handling in the elasmobranch gut: A role for microbial urease. *Journal of Experimental Biology* 222(3):jeb194787.

Wilson. S. G., Stewart, B. S., Polovina, J. J., et al. 2007. Accuracy and precision of archival tag data: A multiple-tagging study conducted on a whale shark (*Rhincodon typus*) in the Indian Ocean. *Fisheries Oceanography* 16(6):547–554.

Wilson, S. G., Polovina, J. J., Stewart, B. S., and M. G. Meekan. 2005. Movements of whale sharks (*Rhincodon typus*) tagged at Ningaloo Reef, Western Australia. *Marine Biology* 148(5):1157–1166.

CHAPTER 5

Genetic Population Structure of Whale Sharks

Jennifer V. Schmidt
Shark Research Institute

CONTENTS

5.1 WHAT IS A POPULATION?

To discuss population structure, it is necessary to establish what is meant by a *population*, but this is not as precise a term as it might appear to be. All known members of a species may be considered a population; for example, the global population of whale sharks. More commonly in research, however, a population is any specific group of animals under study, typically to some degree isolated in space or time from other groups. Members of a single species may be divided into multiple populations if, for example, they inhabit different locations and are studied separately, or if there are barriers to the free movement of animals between locations. The term population *structure* describes the composition of individual populations, with respect to number of individuals, age and sex, and the extent to which the animals interact with other populations. Using whale sharks as an example, animals may be grouped into those inhabiting different regions, for example, the Djibouti population or the Australia population, or those inhabiting different oceans, e.g., the Atlantic Ocean population or the Pacific Ocean population. Specific details and context make clear how the term is being used in a given situation. Population *connectivity* is a term used to refer to the degree to which different populations are in contact and exchange individual animals or genetic material through breeding.

5.1.1 Why Population Structure Matters

Determining population structure helps researchers understand the biology of a species, learn about their reproductive strategies, determine if they require protection from environmental or human-made threats, and devise appropriate conservation plans. Knowing the age and sex of animals in a population, for example, can help researchers predict the social structure of the species and its breeding strategies. In determining conservation strategies for threatened species, isolated breeding populations of resident animals have different needs than a species with a single mixed population that undertakes long-range migrations. Resident animals or short-distance migrants can be protected with regional approaches, while highly migratory animals require international collaborations to assure their safety over a large range. These distinctions are particularly important for marine species, which have fewer boundaries and greater migratory capabilities than terrestrial animals (Avise 1998), and especially important for the whale shark, which is regarded as a highly migratory species.

Whale sharks are found in all tropical and warm temperate marine waters, but they are most frequently encountered at coastal feeding aggregations (Colman 1997, Norman et al. 2017). If each whale shark aggregation is considered a population, then it is important to ask if these are isolated, with no interaction between geographically separate locations or if migration brings different populations into contact.

A better understanding of whale shark population genetics could lead to improved conservation interventions for this species. Whale sharks are listed as Endangered by the International Union for the Conservation of Nature and are protected within the waters of many countries, but there remains an active trade in whale shark products (Jabado et al. 2015, Steinke et al. 2017). Knowing whether whale sharks remain local or migrate long distances will determine if national regulations can properly protect these animals or if all members are threatened as they move through unsafe waters.

5.1.2 Determining Population Structure

To define the structure of any given population requires data on the number of individuals present, their demographics, their relatedness to each other and to animals in different populations, and the manner and frequency with which populations interact. These data are typically gathered through field work at each population site and from genetic analysis of DNA samples collected across the range of the species.

5.1.2.1 Field Studies of Populations

Field work is essential to surveying wild populations, particularly the ability to distinguish individual animals and track their habitat usage and interactions with other populations. Boat-based and aerial surveys can count the number of individuals in a population, while observation or physical examination of individuals can determine demographic characteristics of age and sex. Collection of tissue biopsies, or of hair or feces, allows for DNA isolation from individuals. Capture–mark–recapture is a technique that allows for marking of individuals with numbered tags that can be recorded if the animal is re-sighted, or more commonly now with satellite tags that report the animal's location. For animals such as the whale shark, that carry unique skin patterns, photo-identification greatly increases the number of animals that can be followed. See Chapter 7 for more on photo-identification.

5.1.2.2 Biochemical and Genetic Analysis of Populations

Field studies provide information about one group of animals during one period of time, but larger-scale information can be gained by biochemical and genetic analyses. Early studies used differences in the structures of certain proteins to provide a measure of genetic diversity and differentiation within a population. These studies were cumbersome, however, and the degree of variability in proteins was not sufficient for high-resolution analysis. The ability to work with DNA instead of proteins, largely as a result of the invention of the polymerase chain reaction (PCR) that allows DNA recovery from very small amounts of tissue, revolutionized population genetics (Saiki et al. 1985). DNA sequences mutate, or change, naturally over time, and populations that do not interact will accumulate a profile of changes unique to each group. Typically, the longer two populations are separated the more genetically distinct they will become. If different populations do interact and interbreed, the DNA changes arising in each group will become mixed, rendering them genetically similar. A frequently used term is gene flow: the movement of genetic information between populations by interbreeding. Breeding between populations is required for gene flow, where genetic changes are passed from parent to offspring, and it is important to remember that a migratory event that does not result in breeding is unrecorded on a genetic level. Population genetics determines the degree to which two or more populations interact by measuring differences in their genetic profiles.

The typical goals of population genetic studies are threefold. First, to determine the genetic diversity of the species, both within discrete populations and globally. Genetic diversity is a general measure of the health of a population that reflects both its size and its ability to respond to environmental changes. Second, to measure the level of genetic difference (differentiation) between populations of animals found in different parts of the species' range. Third, to use the amount of genetic differentiation between populations to estimate their degree of connectivity, which is the extent to which these different groups of animals interact. These ideas are discussed in more detail through the rest of this chapter and in Chapters 6 and 7. No single approach can fully define the population structure for a species. Rather, it is essential to integrate data from field studies and laboratory genetic analyses when evaluating population structure (Lowe and Allendorf 2010).

5.1.2.3 Using Genetics to Study Wild Populations

Genetic analysis involves the isolation and comparison of DNA sequences, typically across multiple individuals or multiple populations. Varied types of genetic analyses provide different types of population information, and each has its advantages and disadvantages with respect to the technology required and the data obtained.

In human biology, genetic analysis is often used to identify genes that, when mutated, lead to the development of a certain disease (Eichler 2019). Initial genetic applications were therefore targeted towards humans and model animal species that are used in studying human health, such as mice and rats (Altschuler et al. 2008, Lehner 2013). Eventually, genetic technologies began to be applied to other animal species, and are now used broadly in both captive and wild populations (DeSalle and Amato 2004).

Genetic analysis of wild organisms can provide information about the health of populations, their movements and migrations, their reproductive strategies, and the best approach for protecting animals that may be threatened with extinction. The term *conservation genetics* has been used to denote the genetic study of wild populations with the explicit goal of informing their conservation and management. Genetic analysis gives a different view of animal populations than does traditional field work. Genetic population structure measures the history of the species, with the timeframe visualized being determined by the longevity of the animals, their age at maturity, and the rate of genetic change inherent to the species. Genetic data are not affected by recent changes to animal numbers and/or behavior. They are insulated from spurious results that may be caused by temporary population aberrations, but they also do not reflect recent actual changes in population size or behavior.

5.1.2.4 Genetic Analysis of Whale Shark Populations

Genetic analysis of marine species has lagged behind studies of terrestrial animals (Kelley et al. 2016). Difficulties in locating and sampling oceanic species make marine genetic research logistically complex. For whale sharks, early research relied on individual animals discovered opportunistically (White 1930, Denison 1937), but once it was found that whale sharks aggregate predictably in certain coastal areas, large-scale whale shark research, including genetic analysis, became possible. Several research groups initiated genetic studies of whale sharks, realizing that the skin of these large but docile animals was easily biopsied by free divers. Initial research was directed at determining the relatedness of whale sharks observed at feeding aggregations in different oceans. Field studies suggested that whale sharks moved away from seasonal coastal aggregation sites, and migrated regionally, but it was unclear if they completed trans-oceanic migrations, allowing for global mixing of populations. Genetic studies could help answer this question by measuring genetic differentiation between populations. With legal and illegal fisheries targeting whale sharks in several places, it was important to know whether animals sampled in protected areas were likely to migrate into areas that contain active fisheries. Genetics could also help investigate the breeding habits of these animals, about which little is known.

5.2 GENETIC MARKERS FOR POPULATION ANALYSIS

Genetic analysis can utilize a variety of DNA sequences as markers for comparative studies. Two main types of DNA elements are most often used in genetic studies: nucleotide sequencing of the highly variable control region of the mitochondrial genome; and determining the nucleotide repeat number of a panel of DNA elements called microsatellites. To perform genetic analysis, tissue is obtained from individuals of the species under study (by tissue biopsy, blood sampling, etc.), DNA is chemically extracted, and the chosen region of DNA is amplified by PCR to yield many billions of copies for analysis. The PCR product DNA is either subjected to DNA sequencing (for mitochondrial DNA) or repeat number analysis (for microsatellites). The choice of methodology is determined by the goals of the project, the amount and condition of tissue/DNA available, the existing genetic information for the species, and the technology available to the laboratory. Poor-quality tissues, for example, lend themselves better to the isolation of small microsatellites than to large regions of mitochondrial DNA. On the other hand, mitochondrial sequences are more similar across species, and so may be preferable where little prior sequence information is available.

5.2.1 Genetic Diversity and Differentiation

Two key ideas in population genetic analysis are genetic diversity and genetic differentiation. These similar sounding terms actually convey very different information (see Glossary). Genetic diversity can be assessed within or across populations and is a measure of the total amount of genetic information present in the animals surveyed, for example, the number of mtDNA haplotypes (sequence variants) or microsatellite alleles (different repeat numbers) found. It is generally accepted that high genetic diversity is beneficial to a species, as it provides the genetic raw material that allows populations to respond more effectively to the selection pressures of disease threats or habitat changes (Reed and Frankham 2003). High genetic diversity usually indicates a healthy population, whereas low genetic diversity may, but does not always, indicate a population at greater risk of decline or extinction. Genetic differentiation, on the other hand, refers to the degree of genetic difference between two or more populations. It is a measure of the degree to which two populations have accrued unique DNA sequence changes through isolation without gene flow, i.e. lack of migration and genetic mixing through interbreeding. Genetic differentiation is not inherently good or bad for a species, but conveys important information for understanding the behavior and population structure of that species.

5.2.2 Mitochondrial DNA Analysis

Mitochondrial DNA analysis involves isolating and determining the DNA sequence of one or more regions of the mitochondrial genome. The mitochondrial control region is typically used for comparisons between individuals of the same species, while the *Cytochrome oxidase I (COI)* gene is used for comparison across species, or for species identification of wildlife products. The mitochondrial control region has a regulatory function for the chromosome, but does not code for proteins. This allows it to mutate over time with little impact on the fitness of the individual, which increases the chance of sequence differences between even closely related animals. Protein coding genes such as *COI* are constrained from rapid change by the need to direct protein production. A unique mitochondrial DNA sequence variant is referred to as a haplotype. Each individual carries a single mitochondrial haplotype matching that of its mother, unless a mutation has occurred in one of her gametes (egg cells). The full control region in most sharks is roughly 1200 nucleotides, for example, but a haplotype need not contain this entire region. A series of haplotypes can be generated from any length of sequence as long as the regions compared across individuals are the same. Within a given study, many haplotypes may be found that differ by as little as a single nucleotide in closely related animals, or by many changes in distantly related animals. Tracking haplotypes across generations allows researchers to follow the genetic relationships between animals.

5.2.3 Microsatellite Analysis

Microsatellites are repetitive DNA elements located throughout the nuclear genome. They contain strings of two, three, or four nucleotide repeats that take forms such as CACACA…, CAGCAGCAG…, CAGTCAGTCAGT…, etc. The high mutation rate of microsatellites makes them useful for genetic studies, with the ability to identify differences between even closely related individuals. The rate of microsatellite change varies both between and within species, but an average rate of change across all loci is believed to be relatively consistent for a given species (Martin et al. 1992). If the average rate is not known for a species being studied, it must be inferred from the most closely related available species.

A typical nuclear genome contains thousands of microsatellite repeats. Since the nuclear genome carries two copies of each chromosome, it also carries two copies of each microsatellite. Any specific position in a genome is called a locus (plural loci), and the exact sequence at any one locus is called an allele. A given locus has the potential to carry multiple variant alleles, and the two alleles found in any one individual can be the same or different. The data for any microsatellite locus therefore takes the form of two numbers, which indicate the number of repeat nucleotides of the two alleles at that locus in that animal. The data for one tetra- (four) nucleotide repeat locus in one animal, for example, might take the form 120/124, with the 124 allele having one more repeat than the 120 allele. (Note the different format of microsatellite data from that of mitochondrial data, which is given as a long string of DNA sequence.) Microsatellite-based genetic analysis examines multiple different microsatellite loci, assembling DNA profiles that can have sufficient resolution to identify individual animals genetically.

One particularly effective way of analyzing population structure using genetics is to perform both mitochondrial sequencing and microsatellite analysis on the same individuals. Genetic data from mitochondrial haplotypes gauge inheritance through the female line, while the nuclear microsatellites track gene flow through both maternal and paternal inheritance. A specialized behavior called reproductive philopatry is found in some species, where one sex, typically the female, either remains in one area or returns to a specific area to breed, while males migrate more broadly. Reproductive philopatry can be detected by field analysis, for example, from tagging of animals that shows them returning to certain areas, but it also leaves a genetic signature that can be found through laboratory analysis. If both males and females of a species associate and breed similarly, genetic data from mitochondrial sequences will be similar to that from nuclear DNA markers such as microsatellites. Under circumstances of female reproductive philopatry, however, microsatellite markers often show no or decreased genetic differentiation between populations, while the maternally inherited mitochondrial data show more significant levels of differentiation. Combined field work and genetic studies of white shark populations in South Africa and Australia, for example, have shown evidence for a high degree of philopatry. Male white sharks in these regions appear to migrate and breed between populations, while females either remain localized or return to their original populations to breed (Pardini et al. 2001, Bonfil et al. 2005).

5.2.4 Visual Representations of Population Data

Population genetics data, i.e. a large set of mitochondrial haplotypes and/or microsatellite alleles, are analyzed in numerous ways to determine levels of genetic diversity and differentiation within and between populations. One method of representing genetic differentiation is to generate a network of mtDNA haplotypes, producing a visual display of the relatedness of animals. Figure 5.1 illustrates the mitochondrial haplotype network from a whale shark study by Castro et al. (2007). The data implications of this figure will be described in more detail below, but it is shown here to illustrate the concept of genetic distance between haplotypes. Each circle in the image represents a specific haplotype, with one or more colors indicating the geographic regions where animals were found to carry that haplotype. Two circles connected by a single line indicate that these haplotypes differ by only one sequence change. One or more black dots between haplotypes indicates that there are intermediate haplotypes that theoretically exist, but were not found in the animals studied. Haplotype H6 differs by one sequence change from H4, for example, but H4 is two sequence changes from H30. H6, H4, and H30 are found only in the Atlantic Ocean, while closely related H17 is found in the eastern Indian Ocean. The size of each circle indicates how common the haplotype is, larger circles mean a greater number of that haplotype found, and the total number of steps gives a measure of the amount of genetic change within the species. A comprehensive general reference for genetic analysis of marine species can be found in Hellberg et al. (2002).

5.3 POPULATION ANALYSIS OF WHALE SHARKS

5.3.1 Field Studies of Whale Shark Populations

Research on whale sharks accelerated greatly once they were found to occur reliably at coastal feeding aggregations (Taylor 1996, Clark and Nelson 1997, Colman 1997, Norman et al. 2017).

Aerial surveys determined that the aggregations occurred seasonally and counted numbers of individuals with greater accuracy than could be done from the surface (Cliff et al. 2007, Rowat et al. 2009a, de la Parra Venegas et al. 2011). Rowat et al. (2009a), for example, found that whale shark residency in the Seychelles was highest in September and October, with a peak sighting of 38 individuals. Boat-based studies that allow access to animals in the water determined the overall size and sex composition of these aggregations, finding that many of them are composed primarily of juvenile male animals. Plankton analysis showed that many aggregations are driven by the presence of a transient but abundant food source such as a spawning event (Colman 1997, Gunn et al. 1999, Heyman et al. 2001, de la Parra et al. 2011, Robinson et al. 2013). Tracking studies of whale sharks, initially with numeric marker tags and later with satellite and acoustic tags, found that most animals carry out short range movements, sometimes between aggregation sites within the same region (Wilson et al. 2006, Gifford et al. 2007, Graham and Roberts 2007, Hsu et al. 2007, Sleeman et al. 2010, Acuña-Marrero et al. 2014, Berumen et al. 2014, Robinson et al. 2017, Ryan et al. 2017, Rohner et al. 2018). Whale sharks tagged off Ningaloo Reef, Western Australia, for example, moved into the northeastern Indian Ocean, towards Indonesia (Wilson et al. 2006, Sleeman et al. 2010). Juvenile male and female sharks tagged in the Red Sea stayed largely within the Red Sea basin, with a few animals moving into the Gulf of Aden and the northwestern Indian Ocean (Berumen et al. 2014).

Occasionally, however, satellite tags have captured long-range migrations that carried sharks across oceans (Eckert and Stewart 2001, Hueter et al. 2013, Guzman et al. 2018). Two whale sharks tagged in the eastern Pacific Ocean, one in the Gulf of California and the other off Panama, are postulated to have moved 13,000 and 20,000 km, respectively, to the northwestern Pacific (Eckert and Stewart 2001, Guzman et al. 2018). A female whale shark tagged off the Atlantic coast of Mexico swam more than 7,000 km over 150 days to the Mid-Atlantic Ridge, where the satellite tag detached (Hueter et al. 2013). It was suggested that this shark was headed for the western African coast, although populations of adults, including mature females, have been found at other mid-Atlantic sites that could also have been a final destination, such as the archipelago of St Peter & St Paul Rocks, and the UK overseas territory of St Helena (Macena and Hazin 2016, Perry et al. 2020). This shark was not seen again for 4 years. In 2011, it was re-sighted at the original tagging location in Mexico and has returned each year since (www.whaleshark.org). Several adult female whale sharks tagged near the Galapagos Islands undertook long movements into the eastern Pacific, the most far-ranging of these being over 7,000 km total distance (Hearn et al. 2016). Satellite telemetry provides high-resolution data on whale shark migration, but the high cost of this approach prevents employment at large scales. Photo-identification enables the tracking of thousands of individuals from many geographic locations much more cheaply, using spot pattern "fingerprints" that persist for the life of the animal. This approach has provided information about aggregation site residency rates and return time, and has documented regional and occasional long-range migrations (Arzoumanian et al. 2005, Holmberg et al. 2009, Rowat et al. 2009b, Norman et al. 2017).

5.3.2 Mitochondrial DNA Analysis of Whale Shark Populations

The first global population genetic analysis of whale sharks was published in 2007 (Castro et al. 2007). This work described the sequencing of the complete mitochondrial control region from 70 whale shark samples from aggregation sites around the world. Among these animals, the authors identified 44 haplotypes, an indication of high genetic diversity within the species. The researchers stated that this genetic diversity indicates a "relatively large, stable population". They found that the haplotypes did not cluster geographically, but were well distributed across the Atlantic, Indian, and Pacific Oceans. Their analysis of genetic differentiation showed that animals from the Indian and Pacific Oceans were genetically indistinguishable from each other. They did find statistically significant genetic differentiation between animals from the Atlantic Ocean and those from the Indo-Pacific. The mitochondrial haplotype network in Figure 5.1 presents data for

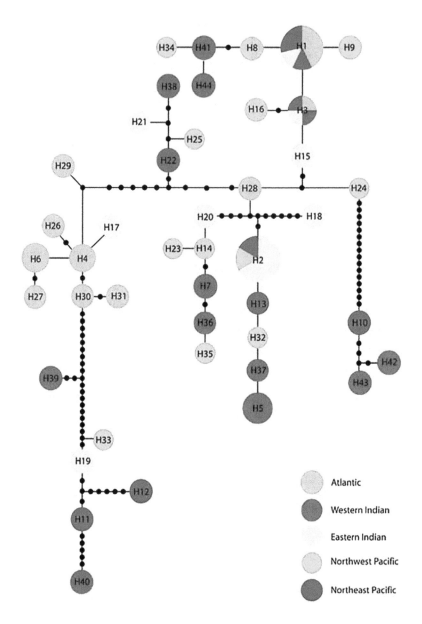

Figure 5.1 The mitochondrial haplotype network from a whale shark study by Castro et al. (2007). (© John Wiley & Sons, reproduced with permission.)

the whale shark from Castro et al. (2007); in this network, Atlantic Ocean animals are colored light blue and can be seen to carry haplotypes such as H1, H3, and H8, which are also shared with sharks from the Indian and Pacific Oceans. There is another group of light blue haplotypes that are not shared, however, including H4, H6, and others. Atlantic Ocean whale sharks therefore share some haplotypes with Indo-Pacific sharks, but also have haplotypes that are unique to the Atlantic. Other ocean basins also carry unique haplotypes, as shown by the red, yellow, green, and dark blue circles, but those from the Atlantic are more numerous and more divergent than in other ocean basins. That paper provided the first data to suggest that whale sharks move freely between the Indian and Pacific Oceans, but that movement of Atlantic sharks to other ocean basins is more restricted.

Castro et al (2007) also used their genetic data to estimate the effective population size of whale sharks, i.e. the number of breeding animals in the entire global whale shark population. Measures of genetic diversity can provide a statistical estimate for how many breeding animals are likely present to generate the pattern found. One qualifier of this calculation is that it is dependent on the rate at which DNA sequences change, a value that is unknown for whale sharks and must be extrapolated from related species. All breeding population estimates should be considered rough approximations without a true value for whale sharks being known. The breeding population estimate from the Castro et al. (2007) dataset was 119,000–238,000 females or, as it as derived from maternally inherited mitochondrial data, 238,000–476,000 total animals.

Ramírez-Macías et al. (2007) used a regional mtDNA analysis to study whale sharks from the Gulf of California, adjacent to the Baja Peninsula of Mexico (Ramírez-Macías et al. 2007). The Gulf of California is one of only a few places where adult female whale sharks are regularly found, and most of these animals show distended abdomens that may indicate pregnancy (Ramírez-Macías et al. 2012, Ketchum et al. 2013). This research group sampled sharks from three regions along the length of the Gulf, juveniles of both sexes and adult females, and sequenced a portion of the mitochondrial control region. They found high genetic diversity, with 14 haplotypes among 36 animals, but no genetic differentiation was seen between the different Gulf regions. Although a small study, these data suggest that animals found in different areas of the Gulf of California belong to a single population.

The largest genetic study published to date was performed by Vignaud et al. (2014), who sampled 635 whale sharks across nine global locations and performed both mitochondrial control region sequencing and microsatellite analysis (Vignaud et al. 2014). Mitochondrial DNA analysis found high levels of genetic diversity globally, but with some reduction in the sharks from the Atlantic Ocean. This result indicates that Atlantic Ocean animals may have reduced ability to tolerate population stressors, an area that requires further study. The mitochondrial DNA analysis found little genetic differentiation between sharks from the Indian and Pacific Oceans, but statistically significant differences between the Indo-Pacific and Atlantic Oceans. Extending the results of Castro et al. (2007), these data support reduced migration and diminished gene flow between Atlantic and Indo-Pacific whale sharks. Vignaud et al. (2014) also used microsatellite analysis to study their whale shark samples, which is discussed below.

Toha et al. (2016) used mitochondrial DNA analysis to examine 31 whale sharks sampled in Cenderawasih Bay, Indonesia, where the sharks come to feed at elevated fishing platforms (Toha et al. 2016, Himawan et al. 2015). Indonesia was not included in other genetic studies, so this work represents the first genetic characterization of this population. Rather than using the mitochondrial control region, this group sequenced a portion of a mitochondrial protein coding gene, *Cytochrome oxidase I* (*COI*). Toha et al. found seven *COI* haplotypes among the Indonesian animals, and the most common haplotypes were shared with those of Indo-Pacific reference animals. The use of a different mitochondrial region means that the results cannot be directly compared with studies that utilized the control region, but this work does suggest that Cenderawasih whale sharks belong to the greater Indo-Pacific population.

5.3.3 Microsatellite Analysis of Whale Shark Populations

Ramírez-Macías et al. (2009) first identified whale shark microsatellites, finding nine repeat loci that were amenable to analysis, with between 2 and 17 alleles present in a panel of samples from 50 sharks. Schmidt et al. (2009) published a panel of eight whale shark microsatellite loci, using these to analyze 68 whale sharks with a global distribution (Schmidt et al. 2009). This study found little genetic differentiation between Indian and Pacific Ocean populations, either when comparing all individuals or when grouping sharks into ocean-specific populations. The data showed that Atlantic Ocean animals had subtle genetic differentiation when compared to Indian and Pacific

Ocean animals, but the result was not statistically significant, perhaps because of the relatively small number of samples (statistical significance refers to a calculation that scientists use to determine when an effect is strong enough to be unlikely to occur by chance and therefore represents a true result). The whale shark microsatellite data were used to estimate effective population size, giving a range of 27,000–180,000 breeding animals globally.

In addition to the mtDNA sequencing already discussed, Vignaud et al. (2014) also applied microsatellite analysis to their whale shark samples. They found high levels of genetic diversity in the Indo-Pacific, with somewhat lower values in sharks from Atlantic Mexico. Little genetic differentiation was found between Indo-Pacific locations, but significant differentiation was seen between the Atlantic Ocean samples and all other locations. Figure 5.2 shows a graph of this genetic differentiation using a methodology called principal components analysis (PCA). In this graph, each individual whale shark is represented by a dot, which is colored to indicate where that animal was sampled. Similarly colored ellipses represent the mean of each location and encompass 67% of the relevant individuals. Using PCA, if individuals can be grouped into genetically distinct classes, then they will sort into separate areas of the graph. In Figure 5.2, the groupings of Indian and Pacific Ocean sharks (all colors except yellow) are largely overlapping, indicating little genetic differentiation. The Atlantic animals, however, indicated by yellow dots, overlap partially with the other colors, but also spread out to the right side of the chart. This type of visual representation helps researchers understand genetic differentiation between populations.

The availability of both mitochondrial and microsatellite (nuclear) data for a single set of individuals allowed these authors to ask about female philopatry in this species. If females of a species return to a specific region to breed, this mating strategy leaves a genetic signature that is best seen when comparing mitochondrial and nuclear markers. Among the whale sharks studied in this work, the levels of genetic differentiation observed for mitochondrial haplotypes and microsatellite markers were quite similar, suggesting whale sharks do not display female reproductive philopatry.

5.3.4 Emerging Genetic Technologies

Newly developed technologies are making it possible to perform genetic analysis without sampling of individuals, even without observing the animals directly. Studies have shown that animal DNA can be recovered from feces, hair, body mucus, and even from parasites that feed on skin or blood (Calvignac-Spencer et al. 2013, Kocher et al. 2017). DNA recovered from parasites that feed on multiple animals could give a picture of local biodiversity, whereas DNA from invertebrates that parasitize single animals could provide individual genetic profiles. This is a promising approach for pelagic species that can be difficult to locate or cannot be easily sampled. Recent work showed that whale shark mitochondrial DNA can be isolated from parasitic copepods that were collected from free-swimming whale sharks, an approach termed iDNA, for invertebrate-derived DNA (Austin et al. 2016, Meekan et al. 2017). Meekan et al. (2017) used iDNA recovered from parasites of Indian Ocean sharks as a source to amplify the whale shark mitochondrial control region. Each individual copepod was found to have a single mitochondrial haplotype (i.e. to likely have fed on a single shark) and multiple copepods recovered from the same shark carried the same haplotype. Comparing copepod-derived sequences from 31 sharks with the known whale shark haplotypes, the authors identified 4 previously unknown mtDNA haplotypes among the iDNA samples (Figure 5.3). Population genetic analysis of these data was consistent with previous studies from tissue biopsies: Indo-Pacific whale sharks showed no genetic differentiation between populations, whereas Atlantic Ocean sharks were significantly differentiated from the Indo-Pacific. This type of analysis offers a potential non-invasive method of obtaining whale shark DNA for genetic studies.

Environmental DNA (eDNA) analysis uses traces of DNA left behind in the environment to identify the species or individuals currently or recently present in that environment. Applied to

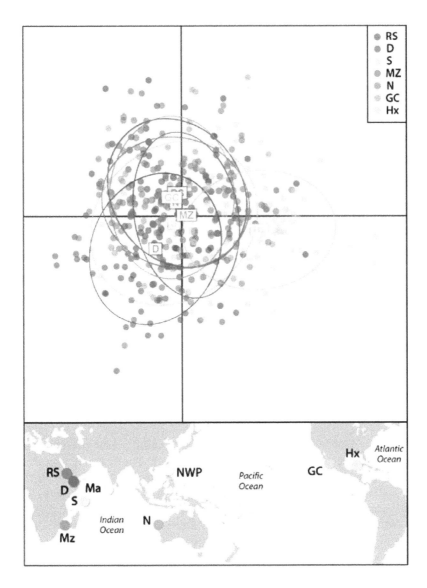

Figure 5.2 Genetic differentiation of different whale shark subpopulations from Vignaud et al. (2014). (© John Wiley & Sons, reproduced with permission.)

marine animals, eDNA analysis involves collecting water samples from a region and then analyzing filtered material by PCR to find DNA sequences from the desired species. Sigsgaard et al. (2016) collected seawater within a whale shark aggregation in the Arabian Gulf. Using PCR to search for whale shark sequences, they amplified small segments of the mitochondrial control region. Tissue biopsies from sharks in the same aggregation allowed comparison between animal DNA and sea-water eDNA, with many identical mtDNA haplotypes found. This comparison provided proof of principle that eDNA could be used for whale shark population genetic analysis. They were able to confirm that Arabian Gulf sharks are part of the Indo-Pacific population and are genetically differentiated from Atlantic Ocean animals. The authors also used their data to predict the whale shark effective population size, arriving at 43,618–183,526 females or 87,236–367,052 total individuals. Analysis of water near shark aggregations can serve as a proxy for individual sampling of sharks to answer some population genetics questions.

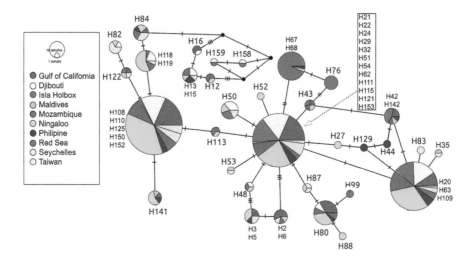

Figure 5.3 Haplotype network diagram from Meekan et al. (2017), generated using samples isolated from parasitic copepods, *Pandarus rhincodonicus* (see Chapter 4). (Picture credit: Frontiers in Marine Science.)

5.4 CONCLUSIONS FROM WHALE SHARK GENETIC ANALYSES

With several studies of whale shark population genetics completed, some common themes have emerged. First, all analyses found evidence of high genetic diversity in whale shark populations, though with potentially lower diversity in Atlantic Ocean animals. This result could be due to a smaller number of sharks in the Atlantic than in the Indo-Pacific region, either the initial founding animals or the current population, or could be a result of a past dramatic reduction in the size of the population with later recovery, an event called a bottleneck. The health of a species is not tied solely to genetic diversity, but strong diversity is known to promote adaptation and survival (Reed and Frankham 2003). The lower level of diversity seen among Atlantic Ocean sharks is difficult to interpret in the absence of reliable population counts, and the number of Atlantic individuals in most studies is low and quite geographically restricted. Further analysis of whale sharks from different parts of the Atlantic is needed to better understand this significance of this result.

Second, current evidence suggests that whale sharks comprise a single breeding population across the Indian and Pacific Oceans. This is not too surprising given the relative ease with which the animals can cross this region while remaining in the tropical waters they favor. It is a matter of debate among geneticists just how many animals must move to normalize genetic structure across a species' range, though all agree it is only a fraction of animals at each generation. From a conservation standpoint, this finding means that even sharks identified in protected regions are potentially at risk from fisheries during their migrations and it is a strong argument for seeking and enforcing international protection for whale sharks.

Third, whale sharks found in the Atlantic Ocean show greater genetic differentiation when compared to Indian or Pacific Ocean sharks than do those populations when compared to each other. The statistical significance of this effect varied among the three global studies. This is likely a result of differing sample sizes and methodologies, as a greater number of animals will allow more subtle differentiation to be visualized. These data indicate reduced gene flow between the Indo-Pacific and Atlantic basins. It is not possible to gauge the magnitude of this effect from the existing data, although a strict interpretation would be that no sharks migrate and gene flow is zero. Cold temperatures and strong currents near South Africa that should constrain whale shark migration would seem to support this idea, but movements of warmer water from the Indian Ocean are documented along the coast and whale sharks are frequently seen here (Beckley et al. 1997, Cliff et al. 2007). Field studies are therefore inconsistent with a complete lack of population interaction. If the most likely situation is reduced yet ongoing gene flow, how does this occur? Possible explanations are considered in the final section of this chapter.

Fourth, the availability of both mitochondrial haplotype and microsatellite data for a single large group of whale sharks made it possible to investigate the reproductive philopatry of these animals. Finding a lack of genetic differentiation using microsatellite markers, but a greater degree of differentiation with mtDNA haplotypes, would be characteristic of female philopatry. The fact that quite similar results were obtained by these two methods suggests that whale sharks do not practice female reproductive philopatry.

Fifth, estimates of effective (breeding) population size from the different studies show large but overlapping ranges. Combined, the reports suggest a total breeding population of between 27,000 and 476,000 animals. The actual population size of a species is typically much larger than its breeding population, particularly for late-maturing species with many juvenile animals. Even the smallest of these estimates is much larger than the roughly 12,000 whale sharks currently listed in the Wildbook for Whale Sharks (https://www.whaleshark.org/). It should be considered that genetic calculation of effective population size is only an estimate based on measures of genetic diversity, and many physiological and ecological factors may have effects that are not accounted for. For example, a recent decline in whale shark numbers may not yet be reflected in genetic data. Despite these uncertainties, such estimates are often the best available to conservation managers for species that are difficult or impossible to census in the wild. Taken at face value, the available information suggests that there is hope for a healthy whale shark population, with large numbers of unrecorded animals as yet.

5.5 INTEGRATING FIELD AND GENETIC STUDIES OF POPULATION STRUCTURE

Is it possible to reconcile field studies of whale shark populations with the genetic data? Data from both approaches make clear the many gaps in our knowledge about these animals. Of foremost interest is the degree to which whale sharks undertake intra- and inter-oceanic migrations. The genetic data show little population differentiation within the Indo-Pacific and a degree of differentiation that indicates reduced migration between the Atlantic and the Indo-Pacific. Tracking data that demonstrate long movements, however, have been difficult to obtain. Satellite tracking has recorded several individual whale sharks carrying out long distance movements: from the eastern to the western Pacific (Eckert and Stewart 2001, Guzman et al. 2018), and from the Western Atlantic to the Mid-Atlantic Ridge (Hueter et al. 2013). Far more commonly, tagged animals remain close to the area where they are first recorded (Wilson et al. 2006, Gifford et al. 2007, Hsu et al. 2007, Sleeman et al. 2010, Berumen et al. 2014, Hearn et al. 2016, Rohner et al. 2018). Photo-identification has also shown largely local movements and annual returns to feeding aggregations, with few sharks making regional migrations (Andrzejaczek et al. 2016, Araujo et al. 2017, Norman et al. 2017).

Genetic data track the history of entire populations and record migratory events that lead to breeding at any point in the life of an animal. Satellite tagging marks a small number of animals for months, or at most for a year or so. After many years of research, only a small number of whale sharks have been tagged. The majority of animals tracked by both methodologies are the juvenile males found at coastal aggregations and a few adult females from the Gulf of California, the Galapagos, St. Helena, and the Mexican Caribbean. Assuming only a percentage of animals migrate long distances, and only infrequently or at certain stages of their lives, capturing this event in a tagged animal is unlikely. One possible behavior, for example, is that whale sharks undertake long migrations during the transition from juveniles to adults, perhaps only once in their lives.

Ongoing satellite tracking is needed to capture long-distance movements, particularly those of adult and near-adult sharks. Continued genetic analysis is also necessary, to refine the relationship between Atlantic and Indo-Pacific animals. Additional questions and pressing research needs are discussed in the final section of this chapter.

5.6 GENOMIC ANALYSIS IN WHALE SHARKS

The term *genomics*, related to and yet distinct from genetics, refers to isolating and characterizing the entire DNA complement of a species. This includes not only the portions of the genome that contain genes and code for proteins, but also the larger amount of non-gene sequence that controls many other aspects of cellular function or has no known function to date. Access to the full genome of a species allows for detailed study of its evolution, adaptation, physiology, and reproduction. New avenues of research are constantly emerging as our understanding of DNA function and regulation grows. Essential to genomics is obtaining detailed sequence of large regions of the genome of the species under study. "Sequencing a genome" has become a common term in research and medicine, and even in popular culture, but it is a highly technical endeavor requiring fast computers and advanced bioinformatics software to assemble and make sense of the billions of nucleotides that make up a genome. Applying genomic information to population study provides for higher resolution genetic markers, even up to the comparison of populations at the level of entire genomes. Genomic comparison across populations has been shown to highlight differences in gene function and regulation correlated with micro-adaptation to ecological niches and to suggest the presence of subspecies. This is a rapidly growing area of research, but presently, there is relatively little genomic data published for sharks. Currently genomic sequences are available for four elasmobranch species, the whale shark (Read et al. 2017, Weber et al. 2020), the white shark *Carcharodon carcharias* (Marra et al. 2019), the brown-banded bamboo shark *Chiloscyllium punctatum* and the cloudy catshark (*Scyliorhinus torazame*) (Hara et al. 2018), and for one chimaera (a chondrichthyan that is related to true sharks and rays): the Australian ghost shark *Callorhinchus milii* (Venkatesh et al. 2014). A genome sequencing project for the little skate *Leucoraja erinacea* is in process (Wyffels et al. 2014), and likely other efforts are under way as well. Given the vast diversity of elasmobranch species, additional genomes are needed for comparative study.

Efforts to sequence a genome often begin with the mitochondrial chromosome, which is far smaller than the nuclear genome and occurs in much higher copy number per cell due to the presence of only one nucleus but many mitochondria. The sequencing of the whale shark mitochondrial genome has been completed for two different individuals from Taiwan (Alam et al. 2014, Chen et al. 2016). The whale shark mitochondrial genome was found to be slightly longer that of other shark species, largely due to the presence of a large insertion in the D-loop control region relative to other species, with the sequence otherwise most closely related to other members of the order Orectolobiformes (several of the bamboo sharks, and the Japanese wobbegongs) (Figure 5.4). Sequencing of the whale shark nuclear genome was completed in 2017 (Read et al. 2017), using DNA isolated from a male shark collected in Taiwan and formerly kept at the Georgia Aquarium; this was the same individual that was used for the mitochondrial genome described in Alam et al. (2014). Read et al. (2017) found a genome size of 3.44 GB (1 GB is equal to one billion nucleotides), within the average size range for vertebrate genomes, although several times larger than the unusually small genome of the Australian ghost shark (Venkatesh et al. 2014) and half the size of that in the cloudy catshark (Hara et al. 2018). Preliminary characterization of the whale shark genome sequence found loss of proteins involved in bone formation, which is expected for a fish with a cartilaginous skeleton, and identified novel genes that may be involved in immune system function (Read et al. 2017). Whale shark, white shark, and Australian ghost shark genomes all show adaptations that may allow for rapid wound healing, an interesting phenomenon that is well documented

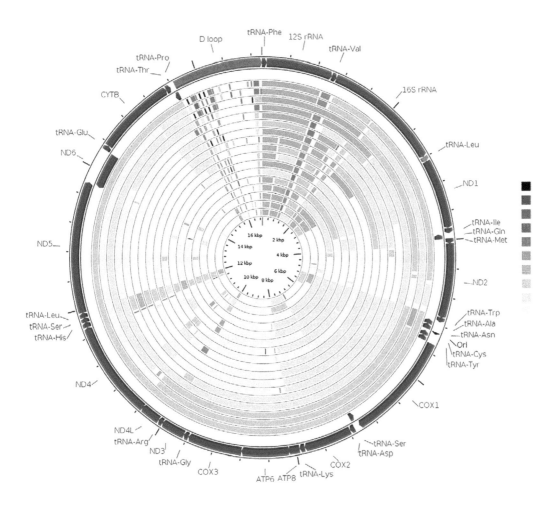

Figure 5.4 Mitochondrial genome structure in whale sharks and 17 other shark species, from Alam et al. (2014), showing large insertion in the control region.

in a number of elasmobranch species (Read et al. 2017, Marra et al. 2019, Venkatesh et al. 2014). Hara et al. (2018) reassembled the genomic data of Read et al. (2017) and included it in an analysis of genomic trends in fishes, and they found a reduced mutation rate in all the chondrichthyan (shark and ray) species, compared to the teleosts or bony fishes. More recently, a second genome assembly for a different whale shark was published by Weber et al. (2020), which, when compared with 84 other genomes across a wide range of the tree of life, showed a curious relationship between the length of genes and the longevity of the animal in question. Clearly, much more information will come from future characterization of the two current assemblies of the whale shark genome.

5.7 FUTURE CHALLENGES IN WHALE SHARK POPULATION ANALYSIS

Despite the accelerating pace of whale shark research over the last 20 years, many important aspects of their biology remain unknown. One of the most pressing questions is to identify the habitat(s) used by the age and sex groups that are missing from current datasets: adult males and females, juvenile females, and very young sharks less than 3 m in length. Some progress had been made recently in identifying oceanic regions that host seemingly adult females, such as the Gulf of California, the Galapagos, and St.

Helena island (Ramírez-Macías et al. 2012, Acuña-Marrero et al. 2014, Perry et al. 2020, respectively). These female sharks are transient and seasonal, however, and their broader range is unknown. While the juvenile males studied in the genetic analyses discussed here are representative of global whale shark diversity in the broadest sense, understanding how the observed patterns of genetic diversity and differentiation arise requires a great deal more data on migratory and reproductive behavior. This will require field ecology studies, with perhaps a bit of luck in discriminating long-range migrants, followed by genetic analysis of newly identified populations (Lowe and Allendorf 2010).

It is also important to refine current evidence for the genetic differentiation of Atlantic Ocean whale sharks, by identifying and analyzing populations from the many unsampled areas of the Atlantic basin to determine levels of ongoing gene flow. Of great interest is whether the gene flow is episodic, with some individuals performing long-range migrations within each generation. One female shark, as already discussed, migrated halfway across the Atlantic and back (Hueter et al. 2013), and recent work has begun to characterize whale sharks found at other potential Atlantic migratory waypoints such as the Azores, Saint Peter & Paul Archipelago and Saint Helena (Afonso et al. 2014, Macena and Hazin 2016, Perry et al. 2020, respectively). Alternatively, the gene flow could be sequential or "stepping stone" in nature, with animals migrating throughout the Atlantic, breeding at oceanic islands and/or along the western coast of Africa, but not making full inter-oceanic movements. Separate regional breeding events would then perpetuate the gene flow southwards along the African coast and around South Africa. This hypothesis doesn't require single animals to cross from the Atlantic to the Indian Ocean. Rather, multiple breeding events combine to ensure gene flow. A few reports indicate the presence of whale sharks along the western and southern African coast, suggesting regional movements could convey genetic information across long distances (Beckley et al. 1997, Cliff et al. 2007, Weir 2011). Isolation by distance (IBD) is a term used to indicate contiguous populations of animals that breed only across segments of their range. IBD can establish a genetic gradient, with the extremes of the range genetically differentiated from each other (Slatkin 1993, Irwin et al. 2005). Whether IBD applies to whale sharks is unclear, as little differentiation is seen across the larger Indo-Pacific range under similar analytical conditions. Unstudied populations in the southern and eastern Atlantic must be incorporated into ongoing genetic studies to allow for Atlantic-wide mapping of population differentiation.

Sampling of the Indo-Pacific region has been far more extensive than that of the Atlantic, but it is a vast area and there are hints of subtle population differentiation at its eastern and western extremes (Vignaud et al. 2014). Further genetic studies should compare animals at the far western and eastern Indo-Pacific; analysis using more individuals and advanced marker sets may uncover genetic differentiation across this seemingly homogeneous range. The availability of the whale shark genome will help in these efforts by allowing the development of genetic markers for high-resolution analysis. All genetic work would be improved by the establishment of the average DNA mutation rate in whale sharks, although this rate is likely to be sufficiently similar to related species that it is not the highest priority for research.

In the absence of detailed information about large-scale demographics of the whale shark, computer-assisted modeling has filled some gaps in our understanding of their behaviors. Sequeira et al. (2013), for example, compiled all known information on whale shark populations and movements, suggesting that they comprise a single highly migratory population globally (Sequeira et al. 2013). While genetic data indicate that mobility is reduced with respect to Atlantic sharks, the concept of a single connected population is supported on many levels. Improved software for ecological modeling will allow researchers to continue to study animals they cannot find and hopefully to understand how the stressors of climate change, fisheries, and other anthropogenic activities will impact the future of the species. All research questions lead ultimately towards establishing the population status of whale sharks and devising conservation schemes that are supported by scientific data.

ACKNOWLEDGMENTS

The author thanks Kevin Feldheim, Fusun Ozer, and Thomas Vignaud for comments on the manuscript.

GLOSSARY

Allele: A unique or recognizable *DNA* sequence variant at a given *gene* or *locus.*

Amplify: In genetics, to make many copies of a sequence, usually by *PCR.*

Bases: Four chemicals that make up the letters of the universal genetic code contained in *DNA*: adenine (A), cytosine (C), thymine (T), and guanine (G).

Chromosome: A molecular structure composed of a long and highly coiled *DNA* strand. Cells have a chromosome number that is characteristic of the species.

Conservation genetics: the genetic analysis of wild populations, with the explicit goal of informing their conservation and/or population management.

Conserved: In genetics, refers to a slow rate of accumulated *sequence* changes as a result of the essential functions of the *gene* in question.

Control region: A region of the *mitochondrial genome* that is often used for population genetic studies because it is highly variable in sequence.

DNA: The chemical polymer that carries the hereditary information of a cell.

eDNA: An abbreviation for Environmental DNA, which is trace organismal *DNA* found free in environments currently or recently occupied by that organism.

iDNA: An abbreviation for invertebrate-derived DNA, which is when host *DNA* is collected from invertebrate parasites of that animal, rather than from the animal itself

Fitness: The ability of a population of animals to adapt to changes in their environment.

Gene: A segment of *DNA* that directs the production of a specific *protein.*

Gene flow: The spread of genetic information between geographic regions by the movement and interbreeding of animals.

Genetic analysis: The isolation and comparison of DNA sequences, typically across multiple individuals or multiple populations of one or more species.

Genetic differentiation: A measure of the degree of genetic difference between populations of animals. Greater genetic differentiation indicates that the populations are less closely related.

Genetic diversity: A measure of the total amount of genetic information in a population. High genetic diversity is correlated with increased fitness.

Genome: The entire *DNA* complement of a cell. All animal cells contain two genomes, a large genome in the *nucleus* and a smaller, circular genome in each *mitochondrion.*

Haplotype: A particular variant of a region of *DNA* sequence. In population genetics, haplotype usually refers to a specific mitochondrial *control region* sequence.

Haplotype network: A graphical depiction of the relationships between all haplotypes in a population.

Locus: A specific position in a genome that is occupied by one of many possible *alleles.*

Microsatellite: A short, repetitive *DNA* sequence, the number, or sequence of which can be used for genetic studies of populations.

Mitochondria: A cellular organelle that contains its own small, circular *genome.* Mitochondria, and their genomes, are inherited only from the mother.

mtDNA: An abbreviation for mitochondrial *DNA.*

Mutation: In this chapter, any heritable change in a *DNA* sequence. Unlike the usual medical connotation, a mutation is not inherently helpful or harmful, just different.

Nuclear: Relating to the nucleus of the cell.

Nucleotides: The individual chemical structures or *bases* that make up the *sequence* of a *DNA* strand.

PCR: Abbreviation for the polymerase chain reaction, a technique for amplifying a target section of DNA billions of times using polymerase enzymes isolated from bacteria.

Population genetics: The use of DNA sequence analysis to compare populations of animals to determine the relationships between them.

Protein: One of many different molecules that carry out major functions within a cell.

Sequence: The particular arrangement of nucleotide bases (A, C, T, or G) at a given *locus*. Also, the verb used to describe the laboratory method used to determine a sequence.

REFERENCES

Acuña-Marrero, D., Jiménez, J., Smith, F. et al. 2014. Whale shark (*Rhincodon typus*) seasonal presence, residence time and habitat use at Darwin Island, Galapagos Marine Reserve. *PLoS One* 9:e115946. doi:10.1371/journal.pone.0115946.

Afonso, P., McGinty, N., and M. Machete. 2014. Dynamics of whale shark occurrence at their fringe oceanic habitat. *PLoS One* 9:e102060. DOI:10.1371/journal.pone.0102060.

Alam, M. T., Petit III, R. A., Read, T. D., and A. D. M. Dove. 2014. The complete mitochondrial genome sequence of the world's largest fish, the whale shark (*Rhincodon typus*), and its comparison with those of related shark species. *Gene* 539:44–49. doi:10.1016/j.gene.2014.01.064.

Altschuler, D., Daly, M. J., and E. S. Lander. 2008. Genetic mapping in human disease. *Science* 322:881–888. doi:10.1126/science.1156409.

Andrzejaczek, S., Meeuwig, J., Rowat, D., et al. 2016. The ecological connectivity of whale shark aggregations in the Indian Ocean: A photo-identification approach. *Royal Society Open Science* 3:160455. doi:10.1098/rsos.160455.

Araujo, G., Snow, S., So, C. L., et al. 2017. Population structure, residency patterns and movements of whale sharks in Southern Leyte, Philippines: Results from dedicated photo-ID and citizen science. *Aquatic Conservation* 27:237–252. doi:10.1002/aqc.2636.

Arzoumanian, Z., Holmberg, J., and B. Norman. 2005. An astronomical pattern-matching algorithm for computer-aided identification of whale sharks *Rhincodon typus*. *Journal of Applied Ecology* 42:999–1011. doi:10.1111/j.1365–2664.2005.01117.x.

Austin, C. M., Tan, M. H., Lee, Y. P., et al. 2016. The complete mitogenome of the whale shark parasitic copepod *Pandarus rhincodonicus* Norman, Newbound & Knott (Crustacea: Siphonostomatoida: Pandaridae) – a new gene order for the copepod. *Mitochondrial DNA Part A* 27:694–695. doi:10.310 9/19401736.2014.913147.

Avise, J. C. 1998. Conservation genetics in the marine realm. *Journal of Heredity* 89:377–382. doi:10.1093/jhered/89.5.377.

Beckley, L. E., Cliff, G., Smale, M. J., and L. J. V. Compagno. 1997. Recent strandings and sightings of whale sharks in South Africa. *Environmental Biology of Fishes* 50:343–348. doi:10.1023/A:1007355709632.

Bonfil, R., Myer, M., Scholl, M., et al. 2005. Transoceanic migration, spatial dynamics, and population linkages of white sharks. *Science* 310:100–103.

Berumen, M. L., Braun, C. D., Cochran, J. E. M., Skomal, G. B., and S. R. Thorrold. 2014. Movement patterns of juvenile whale sharks tagged at an aggregation site in the Red Sea. *PLoS One* 9:e103536. doi:10.1371/journal.pone.0103536.

Calvignac-Spencer, S., Leendertz, F. H., Gilbert, M. T., and G. Schubert. 2013. An invertebrate stomach's view on vertebrate ecology. *BioEssays* 35:1004–1013. doi:10.1002/bies.201300060.

Castro, A. L. F., Stewart, B. S., Wilson, S. G., et al. 2007. Population genetic structure of Earth's largest fish, the whale shark (*Rhincodon typus*). *Molecular Ecology* 16:5183–5192. doi:10.1111/j.1365-294X.2007.03597.x.

Chen, X., Ai, W., Pan, L., and X. Shi. 2016. Complete mitochondrial genome of the largest living fish: Whale shark *Rhincodon typus* (Orectolobiformes: Rhincodontidae). *Mitochondrial DNA Part A* 27:425–426. doi:10.3109/19401736.2014.898288.

Clark, E., and D. R. Nelson. 1997. Young whale sharks, *Rhincodon typus*, feeding on a copepod bloom near La Paz, Mexico. *Environmental Biology of Fishes* 50:63–73. doi:10.1023/A:1007312310127.

Cliff, G., Anderson-Reade, M. D., Aitken, A. P., Charter, G. E., and V. M. Peddemors. 2007. Aerial census of whale sharks (*Rhincodon typus*) on the northern KwaZulu-Natal coast, South Africa. *Fisheries Research* 84:41–46. doi:10.1016/j.fishres.2006.11.012.

Colman, J. G. 1997. A review of the biology and ecology of the whale shark. *Journal of Fish Biology* 51:1073–1275. doi:10.1111/j.1095–8649.1997.tb01138.x.

de la Parra Venegas, R., Hueter, R., González Cano, J., et al. 2011. An unprecedented aggregation of whale sharks, *Rhincodon typus*, in Mexican coastal waters of the Caribbean Sea. *PLoS One* 6:e18994. doi:10.1371/journal.pone.0018994.

Denison, R. H. 1937. Anatomy of the head and pelvic fin of the whale shark, *Rhineodon*. *Bulletin of the American Museum of Natural History* 73:477–515.

DeSalle, R., and G. Amato. 2004. The expansion of conservation genetics. *Nature Reviews Genetics* 5:702–712. doi:10.1038/nrg1425.

Eckert, S. A., and B. S. Stewart. 2001. Telemetry and satellite tracking of whale sharks, *Rhincodon typus*, in the Sea of Cortez, Mexico, and the North Pacific Ocean. *Environmental Biology of Fishes* 60:299–308. doi:10.1023/A:1007674716437.

Eichler, E. E. 2019. Genetic variation, comparative genomics, and the diagnosis of disease. *New England Journal of Medicine* 381:64–74. doi:10.1056/NEJMra1809315.

Gifford, A., Compagno, L. J. V., Levine, M., and A. Antoniou. 2007. Satellite tracking of whale sharks using tethered tags. *Fisheries Research* 84:17–24. doi:10.1016/j.fishres.2006.11.011.

Graham, R. T., and C. M. Roberts. 2007. Assessing the size, growth rate and structure of a seasonal population of whale sharks (*Rhincodon typus* Smith 1828) using conventional tagging and photo identification. *Fisheries Research* 84:71–80. doi:10.1016/j.fishres.2006.11.026.

Gunn, J. S., Stevens, J. D., Davis, T. L. O., and B. M. Norman. 1999. Observations on the short-term movements and behaviour of whale sharks (*Rhincodon typus*) at Ningaloo Reef, Western Australia. *Marine Biology* 135:553–559. doi:10.1007/s002270050656.

Guzman, H. M., Gomez, C. G., Hearn, A., and S. A. Eckert. 2018. Longest recorded trans-Pacific migration of a whale shark (*Rhincodon typus*). *Marine Biodiversity Records* 11:8. doi:10.1186/s41200-018-0143-4.

Hara, Y., Yamaguchi, K., Onimaru, K., et al. 2018. Shark genomes provide insights into elasmobranch evolution and the origin of vertebrates. *Nature Ecology & Evolution* 2:1761–1771. doi:10.1038/s41559-018-0673-5.

Hearn, A. R., Green, J., Román, M. H., Acuña-Marrero, D., Espinoza, E., and A. P. Klimley. 2016. Adult female whale sharks make long-distance movements past Darwin Island (Galapagos, Ecuador) in the Eastern Tropical Pacific. *Marine Biology* 163:214. doi:10.1007/s00227-016-2991-y.

Hellberg, M. E., Burton, R. S., Neigel, J. E., and S. R. Palumbi. 2002. Genetic assessment of connectivity among marine populations. *Bulletin of Marine Science* 70:273–290.

Heyman, W. D., Graham, R. T., Kjerfve, B., and R. E. Johannes. 2001. Whale sharks *Rhincodon typus* aggregate to feed on fish spawn in Belize. *Marine Ecology Progress Series* 215:275–282. doi:10.3354/meps215275.

Himawan, M. R., Tania, C., Noor, B. A., Wijonarno, A., Subhan, B., and H. Madduppa. 2015. Sex and size range composition of whale shark (*Rhincodon typus*) and their sighting behaviour in relation with fishermen lift-net within Cenderawasih Bay National Park, Indonesia. *AACL Bioflux* 8:123–133.

Holmberg, J., Norman, B., and Z. Arzoumanian. 2009. Estimating population size, structure, and residency time for whale sharks *Rhincodon typus* through collaborative photo-identification. *Endangered Species Research* 7:39–53. doi:10.3354/esr00186.

Hueter, R. E., Tyminski, J. P., and R. de la Parra. 2013. Horizontal movements, migration patterns, and population structure of whale sharks in the Gulf of Mexico and Northwestern Caribbean Sea. *PLoS One* 8:e71883. doi:10.1371/journal.pone.0071883.

Hsu, H. H., Joung, S., Liao, Y. Y., and K. M. Liu. 2007. Satellite tracking of juvenile whale sharks, *Rhincodon typus*, in the Northwestern Pacific. *Fisheries Research* 84:25–31 doi:10.1016/j.fishres.2006.11.030.

Irwin, D. E., Bensch, S., Irwin, J. H., and T. D. Price. 2005. Speciation by distance in a ring species. *Science* 307:414–416. doi:10.1126/science.1105201.

Jabado, R. W., Al Ghais, S. M., Hamza, W., et al. 2015. The trade in sharks and their products in the United Arab Emirates. Biological Conservation 181:190–198 DOI:10.1016/j.biocon.2014.10.032

Kelley, J. L., Brown, A. P., Therkildsen, N. O., and A. D. Foote. 2016. The life aquatic: Advances in marine vertebrate genomics. *Nature Reviews Genetics* 17:523–534. doi:10.1038/nrg.2016.66.

Ketchum, J. T., Galván-Magaña, F., and P. Klimley. 2013. Segregation and foraging ecology of whale sharks, *Rhincodon typus*, in the southwestern Gulf of California. *Environmental Biology of Fishes* 96:779–795. doi: 10.1007/s10641-012-0071-9.

Kocher, A., De Thoisy, B., Catzeflis, F., Valiere, S., Bañuls, A. L., and J. Murienne. 2017. iDNA screening: Disease vectors as vertebrate samplers. *Molecular Ecology* 26:6478–6486. doi:10.1111/mec.14362.

Lehner, B. 2013. Genotype to phenotype: Lessons from model organisms for human genetics. *Nature Reviews Genetics* 14:168–178. doi:10.1038/nrg3404.

Lowe, W. H., and F. W. Allendorf. 2010. What can genetics tell us about population connectivity? *Molecular Ecology* 19:3038–3051. doi:10.1111/j.1365-294X.2010.04688.x.

Macena, B. C. L., and F. H. V. Hazin. 2016. Whale Shark (*Rhincodon typus*) seasonal occurrence, abundance and demographic structure in the mid-equatorial Atlantic Ocean. *PLoS One* 11:e0164440. doi:10.1371/journal.pone.0164440.

Marra, N. J., Stanhope, M. J., Jue, N. K., et al. 2019. White shark genome reveals ancient elasmobranch adaptations associated with wound healing and the maintenance of genome stability. *Proceedings of the National Academy of Sciences* 116:4446–4455. doi:10.1073/pnas.1819778116.

Martin, A. P., Naylor, G. J. P., and S. R. Palumbi. 1992. Rates of mitochondrial DNA evolution in sharks are slow compared with mammals. *Nature* 357:153–155. doi:10.1038/357153a0.

Meekan, M., Austin, C. M., Tan, M. H., et al. 2017. iDNA at sea: Recovery of whale shark (*Rhincodon typus*) mitochondrial DNA sequences from the whale shark copepod (*Pandarus rhincodonicus*) confirms global population structure. *Frontiers in Marine Science* 4:420. doi:10.3389/fmars.2017.00420.

Norman, B. M., Holmberg, J. A, Arzoumanian, Z., et al. 2017. Undersea constellations: The global biology of an endangered marine megavertebrate further informed through citizen science. *BioScience*. doi:10.1093/biosci/bix127.

Pardini, A. T., Jones, C. S., Noble, L. R., et al. 2001. Sex-biased dispersal of great white sharks. *Nature* 412:139. doi:10.1038/35084125.

Perry, C. T., Clingham, E., Webb, D. H., et al. 2020. St. Helena: An important reproductive habitat for whale sharks (*Rhincodon typus*) in the South Central Atlantic. *Frontiers in Marine Science* 7:576343. doi:10.3389/fmars.2020.576343.

Ramirez-Macias, D., Shaw, K., Ward, R., Galvan-Magana, F., and R. Vazquez-Juarez. 2009. Isolation and characterization of microsatellite loci in the whale shark (*Rhincodon typus*). *Molecular Ecology Resources* 9:798–800. doi:10.1111/j.1755-0998.2008.02197.x.

Ramirez-Macias, D., Vazquez-Juarez, R., Galvan-Magana, F., and A. Munguia-Vega. 2007. Variations of the mitochondrial control region sequence in whale sharks (*Rhincodon typus*) from the Gulf of California, Mexico. *Fisheries Research* 84:87–95. doi:10.1016/j.fishres.2006.11.038.

Ramírez-Macías, D., Vázquez-Haikin, A., and R. Vázquez-Juárez. 2012. Whale shark *Rhincodon typus* populations along the west coast of the Gulf of California and implications for management. *Endangered Species Research* 18:115–128. doi:10.3354/esr00437.

Read, T. D., Petit 3rd, R. A., Joseph, S. J., et al. 2017. Draft sequencing and assembly of the genome of the world's largest fish, the whale shark: *Rhincodon typus* Smith 1828. *BMC Genomics* 18:532. doi:10.1186/s12864-017-3926-9.

Reed, D. H., and R. Frankham. 2003. Correlation between fitness and genetic diversity. *Conservation Biology* 17:230–237. doi:10.1046/j.1523-1739.2003.01236.x.

Robinson, D. P., Jaidah, M. Y., Jabado, R. W., et al. 2013. Whale sharks, *Rhincodon typus*, aggregate around offshore platforms in Qatari waters of the Arabian Gulf to feed on fish spawn. *PLoS One* 8:e58255. doi:10.1371/journal.pone.0058255.

Robinson, D. P., Jaidah, M. Y., Bach, S. S., et al. 2017. Some like it hot: Repeat migration and residency of whale sharks within an extreme natural environment. *PLoS One* 12:e0185360. doi:10.1371/journal.pone.0185360.

Rohner, C. A., Richardson, A. J., Jaine, F. R. A., et al. 2018. Satellite tagging highlights the importance of productive Mozambican coastal waters to the ecology and conservation of whale sharks. *PeerJ* 6:e4161. doi:10.7717/peerj.4161.

Rowat, D., Gore, M., Meekan, M. G., Lawler, I. R., and C. J. A. Bradshaw. 2009a. Aerial survey as a tool to estimate whale shark abundance trends. *Journal of Experimental Marine Biology and Ecology* 368:1–8. doi:10.1016/j.jembe.2008.09.001.

Rowat, D., Speed, C., Meekan, M., Gore, M., and C. Bradshaw. 2009b. Population abundance and apparent survival of the vulnerable whale shark *Rhincodon typus* in the Seychelles aggregation. *Oryx* 43:591–598. doi:10.1017/S0030605309990408.

Ryan, J. P., Green, J. R., Espinoza, E. and A. R. Hearn. 2017. Association of whale sharks (*Rhincodon typus*) with thermo-biological frontal systems of the eastern tropical Pacific. *PLoS One* 12:e0182599. doi:10.1371/journal.pone.0182599.

Saiki, R. K., Scharf, S., Faloona, F., et al. 1985. Enzymatic amplification of beta-globin genomic sequences and restriction site analysis for diagnosis of sickle cell anemia. *Science* 230:1350–1354. doi:10.1126/science.2999980.

Schmidt, J. V., Schmidt, C. L., Ozer, F., et al. 2009. Low genetic differentiation across three major ocean populations of the whale shark, *Rhincodon typus*. *PLoS One* 4:e4988. doi:10.1371/journal.pone.0004988.

Sequeira, A. M., Mellin, C., Meekan, M. G., Sims, D. W., and C. J. A. Bradshaw. 2013. Inferred global connectivity of whale shark *Rhincodon typus* populations. *Journal of Fish Biology* 82:367–389. doi:10.1111/jfb.12017.

Sigsga ard, E. E., Nielsen, I. B., Bach, S. S., et al. 2016. Population characteristics of a large whale shark aggregation inferred from seawater environmental DNA. *Nature Ecology & Evolution* 1:0004. doi:10.1038/s41559-016-0004.

Slatkin, M. 1993. Isolation by distance in equilibrium and non-equilibrium populations. *Evolution* 47:264–279. doi:10.1111/j.1558–5646.1993.tb01215.x.

Sleeman, J. C., Meekan, M. G., Wilson, S. G., et al. 2010. To go or not to go with the flow: Environmental influences on whale shark movement patterns. *Journal of Experimental Marine Biology and Ecology* 390:84–98. doi:10.1016/j.jembe.2010.05.009.

Steinke, D., Bernard, A. M., Horn, R.L., et al. 2017. DNA analysis of traded shark fins and mobulid gill plates reveals a high proportion of species of conservation concern. *Scientific Reports* 7:9505. doi:10.1038 /s41598-017-10123-5.

Taylor, J. G. 1996. Seasonal occurrence, distribution and movements of the whale shark, *Rhincodon typus*, at Ningaloo Reef, Western Australia. *Marine & Freshwater Research* 47:637–642. doi:10.1071/MF9960637.

Toha, A. H., Widodo, N., Subhan, B., et al. 2016. Close genetic relatedness of whale sharks, *Rhincodon typus* in the Indo-Pacific region. *AACL Bioflux* 9:458–465.

Venkatesh, B., Lee, A. P., Ravi, V., et al. 2014. Elephant shark genome provides unique insights into gnathostome evolution. *Nature* 505:174–179. doi:10.1038/nature12826.

Vignaud, T. M., Maynard, J. A., Leblois, R., et al. 2014. Genetic structure of populations of whale sharks among ocean basins and evidence for their historic rise and recent decline. *Molecular Ecology* 23:2590–2601. doi:10.1111/mec.12754.

Weber, J. A., Park, S. G., Luria, V., et al. 2020. The whale shark genome reveals how genomic and physiological properties scale with body size. *Proceedings of the National Academy of Sciences* 117:20662–20671. doi:10.1073/pnas.1922576117.

Weir, C. 2011. Sightings of whale sharks (*Rhincodon typus*) off Angola and Nigeria. *Marine Biodiversity Records* 3:e50. doi:10.1017/S1755267209990741.

White, E. G. 1930. The whale shark, *Rhineodon typus*. Description of the skeletal parts and classification based on the Marathon specimen captured in 1923. *Bulletin of the American Museum of Natural History* 61:129–160.

Wilson, S. G., Polovina, J., Stewart, B., and M. Meekan. 2006. Movements of whale sharks (*Rhincodon typus*) tagged at Ningaloo Reef, Western Australia. *Marine Biology* 148:1157–1166. doi:10.1007/s00227-005-0153-8.

Wyffels, J., King, B. L., Vincent, J., et al. 2014. SkateBase, an elasmobranch genome project and collection of molecular resources for chondrichthyan fishes. *Faculty of 1000 Research* 3:191. doi:10.12688/f1000research.4996.1.

CHAPTER 6

Whale Shark Movements and Migrations

Alex R. Hearn
Universidad San Francisco de Quito

Jonathan R. Green
Galapagos Whale Shark Project

César Peñaherrera-Palma
Pontificia Universidad Católica del Ecuador

Samantha Reynolds
The University of Queensland

Christoph A. Rohner
Marine Megafauna Foundation

Marlon Román
Inter-American Tropical Tuna Commission

Ana M. M. Sequeira
University of Western Australia

CONTENTS

6.1 INTRODUCTION

The geographic range of whale sharks encompasses most tropical and warm temperate waters (Compagno 2001, Last and Stevens 2009), but information on geographic connectivity, migration and demography of this species is still limited. For example, the first quantitative approach to understanding the pelagic distribution of whale sharks was carried out in the Indian Ocean as recently as 2012 (Sequeira et al. 2012). Genetic studies suggest at least two populations that rarely mix: an Atlantic and an Indo-Pacific population (Castro et al. 2007; Vignaud et al. 2014). Whale sharks are generally considered to be mostly solitary animals that spend their time in the open ocean, but occasionally form groups (Gudger 1941). In the last 30 years, several locations have been identified globally where constellations occur on a seasonal basis to take advantage of a predictable food supply (Colman 1997; Stevens 2007; Rowat and Brooks 2012; Sequeira et al. 2013a; Norman et al. 2017a). Understanding whether or not the sharks migrate between these constellation sites is central to our knowledge of their population ecology and thus to our efforts to protect them.

The first two decades of the 21st century have seen a dramatic increase in our knowledge of both the vertical and horizontal movement behavior of whale sharks globally due to an increase in the use of photo-identification techniques and advances in tracking technology available for the marine environment. It is currently thought that there is little movement of individuals between constellation sites over medium-term time scales (decades) (Andrzejaczek et al. 2016). Even though whale sharks can travel thousands of kilometers per year, those visiting constellation sites (mostly males, ranging from 3 to 9 m in total length) seem to largely stay in the broader area of those sites (Norman et al. 2017a; Prebble et al. 2018). In contrast, large females appear to range much further, as outlined below.

6.2 TRACKING ANIMALS IN THE OCEAN

Tracking animal movements in the ocean is challenging for scientists. In many studies, there may be only a single, fleeting physical interaction with the subject (e.g. tagging), so we must rely on technology to provide us with information on their movements and whereabouts. Biotelemetry involves the attachment of electronic devices to animals, devices which transmit information to researchers by different means (see Lennox et al. 2017 for an in-depth review of marine tracking), allowing scientists to recreate the paths taken by their subjects. Tracking marine animals using very-high-frequency (VHF) radio transmitters, GPS and satellite-linked tags (Figure 6.1) is possible when they routinely spend time at the surface (Hooker et al. 2002). For other animals that seldom or never spend time at the surface, light-based or even geomagnetic-based (Klimley et al. 2017) archival tags can store data that are later transmitted to satellites once the tag has detached (Figure 6.1, Argos 2011) and used to calculate daily position estimates. These archival tags are needed because transmission signals cannot propagate through seawater and so cannot be successfully sent to a receiver whenever animals are below the surface. Acoustic tags emit coded ultrasonic signals, which can be used to actively track the animals in real time by using arrays of underwater or vessel-based hydrophones (receivers). These tags are useful to understand residency at particular sites and movement patterns between receiver stations. Both acoustic and satellite tags can be fitted with sensors to provide information on depth, temperature and other variables. Dead reckoning techniques have also been employed, which involves the reconstruction of tracks from a known starting point by using continuous depth, bearing, and speed data. This approach requires either remote transmission of the data or recovery of the tag, which results in short but highly detailed tracks (Wilson et al. 2007). The use of fine scale accelerometers has allowed scientists to differentiate between depth occupancy and locomotor activity (Gleiss et al. 2009, 2011, 2013; Cade et al. 2020). These instruments are attached to the first or second dorsal

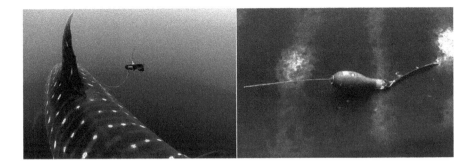

Figure 6.1   (left) Towed satellite positioning "smart position and temperature" (SPOT) tag attached to a whale shark on a 1.5 m leader. This near-real time tag will transmit to the Argos satellite network whenever it is at the surface. (right) MiniPAT tag attached to a whale shark on a short leader. This archival tag will detach after a pre-programmed period, float to the surface and transmit its information via satellite. (Photos: Jonathan R. Green.)

fin of the sharks, where they record body orientation in three dimensions and thus allow scientists to calculate the precise depth, body movements and orientation, speed and acceleration of the sharks. Increasingly, these tags are being combined with cameras to provide more insight into particular behaviors, and their timing and frequency. All of these techniques have been applied to the tracking of whale sharks and are helping researchers to follow their movements in the world's vast oceans.

Tracking data may be supplemented by the use of photo-identification, particularly in locations frequented by tourists and/or scientists. This technique is based on evidence that the spot patterns on whale sharks are unique for each individual and do not change as they grow (Arzoumanian et al. 2005; Pierce et al. 2019). Thus, through the global Wildbook for Whale Sharks database (www. whaleshark.org), which encourages contributions from both scientists and tourists, it is possible to obtain information on the residency of individuals at particular constellation sites and whether individuals move between sites (Norman et al. 2017b, and see Chapter 7).

6.3 VERTICAL MOVEMENTS OF WHALE SHARKS

Whale sharks are a mostly epipelagic species, meaning that they spend much of their time in the top, sunlit layer (<200 m) of the ocean (Tyminski et al. 2015). They spend a significant amount of time at or near the surface while at feeding constellation sites, such as in the Gulf of California (>80% of time above 10 m), at Ningaloo Reef in Western Australia (50% at surface), or off the Yucatan peninsula (7%–72% of time in the top 3 m; Eckert and Stewart 2001; Motta et al. 2010; Gleiss et al. 2013; Tyminski et al. 2015; Cade et al. 2020). Despite this, whale sharks can occupy an extremely wide vertical depth range, from surface waters down to a maximum recorded depth of 1,928 m (Tyminski et al. 2015), which makes the whale shark the deepest-diving fish known to science. This observation is right at the limits that the current generation of PATs can record. We don't actually know how deep whale sharks are capable of diving yet, but it could be substantially more than 1,900 m. We remain technologically limited, and so this remains an outstanding question for this species. Studies from coastal constellation sites and oceanic waters around the world have shown that both their vertical occupancy and their diving behaviors are complex and change depending on habitat, location, bathymetry and time of day, among other factors. Various non-exclusive explanations for their diving behaviors have been suggested, including thermoregulation, feeding, foraging, navigation and minimizing the energetic cost of transport (Gunn et al. 1999;

Graham et al. 2006; Rowat and Gore 2007; Thums et al. 2012; Gleiss et al. 2013; Rohner et al. 2013; Tyminski et al. 2015). Let us explore these in more detail.

Due to their ectothermic physiology (i.e. their inability to regulate their internal body temperature), whale sharks are only able to control their body temperature behaviorally, meaning that they move to waters with different temperatures to meet their thermo-physiological needs. For example, whale sharks feeding in surface waters of Al Shaheen in the Arabian Gulf at temperatures of up to 35°C had a shallower mean depth during the morning (6:00–12:00) than during the remainder of the day and nighttime, suggesting that they cool down at depth after feeding at the surface in the morning (Robinson et al. 2017). Conversely, at Ningaloo Reef and at Christmas Island, where sharks performed single deep dives during daytime hours, researchers found a negative relationship between minimum temperature and the duration of the interval between dives, leading them to suggest that the sharks warm up at the surface following deep foraging dives below the thermocline (Thums et al. 2012). However, this relationship broke down in waters warmer than 25°C, and deep diving also occurs in poorly stratified waters (Gleiss et al. 2011), so thermoregulation evidently does not fully explain their diving behavior.

Nine whale sharks were tracked using accelerometers over a combined period of 171.5 hours at Ningaloo Reef (Gleiss et al. 2011). These sharks adopted a highly efficient mode of swimming, which involves using their negative buoyancy to "glide" down during descents, thus harnessing the potential energy from gravity and converting it into both horizontal and vertical movements (kinetic energy) without beating their caudal fin. These observations indicate that whale sharks have optimized their gliding performance, achieved by their angle of descent in combination with the lift produced by their broad head and wide pectoral fins. Ascents are then made by slowly beating the caudal fin.

Based on their geometry, duration and in particular the angle of ascent, Gleiss et al. (2011) described five different whale shark dive types. In bounce (or yo-yo) dives (Figure 6.2a), the shark moves up and down in the water column over vertical ranges of >10 m, usually beginning within 15 m of the surface, without any protracted time between dives. Bounce dives are also fairly common in other elasmobranchs, such as scalloped hammerhead sharks *Sphyrna lewini* (Klimley 1993; Ketchum et al. 2014), white sharks *Carcharadon carcharias* (Weng et al. 2007) and blue sharks *Prionace glauca* (Carey and Scharold 1990). V-dives (Figure 6.2b) are similar to the typical dive pattern of many air-breathing marine animals but, in whale sharks, they were not constrained to starting at the surface, but rather in shallow water. V-dives can be either single dives with at least 30 minutes of surface time intervals (Type 2) or a series of consecutive dives (Type 3). These types of dives are presumably a foraging strategy, to search for prey in a food scarce, environment to maximize vertical distance searched (see Chapter 8). Bottom bounce dives (Figure 6.2c) are U-shaped dives with small (<10 m) oscillations occurring between the descent and ascent phases. Both bounce dives (Type 1) and bottom bounce dives (Type 4) are the most efficient for horizontal travel, although if over large enough vertical distances, they may also serve as a strategy for searching for prey in different water layers (Gleiss et al. 2011). U-dives (Figure 6.2d) are characterized by a protracted bottom time before the ascent, likely associated with feeding at depth. These dives display a descent pitch, which is shallower than the other dive types (−8° as opposed to −13°).

The depth occupancy of whale sharks also appears to be linked to the distribution and movements of their prey. At Gladden Spit (Belize), vertical movements of whale sharks coincided with spawning aggregations of snappers (*Lutjanus* spp.) occurring at the surface in the late evenings, after full moon. Sharks tended to be shallower during the night, feeding on snapper spawn, while making deep dives during the day. Furthermore, deep diving occurred after the peak spawning season, presumably as part of a foraging strategy employed once their surface food source was no longer available (Graham et al. 2006). Many potential prey species undertake a diel vertical migration (DVM), staying deep during the day and feeding in the euphotic zone (the ocean layer that

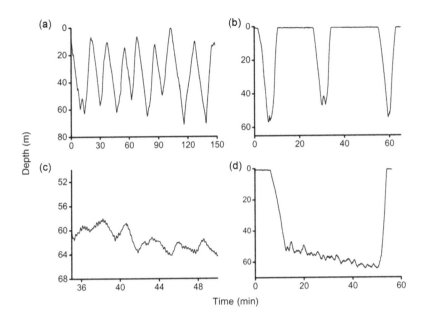

Figure 6.2 Dive types distinguished from time–depth profiles. (a) Bounce or yo-yo diving (Type 1), (b) V-dives (Types 2 and 3), which can further be classified based on surface time between dives (see main text), (c) bottom bounce dives (Type 4) and (d) U-dives (Type 5). (Reproduced from Gleiss et al. (2011) with permission.)

receives enough light for photosynthesis to occur) at night to avoid visual predators (Brierley 2014). This seems to indicate that the sharks may be targeting these mesopelagic prey.

Reverse DVM is more common at daytime surface feeding constellations. One of the largest known feeding "constellations" occurs off the Yucatan Peninsula in Mexico, where whale sharks gather to feed on tuna eggs at the surface from May to September (de la Parra et al. 2011; Hueter et al. 2013). Whale sharks tracked here spent 96.5% of their time in the epipelagic zone (0–200 m depth) and >30% of their time in the top 3 m of the water column (Motta et al. 2010; Tyminski et al. 2015). While at this feeding constellation, they occupy slightly shallower depths during daytime (1.9–15.4 m) than at night (2.4–20.7 m), probably due to feeding at the surface during daytime. They make more vertical excursions to below the thermocline at night – a behavior that has been linked to thermoregulation or to feeding at depth (Tyminski et al. 2015). Their vertical profile also changed within the day, with more surface suction feeding behavior (see Chapter 8) observed in the morning and afternoon hours, with a dip around midday and little surface activity displayed at night (Motta et al. 2010).

At Ningaloo Reef, Western Australia, short-term active tracking of whale sharks using acoustic tags and a hydrophone mounted on a small boat showed that there was considerable variation between individuals, in terms of diving behavior and vertical occupancy, although in general sharks were closer to the surface at night (Gunn et al. 1999). In a subsequent study at the same location, sharks were shallower during daytime hours, likely due to feeding on surface schools of tropical krill *Pseudeuphasia latifrons* (Wilson et al. 2006). Whale sharks exhibited a moderate reverse DVM, but clear increased locomotory activity around dawn and dusk, and the greatest ascent pitch angles just before dusk (Gleiss et al. 2013). In other words, they did most of their diving around dawn and dusk.

A female whale shark tagged at the Tofo constellation site in Mozambique moved to Madagascar over a period of 87 days, a distance of at least 1,200 km (Brunnschweiler et al. 2009). While in coastal waters, this shark spent 64% of its time in shallow water (<10 m) and carried out occasional deeper dives (to 108 m), mostly at night. Once in the oceanic waters of the Mozambique Channel, however, it switched to a DVM pattern, spending daytime hours at greater depths (118 m on average), rising to shallow waters (41 m average) at night and making regular deep dives to below 1,000 m (maximum recorded depth 1,286 m).

Once whale sharks move away from their coastal feeding constellations to oceanic waters, they are less constrained by bathymetry and more challenged with lower prey availability at the surface. Consequently, their vertical movements change. Here, as well as optimizing the efficiency of their horizontal displacement, vertical excursions might also be driven by foraging, as whale sharks cross different water layers in which they can potentially detect different cues for a feeding opportunity (Gunn et al. 1999; Wilson et al. 2006; Brunnschweiler and Sims 2012). Biochemical evidence suggests that whale sharks feed at depth as well as in surface waters (Rohner et al. 2013). Prey items in the deep scattering layer, a deep midwater layer of the ocean with a high biomass of living organisms such as small fishes (e.g. myctophids) and crustaceans (e.g. euphausiids), can be targeted by whale sharks away from shallow coastal waters as they vertically migrate to deep waters during day time (Brierley 2014).

During deep dives, whale sharks enter a habitat with very different environmental parameters than those found in the epipelagic zone. The hydrostatic pressure increases with depth, temperature decreases rapidly beneath the well-mixed surface layer, and natural light disappears past the euphotic zone (about 200 m). Whale sharks have been recorded in almost 200 atmospheres of pressure at 1,928 m depth and in water as cold as 3.4°C (Brunnschweiler et al. 2009; Tyminski et al. 2015), where their large size prevents heat loss in their bodies, even as the surrounding temperature can decline by over 20°C (Nakamura et al. 2020); this allows them to act as functionally warm-blooded since, for a while at least, they are much warmer than their prey and therefore operating at a higher metabolic rate (see Chapter 1). The mesopelagic zone also often contains the oxygen minimum zone, where limited mixing and lack of photosynthesis result in low levels of dissolved oxygen (Stramma et al. 2008). These short, but deep dives are type 2 "V-shaped" dives described above and seem to be part of their open-ocean foraging behavior in the absence of predictable high-density patches of prey at the surface, although they may also play another role: Graham et al. (2006) suggest that the rapid ascents might be undertaken to re-oxygenate the gills following time spent in these low oxygen layers.

6.4 HORIZONTAL MOVEMENTS OF WHALE SHARKS

Our knowledge about whale shark movements has vastly increased over the last century since Gudger (1932) proposed that the Sulu Sea, between southern Philippines and northern Borneo, was the single point of origin from which whale sharks dispersed throughout the oceans depending on environmental conditions.

Sequeira et al. (2013a) explored the influence of climate signals on the temporal distribution of whale sharks. There is some evidence that sighting probability varies in long-term cycles, and three competing hypotheses were put forward to explain the observed trend. First, it may be driven by a potentially inter-decadal cyclical pattern in relative abundance associated with the migratory behavior of the species (Sequeira et al. 2013a). Similar cycles have been observed in fish catches (Jury et al. 2010) and in stranding of other large marine species (Evans et al. 2005). Second, the trend might reflect population declines resulting from overfishing and other threats (Bradshaw et al. 2007, 2008; and see Chapter 11). Third, there may be a distributional shift in whale shark occurrence related to changes in habitat suitability as a result of climate change and its impact on thermal regimes, productivity and current patterns in the pelagic zone.

Tracking studies and complementary work carried out at constellation sites, where researchers can easily interact with whale sharks, have provided a fair understanding of the spatial dynamics of the sharks present at specific locations. Compilations of tracks at a global scale (Sequeira et al. 2013b), however, highlight that the available information on whale shark movement, range and habitat use on a global scale is still fragmentary and scattered across their vast range.

The first published whale shark tracks were from Ningaloo, Western Australia, where modern whale shark research first began. Here, in the 1990s, three whale sharks were fitted with acoustic tags and tracked for 3–26 hours from a small boat, while an early model of geolocation tag provided 24 hours of data for a fourth shark (Gunn et al. 1999). A second study in the Gulf of California showed the movements of 17 whale sharks (Eckert and Stewart 2001). By 2013, 109 whale shark tracks from 15 locations were publicly available, though only 75 had been published in peer-reviewed literature (Sequeira et al. 2013b). The number of published tracks had increased fourfold to 342 by 2018 (Table 1 and Figure 6.3), encompassing recent studies in the Galapagos Islands (Hearn et al. 2013, 2016), Red Sea (Berumen et al. 2014), Arabian Gulf (Robinson et al. 2017), Mozambique (Rohner

Table 6.1 Summary of Published Tracking Studies on Whale Sharks

Location	N (F,M,U)	TL (m) (min–max)	Duration (days)	Distance (km)	References
Belize	11 (0,8,3)	6 (3.6–9.7)	125 (14–249)		Graham et al. (2006)
China	1 (0,1,0)		33		Marine Conservation Society Seychelles, cited in Sequeira et al. (2013b)
Christmas Island	1 (0,1,0)	5	88		Australian Institute of Marine Sciences, cited in Sequeira et al. (2013b)
Djibouti	1 (0,1,0)	3	9	81	Rowat et al. (2007)
Dubai	1 (1,0,0)	4.6	33	348	MML 2010 (in Sequeira et al. 2013b)
Galapagos	51 (50,1,0)	10.7 (4–13.4)	48 (1–178)	4,155 (52–16,407)	Hearn et al. (2013, 2016, 2017)
Gulf of California	23 (12,3,8)	7.4 (3–18)	112 (1–1,144)	1,812 (9–12,620)	Eckert and Stewart (2001), Ramirez-Macias et al. (2017)
Honduras	2 (0,2,0)	8	82 (31–132)		Gifford et al. (2007)
Madagascar	13 (3,7,3)	6.8 (4–8)	110 (9–199)	1,958 (191–4,275)	Graham et al. (2008) (in Sequeira et al. 2013b), Diamant et al. (2018)
Malaysia	3 (0,0,3)	7	48 (6–127)	3,048 (220–8,025)	Eckert et al. (2002)
Mozambique	17 (4,13,0)	6.5 (5.4–8.7)	29 (2–88)	722 (10–2,327)	Brunnschweiler et al. (2009), Rohner et al. (2018)
Panama	1 (1,0,0)	7	841	20,142	Guzman et al. (2018)
Philippines	20 (4,11,5)	5.5 (3–7)	59 (6–126)	880 (54–2,580)	Eckert et al. (2002), Araujo et al. (2018)
Qatar	28 (11,17,0)	7 (5–9)	69 (1–227)	378 (0–2,613)	Robinson et al. (2017)
Red Sea	59 (21,18,20)	4 (2.5–7)	117 (0–315)	509 (5–2,950)	Berumen et al. (2014)
Seychelles	3 (0,1,2)	6.2 (5–7)	42 (19–60)	1,769 (502–3,383)	Rowat and Gore (2007)
South Africa	3 (1,2,0)	7.3 (7–8)	9 (3–17)	118 (65–199)	Gifford et al. (2007)
St Helena	30 (14, 11, 5)	8.1 (4–12)	75 (1–432)	465 (3–3,396)	Perry et al. (2020)
Taiwan	4 (1,3,0)	4.2 (4–4.5)	108 (0–208)	4,250 (3,121–5,896)	Hsu et al. (2007)
USA	2 (1,1,0)	7.5	96 (84–108)		MML (2010) (in Sequeira et al. 2013b)
Western Australia	57 (25,25,7)	6.4 (3–11)	81 (0–261)	1,776 (4–6,157)	Wilson et al. (2006, 2007), Reynolds et al. (2017)
Yucatan Peninsula	35 (22,12,1)	7.3 (4.5–9)	69 (0–190)	699 (20–7,213)	Hueter et al. (2013)

N, number of individuals; F, female; M, male; and U, unknown.
Note that the distance traveled (mean and range) was calculated differently among studies.

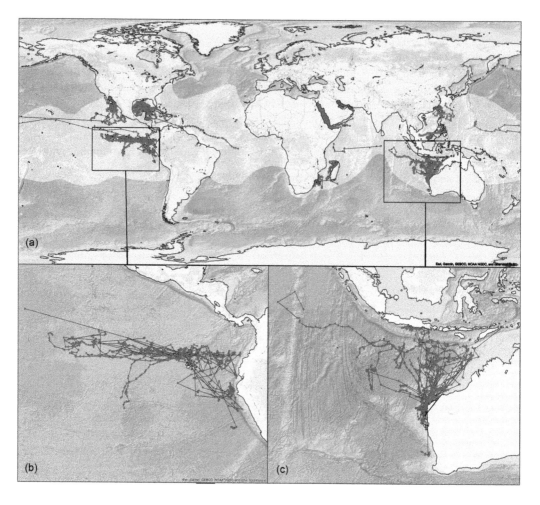

Figure 6.3   (a) Global summary of whale shark movements based on tracks summarized in Table 1, with geographic range shown in green; (b) close-up of tracks from Eastern Tropical Pacific; and (c) close-up of tracks from Western Australia

et al. 2018), Madagascar (Diamant et al. 2018), Ningaloo Reef (Reynolds et al. 2017), Philippines (Araujo et al. 2018), St. Helena (Perry et al. 2020), and the Yucatan Peninsula (Hueter et al. 2013) and Baja California in Mexico (Ramirez-Macías et al. 2017).

The size structure of tagged whale sharks to date ranges from 2.5 to 10 m for males and from 3 to 18 m for females (Figure 6.4). All but one of the large females were tagged in either Baja California or the Galapagos Islands, both in the Eastern Pacific. The majority of the remaining tracks have been of immature males, reflecting the biased mean size and sex structure of most constellations (see Chapter 7). Sightings of neonate and very young (<2 m) whale sharks are rare globally, and their movements are not yet known. Similarly, little is known about the movements of sexually mature males.

What is striking from the compilation of all tracking results is that only 15 individuals moved more than 5,000 km from their tagging location (Figure 6.3). This may be partly due to issues of tag retention, with average track durations varying from 9 to 125 days between locations. In the Galapagos Islands, for example, a significant proportion of towed tags are shed the same day, with evidence pointing to blacktip sharks *Carcharhinus limbatus* and silky sharks (*C. falciformis*) actively ripping them out (author's unpublished data). Only two transoceanic movements have been

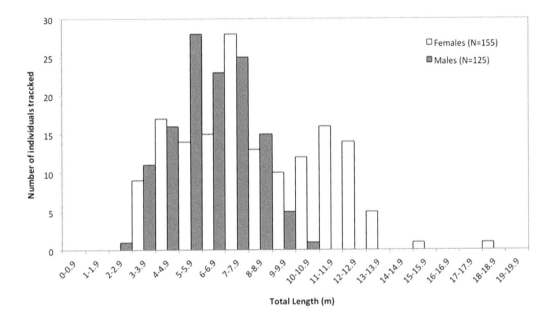

Figure 6.4 Composition (size and sex) of whale sharks used in tracking studies globally. NB: numbers may not coincide with Table 6.1, because for some individuals, either sex or size data were lacking.

reported: one individual tagged in the Gulf of California, which apparently made a transoceanic movement of 12,620 km over a 3-year period (Eckert and Stewart 2001), and one tagged in Panama, which moved past Hawai'i as far as the region of the Mariana Trench in the western Indo-Pacific (20,142 km), over a period of 841 days (Guzman et al. 2018). Given that these datasets lack depth information, it is difficult to ascertain whether these are bona fide tracks or detached floating tags (Hearn et al. 2013; Sequeira et al. 2013b; Andrzejaczek et al. 2016). Other long-distance movements include those of large female whale sharks tracked at the Galapagos Islands (outlined in detail below), and a 7.5 m female whale shark tagged in the Gulf of Mexico which was tracked over 7,213 km to near the St. Peter and St. Paul Archipelago in the mid-Atlantic in 2007 (Hueter et al. 2013), and was subsequently identified back in the Gulf of Mexico in 2011 (R. de la Parra, pers. comm.).

The most whale shark tracking and photo-identification data have been collected in the Indian Ocean (including the Red Sea). Some movements between regions within this area have been observed, such as one shark swimming between Mozambique and Seychelles (Rowat and Gore 2007), two between Mozambique and Tanzania (Norman et al. 2017b), and others between Ningaloo and Indonesia (Norman et al. 2017b; Reynolds et al. 2017). Despite the relatively large number of constellation locations known in the Indian Ocean, no definitive breeding area has been identified there (or indeed, anywhere), and large females are not commonly seen at constellation sites, although one 9 m female shark was tagged off Qatar (Robinson et al. 2017). While most constellations appear to be seasonal, whale sharks are seen year round at some locations such as the Maldives, Thailand, the Red Sea, Mozambique, and Honduras (Norman et al. 2017b) and recent evidence suggests that cryptic residency could be more common than previously realized. In Tanzania, for example, acoustically tagged whale sharks were detected throughout the year, despite not often being seen at the surface across cooler months (Cagua et al. 2015; Rohner et al. 2020), while in Cenderawasih Bay (Indonesia), individual sharks display a wide range of behaviors, including long-term residency (Meyers et al. 2020).

Whale sharks often appear at constellation sites coinciding with ephemeral high prey availability, such as fish spawning events (e.g. in Yucatan Mexico, Qatar and Belize), but when they leave, whale sharks generally disperse in multiple directions, often linked with prevailing currents (Eckert and Stewart 2001; Eckert et al. 2002; Hsu et al. 2007; Hueter et al. 2013; Berumen et al. 2014). Whale sharks tagged at Ningaloo Reef generally move northeast into the Indian Ocean (Wilson et al. 2006; Reynolds et al. 2017), perhaps due to cold temperatures outside their thermal range to the south. There is some evidence linking whale sharks' oceanic movements with large-scale oceanographic features, such as the Kuroshio Current (Iwasaki 1970; Hsu et al. 2007), the Equatorial Front (EF) in the Pacific (Hearn et al. 2016; Ryan et al. 2017), and the Yucatan and Loop currents in the Gulf of Mexico region (Hueter et al. 2013). Although movements in response to currents have been reported, it is unclear whether these associations are to aid movements or navigation between distant locations, or are destinations in themselves, perhaps as offshore foraging areas where entrainment processes result in greater densities of zooplankton prey.

6.4.1 Case Study: Ningaloo Reef

Ningaloo Reef, Australia's largest fringing coral reef, stretches 260 km along the west coast of the Cape Range Peninsula, Western Australia (WA). It was one of the first-known constellation sites for whale sharks, and a lucrative tourism industry has developed since the world's first in-water experiences with whale sharks were offered here in 1989 (Catlin and Jones 2010). Whale sharks aggregate at Ningaloo Reef in response to the mass spawning of corals in March and April (Simpson et al. 1991; Taylor 1996). The tourism industry operates during the austral autumn and winter, and this has become known as "whale shark season" because of the reliability of encountering whale sharks at this time of year.

As at many of the known coastal constellation sites for whale sharks, the majority of those found at Ningaloo Reef are juvenile males (Meekan et al. 2006; Norman and Stevens 2007; Norman et al. 2017b). Although large, mature sharks of both sexes are sometimes sighted there, to date there has been no evidence of mating. Ningaloo Reef is, however, an important feeding area for whale sharks, and their daily movements within the area likely reflect foraging and feeding behaviors (Gleiss et al. 2013). Whale sharks stay close to the reef edge, moving north and south, parallel with the reef (Gunn et al. 1999), and vertically in the water column (Gleiss et al. 2013) in their search for food. They are often found in higher abundance close to passages in the reef where their zooplankton prey are concentrated by currents (Anderson et al. 2014). They also feed on subtropical krill during the day and night (Taylor 2007). Oceanographic and atmospheric conditions influence the presence and abundance of whale sharks at Ningaloo Reef, with more sharks sighted when the Southern Oscillation Index is stronger (Wilson et al. 2001; Sleeman et al. 2010a) and along-shelf winds more prevalent, perhaps making prey more available (Sleeman et al. 2010a).

More than 1,770 individual sharks had been identified at Ningaloo Reef, at the time of writing, through the global citizen science photo-identification database (Wildbook 2020). This long-term monitoring has revealed that individual whale sharks can be seen multiple times throughout each whale shark season, with an average residency rate of 33 days (Holmberg et al. 2009). There is evidence of site fidelity at Ningaloo Reef, just as there is at other coastal constellation sites, with 40% of identified sharks seen there in two or more years, and some encountered year after year for over 20 years (Norman and Morgan 2016; Norman et al. 2017b). However, the movements and whereabouts of sharks outside whale shark season remain unclear. The timings of peak occurrences of whale sharks at other constellation sites around the Indian Ocean led to the hypothesis that at least some sharks could be making a clockwise migration around the ocean basin over a period of 2 years (Sequeira et al. 2013b). However, photo-identification studies have shown that few sharks move between countries that are separated by long distances, and only one shark identified at Ningaloo Reef (i.e. A-424, which was also seen off the coast of East Kalimantan in Indonesian Borneo) has been sighted in the waters of another country to date (Norman et al. 2016; Norman et al. 2017b; Wildbook 2020).

The first satellite tracking studies of whale sharks tagged at Ningaloo Reef (Figure 6.3c) showed individuals moving to the north, north-east and north-west of Ningaloo after the whale shark season (Wilson et al. 2006; Sleeman et al. 2010b). More recent satellite tracking has documented the first fully tracked return movements of whale sharks to Ningaloo Reef, with sharks making relatively short forays away from the coast before returning to the area within a few months (Reynolds et al. 2017). Photo-identification has also revealed the return of other sharks after their tracking periods had ended (Norman et al. 2016; Reynolds et al. 2017). Sharks returned to Ningaloo Reef outside the whale shark season and appeared to be using an area south of where they are typically found during the season (Reynolds et al. 2017). Whale sharks have been recorded at Ningaloo Reef by satellite tracking, photo-identification, and acoustic tracking in all months of the year (Norman et al. 2017a), and habitat selectivity modeling has shown that Ningaloo Reef provides suitable conditions for whale sharks throughout the year (Reynolds et al. 2017). This recent research suggests that, rather than hosting a seasonal constellation of whale sharks in the austral autumn and winter, Ningaloo Reef is an important aggregation site for whale sharks all year round.

It seems, however, that the whale sharks that aggregate at Ningaloo Reef also use other areas along the WA coast, possibly to take advantage of ephemeral sources of prey, such as western rock lobster larvae off Geraldton (Norman et al. 2016). Photo-identification and satellite tracking, along with anecdotal reports from members of the public, have shown the presence of whale sharks at many places along the coast, from as far north as Ashmore Reef (in the Timor Sea) and Broome, to as far south as Albany on the Southern Ocean coast of WA, at different times of the year (Norman et al. 2016). Whale sharks tagged at Ningaloo Reef have been tracked to and from Shark Bay (Norman et al. 2016; Reynolds et al. 2017; S. Reynolds unpublished data available on www.zoatrack. org), and all four whale sharks tagged in the first satellite-tagging program at Shark Bay in January 2018, moved north to Ningaloo Reef (S. Reynolds unpublished data available on www.zoatrack. org). From Ningaloo, tagged sharks have also been tracked moving further than 1,200 km south to areas off the coast of Perth and Bunbury (Norman et al. 2016; Reynolds et al. 2017; S. Reynolds unpublished data available on www.zoatrack.org). As well as these movements along the WA coast, whale sharks have been tracked from Ningaloo Reef moving north to Indonesia (Sleeman et al. 2010b; Reynolds et al. 2017), and north-west towards Christmas Island (Sleeman et al. 2010b) and Cocos (Keeling) Islands (S. Reynolds unpublished data available on www.zoatrack.org). In 2017 and 2018, a tagged whale shark (A-496) traveled further than 11,500 km in a loop from Ningaloo Reef north-west towards Cocos (Keeling) Islands, south-east to an area off the coast of Perth and then north again to an area close to Ningaloo Reef (S. Reynolds unpublished data available on www. zoatrack.org). This is the longest recorded movement of a whale shark from Ningaloo Reef and from the entire Indian Ocean to date. Many sharks tracked from Ningaloo Reef move thousands of kilometers, and some spend time in offshore areas of the Indian Ocean where water depths exceed 6,000 m. However, 56% of the locations transmitted from tagged sharks were in coastal waters where depths were ≤200 m (Reynolds et al. 2017). This is similar to whale sharks tracked in Mozambique, which spent a lot of their time in coastal waters, even though the continental shelf is narrow and deeper waters are within easy reach (Rohner et al. 2018). It appears that some sharks prefer to remain in shallower waters closer to the coast, perhaps selecting areas of higher productivity, while others spend more time in the open ocean.

6.4.2 Case Study: Galapagos

The Galapagos Islands of Ecuador provide a unique opportunity to study the movement patterns of a hitherto understudied segment of the whale shark population in an open ocean environment: adult females. This stands in contrast to Ningaloo and almost all the other known coastal constellation sites, where mostly immature males come to feed seasonally

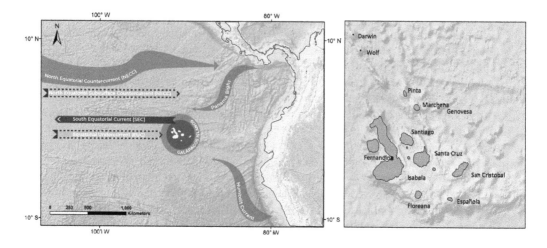

Figure 6.5   (left) Eastern Tropical Pacific showing Galapagos Islands and main ocean currents (solid lines: surface currents, dashed lines: subsurface currents). (right) the Galapagos Archipelago, showing main islands.

Situated in the Eastern Tropical Pacific, straddling the equator approximately 1,000 km off the coast of mainland Ecuador, the Galapagos Islands are made up of 13 major islands (Figure 6.5) and over 100 islets and emergent rocks (Snell et al. 1996). The main island archipelago sits on a platform with an average depth of 200 m, which drops rapidly down to >2,000 m in the surrounding ocean, while the remote northernmost islands of Darwin and Wolf rise straight from the seabed (Geist 1996). The archipelago sits in the South Equatorial Current, which flows westwards along the equator, and is fed from the northeast by the warm Panama Bight, and from the southeast by the cool Humboldt Current (Figure 6.5). The deep subsurface Equatorial Undercurrent, also known as the Cromwell Current, flows from west to east and is blocked by the islands, causing upwelling and high productivity in the western region of the archipelago (Karnauskas et al. 2007). This results in one of the largest oceanographic features on the planet – the Equatorial cold tongue, which extends for thousands of kilometers into the central Pacific Ocean and plays a key role in global climate and productivity patterns (Karnauskas et al. 2007).

The first record of a whale shark at the Galapagos Islands was of a large female to the north of Fernandina Island in June 1925 (Beebe 1926; Gudger 1927). A second small individual was reported at Isabela Island in spring of 1933 (Gudger 1933). Whale sharks continued to be reported sporadically over the next 60 years, and four large females were observed south of the island of Isabela in March and April 1985 (Arnbom and Papastavrou 1988). Whale sharks are now reported occasionally throughout the year at dive sites in the central archipelago, but especially in the months of March through May (Hearn et al. 2014). However, since the advent of dive-based tourism in the 1990s, they have become an important attraction at the northernmost island of Darwin.

Darwin is the remotest of the islands – almost 300 km from the central point of the archipelago, and 40 km northwest of its closest neighbor, Wolf. It is also the smallest of the Galapagos Islands, barely over 1 km², and it is surrounded by steep cliffs and a shallow (<10 m) rocky platform to the southeast, which ends at Darwin's Arch, a rocky outcrop which signals the edge of the platform. The prevailing current impinges the island here and splits around to the north and south. Large aggregations of scalloped hammerhead sharks occur at the Arch, alongside Galapagos, silky and blacktip sharks, jacks, tuna and bottlenose dolphins (Hearn et al. 2017). This is also where whale sharks can be found during the cool season from July through early December (Zarate 2002; Acuña-Marrero et al. 2014; Hearn et al. 2014).

The whale sharks at Darwin Island are predominantly large adult females (Acuña-Marrero et al. 2014; Hearn et al. 2016; Norman et al. 2017b), many of which display distended abdomens, which has been suggested to be an indication of pregnancy (Ramirez-Macías et al. 2012). The site is small, with sharks appearing to utilize an area of only 2.4 km² (Acuña-Marrero et al. 2014) and there may be on average only up to four individuals present on any given day, although over the season, up to 700 individuals may visit the island (Acuña-Marrero et al. 2014). Their rapid turnover rate – an average residency of only 2 days – compares with average residencies of 19 days at St. Helena (Perry et al. 2020) 33 days at the Ningaloo constellation (Holmberg et al. 2009), 60 days in Baja California (Ramirez-Macías et al. 2012) and approximately 1 month at Holbox/Isla Mujeres (Hueter et al. 2013). If Darwin Island were an important feeding area, attracting individuals from the surrounding ocean, then we would expect to see longer residency and larger numbers of sharks at any given time. In fact, the only feeding event reported for Galapagos occurred at neighboring Wolf Island, when a large female was seen swimming through snapper spawning aggregation with her mouth open wide (Hearn et al. 2016). In 2013, four large females were fitted with continuous tags with depth sensors, which sent acoustic signals every 3 seconds. These signals were detected by a hydrophone mounted on a small boat used to track their vertical and horizontal movements for 3–25 hours (Acuña-Marrero et al. 2014). The sharks restricted their movements to within 2.4 km of the island and its shallow platform, with the longest occupancy (>50%) on the windward side of Darwin's Arch to the southeast of the island, where the prevailing current impinges upon the island platform. The sharks did not utilize the lee of the island. Their average depth was around 20 m, and only a single shark made one dive to >200 m. Two of the sharks were tracked for a short distance as they left the island, and during this movement, as the water depth dropped below 60 m, they spent more time in surface waters. This contrasts with the behavior displayed off Ningaloo Reef, where sharks carry out long-shore movements along the inner portion of the continental shelf, along with long series of bounce dives to the bottom, and spent a significant portion of their time at the surface, especially at night, seeming to match DVM of zooplankton (Gunn et al. 1999). If they are not feeding and are not interacting socially, their presence at Darwin may be to use the island as navigational waypoint on a large-scale migratory route.

Long-distance tracking studies were carried out at the island from 2011–2017 (Hearn et al. 2013, 2016, 2017; Ryan et al. 2017) by using spear guns and fin clamps to attach towed satellite transmitters to sharks (Figure 6.6). Fifty-one sharks, all adult females with the exception of one 8 m male and three immature females (4–6 m), were tracked as they moved away from the island (Figure 6.3b). None of the sharks remained at Darwin Island. In all cases where sharks were tagged in July, they moved in a westwards direction to the EF, with the exception of the male, which spent 6 months moving in a haphazard manner in the open ocean between Galapagos and mainland Ecuador. The adult females then continued along the EF for up to 2,000 km (Ryan et al. 2017). The immature

Figure 6.6 Whale shark being tagged by diver at Darwin, Galapagos Islands. (Photo: Simon J Pierce.)

females, in contrast, veered to the southwest after only 500 km, and their tags detached in almost the same location, approximately 2,000 km southwest of Galapagos.

During their movements along the EF, whale sharks appeared to track tropical instability waves (Ryan et al. 2017), which are generated seasonally by shearing of the South Equatorial Current and the North Equatorial Countercurrent (Sweet et al. 2009). These instability waves propagate along the front and accumulate both plankton and their predators, such as planktivorous seabirds (Spear et al. 2001). The EF is the largest, but not the only, upwelling system in the region. Other upwelling systems include the Peru-Humboldt and the region southwest of Galapagos (Ryan et al. 2017). Indeed, by September, the adult females turned back along the EF towards Galapagos, with at least five individuals being either tracked or photographed near the island later in the same season (Hearn et al. 2016). These sharks, along with those tagged at Darwin Island in September–October, continued to move east, and by the end of December and early January each year, all individuals whose tags had not detached were in the upwelling system along the shelf break of southern Ecuador and northern Peru (Figure 6.7). Their mean speed was >40 km each day, far greater than that of most sharks tagged at feeding constellations (~10–30 km per day; see Table 6.1).

Sharks have not been tracked beyond this period due to tag shedding, but a dataset of whale shark encounters with the tuna purse-seine fishery, provided by the onboard observer program of the Inter-American Tropical Tuna Commission (IATTC), both validates the movement results and suggests a hitherto unknown important area for whale sharks (Figure 6.8; Román et al. 2018).

From July through September, as with the satellite tracks, whale sharks were encountered along the EF, coinciding with the period of the highest surface water chlorophyll levels in this region and the lowest levels in coastal waters of Ecuador (Figure 6.8a). At the end of the year, whale sharks were reported along the continental shelf, again coinciding with movement data and with local increased productivity in surface waters (Figure 6.8b). However, in the first quarter, for which no tracks exist, whale shark encounters were almost exclusively limited to the coastal region and to a

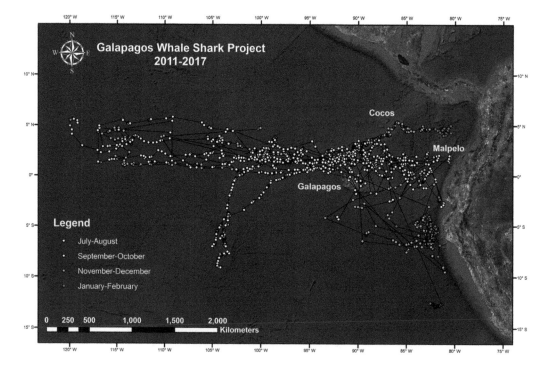

Figure 6.7 Annual tracks from whale sharks tagged at Darwin Island, Galapagos Marine Reserve.

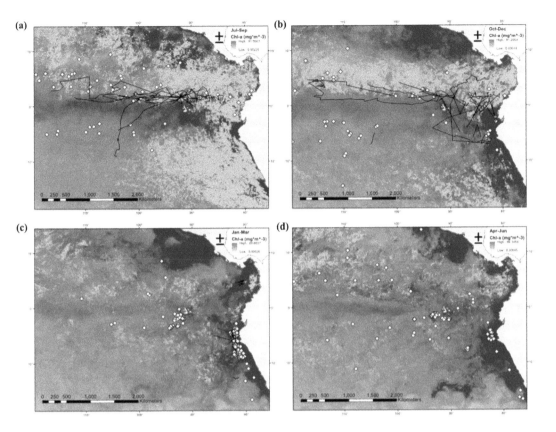

Figure 6.8 Aggregated (2003–2014) seasonal record of whale shark encounters (white circles) reported by the onboard observer program of the Inter-American Tropical Tuna Commission, IATTC (Román et al. 2018), overlaid on average chlorophyll a concentration. Black lines represent tracks from Galapagos within each quarter.

highly productive upwelling area to the southwest of Galapagos (Figure 6.8c). This pattern is similar although somewhat more dispersed in the second quarter (Figure 6.8d). Another piece of the puzzle was provided in February 2018, when an aerial patrol survey of the Galapagos National Park Directorate reported an aggregation of at least 18 large whale sharks in the area between Isabela and Floreana islands, approximately along the platform margin (DPNG 2018).

Whale sharks of both sexes and at sizes ranging from 4 to 13 m are reported occasionally at locations along the coast of Ecuador and Peru (Ramirez 1995; Torres et al. 2018). Similarly, a wide range of sizes may be seen at certain times of year by divers at the oceanic islands of Cocos and Malpelo (Sibaja-Cordero 2008; Sandra Bessudo pers. comm.). In contrast, it seems that it is almost entirely the adult females that make the movement to Darwin Island and beyond. These movements may reflect a large-scale commute between feeding grounds. Whale sharks appear to shift their diet as they grow, so may be expected to occupy different habitats based on their size (Borrell et al. 2011), but this would not explain the highly skewed sex ratio.

There has been much speculation in the literature that large females with distended abdomens in Philippines, Baja California, Qatar and Galapagos may be pregnant (Eckert and Stewart 2001; Ramirez-Macías et al. 2012, 2017; Rowat and Brooks 2012; Ketchum et al. 2013; Acuña-Marrero et al. 2014; Robinson et al. 2017). However, determining the gestation state of large, free-swimming whale sharks at depths of >20 m and currents often exceeding two knots is no mean feat. In 2017 and 2018, researchers from the Okinawa Churashima Foundation were able to take blood samples and

Figure 6.9 Rui Matsumoto from the Okinawa Churashima Foundation carries out an ultrasound examination of a female whale shark at Darwin, Galapagos Islands, in 2018. (Photo: Jonathan R Green.)

carry out ultrasound recordings from large female whale sharks in the wild for the first time (Green and Hearn 2018; Figure 6.9). As yet, no evidence of pregnancy has been found (see Chapter 2).

6.5 CLIMATE CHANGE

Whale sharks mainly occur within sea surface temperatures ranging between 23°C and 30°C across the world's oceans (Eckert and Stewart 2001; Hsu et al. 2007; Rowat and Gore 2007; Sequeira et al. 2012; Reynolds et al. 2017). Although they have also been reported from locations outside this range of temperatures (e.g. Turnbull & Randell 2006; Tomita et al. 2014; Norman et al. 2016; Robinson et al. 2017), whale sharks appear to avoid prolonged exposure to excessively low or high temperatures that may inhibit metabolic function. A 17-year dataset of whale shark occurrences, gleaned in logbooks from tuna purse-seine fisheries (Sequeira et al. 2012), was used to predict their pelagic distribution. Bearing in mind that the data were generated by fishing vessels actively seeking tuna, the main results suggested that whale shark habitat suitability, at least in the Indian Ocean, may indeed be mainly correlated with spatial variation in sea surface temperature, with 90% of sightings occurring between 26.5°C and 30°C, and that ocean temperatures can therefore provide insights into their migratory behavior.

A global analysis of a 30-year dataset of whale shark observations in the Atlantic, Indian and Pacific Oceans led to the understanding that the predicted increase in sea surface temperature by 2070 will lead to an overall redistribution of the species towards the poles, with reductions in occurrence in many of the currently known constellations (Sequeira et al. 2014a). These changes will likely be the result of whale sharks adapting to new temperature regimes and to associated changes in prey distributions (Gutierrez et al. 2008; Beaugrand et al. 2009). The predicted pole-ward shift in whale shark distributions echoes predictions made for other marine species (Perry et al. 2005; Dulvy et al. 2008). Indeed, this shift appears to be underway, with previously sporadic sightings of whale sharks at the Azores (Portugal), at the northern limit of their range, showing a dramatic increase since 2008, coinciding with a poleward shift of the seasonal 22°C isotherm (Afonso et al. 2014).

6.6 FUTURE PERSPECTIVES

While the number of published studies on whale shark movements, both vertical and horizontal, has increased dramatically over the past decade, many questions remain to be answered. The patterns and drivers of whale shark movements from constellation sites around the world have not yet been fully elucidated. Do juveniles stay closer to the coast while mature adults move offshore (as seen off the Pacific coast of Mexico) (Ketchum et al. 2013; Ramirez-Macías et al. 2017), possibly to find mates or different food sources (Figure 6.10)? Do different sexes and life stages have different habitat requirements and movement patterns? As yet, there is little to no information about the location of neonatal whale sharks, let alone their movements. And while the Galapagos studies have begun to shed light on the regional movements of adult females, few sexually mature males have been tracked to date. The lack of understanding of the distribution, connectivity and migration pathways of whale sharks continues to limit the development of international conservation plans for the species.

Tag retention has consistently been a limiting factor in these studies; a significant proportion of tags are shed almost immediately and those that do remain attached are often shed within 3–6 months (Table 6.1). New attachment techniques must seek to solve this issue, ideally to fulfill the battery potential of the tags and to track individuals for at least a year. Currently, some progress has been made with the use of clamps to attach tags to the dorsal fins of sharks (Norman et al. 2016; Green and Hearn 2018; Meyers et al. in 2020), and we expect mean retention time to increase as the technique is perfected (Figure 6.6). In addition to improved retention, increased satellite coverage through initiatives such as the Max Planck Institute's Icarus program for global monitoring of animals (www.icarus.mpg.de/en) should allow for improved detection, especially at lower latitudes where ARGOS coverage is spotty. This, along with greater transmission capabilities of future generations of tags, should increase the amount of information we are able to collect remotely about the behavior of the sharks. It's really our only hope, because we are unable to follow them in person to the depths and places they are capable of visiting.

Improved tagging technology, combined with recent efforts to provide a unified theoretical framework for animal movements (e.g. Nathan et al. 2008; Papastamatiou and Lowe 2012) and with

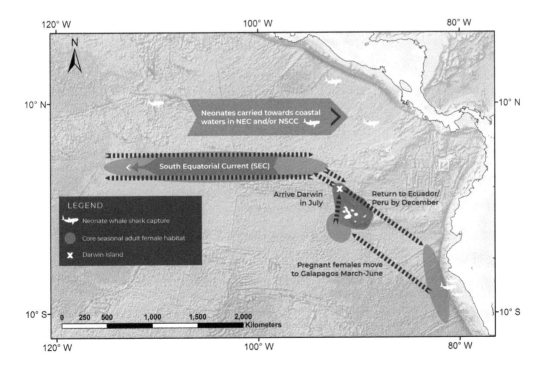

Figure 6.10 Proposed movement scheme of whale sharks in the Eastern Tropical Pacific.

advances in spatial analytical statistics should result in a shift from descriptive tracking studies to a more hypothesis-based approach. For example, Rohner et al. (2018) compared random model tracks based on characteristics of real sharks with tracks from juvenile sharks tagged in Mozambique and showed that they prefer shallow, highly productive waters. This kind of study will help elucidate the drivers behind whale shark long-distance movements and their search strategies in both a vertical context and a horizontal context.

The recent discovery of a predominantly adult whale shark constellation at St. Helena in the Atlantic, with a sex ratio of approximately 1:1, may provide the opportunity to address the location of putative mating grounds and track adult males (Perry et al. 2020). Partnerships with offshore fishing vessels may also provide opportunities to track the offshore movements of whale sharks. The drive to understand the movements of these sharks has been taken up by new research groups, and we can expect results from tagging studies at the offshore islands of Revillagigedo (Mexico) and Malpelo (Colombia) (S. Bessudo pers. comm) and Ascension (A. Dove, pers. comm.), along with studies at coastal locations such as Peru and Panama, in the next few years.

With continued increases in photo-identification monitoring of global whale shark populations, further genetic studies and improvements in tracking technology, the true extent of connections between whale shark populations across ocean basins, and how these connections may be affected by climate change, may finally be revealed.

ACKNOWLEDGMENTS

IATTC data were collected thanks to the on-board PICD observers managed by IATTC and National Observer programs. Thanks to Cassandra Garduño Mendoza for help with the figures. AMMS was supported by the Australian Research Council (Grant no.: DE170100841). AH was supported by the Universidad San Francisco de Quito, Ecuador.

REFERENCES

Acuña-Marrero, D., J. Jimenez, F. Smith, et al. 2014. Whale shark (*Rhincodon typus*) seasonal presence, residence time and habitat use at Darwin Island, Galapagos Marine Reserve. *PLoS One* 9(12):e1159456.

Afonso, P., N. McGinty, and M. Machete. 2014. Dynamics of whale shark occurrence at their fringe oceanic habitat. *PLoS One* 9(7):e102060.

Anderson, D. J., H. T. Kobryn, B. M. Norman, et al. 2014. Spatial and temporal patterns of nature-based tourism interactions with whale sharks (*Rhincodon typus*) at Ningaloo Reef, Western Australia. *Estuarine, Coastal and Shelf Science* 148:109–19.

Andrzejaczek, S., J. Meeuwig, D. Rowat, et al. 2016. The ecological connectivity of whale shark aggregations in the Indian Ocean: A photo-identification approach. *Royal Society Open Science* 3:160455.

Araujo, G., C. A. Rohner, J. Labaja, et al. 2018. Satellite tracking of juvenile whale sharks in the Sulu and Bohol Seas, Philippines. *PeerJ* 6:e5231.

Argos. 2011. Argos User's Manual. Worldwide tracking and environmental monitoring by satellite. http://www.argos-system.org/wp-content/uploads/2016/08/r363_9_argos_users_manual-v1.6.6.pdf (accessed March 6, 2013).

Arnbom, T., and V. Papastavrou. 1988. Fish in association with whale sharks *Rhiniodon typus* near the Galapagos Islands. *Noticias de Galapagos* 46:13–15.

Arzoumanian, Z., J. Holmberg, and B. Norman. 2005. An astronomical pattern-matching algorithm for computer-aided identification of whale sharks *Rhincodon typus. Journal of Applied Ecology* 42:999–1011.

Beaugrand, G., C. Luczak, and M. Edwards. 2009. Rapid biogeographical plankton shifts in the North Atlantic Ocean. *Global Change Biology* 15:1790–803.

Beebe, W. 1926. *The Arcturus Adventure*. New York: Putnam.

Berumen, M.L., C. D. Braun, J. Cochran, et al. 2014. Movement patterns of juvenile whale sharks tagged at an aggregation site in the Red Sea. *PLoS One* 9(7):e103536.

Borrell, A., A. Aguilar, M. Gazo, et al. 2011. Stable isotope profiles in whale shark (*Rhincodon typus*) suggest segregation and dissimilarities in the diet depending on sex and size. *Environmental Biology of Fishes* 92:559–67.

Bradshaw, C. J. A., B. M. Fitzpatrick, C. C. Steinberg, et al. 2008. Decline in whale shark size and abundance at Ningaloo Reef over the past decade: The world's largest fish is getting smaller. *Biological Conservation* 141:1894–905.

Bradshaw, C. J. A., H. F. Mollet, and M. G. Meekan. 2007. Inferring population trends for the world's largest fish from mark–recapture estimates of survival. *Journal of Animal Ecology* 76:480–89.

Brierley, A. S. 2014. Diel vertical migration. *Current Biology* 24(22):R1074–76.

Brunnschweiler, J. M., H. Baensch, S. J. Pierce, and D. W. Sims. 2009. Deep-diving behaviour of a whale shark *Rhincodon typus* during long-distance movement in the Western Indian Ocean. *Journal of Fish Biology* 74(3):706–14.

Brunnschweiler, J. M., and D.W. Sims. 2012. Diel oscillations in whale shark vertical movements associated with meso-and bathypelagic diving. *American Fisheries Society Symposium* 76:457–69.

Cade, D. E., J. J. Levenson, R. Cooper, et al. 2020. Whale sharks increase swimming effort while filter feeding, but appear to maintain high foraging efficiencies. *Journal of Experimental Biology* 223:jeb224402.

Cagua, E. F., J. E. M. Cochran, C. A. Rohner, et al. 2015. Acoustic telemetry reveals cryptic residency of whale sharks. *Biology Letters* 11:20150092.

Carey, F. G., and J. V. Scharold. 1990. Movements of blue sharks (*Prionace glauca*) in depth and course. *Marine Biology* 106:329–42.

Castro, A. L. F., B. S. Stewart, S. G. Wilson, et al. 2007. Population genetic structure of Earth's largest fish, the whale shark (*Rhincodon typus*). *Molecular Ecology* 16:5183–92.

Catlin, J., and R. Jones. 2010. Whale shark tourism at Ningaloo Marine Park: A longitudinal study of wildlife tourism. *Tourism Management* 31:386–94.

Colman, J. G. 1997. A review of the biology and ecology of the whale shark. *Journal of Fish Biology* 51:1219–34.

Compagno, L. J. V. 2001. *Sharks of the world: An annotated and illustrated catalogue of shark species known to date*, Vol. 2: *Bullhead, Mackerel and Carpet Sharks (Heterodontiformes, Lamniformes and Orectolobiformes)*. Rome: FAO Species Catalogue for Fishery Purposes.

De la Parra Venegas, R., R. Hueter, J. Gonzalez, et al. 2011. An unprecedented aggregation of whale sharks *Rhincodon typus*, in Mexican coastal waters of the Caribbean Sea. *PLoS One* 6(4):e18994.

Diamant, S., C. A. Rohner, J. J. Kiszka, et al. 2018. Movements and habitat use of satellite-tagged whale sharks off western Madagascar. *Endangered Species Research* 36:49–58.

DPNG. 2018. *Agregación de tiburones ballena reportada en el sur de Isabela*. Puerto Ayora: Dirección del Parque Nacional Galápagos.

Dulvy, N. K., J. K. Baum, S. C. Clarke, et al. 2008. You can swim but you can't hide: The global status and conservation of oceanic pelagic sharks and rays. *Aquatic Conservation: Marine and Freshwater Ecosystems* 18:459–82.

Eckert, S., L. Dolar, G. Kooyman, et al. 2002. Movements of whale sharks (*Rhincodon typus*) in South East Asian waters as determined by satellite telemetry. *Journal of Zoology* 257:111–15.

Eckert, S. A., and B. S. Stewart. 2001. Telemetry and satellite tracking of whale sharks, *Rhincodon typus*, in the Sea of Cortez, Mexico, and the north Pacific Ocean. *Environmental Biology of Fishes* 60(1–3):299–308.

Evans, K., R. Thresher, R. M. Warneke, et al. 2005. Periodic variability in cetacean strandings: Links to large-scale climate events. *Biology Letters* 1:147–50.

Geist, D. 1996. On the emergence and submergence of the Galapagos Islands. *Noticias de Galapagos* 56:5–9.

Gifford, A., L. J. V. Compagno, M. Levine, and A. Antoniou. 2007. Satellite tracking of whale sharks using tethered tags. *Fisheries Research* 84:17–24.

Gleiss, A. C., B. Norman, N. Liebsch, et al. 2009. A new prospect for tagging large free-swimming sharks with motion-sensitive data-loggers. *Fisheries Research* 97:11–16.

Gleiss, A. C., B. Norman, and R. P. Wilson. 2011. Moved by that sinking feeling: Variable diving geometry underlies movement strategies in whale sharks. *Functional Ecology* 25(3):595–607.

Gleiss, A. C., S. Wright, N. Liebsch, et al. 2013. Contrasting diel patterns in vertical movement and locomotor activity of whale sharks at Ningaloo Reef. *Marine Biology* 160:2981–92.

Graham, R. T., C. M. Roberts, and J. C. Smart. 2006. Diving behaviour of whale sharks in relation to a predictable food pulse. *Journal of the Royal Society Interface* 3(6):109–16.

Green, J., and A. Hearn. 2018. Galapagos Whale Shark Project, Fieldwork Report 2018. Galapagos Whale Shark Project, Ecuador.

Gudger, E. W. 1927. The whale shark *Rhiniodon typus* at the Galápagos Islands – a new faunal record. *Science* 65:545.

Gudger, E. W. 1932. The whale shark, *Rhineodon typus*, among the Seychelles islands. *Nature* 130:169.

Gudger, E. W. 1933. A second whale shark *Rhineodon typus* at the Galapagos Islands. *Nature* 132:569.

Gudger, E. W. 1941. The whale shark unafraid: The greatest of the sharks, *Rhineodon typus*, fears not shark, man nor ship. *The American Naturalist* 75(761):550–68.

Gunn, J. S., J. D. Stevens, T. L. O. Davis, and B. M. Norman. 1999. Observations on the short-term movements and behaviour of whale sharks (*Rhincodon typus*) at Ningaloo Reef, Western Australia. *Marine Biology* 135(3):553–59.

Gutierrez, A. P., L. Ponti, T. d'Oultremont, and C. K. Ellis. 2008. Climate change effects on poikilotherm tritrophic interactions. *Climatic Change* 97(1):S167–92.

Guzman, H. M., C. G. Gomez, A. Hearn, and S. A. Eckert. 2018. Longest recorded trans-Pacific migration of a whale shark (*Rhincodon typus*). *Marine Biodiversity Records* 11:8.

Hearn, A., E. Espinoza, J. Ketchum, et al. 2017. Ten years of tracking shark movements highlights the ecological importance of the northern islands: Darwin and Wolf. In *Galapagos Report 2015–2016*, ed. GNPD, GCREG, CDF, and GC, 130–139. Puerto Ayora.

Hearn A., J. R. Green, E. Espinoza, et al. 2013. Simple criteria to determine detachment point of towed satellite tags provide first evidence of return migrations of whale sharks (*Rhincodon typus*) at the Galapagos Islands, Ecuador. *Journal of Animal Biotelemetry* 1:11.

Hearn, A. R., D. Acuña, J. T. Ketchum, et al. 2014. Elasmobranchs of the Galapagos Marine Reserve. In *The Galapagos Marine Reserve: Social and Ecological Interactions in the Galapagos Islands*, eds. J. Denkinger, and L. Vinueza, 23–59. New York: Springer Science.

Hearn, A. R., J. Green, M. H. Roman, et al. 2016. Adult female whale sharks make long-distance movements past Darwin Island (Galapagos, Ecuador) in the Eastern Tropical Pacific. *Marine Biology* 163:214.

Holmberg, J., B. Norman, and Z. Arzoumanian. 2009. Estimating population size, structure, and residency time for whale sharks *Rhincodon typus* through collaborative photo-identification. *Endangered Species Research* 7:39–53.

Hooker, S. K., H. Whitehead, S. Gowans, and R. W. Baird. 2002. Fluctuations in distribution and patterns of individual range use of northern bottlenose whales. *Marine Ecology Progress Series* 225:287–97.

Hsu, H. H., S. J. Joung, Y. Y. Liao, and K. M. Liu. 2007. Satellite tracking of juvenile whale sharks, *Rhincodon typus*, in the northwestern Pacific. *Fisheries Research* 84:25–31.

Hueter, R. E., J. P. Tyminski, and R. de la Parra. 2013. Horizontal movements, migration patterns and population structure of whale sharks in the Gulf of Mexico and northwestern Caribbean Sea. *PLoS One* 8(8):e71883.

Iwasaki, Y. 1970. On the distribution and environment of the whale shark *Rhincodon typus*, in skipjack fishing grounds in the western Pacific Ocean. *Journal of Marine Science and Technology* 4:37–51.

Jury, M., T. McClanahan, and J. Maina. 2010. West Indian Ocean variability and East African fish catch. *Marine Environmental Research* 70:162–70.

Karnauskas, K. B., R. Murtugudde, and A. J. Busalacchi. 2007. The effect of the Galapagos Islands on the equatorial pacific cold tongue. *Journal of Physical Oceanography* 37:1266–81.

Ketchum, J. T., F. Galván-Magaña, and A. P. Klimley. 2013. Segregation and foraging ecology of whale sharks *Rhincodon typus*, in the southwestern Gulf of California. *Environmental Biology of Fishes* 96 (6):779–95.

Ketchum, J. T., A. Hearn, A. P. Klimley, et al. 2014. Interisland movements of scalloped hammerhead sharks (*Sphyrna lewini*) and seasonal connectivity in a marine protected area of the eastern tropical Pacific. *Marine Biology* 161(4):939–51.

Klimley, A. P. 1993. Highly directional swimming by scalloped hammerhead sharks, *Sphyrna lewini*, and subsurface irradiance, temperature, bathymetry, and geomagnetic field. *Marine Biology* 117(1):1–22.

Klimley, A. P., M. Flagg, N. Hammerschlag, and A. Hearn. 2017. The value of using measurements of geomagnetic field in addition to irradiance and sea surface temperature to estimate geolocations of tagged aquatic animals. *Journal of Animal Biotelemetry* 5:19.

Last, P. R., and J. D. Stevens. 2009. *Sharks and Rays of Australia*. Second edition. Collingwood: CSIRO Publishing.

Lennox, R. J., K. Aarestrup, S. J. Cooke, et al. 2017. Envisioning the future of aquatic animal tracking: Technology, science and application. *BioScience* 67(10):884–96.

Meekan, M. G., C. J. A. Bradshaw, M. Press, et al. 2006. Population size and structure of whale sharks *Rhincodon typus* at Ningaloo Reef, Western Australia. *Marine Ecology Progress Series* 319:275–85.

Meyers, M. M., M. P. Francis, M. Erdmann, et al. 2020. Movement patterns of whale sharks in Cenderawasih Bay, Indonesia, revealed through long-term satellite tagging. *Pacific Conservation Biology*. doi:10.1071/PC19035.

Motta, P. J., M. Maslanka, R. E. Hueter, et al. 2010. Feeding anatomy, filter-feeding rate, and diet of whale sharks *Rhincodon typus* during surface ram filter feeding off the Yucatan Peninsula, Mexico. *Zoology* 113:119–212.

Nakamura, I., R. Matsumoto, and K. Sato. 2020. Body temperature stability in the whale shark, the world's largest fish. *Journal of Experimental Biology* 223:jeb210286.

Nathan, R., W. M. Getz, E. Revilla, et al. 2008. A movement ecology paradigm for unifying organismal movement research. *PNAS* 105(49):19052–59.

Norman, B. M., J. A. Holmberg, Z. Arzoumanian, et al. 2017b. Undersea constellations: The global biology of an endangered marine megavertebrate further informed through citizen science. *BioScience* 67:1029–43.

Norman, B. M., and D. L. Morgan. 2016. The return of "Stumpy" the whale shark: Two decades and counting. *Frontiers in Ecology and the Environment* 14:449–50.

Norman, B. M., S. Reynolds, and D. L. Morgan. 2016. Does the whale shark aggregate along the Western Australian coastline beyond Ningaloo Reef? *Pacific Conservation Biology* 22:72–80.

Norman, B. M., and J. D. Stevens. 2007. Size and maturity status of the whale shark (*Rhincodon typus*) at Ningaloo Reef in Western Australia. *Fisheries Research* 84:81–86.

Norman, B. M., J. M. Whitty, S. J. Beatty, et al. 2017a. Do they stay or do they go? Acoustic monitoring of whale sharks at Ningaloo Marine Park, Western Australia. *Journal of Fish Biology* 91:1713–20.

Papastamatiou, Y. P., and C. G. Lowe. 2012. An analytical and hypothesis-driven approach to elasmobranch movement studies. *Journal of Fish Biology* 80:1342–60.

Perry, A. L., P. J. Low, J. R. Ellis, and J. D. Reynolds. 2005. Climate change and distribution shifts in marine fishes. *Science* 308:1912–15.

Perry, C. T., E. Clingham, D. H. Webb, et al. 2020. St. Helena: An important reproductive habitat for whale sharks (*Rhincodon typus*) in the south central Atlantic. *Frontiers in Marine Science* 7:576343.

Pierce, S. J., J. Holmberg, A. A. Kock, and A. D. Marshall. 2019. Photographic identification of sharks. In *Shark Research: Emerging Technologies and Applications for the Field and Laboratory*, eds. J. C. Carrier, M. R. Heithaus, and C. A. Simpfendorfer, 219–34. Boca Raton, FL: CRC Press.

Prebble, C. E. M., C. A. Rohner, S. J. Pierce, et al. 2018. Limited latitudinal ranging of juvenile whale sharks in the Western Indian Ocean suggests the existence of regional management units. *Marine Ecology Progress Series* 601:167–83.

Ramirez, P. 1995 Observaciones de tiburón ballena, *Rhineodon typus*, frente a Paita, Peru. In *Biología y pesquería de tiburones de las islas Lobos, Peru*, eds. W. Elliot, F. Paredes, and M. Bustamante, 23–29. Lima: Instituto del Mar de Peru.

Ramirez-Macías, D., N. Queiroz, S. J. Pierce, et al. 2017. Oceanic adults, coastal juveniles: Tracking the habitat use of whale sharks off the Pacific coast of Mexico. *PeerJ* 5:e3271.

Ramirez-Macías, D., A. Vazquez-Haikin, and R. Vazquez-Juarez. 2012. Whale shark *Rhincodon typus* populations along the west coast of the Gulf of California and implications for management. *Endangered Species Research* 18(2):115–28.

Reynolds, S. D., B. M. Norman, M. Beger, et al. 2017. Movement, distribution and marine reserve use by an endangered migratory giant. *Diversity and Distributions* 23:1268–79.

Robinson, D. P., M. Y. Jaidah, S. S. Bach, et al. 2017. Some like it hot: Repeat migration and residency of whale sharks within an extreme natural environment. *PLoS One* 12(9):e0185360.

Rohner, C. A., J. E. M. Cochran, E. F. Cagua, et al. 2020. No place like home? High residency and predictable seasonal movement of whale sharks off Tanzania. *Frontiers in Marine Science* 7: 423.

Rohner, C. A., L. I. E. Couturier, A. J. Richardson, et al. 2013. Diet of whale sharks *Rhincodon typus* inferred from stomach content and signature fatty acid analyses. *Marine Ecology Progress Series* 493:219–35.

Rohner, C. A., A. J. Richardson, F. R. A. Jaine, et al. 2018. Satellite tagging highlights the importance of productive Mozambican coastal waters to the ecology and conservation of whale sharks. *PeerJ* 6:e4161.

Román, M. H., A. Aires-da-Silva, and N. W. Vogel. 2018. *Whale shark interactions with the tuna purse-seine fishery in the Eastern Pacific Ocean: Summary and analysis of available data.* 8th Meeting of the Working Group of Bycatch. La Jolla, California (USA). 10–11 May, 2018. IATTC Document BYC-08 Inf-A.

Rowat, D., and K. S. Brooks. 2012. A review of the biology, fisheries and conservation of the whale shark *Rhincodon typus. Journal of Fish Biology* 80:1019–56.

Rowat, D., and M. Gore. 2007. Regional scale horizontal and local scale vertical movements of whale sharks in the Indian Ocean off Seychelles. *Fisheries Research* 84(1):32–40.

Ryan, J. P., J. R. Green, E. Espinoza, and A. R. Hearn. 2017. Association of whale sharks (*Rhincodon typus*) with thermo-biological frontal systems of the eastern tropical Pacific. *PLoS One* 12(8):e0182599.

Sequeira, A., C. Mellin, D. Rowat, et al. 2012. Ocean-scale prediction of whale shark distribution. *Diversity and Distributions* 18:504–18.

Sequeira, A. M. M., C. Mellin, S. Delean, et al. 2013a. Spatial and temporal predictions of inter-decadal trends in Indian Ocean whale sharks. *Marine Ecology Progress Series* 478:185–95.

Sequeira, A. M. M., C. Mellin, D. A. Fordham, et al. 2014a. Predicting current and future global distributions of whale sharks. *Global Change Biology* 20:778–89.

Sequeira, A. M. M., C. Mellin, M. G. Meekan, et al. 2013b. Inferred global connectivity of whale shark *Rhincodon typus* populations. *Journal of Fish Biology* 82:367–89.

Sibaja-Cordero J. A. 2008. Tendencias espacio-temporales de los avistamientos de fauna marina en los buceos turísticos (Isla del Coco, Costa Rica). *Revista de Biología Tropical* 56(2):S113–32.

Simpson, C., A. Pearce, and D. I. Walker. 1991. Mass spawning of corals on Western Australian reefs and comparisons with the Great Barrier Reef. *Journal of the Royal Society of Western Australia* 74:85–91.

Sleeman, J. C., M. G. Meekan, B. J. Fitzpatrick, et al. 2010a. Oceanographic and atmospheric phenomena influence the abundance of whale sharks at Ningaloo Reef, Western Australia. *Journal of Experimental Marine Biology and Ecology* 382:77–81.

Sleeman, J. C., M. G. Meekan, S. G. Wilson, et al. 2010b. To go or not to go with the flow: Environmental influences on whale shark movement patterns. *Journal of Experimental Marine Biology and Ecology* 390:84–98.

Snell, H. M., P. A. Stone, and H. L. Snell. 1996. A summary of geographical characteristics of the Galapagos Islands. *Journal of Biogeography* 23(5):619–24.

Spear, L. B., L. T. Balance, and D. G. Ainley. 2001. Response of seabirds to thermal boundaries in the tropical Pacific: The thermocline versus the Equatorial Front. *Marine Ecology Progress Series* 219:275–89.

Stevens, J. D. 2007. Whale shark (*Rhincodon typus*) biology and ecology: A review of the primary literature. *Fisheries Research* 84:4–9.

Stramma, L., G. C. Johnson, J. Sprintall, and V. Mohrholz. 2008. Expanding oxygen-minimum zones in the tropical oceans. *Science* 320:655–59.

Sweet, W. V., J. M. Morrison, K. M. Liu, et al. 2009. Tropical instability wave interactions within the Galapagos archipelago. *Deep-Sea Research, Part I* 56:1217–29.

Taylor, J. G. 1996. Seasonal occurrence, distribution and movements of the whale shark, *Rhincodon typus*, at Ningaloo Reef, Western Australia. *Marine and Freshwater Research* 47:637–42.

Taylor, J. G. 2007. Ram filter feeding and nocturnal feeding of whale sharks (*Rhincodon typus*) at Ningaloo Reef, Western Australia. *Fisheries Research* 84:65–70.

Thums, M., M. Meekan, J. Stevens, et al. 2012. Evidence for behavioural thermoregulation by the world's largest fish. *Journal of the Royal Society Interface* 10(78):20120477.

Tomita, T., T. Kawai, H. Matsubara, et al. 2014. Northernmost record of a whale shark *Rhincodon typus* from the Sea of Okhotsk. *Journal of Fish Biology* 84:243–46.

Torres, K. A., K. Ramirez-Pozo, and A. Hearn. 2018. *Línea base de conocimiento sobre la megafauna marina (tiburon ballena, manta gigante y pez luna) en la costa de Ecuador.* Quito: WWF-Ecuador / Fundacion Megafauna Marina del Ecuador.

Turnbull S. D., and J. E. Randell. 2006. Rare occurrence of a *Rhincodon typus* (whale shark) in the Bay of Fundy, Canada. *Northeastern Naturalist* 13(1):57–58.

Tyminski, J. P., R. de La Parra-Venegas, J. González Cano, and R. E. Hueter. 2015. Vertical movements and patterns in diving behavior of whale sharks as revealed by pop-up satellite tags in the eastern Gulf of Mexico. *PLoS One* 10(11):e0142156.

Vignaud, T. M., J. A. Maynard, R. Leblois, et al. 2014. Genetic structure of populations of whale sharks among ocean basins and evidence for their historic rise and recent decline. *Molecular Ecology* 23:2590–601.

Weng, K. C., J. B. O'Sullivan, C. G. Lowe, et al. 2007. Movements, behavior and habitat preferences of juvenile white sharks *Carcharadon carcharias* in the eastern Pacific. *Marine Ecology Progress Series* 338:211–24.

Wildbook. 2020. Wildbook for Whale Sharks. https://www.whaleshark.org (accessed August 8, 2020).

Wilson, R. P., N. Liebsch, I. M. Davies, et al. 2007. All at sea with animal tracks: Methodological and analytical solutions for the resolution of movement. *Deep-Sea Research II* 54:193–210.

Wilson, S. G., J. G. Taylor, and A. F. Pearce. 2001. The seasonal aggregation of whale sharks at Ningaloo Reef, Western Australia: Currents, migrations and the El Niño/ Southern Oscillation. *Environmental Biology of Fishes* 61:1–11.

Wilson, S. G., J. J. Polovina, B. S. Stewart, and M. G. Meekan. 2006. Movements of whale sharks (*Rhincodon typus*) tagged at Ningaloo Reef, Western Australia. *Marine Biology* 148(5):1157–66.

Zarate, P. 2002. Tiburones. In *Reserva Marina de Galapagos. Linea Base de la Biodiversidad,* eds. E. Danulat, and G. Edgar, 361–75. Santa Cruz: Fundacion Charles Darwin/Servicio Parque Nacional Galapagos.

CHAPTER 7

Population Ecology of Whale Sharks

Christoph A. Rohner
Marine Megafauna Foundation

Bradley M. Norman and Samantha Reynolds
The University of Queensland ECOCEAN

Gonzalo Araujo
Large Marine Vertebrates (LAMAVE) Research Institute Philippines

Jason Holmberg
Wild Me (Wildbook)

Simon J. Pierce
Marine Megafauna Foundation

CONTENTS

7.1 INTRODUCTION

Whale sharks are the world's largest extant fish, and potentially the largest fish species that has ever lived (Chapter 1). The largest whale sharks on record were around 18–20 m in length, while the smallest free-swimming neonate was 46 cm long (Chen et al. 1997; Aca and Schmidt 2011; Borrell et al. 2011). Despite their enormous adult size, however, whale shark sightings were rarely recorded in the first 150 years following the species' scientific description in 1828. Studying large, endangered and wide-ranging marine animals, such as whale sharks, is a significant challenge. We speak from some experience on this. We do now have a much-improved understanding of where whale sharks are likely to be found, and have identified ~30 global hotspots for the species (Norman et al. 2017), but our observations are heavily biased towards individuals between 3 and 9 m long, and particularly male sharks, because these are the animals that predominate at the known aggregation sites. The obvious question – where do the rest of the whale sharks live? – is largely informed by where we *don't* see them, rather than our knowledge of their specific ecology at different life stages.

As we outline in this chapter, studying the population ecology of whale sharks is a highly collaborative endeavor. Researchers are sharing their data across the globe, modern computer science is being integrated into shark identification, and thousands of citizen scientists each year are joining the quest to discover more about the world's largest fish. The changing ecology and habitat use of whale sharks as they grow is starting to become more apparent. Nevertheless, some interesting mysteries still remain.

7.2 IDENTIFYING INDIVIDUAL SHARKS

7.2.1 Photographic Identification

Photographic identification (photo-ID), in which the natural markings on the sharks' skin are photographed to aid the recognition of individuals (Pierce et al. 2019; Taylor 1994, 2019), has become a foundational technique in whale shark research. Two assumptions must be met for successful photo-ID studies: (a) Individuals can be reliably distinguished from others, and (b) they can then be re-identified over time (Pierce et al. 2019). Both assumptions are demonstrably true for whale sharks, with individuals having been re-identified >20 years apart at Ningaloo Reef in Western Australia (Norman and Morgan 2016), and genetic analyses confirming the uniqueness of individual spot patterns (Meenakshisundaram et al. In press). As yet, no significant issues with this approach have been documented, although a single sighting of one completely white albino or leucistic whale shark in the Galápagos Islands (Figure 7.1) does prove that exceptions are technically

Figure 7.1   A white whale shark sighted at Darwin Island, Galápagos, on 25 August 2007. (Photos: Antonio Moreano.)

possible! The lack of observations of very small whale sharks means we do not yet know if, or how, their spot pattern may change in the first months to years of a shark's life when they grow from ~60 cm to several meters long. Fortunately, the bias towards medium to large juvenile sharks at most hotspots (i.e., the 3–9 m range, as discussed below) has rendered this point moot in practical terms, as there is no evidence to date that their patterns do not remain stable for the rest of their lives.

The use of photo-ID to study whale sharks provides several advantages over the application of external marker tags which, as a common technique in shark research, was often used as a default method in earlier population studies of the species (Graham and Roberts 2007; Rowat et al. 2009). These marker tags can become biofouled and unreadable, or be shed or break, with estimated annual retention for marker tags of only 43%–69% in Seychelles whale sharks (Rowat et al. 2009). The unreliability of re-identifying "known" animals, and the potential overestimation of population size that can result from these being assigned as "new" sharks, has led to most subsequent projects exclusively using photo-ID as the basis for whale shark population studies (Norman et al. 2017).

Photo-ID provides a few other practical advantages. Individual sharks can be identified from a distance, without any physical interaction or disturbance, and thereby adhering to the "codes of conduct" which have been developed at many whale shark tourism destinations to avoid negative impacts on the sharks (Chapter 10). This often allows for researchers to work from tourism vessels as a way to maximize field time while minimizing expenses. By contrast, more invasive methods such as external tags can have real (such as superficial tissue damage) or perceived detrimental impacts – even if there is no behavioral change in the sharks (Rohner et al. 2020) – which can lead to negative opinions of researchers among stakeholders. Maintaining public support for research is often an important consideration, particularly for conservation and outreach activities, and can require a high degree of effort to achieve.

Photo-ID, however, can also have a significant downside: the time required to match photos and identify sharks as either new or a resighting. To overcome this limitation, two main computer-based algorithms (*I3S* and *modified Groth*) have been developed to semi-automate photo comparisons among individual whale sharks (Arzoumanian et al. 2005). While some researchers do still maintain personal or organizational photo databases for shark identification, most now use the global, collaborative database at Wildbook for Whale Sharks (located online at www.whaleshark.org) to submit, process, and store their photo-ID records of whale sharks. Individual sharks are identified by their unique spot patterns on the flanks, behind the 5th gill and above the pectoral fin, and are compared to all other whale sharks in the database (Arzoumanian et al. 2005). The left and right sides of the shark's flanks have different spot patterns. Unmatched sharks are assigned a new individual number (i.e. MZ-100 for the 100th shark identified from Mozambique), as long as a high-quality photo of the left flank is available. The left side was chosen arbitrarily as the standard for assigning a new identification number. Lower-quality left-side photos or previous photos of the right flank can often be retrospectively matched if photos of both sides exist from an encounter of an individual. A new sighting of a previously identified individual is added to that shark's list of encounters. Encounter information also includes the date and location of the sighting, as well as the sex, size, behavior, and description and photos of any scars. Length estimates are often done visually by comparing the length of a swimmer or a boat to the length of the whale shark. While visual estimates are not accurate or precise, particularly from untrained observers, they are still useful to differentiate between small and large whale sharks. First laser photogrammetry and then stereo-videography have improved the accuracy of in-water methods, and the relatively low cost of these techniques means that they are widely used by researchers (Rohner et al. 2015b; Meekan et al. 2020).

7.2.2 Citizen Science

In the age of high public awareness for scientific projects and accessible, affordable and mobile digital technology, scientists are able to harness the observations of millions of "citizen scientists" to leverage the study of biodiversity and species distributions at global scales (Bonney et al. 2009; Newman et al. 2012; Bird et al. 2014). Whale shark tourism has become increasingly popular over the past decade (Chapter 10), as has the use of underwater digital cameras, which means that a large pool of whale shark identification photographs and videos is now available to be harnessed for research. Indeed, successful identifications can be routinely extracted from public photographs, taken without prior knowledge of the technique being required by the photographer (Davies et al. 2013). The online platform www.whaleshark.org facilitates data gathering by supporting a common information architecture, and making it easy for citizen scientists to upload ID photos from a browser and share their encounter information, while keeping the data validation and spot pattern matching to the researchers who manage the data from a particular aggregation site. Once the shark has been identified, citizen scientists automatically find out which individual they saw, and receive updates via email whenever that shark they helped identify is seen again by others. To date, over 7,000 different submitters (by counting unique email addresses) have contributed to the database. Citizen science has thus played a significant part in advancing our understanding of the basic biology and ecology of whale sharks. In particular, citizen scientists have often provided useful sighting data to fill in gaps between areas where active research programs are conducted on whale sharks, enabling shark movements to be documented. For example, a 2,700 km movement from Western Australia to Sangalaki Island in Indonesia (Norman et al. 2017) was determined with the help of citizen scientists. The inclusion of citizen scientists facilitates the creation of cost-effective, long-term monitoring programs to measure whale shark abundance and helps to engender public awareness, engagement, and support.

7.2.3 Data Mining with Artificial Intelligence

The Wildbook for Whale Sharks database is truly global. Access is available to many collaborating research groups, allowing for unprecedented data-sharing and insights into whale shark movement ecology (Norman et al. 2017). This database is still rapidly growing. Between 1992 and 2014, when the whole database was last fully assessed, ~28,000 encounters had been recorded (Norman et al. 2017). As of October 2020, the number of encounters has more than doubled to 77,372 encounters and 12,241 identified sharks, with 7,006 contributors. The growth of the Wildbook collaborative global database has also facilitated, and in turn been accelerated by, the incorporation of artificial intelligence (AI; e.g., computer vision trained by deep convolutional neural networks) to minimize the need for researchers to manually enter and process photographic data (Pierce et al. 2019). Manual efforts do not scale well. As Duyck et al. (2015) pointed out, "at 10 seconds per comparison, a 10,000-sized catalog will take approximately 15 person-years to analyze." The use of AI is enabling the data mining of massive online public video archives such as YouTube.com, which as of May 2019 had more than 720,000 hours of video content being uploaded every day (Hale 2019). Wild Me, the non-profit that hosts and leads development on Wildbook, released an "intelligent agent" in 2018 that applies multiple forms of machine learning (e.g., computer vision, natural language processing, machine translation) to transform social media-sourced data into curated data for whale shark researchers. The agent data-mines YouTube nightly to identify wild whale shark sightings and initiates a conversation with the video poster to gather additional metadata, if missing, such as the geographical location of the sighting. It also retrospectively searches for older whale shark sightings. So far, the agent has engaged over 2,200 video posters, an audience that had evidently been largely missed by regular community outreach efforts. The agent was the second-highest contributor of whale shark encounters to Wildbook in 2018, with 91% of the agent's data submissions not being duplicated by human contributors. In 2019, it uploaded more encounters than human contributors (Figure 7.2).

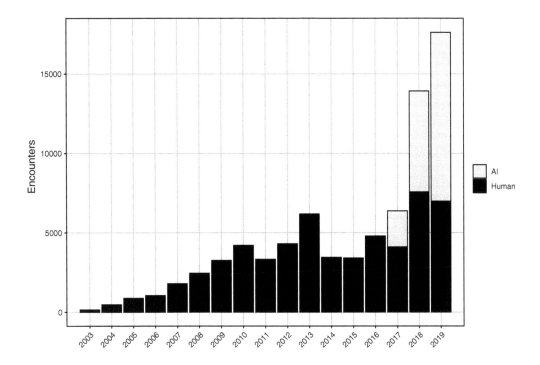

Figure 7.2 The yearly number of encounters uploaded to the global photo-ID database www.whaleshark.org by human and AI contributors.

This flood of new data will require some work to benchmark human-curated population esti-mates, and the bias and error rate therein, with automated computer estimates from the same datasets (Pierce et al. 2019). A caveat is that, because citizen science and AI-curated photos and videos are not being collected with science in mind, they are often suboptimal as identification photos. Dealing with this much incoming data can be rather overwhelming for individual researchers, so ongoing improvements in algorithm development to accurately distinguish new sharks from resightings, par-ticularly in sites that have a high level of tourism activity, will also be hugely useful. Additionally, citizen science and AI-curated data are collected continuously, rather than in a pre-defined sam-pling season which dedicated research programs often use to conduct population models that rely on such a study design. At an analytical level, opportunistic photo-ID data can be used in generalized additive (GAM) or linear (GLM) models to examine temporal trends in sightings and taking a suite of environmental, meteorological, temporal, and biological predictors that may affect sightability into account (Rohner et al. 2013). Citizen science data are typically "presence only" (Williams et al. 2015); i.e., photos are submitted when sharks are seen, but often no records are kept of unsuccess-ful trips. This makes it challenging to calculate the search effort for hotspots with mixed datasets. Estimating abundance and demographic parameters, such as apparent survival or the probability of entry into the study area, usually require capture–mark–recapture models (technically sight–resight models, when applied to photo-ID datasets). Current analyses require a carefully designed study with short sampling periods and longer time periods in between, ignoring other data that may be continuously or opportunistically obtained across the year. Some alternative analytical techniques to model demographic parameters from irregular photo-ID data have been developed to partially address this issue, which is common to many marine species (e.g. Whitehead 2001; Fox et al. 2013). Further improvements to analyses that can estimate sighting or abundance trends over time from opportunistic data would be a major advance for conservation assessment.

7.3 POPULATION ECOLOGY

7.3.1 Constellations

Considering their size and global distribution, whale sharks are a relatively recent subject of scientific inquiry. Many of the early whale shark records were reported by E. W. Gudger, who published 47 manuscripts on whale shark biology, ecology, and sightings over 40 years (Stevens 2007). Gudger claimed to have a copy of every whale shark photo taken in the early 20th century (Stevens 2007). Silas (1986) published a complete summary of biological information from 68 whale sharks caught in Indian fisheries, along with a thorough review of ecological knowledge on the species. Around the same time, F. Wolfson (1986) published a review summarizing all 320 known whale shark sightings that were available to her up to 1985. She concluded her work by noting "… we must move beyond the purely descriptive natural history approach and design and implement sustained programs of investigation, using the most advanced equipment and techniques that are amenable to statistical treatment."

Although Wolfson was not aware of it, she was providing excellent foreshadowing for the start of the "modern" phase of whale shark research – the beginning of field investigations at Ningaloo Reef in Western Australia. This work began in the early 1980s as a pioneering effort by a medical doctor, G. Taylor, based in Exmouth, who enlisted family members, friends, and scientific collaborators to conduct the first field studies on the ecology and population structure of the species. By the early 1990s, this had become a sophisticated project, coupling aerial surveys with boat-based in-water observations, and culminating in the publication of a book (Taylor 1994, 2019) and multiple scientific papers (e.g. Taylor and Grigg 1991; Taylor 2007, 1996). One of the enduring legacies of this work was the start of a photo-identification catalog for whale sharks, which eventually developed into the global Wildbook for Whale Sharks Photo-Identification Library.

Ningaloo Reef was the first site to be formally documented as a whale shark aggregation area, hereafter termed "constellation" (see "Preface"). Additional constellations have been discovered and documented in the years since. Predictable and reliable constellations have provided access for scientists to study the species and for tourism operators to start commercial whale shark snorkeling or diving operations (Chapter 10), which in turn has enabled the bulk of public photo-identification contributions. As a result, these hotspots are the source of most of our current knowledge of whale shark biology and ecology. Around 30 of these sites – here defined as locations with >100 recorded whale shark encounters – had been identified around the world by 2014 (Norman et al. 2017; Figure 7.3 and Table 7.1). Many of these constellations occur seasonally (Norman et al. 2017) and are driven by localized high prey availability (e.g. Nelson and Eckert 2007; Robinson et al. 2013; Rohner et al. 2015a). New sites continue to be discovered, with significant constellations also described from northwest Madagascar (Diamant et al. 2018), St. Helena Island in the mid-Atlantic (Perry et al. 2020), and Puerto Princesa in the Philippines (Araujo et al. 2019). Other sites are known, but dedicated field studies have not taken place or been published as yet, such as around Taiwan, Gabon, off the Gujarat coast of India, and Oman. The largest constellations to date, where hundreds of whale sharks sometimes gather together to feed at the same time, have been recorded north of Isla Mujeres in Mexico (Figure 7.4), at the Al Shaheen area off Qatar (de la Parra Venegas et al. 2011; Robinson et al. 2013), and off Gujarat in India (Hanfee 2001).

Some other hotspots rarely have multiple sharks present at the same time, but rather are distinct locations where individual whale sharks are reliably sighted throughout a season. These include Darwin Island in the Galápagos Islands, the Tubbataha Reefs in the Philippines, and Koh Tao in Thailand (Theberge and Dearden 2006; Acuña-Marrero et al. 2014; Araujo et al. 2018). Feeding does not necessarily take place at these sites, although it may well occur in the vicinity. For convenience's sake, and to maintain coherence with the previous literature, we will refer to both types of hotspots as constellations in this chapter.

Figure 7.3 Global whale shark constellations, defined here as areas where >100 encounters have been reported to the global photo-identification database www.whaleshark.org. Blue spots are sites with 67% or more male whale sharks; red spots are sites with 67% or more female whale sharks. Spots where both colors are used are where neither sex was >66% of the total number of identified individuals. Darker blue or red spots are where the mean size of sharks was >8 m, or where the majority of identified sharks were mature. Data sources are listed in Table 7.1. Design by Madeleine Pierce.

Table 7.1 Population Structure of Whale Sharks at Constellation Sites, with the Mean Length, the Percentage of Male Individuals, and a Primary Reference

Location	% Male	Mean Length (m)	N	References
Al Shaheen, Qatar	68.7	6.9	296	Robinson et al. (2016)
Bahia de La Paz, Mexico	74.8	4.4	101	Ramírez-Macías et al. (2012)
Bahia de los Angeles, Mexico	75.9	5.3	102	Ramírez-Macías et al. (2012)
Belize		7.2	35	Norman et al. (2017)
Christmas Island, Australia		4.6	82	Hobbs et al. (2009), Norman et al. (2017)
Darwin Island, Galapagos, Ecuador	1.4	11.1	89	Norman et al. (2017)
Djibouti	84.7	3.7	232	Rowat et al. (2011)
Donsol, Philippines	87.8	6.5	396	McCoy et al. (2018)
Daymaniyats, Oman	75	5.9	5	Robinson et al. pers. comm.
Honda Bay, Philippines	96.5	4.5	175	Araujo et al. (2019)
India	75	7.4	164	Pravin (2020), Akhilesh et al. (2013)
Indonesia		4.1	45	Norman et al. (2017)
Isla Mujeres, Mexico		7.1	397	Norman et al. (2017)
Leyte, Philippines	84	5.3	193	Araujo et al. pers. comm.
Mafia Island, Tanzania	85.9	5.6	181	Rohner et al. pers. comm.
Musandam, Oman	85.7	5.4	13	Robinson et al. pers. comm.
Ningaloo Reef, Australia		5.3	758	Norman et al. (2017)
Nosy Be, Madagascar	73.2	5.8	238	Diamant et al. pers. comm.
Oslob, Philippines	80	4.9	371	Araujo et al. (2014)
Praia do Tofo, Mozambique	73.5	7.1	630	Prebble et al. pers. comm.
Seychelles	82.3	5.8	549	Rowat et al. (2011)

(Continued)

Table 7.1 (*Continued*)   Population Structure of Whale Sharks at Constellation Sites, with the Mean Length, the Percentage of Male Individuals, and a Primary Reference

Location	% Male	Mean Length (m)	N	References
Shib Habil, Saudi Arabia	47.2	4.3	116	Cochran et al. (2016)
South Africa		6.8	34	Norman et al. (2017)
St. Helena Island, UK	53	8.1	257	Perry et al. (2020)
Taiwan	61.5	4.6	597	Hsu et al. (2014)
Thailand		4.6	118	Norman et al. (2017)
The Maldives	95	6	64	Riley et al. (2009)
Tubbataha Reefs, Philippines	89	5.1	33	Araujo et al. pers. comm.
The United States of America		8	44	Norman et al. (2017)
Utila, Honduras		6.5	119	Norman et al. (2017)

Figure 7.4   Feeding constellation off Isla Mujeres, Mexico. (Photo: Lilia González.)

7.3.2 Segregated Populations

At birth, there are likely to be roughly equal numbers of both female and male whale sharks (Joung et al. 1996). From there, our understanding of their ecology rapidly becomes uneven and, frankly, confused. Most whale shark constellations consist of juvenile individuals between 3 and 9 m in length, with a significant bias towards male sharks (Figure 7.3, Table 7.1). Smaller juveniles (1.5–3.0 m) and neonates (<1.5 m), as well as large, mature whale sharks, are rarely recorded. Considering the possible size spectrum of the species, from approximately 0.5 to 20 m in length (Rowat and Brooks 2012), it is clear that the sharks present at most of these hotspots represent only a subset of a functional population. The few exceptions to the largely uniform population structure of whale shark constellations provide some clues as to where these "missing" sharks probably live (Figure 7.3).

This pattern, where whale sharks segregate by sex and size, is not unique; many other shark species live separately at different life stages (Springer 1967). With whale sharks, though, we have yet to learn where the majority prefer to live, and this *is* rather unique. The main complication is that whale sharks likely spend a lot of their lives away from coastal areas. Whale sharks are effectively unconstrained by distance or habitat within ocean basins, as they can swim a long way quite fast. Additionally, they dive exceptionally deep, resulting in a big vertical extension of their vast habitat. Much of their potential habitat, particularly in the open ocean, is hard for researchers to access, and a tough place to find solitary sharks, irrespective of how large the individuals are. In the next section, we discuss what whale sharks are doing at different sizes. Following that, we consider what females and males do differently. These phenomena are undoubtedly intertwined, but it helps to separate them here to tease out our specific knowledge gaps.

7.3.3 Size Segregation

Sightings of neonates and very small whale sharks are extremely rare. Only six encounters of live whale sharks of 1.5 m or less in total length have been submitted to the Wildbook for Whale Sharks database, as of the time of writing in 2020, and Rowat and Brooks (2012) found only 19 literature records in their global review, mostly of sharks captured in fishing gear. While a number of neonates have been seen since (e.g. Figure 7.5; Hsu et al. 2014; Miranda et al. 2020), and there are additional records of slightly larger individuals, the small juveniles <3 m are heavily underrepresented overall – after all, there will be more whale sharks in this size class than any other. The relative lack of these small whale sharks at constellation sites suggests that these individuals occupy a different habitat to larger (>3 m) juveniles. There are several factors that may drive segregation by size, regardless of sex. First, the swimming ability and speed of sharks may vary with size, leading to size-based groups with similar capabilities (Sims 2005). In whale sharks, neonates (<1.5 m long) have a slightly different body design to larger juveniles and adults. Pups have a large, strongly heterocercal tail, i.e. with a much larger upper lobe of the caudal fin than the lower lobe, with otherwise small fins and a low thrust angle (Martin 2007). This leads to an inefficient "anguilliform" (eel-like) swimming style, where their entire body wiggles from side-to-side during swimming, rather than most of the energy of the tail stroke pushing them forward through the water. This means they are relatively poor swimmers, compared to larger whale sharks. Coupled with their small size, young whale sharks are highly vulnerable to predators, and avoiding being eaten is likely to be the main driver of their habitat choice. This may explain why small whale sharks are largely missing from coastal areas, which tend to have higher predator densities than offshore sites. Unfortunately, the limited data on this size-class precludes a more detailed insight or speculation on their ecology.

Figure 7.5   A 60 cm male whale shark pup from Donsol in the Philippines (see Miranda et al. 2020 for details). (Photo: Large Marine Vertebrates Research Institute, Philippines.)

Figure 7.6   Large predatory shark bite on the head of a ~6 m whale shark from Mozambique. (Photo: Chris Rohner.)

The smallest whale sharks seen in coastal constellations tend to be 3–4 m in length, while the largest are usually 9–10 m (Rohner et al. 2015b). The mean length of sharks in these constellations is typically 4–8 m, with most being juveniles (Norman et al. 2017). At this size, whale sharks have fewer predators than when they are small, and this may increase their choice of habitats. While scars from encounters with predators are still apparent on many large juvenile sharks in some areas (Figure 7.6; Speed et al. 2008), presumably the reward of foraging in reliably productive coastal areas starts to outweigh the perceived predation risk in these habitats for many individuals. Schooling behavior itself can serve as an anti-predator strategy, but whale sharks are rarely encountered in clusters except in the presence of high prey availability, so food seems to be the primary influence on the formation and location of constellations. Size-based segregation can be a strategy to minimize competition for limited food (Martin 2007), but this seems unlikely to be a major factor for the segregation of small, large juvenile, and large adult whale sharks. Within those three main size groups, however, there may be limited competition and further segregation. For example, large juveniles of different sizes sometimes compete at tightly clustered "baitballs" of small fishes, where smaller whale sharks may literally be pushed out of the way by larger individuals (SJP pers. obs.). Whale shark prey, while patchily distributed, are mostly diffuse and abundant enough to allow multiple sharks of different sizes to feed in the same area with no evident hierarchy or disharmony. It is entirely possible, however, that the presence of more sharks leads to a faster reduction of prey density, and whale shark feeding appears to be highly attuned to prey density (Nelson and Eckert 2007; Rohner et al. 2015a).

While the mean size and the upper size range vary among sites, it is rare to see large whale sharks exceeding 9–10 m in coastal constellations (Norman et al. 2017). Male whale shark growth may actually asymptote at around this size in some areas (Matsumoto et al. 2019; Meekan et al. 2020), and individual sharks may continue to frequent constellations for years or decades thereafter (Norman and Morgan 2016; Norman et al. 2017). However, the population structure (number of sharks at a given length) at most individual coastal constellations is similar to a bell curve (i.e. Rowat et al. 2011; Araujo et al. 2014, 2017; Rohner et al. 2015b; Robinson et al. 2016), implying that larger sharks of both sexes are, in fact, leaving these sites. The question, then, is where are they going, and why?

The whale sharks seen at some offshore islands, such as Darwin Island in the Galápagos (Figure 7.7; Acuña-Marrero et al. 2014), and St. Helena Island in the mid-Atlantic (Perry et al. 2020) are distinctly larger than those seen regularly in coastal constellations; the mean length was 11.1 m at Darwin Island (Norman et al. 2017), and 8.1 m at St. Helena. Most data support an inference

Figure 7.7   Adult female whale shark at Darwin Island, Galápagos. (Photo: Simon Pierce.)

that large whale sharks are primarily oceanic, and that this habitat shift coincides with the onset of maturity. Habitat differences between adult and large juvenile whale sharks have been well documented from the Pacific coast of Mexico (Ramírez-Macías et al. 2012; Ketchum et al. 2013) and, in the whale sharks tagged in this region, adults used oceanic waters while juveniles stayed primarily in coastal areas (Eckert and Stewart 2001; Ramírez-Macías et al. 2017). Adult female sharks tagged at Darwin Island further support this hypothesis, with many spending their time in the open ocean of the tropical Eastern Pacific over the duration of tagging (Hearn et al. 2016). Whereas large whale sharks are rarely seen in coastal waters, juvenile sharks are frequently tracked moving through the open ocean (i.e. Araujo et al. 2018; Diamant et al. 2018; Cochran et al. 2019), indicating that smaller sharks are not excluded from these habitats. See Chapter 6 for a fuller discussion of these tracking results. Interestingly, large sharks are seldom reported from offshore islands in the Indian Ocean, such as Christmas Island (Hobbs et al. 2009).

It is unclear why larger whale sharks either prefer an oceanic life or avoid shallower waters. They are unlikely to be greatly influenced by the threat of predation, which probably declines with increasing size. Little is known about the adult whale shark diet at this point, although adult females, at least, appear to forage in productive frontal zones (Ramírez-Macías et al. 2017; Ryan et al. 2017). It is plausible that they additionally feed on small mesopelagic fishes, which are extremely abundant in the open ocean (Irigoien et al. 2014). This would mean that large, pelagic whale sharks have a more reliable food source available, while smaller sharks in coastal constellations target ephemeral bursts of productivity, with the potential for long fasts in between feeding opportunities (Wyatt et al. 2019). Larger size does confer some physical advantages, such as a larger gape and perhaps stronger suction-feeding capability, allowing big whale sharks to target larger, more mobile prey species. Although tagged adult whale sharks from Mexico were primarily epipelagic, i.e. spending most of their time between 0 and 200 m depth, they occasionally dived to >1,000 m (Ramírez-Macías et al. 2017). Larger whale sharks can tolerate colder temperatures for longer due to greater thermal inertia (reduced loss of body heat; Nakamura et al. 2020) and may therefore be able to dive deeper for longer, expanding their habitat vertically (Thums et al. 2013; Nakamura et al. 2020). Large, mature whale sharks may thus be less coastal because their greater deep-diving abilities allow them to access prey at depth. Whale sharks can also save energy on horizontal movements by gliding passively on long descents and saving active swimming for shorter ascents (Gleiss et al. 2011). The higher thermal inertia of larger sharks may allow them to use this strategy more frequently than smaller individuals (Thums et al. 2013; Nakamura et al. 2020), thus making it easier for larger individuals to expand their habitat horizontally. A lot more information on the movement of adult sharks, of both sexes, and their behaviors and diet in oceanic waters are needed for us to clearly explain this habitat shift.

7.3.4 Sexual Segregation

At this point, segregation by sex is almost a cliché in whale shark constellations (see Chapter 13). There is a substantial bias towards male sharks at most coastal locations (Figure 7.3, Table 7.1), with up to 95% male sharks in the Maldives (Riley et al. 2009) and 97% in Honda Bay, Philippines (Araujo et al. 2019). One interesting exception is Shib Habil in the Red Sea, with approximately equal numbers of both sexes present (53% of sexed sharks were female; Cochran et al. 2016). No significant differences in habitat use have been documented between sexes tracked from this site (Cochran et al. 2019). Aside from this one location, though, a male bias is present at all other juvenile coastal constellations (Norman et al. 2017). On a global level, the research focus on constellations means that many more males have been identified than females (Norman et al. 2017) and far more is known about their ecology, at least as juveniles.

Sexual segregation can be subdivided into (a) social segregation, where the sexes share the same space but partition by time or diet choice, or (b) habitat segregation, where each sex prefers a different habitat (Wearmouth and Sims 2008). Most constellations show a fairly well-defined seasonality, often due to the ephemeral abundance of a specific prey or feeding opportunity, and social segregation has not been noted from whale sharks so far. Assuming that whale shark pup survival is fairly even for females and males, this indicates that sex-specific behavior is ubiquitous; they select and use different habitats, at least at most locations.

The high densities of prey that are reliably present at some of these constellation sites, and a high proportion of sharks that are observed feeding, provide a clear rationale for the presence of male sharks. While some female sharks are usually present too, their relative absence is far harder to explain, particularly at sites that are almost entirely dominated by juveniles (i.e. Djibouti, Boldrocchi et al. 2020; Madagascar, Diamant et al. 2018; Maldives, Riley et al. 2009). Adult whale sharks have other drivers, as we will discuss, but why do juvenile females not seem to use coastal feeding aggregations as much as males; are they being excluded, or do they prefer different conditions or food? For now, this is still an open question, and these, and other, factors may be interrelated. Meekan et al. (2020) suggested that productive coastal waters offer the potential for faster growth rates, but higher exposure to potential whale shark predators. Faster-growing males would presumably have a better chance of achieving reproductive success over their lifetimes, providing an evolutionary rationale for "riskier" behaviors. Females may then take a different approach, a "slow but steady" growth to their adult size, minimizing their exposure to risk, and in so doing maximizing their own individual chance of producing offspring.

Biochemical studies on whale sharks from India (Borrell et al. 2011), Mozambique, Qatar, and Tanzania (Prebble et al. 2018), and Western Australia (Marcus et al. 2019), have not identified clear diet differences between the sexes in sharks of similar sizes within the same habitats. A long-term, sighting-independent acoustic study at one of those locations in Tanzania found no evidence for small-scale differences in habitat use between the sexes either (Rohner et al. 2020). All in all, female sharks in coastal constellations appear to behave similarly to the male sharks; there are just a lot fewer of them. It is entirely possible that most juvenile female whale sharks are oceanic, but this inference is based simply on the relative lack of females in coastal waters. A male bias is clear even among the smaller size classes (~3–4 m) occurring at most coastal sites (Araujo et al. 2014, 2017; Rohner et al. 2015b). Ultimately, larger studies involving long-term tracking and habitat selectivity modeling for both sexes, from both coastal and offshore environments, might be the best means to resolve this conundrum.

As sharks of both sexes approach maturity, their priorities are likely to change. As noted previously, few sharks exceeding 9 m in length are seen in coastal waters. Interestingly, length and sex data from some constellations, such as Qatar (Robinson et al. 2016) and Mozambique (Rohner et al. 2015), suggest that female sharks either "leave" these habitats first or an influx of larger males occurs at around 8 m lengths. One potential driver of sexual segregation in sharks is the desire of females to avoid male harassment, i.e. undesired courtship or mating attempts, which has seldom been seen in whale sharks but has been anecdotally reported both in the field and captivity (Chapter 2; Perry et al. 2020; Matsumoto et al. 2019). This has also been documented in other species. For example, it appears that mature male blue sharks *Prionace glauca* will routinely try to mate with smaller immature females (Calich and Campana 2015). Mature males represent 63% of the identified whale shark population in Qatar (Robinson et al. 2016). Tagged adult males at Al Shaheen behaved similarly to juvenile males over the duration of tracking, but the single large (≥9 m) female tagged in that study left the Gulf area and moved into the Indian Ocean (Robinson et al. 2017). The success of breeding females will presumably be increased by considering the habitat needs of their pups, which are largely unknown to us (see Section 7.3.2). Giving birth in areas where few predators are present is likely to be as far as parental care extends in whale sharks. Predator hotspots are predictably located at thermal fronts and near prominent topographic features such as islands, seamounts, and continental shelves (Worm et al. 2003, 2005; Block et al. 2011). These are good places for late-term female whale sharks to avoid. Conversely, mature males will receive an advantage by using similar habitats to reproductive females, so as to maximize their own mating opportunities. Courtship and mating behaviors have been witnessed around equatorial mid-Atlantic islands (Macena and Hazin 2016; Perry et al. 2020). An approximately equal sex ratio, mostly of mature sharks, has been documented from St. Helena Island, where mating has also been reported (Perry et al. 2020). However, mature males are rarely seen at the northern Galapagos Islands, where large female whale sharks predominate (Acuña-Marrero et al. 2014; Hearn et al. 2016); the extremely high abundance of predatory sharks present in this particular area may dissuade late-term pregnant females from using the site.

7.3.5 Transit vs Feeding Areas

Although we have been referring to all whale shark hotspots as constellations, there are two broad sub-classifications based on the primary behaviors at the site: (a) feeding constellations and (b) movement corridors, or transit areas. Feeding areas comprise the majority of whale shark constellations, where many individuals feed in the same area on a common prey type. There may be up to several hundred whale sharks aggregating in an area (de la Parra Venegas et al. 2011) to feed in dense prey patches (Chapter 5). Transit areas are sites where whale sharks are reliably seen, but they are typically solitary and observed while apparently swimming past, rather than feeding there. The frequent sightings at such sites may indicate that these are "waypoints" on defined movement pathways for the species, as has been documented for great white sharks and some marine mammals (Horton et al. 2017). These two site categories are characterized by differences in the time individual sharks spend in the area ("residency"), and whether they are likely to return in subsequent years ("site fidelity"). Residency is often estimated from individuals' sightings histories using the lagged identification rate (LIR), the probability of resighting a shark through time (Whitehead 2001). At whale shark feeding areas, typical residency times derived from LIR range from 12 days in the Red Sea to ~50 days in Donsol, Philippines (Cochran et al. 2016; McCoy et al. 2018). Residency estimates based on photo-identification, such as LIR, rely on the sharks being "sightable" by researchers, which is not always the case. At Mafia Island in Tanzania, whale sharks were frequently sighted at the surface, often feeding, for ~5 months of the year, but acoustic telemetry revealed that many sharks were also present through the rest of the year (Cagua et al. 2015; Rohner et al. 2020). Sightings-independent data from acoustic tagging showed that, on average, tagged whale sharks were present on 39% of days in Tanzania (Rohner et al. 2020) and on 28% of days in the Red Sea (Cochran et al. 2019).

Most whale sharks have only ever been seen at a single constellation, although many repeatedly return to that site (Norman et al. 2017). Most of the exceptions are from the western central Atlantic, where ~5% ($N=1,361$) of all sharks have been recorded from multiple constellation sites that are relatively close together (<1,000 km apart, McKinney et al. 2017). Tracking studies have routinely demonstrated that whale sharks can swim thousands of kilometers over a few weeks (e.g. Hueter et al. 2013; Ryan et al. 2017), so they have the capability to swim among constellations more frequently than we have observed to date. The fact that they often re-visit the same aggregation instead may indicate a learned migration behavior. Simulations showed that low levels of migration among constellations in the Indian Ocean are difficult to detect with photo-ID, given the population size, survey effort, and resighting rates at the time of the study (Andrzejaczek et al. 2016). Conversely, this also means that if an individual is identified in two constellations, the exchange between these two hotspots is likely larger than observed (Andrzejaczek et al. 2016). As more whale sharks are added to the global database, we will have more confidence that the low connectivity among constellations reflects the true situation. Importantly, many or even most whale sharks have only been sighted on a single occasion. While site fidelity is high in some individuals, even across decades, there are clearly a lot of other transient whale sharks that do not appear to regularly visit constellations.

Whale sharks are more resident in feeding areas than in transit areas (Norman et al. 2017), as they can obviously take advantage of a reliable and persistent food source for up to several months (Robinson et al. 2013). They also generally have a higher site fidelity to feeding areas than to transit areas (Norman and Morgan 2016). However, there is substantial individual variation in residency and site fidelity, even in reliable feeding areas. The proportion of transient sharks, often defined as individuals that are only seen in 1 year, is remarkably high in whale shark constellations. Even in one of the most resident whale shark constellations documented to date, at Mafia Island in Tanzania, around one-third of all sharks were only seen in a single year (C. Rohner unpublished data). It seems counter-intuitive to leave a reliable feeding area when prey is abundant, or to not return the following year after a good feeding season, but factors other than food may also play a role (see above). While it is also possible that a proportion of these transient sharks do return, but are missed by researchers, mounting evidence from the global photo-ID database suggests transience is a real characteristic of whale shark populations. A high level of transience has three main implications. First, it complicates research to estimate whale shark abundance. Capture–mark–recapture models that use sightings data to project abundance tend to struggle with this individual variation in resighting rates. Some past studies have separated out returning sharks (those seen across more than one year) from transient sharks, and then modeled abundance and trends only for the returning group (Holmberg et al. 2009; Ramírez-Macías et al. 2012). Second, high transience suggests that there are likely to be many more whale sharks alive than those we see in coastal constellations. However, due to the complications that transience brings to estimating abundance, it is difficult to gauge just how many whale sharks there are; genetic techniques, as described in Chapter 5, could be the best available option at this point. Third, transient individuals are inherently more difficult to study because they are only seen once. Transient whale sharks could plausibly have quite different ecologies to resident individuals in their diet, movements, and behaviors, but we can't really know because we don't get an opportunity to study them. This gap in our understanding of the species severely hampers conservation efforts.

Whale sharks in transit areas are less resident and often have a lower site fidelity than in feeding areas (Norman et al. 2017). For example at Darwin Island in the Galápagos, where almost exclusively large adult females are seen, residency is short (~2 days) and individuals are rarely resighted across years (3.5%; S. Green unpublished data). The low residence time indicates that these sites may act as navigational waypoints, rather than core habitats for feeding or mating, and suggests that these sharks may be exploiting different foraging opportunities. For instance, these transient sharks may be orienting towards dynamic oceanographic fronts or eddies rather than fixed locations (Ryan et al. 2017). To sum up, on top of the habitat segregation based on size and sex we discussed above, whale sharks also appear to segregate by their affinity, or lack thereof, to specific constellations areas. High levels of transience thus further cloud our understanding of whale shark population ecology.

7.4 LIFE CYCLE HYPOTHESIS

There are known knowns; there are things we know we know. We also know there are known unknowns; that is to say we know there are some things we do not know. But there are also unknown unknowns — the ones we don't know we don't know.

- Donald Rumsfield, US Secretary of Defense, 2002

In attempting to construct an overview of whale shark population ecology, and how it changes across their lives, the preceding sections have hopefully made the relevance of the above quote clear. Nevertheless, it is worth a try. Here, we draw together some of the knowledge and ideas discussed previously in this chapter and presented elsewhere through the book.

Our current best guess ("working hypothesis") is that whale sharks mate and give birth in oceanic waters, where the small juveniles also spend their first few months or years (Figure 7.8). At around 3–5 m in length, male sharks start to frequent feeding constellations in coastal areas – the hotspots on which so much of whale shark research is based. Some juvenile females are also

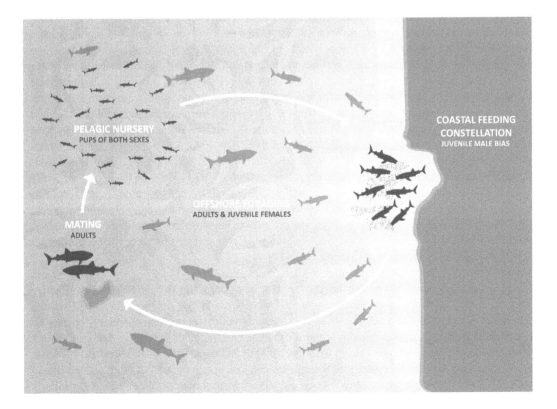

Figure 7.8 A schematic overview of a potential "typical" life cycle for whale sharks. Design by Madeleine Pierce.

present at these constellations, while others stay offshore foraging in frontal zones. Some individuals of both sexes will repeatedly return to predictably located coastal feeding areas over the next few years to decades, while others will be more transient and oceanic. Once they reach around 8–9 m length, the cycle is completed by both sexes of whale sharks becoming primarily oceanic again (Figure 7.8).

7.4.1 "The Lost Years"

Pioneering sea turtle biologist Archie Carr referred to the time between turtles' hatching on beaches and disappearing off into the ocean, and then returning to shallow habitats to feed, as their "lost years" (Carr 1952). We borrow his apt description here, for young whale sharks. Neonate and small juvenile whale sharks, less than 3 m in length, have rarely been encountered, and little information is available on the early stages of their lives. Most, but not all, have been encountered in offshore waters (Rowat and Brooks 2012). They likely have two main goals: to grow fast, and to not get eaten. Whale shark pups can grow impressively quickly in captivity (Chapter 2), but almost nothing is known about their diet and growth in the wild. Predator avoidance is likely the main driver of their distribution and may explain why we rarely see small whale sharks in coastal areas with high predator densities. Small whale sharks will undoubtedly be prey for a number of predators and have been found in the stomachs of blue sharks and blue marlin *Makaira nigricans* (see Rowat and Brooks 2012). This is presumably why the litter size of whale sharks (~300 in the one "megamamma" pregnant female examined so far; Joung et al. 1996) is so large: to maximize the odds that at least some of them will survive.

Many shark species use nursery areas that give the neonates protection from predators and thereby enhance the probability of their reaching maturity (Heupel et al. 2019). Shark nurseries have been characterized as areas where individuals less than 1-year-old are present in a higher density than in other areas, where they spend extended time, and where the area is used repeatedly over the years (Heupel et al. 2007). Pups of both blue sharks and shortfin mako sharks *Isurus oxyrinchus*, which are both oceanic species, probably do use certain ocean areas as "nurseries", though these are vast compared to those used by coastal species (Vandeperre et al. 2014; Nosal et al. 2019). Our limited knowledge on neonate and young whale shark ecology does not allow for much informed speculation on this point, but the existence of "clustered" sightings of neonate whale sharks have been noted from the mid-equatorial Atlantic region (Wolfson 1983; Kukuyev 1996), the northern Indian Ocean (Rowat et al. 2008), the Western Pacific (Aca and Schmidt 2011; Hsu et al. 2014; Miranda et al. 2020; along with the pregnant "megamamma"), and off Peru (Pajuelo et al. 2018). Broader genetic studies across ocean basins will help to identify whether whale sharks from different constellations or regions do swim long distances to use the same pelagic areas for pupping. Such a result would be exciting for conservation, as it would suggest that whale shark constellations have been depleted by overfishing and other human pressures can still be repopulated from other "reservoir" areas over generations to come.

Regardless of whether pelagic nurseries do or do not exist, small whale sharks have few places to hide in the open ocean. They are unlikely to be able to out-swim their predators, so effective camouflage and behavioral avoidance are probably their best adaptive options. They may swim in deeper, dark waters to decrease their encounter probability with visual predators and rely on camouflage in shallower waters. Most camouflage in the marine environment is cryptic, which means that an animal matches its color to the background to blend in and avoid visual detection by a predator. Whale sharks do not use the strategy of a single-color block (i.e. they are not blue, as blue sharks and mako sharks are), but instead their spots and stripes form a disruptive camouflage, similarly to many military uniforms, which conceals their body shape from visual predators (Wilson and Martin 2003; Schaefer and Stobbe 2006). Interestingly, their coloration and pattern do not change as they increase in size and shift to coastal habitats, unlike their relative the zebra shark *Stegostoma tigrinum* (Wilson and Martin 2003). This suggests that whale sharks live predominantly in the sunlit epipelagic zone throughout their lives.

7.4.2 The Hungry Years

Following these early years, characterized by rapid growth and presumably high mortality rates (Chang et al. 1997), the few surviving juvenile whale sharks enter the "hungry years." These years, which actually span decades of their lives (Perry et al. 2018; Meekan et al. 2020; Ong et al. 2020), are focused on feeding and growth. Sexual segregation during the hungry years may be partly explained by the different growth strategies, with males needing to grow to maturity as quickly as possible so they can mate, and females being more cautious to survive to maturity even if that results in slower growth (Meekan et al. 2020). Male sharks thus often seek out ephemeral bursts of productivity in coastal areas and, once located, some of them will return to these sites in future years. They seem to have the capacity to learn and remember how and when to return to these sites (Reynolds et al. 2017). Some feeding areas may be productive enough for whale sharks to stay in the area year-round, as seen in Tanzania (Cagua et al. 2015; Rohner et al. 2020), but for the most part these bursts of productivity are seasonal events. However, we reiterate that most male sharks have only been seen once at these locations, so there is doubtless more nuance to this narrative yet to be uncovered. While many sharks will exploit these reliable feeding opportunities, they tend to forage widely and feed opportunistically on patchily distributed prey through the rest of the year, including in oceanic waters (Chapter 8).

Juvenile females are less inclined to use coastal feeding areas, possibly because of the heightened predation risk in such habitats, and we presume they spend comparatively more time offshore. Here, both sexes are capable of locating oceanographic features, such as eddies and frontal systems, that represent productive feeding areas (Tew Kai et al. 2009; Bailleul et al. 2010; Ryan et al. 2017). This strategy likely means enduring extended periods of fasting when no prey can be found (Wyatt et al. 2019), but individual sharks may improve their foraging ability with experience, as has been suggested for basking sharks (Sims et al. 2006). Less reliable feeding opportunities also likely result in slower growth and coupled with the lower proportion of females in constellations may explain the slower growth in females (Meekan et al. 2020).

7.4.3 Open Water

Whale sharks move to a predominantly offshore lifestyle at or slightly before the onset of reproductive maturity (at ~7–10 m length, depending on sex and geographic area; Chapter 2). This habitat shift may be to exploit offshore prey resources, to breed, or some combination of both. Large female sharks rarely return to coastal waters, particularly once they reach adulthood, and they may leave these areas at a smaller size than do most males, in an effort to reduce unwanted mating attempts. While some areas have comparatively large numbers of mature males, such as Donsol in the Philippines (McCoy et al. 2018) and Al Shaheen off Qatar (Robinson et al. 2016), the scarcity of adult males in most constellations suggest that they rarely use these areas either.

Once offshore, both sexes face the challenge of locating food and mating opportunities in a vast habitat. Oceanic fronts are good feeding grounds in surface waters (Ramírez-Macías et al. 2017; Ryan et al. 2017), which may also allow for opportunistic social behavior to occur in these areas (Sims et al. 2000). They probably also feed on vertically migrating mesopelagic prey. But how do they find a mate? Observations of whale shark reproductive activity have been made off St. Peter and St. Paul Rocks and St. Helena Island in the mid-equatorial Atlantic Ocean (Macena and Hazin 2016; Perry et al. 2020). Whale sharks do feed in both areas (Hazin et al. 2008; Macena and Hazin 2016; Perry et al. 2020), suggesting that either prey availability drives the aggregation behavior, with reproductive opportunities as a bonus, or the inverse. As

Figure 7.9 (Unsuccessful) courtship behavior at St. Helena island. An adult male shark (right) repeatedly approached a female (left) from behind, swam parallel to her course, then peeled off to repeat the activity. This culminated in him making physical contact with her upper caudal fin, as shown. The female shark swam away rapidly, and the male did not follow her. (Photo: Simon Pierce.)

female whale sharks are only likely to mate on a biennial or longer cycle (Chapter 2), some form of signaling is potentially present. Visual and touch cues are a component of pre-courtship behavior in whale sharks (Figure 7.9; Perry et al. 2020), presumably allowing females to accept or reject male overtures. The possibility also remains that adult females release pheromones to indicate readiness to mate, as has been hypothesized for other shark species (Parsons et al. 2008; Whitney et al. 2010). Once pregnant, female whale sharks may move to loosely defined pelagic "nurseries" to give birth. That is likely to be the extent of their maternal care. In the newborn pups, the cycle then begins anew.

7.5 SUMMARY

Whale shark science has come a long way since the 1980s. Laser photogrammetry and stereo-videography, coupled with photo-ID of individual sharks, allows researchers to estimate the age and growth rate of whale sharks through time (Perry et al. 2018; Meekan et al. 2020). Mathematical models based on photo-ID resightings can estimate long-term abundance trends (Holmberg et al. 2008). We can track whale shark movements remotely via satellite, as well as the biological and environmental influences on their movement (Chapter 6). We can even investigate whale shark diet by collecting tiny skin samples (Chapter 8). All of these methods, though, require at least a brief interaction with a shark.

Where are we going with this? Well, we are probably missing a large proportion of the global whale shark population. Perhaps the clearest example is that only one pregnant whale shark has been conclusively documented by scientists (Joung et al. 1996). Whale sharks segregate themselves by size, sex, and movement patterns. If we cannot locate them, we can't use any of the research techniques described above. That means that our broader "understanding" of whale shark ecology is highly influenced by knowledge of the sharks we *can* locate. Constellations are heavily dominated by juvenile male sharks, whereas the females remain elusive. Only a handful of pups are reported worldwide each year. Adult sharks are probably far offshore, where logistical factors and costs make it difficult for us to work, and it would be challenging just to locate individual sharks. Our knowledge of whale shark ecology is partial, at best, and at least some of the inferences we make in this chapter are likely to prove inaccurate in the long run.

The good news is that the enigmatic lives of whale sharks have helped to inspire new research expeditions, tools, and techniques that help to overcome these limitations. Effort and resources are being directed towards constellations where large, mature whale sharks are regularly present, such as St. Helena Island and the Galápagos Islands. Modern computer science technologies are being deployed to crowdsource and process whale shark data from across their distribution, massively leveraging the capabilities of small research teams. Remote research methods such as environmental DNA, i.e. obtaining whale shark samples from the water they swim through (Chapter 5), hold promise going forward. As we enter this next phase of whale shark research, we will undoubtedly see new advances being made to help us unravel some of the mysteries surrounding the world's largest fish.

REFERENCES

Aca, E. Q., and J. V. Schmidt. 2011. Revised size limit for viability in the wild: Neonatal and young of the year whale sharks identified in the Philippines. *Asia Life Sciences* 20:361–68.

Acuña-Marrero, D., J. Jiménez, F. Smith, et al. 2014. Whale shark (*Rhincodon typus*) seasonal presence, residence time and habitat use at Darwin Island, Galápagos Marine Reserve. *PloS One* 9:e115946.

Akhilesh, K.V., Shanis, C.P.R., White, W.T., Manjebrayakath, H., Bineesh, K.K., et al. 2013. Landings of whale sharks *Rhincodon typus* Smith, 1828 in Indian waters since protection in 2001 through the Indian Wildlife (Protection) Act, 1972. *Environmental Biology of Fishes* 96(6):713–722.

Andrzejaczek, S., J. Meeuwig, D. Rowat, S. Pierce, T. Davies, R. Fisher, and M. Meekan. 2016. The ecological connectivity of whale shark aggregations in the Indian Ocean: A photo-identification approach. *Royal Society Open Science* 3:160455.

Araujo, G., A. Agustines, B. Tracey, S. Snow, J. Labaja, and A. Ponzo. 2019. Photo-ID and telemetry highlight a global whale shark hotspot in Palawan, Philippines. *Scientific Reports* 9:17209.

Araujo, G., A. Lucey, J. Labaja, C. L. So, S. Snow, and A. Ponzo. 2014. Population structure and residency patterns of whale sharks, *Rhincodon typus*, at a provisioning site in Cebu, Philippines. *PeerJ* 2:e543.

Araujo, G., C. A. Rohner, J. Labaja, et al. 2018. Satellite tracking of juvenile whale sharks in the Sulu and Bohol Seas, Philippines. *PeerJ* 6:e5231.

Araujo, G., S. Snow, C. L. So, et al. 2017. Population structure, residency patterns and movements of whale sharks in Southern Leyte, Philippines: Results from dedicated photo-ID and citizen science. *Aquatic Conservation: Marine and Freshwater Ecosystems* 27:237–52.

Arzoumanian, Z., J. Holmberg, and B. Norman. 2005. An astronomical pattern-matching algorithm for computer-aided identification of whale sharks *Rhincodon typus*. *The Journal of Applied Ecology* 42:999–1011.

Bailleul, F., C. Cotté, and C. Guinet. 2010. Mesoscale eddies as foraging area of a deep-diving predator, the southern elephant seal. *Marine Ecology Progress Series* 408:251–64.

Bird, T. J., A. E. Bates, J. S. Lefcheck, et al. 2014. Statistical solutions for error and bias in global citizen science datasets. *Biological Conservation* 173:144–54.

Block, B. A., I. D. Jonsen, S. J. Jorgensen, et al. 2011. Tracking apex marine predator movements in a dynamic ocean. *Nature* 475:86–90.

Boldrocchi, G., M. Omar, A. Azzola, and R. Bettinetti. 2020. The ecology of the whale shark in Djibouti. *Aquatic Ecology* 54: 535–51.

Bonney, R., C. B. Cooper, J. Dickinson, et al. 2009. Citizen science: A developing tool for expanding science knowledge and scientific literacy. *Bioscience* 59:977–84.

Borrell, A., A. Aguilar, M. Gazo, R. P. Kumarran, and L. Cardona. 2011. Stable isotope profiles in whale shark (*Rhincodon typus*) suggest segregation and dissimilarities in the diet depending on sex and size. *Environmental Biology of Fishes* 92:559–67.

Cagua, E. F., J. E. M. Cochran, C. A. Rohner, et al. 2015. Acoustic telemetry reveals cryptic residency of whale sharks. *Biology Letters* 11:20150092.

Calich, H. J., and S. E. Campana. 2015. Mating scars reveal mate size in immature female blue shark *Prionace glauca*. *Journal of Fish Biology* 86:1845–51.

Carr, A. 1952. *Handbook of Turtles: The Turtles of the United States, Canada, and Baja California.* Ithaca, NY: Cornell University Press.

Chang, W.-B., M.-Y. Leu, and L.-S. Fang. 1997. Embryos of the whale shark, *Rhincodon typus*: Early growth and size distribution. *Copeia* 1997:444–46.

Chen, C.-T., K.-M. Liu, and S.-J. Joung. 1997. Preliminary report on Taiwan's whale shark fishery. *Traffic Bulletin* 17:53–57.

Cochran, J. E. M., C. D. Braun, E. F. Cagua, et al. 2019. Multi-method assessment of whale shark (*Rhincodon typus*) residency, distribution, and dispersal behavior at an aggregation site in the Red Sea. *PloS One* 14:e0222285.

Cochran, J. E. M., R. S. Hardenstine, C. D. Braun, et al. 2016. Population structure of a whale shark *Rhincodon typus* aggregation in the Red Sea. *Journal of Fish Biology* 89:1570–82.

Davies, T. K., G. Stevens, M. G. Meekan, J. Struve, and J. M. Rowcliffe. 2013. Can citizen science monitor whale-shark aggregations? Investigating bias in mark-recapture modelling using identification photographs sourced from the public. *Wildlife Research* 39:696–704.

de la Parra Venegas, R., R. Hueter, J. González Cano, et al. 2011. An unprecedented aggregation of whale sharks, *Rhincodon typus*, in Mexican coastal waters of the Caribbean Sea. *PloS One* 6:e18994.

Diamant, S., C. A. Rohner, J. J. Kiszka, et al. 2018. Movements and habitat use of satellite-tagged whale sharks off western Madagascar. *Endangered Species Research* 36:49–58.

Duyck, J., C. Finn, A. Hutcheon, P. Vera, J. Salas, and S. Ravela. 2015. Sloop: A pattern retrieval engine for individual animal identification. *Pattern Recognition* 48:1059–73.

Eckert, S. A., and B. S. Stewart. 2001. Telemetry and satellite tracking of whale sharks, *Rhincodon typus*, in the Sea of Cortez, Mexico, and the North Pacific Ocean. *Environmental Biology of Fishes* 60:299–308.

Fox, S., I. Foisy, R. De La Parra Venegas, et al. 2013. Population structure and residency of whale sharks *Rhincodon typus* at Utila, Bay Islands, Honduras. *Journal of Fish Biology* 83:574–87.

Gleiss, A. C., B. Norman, and R. P. Wilson. 2011. Moved by that sinking feeling: Variable diving geometry underlies movement strategies in whale sharks. *Functional Ecology* 25:595–607.

Graham, R. T., Jacquin, S., Bakary, G., et al. 2008. Characteristics of whale shark seasonality and abundance in NW Madagascar and dispersal in the Western Indian Ocean. 2nd International Whale Shark Conference, Holbox, Quintana Roo, Mexico (PowerPoint display only). Available at http://www.domino. conanp.gob.mx/presentaciones.html/

Graham, R. T., and C. M. Roberts. 2007. Assessing the size, growth rate and structure of a seasonal population of whale sharks (*Rhincodon typus* Smith 1828) using conventional tagging and photo identification. *Fisheries Research* 84:71–80.

Hale, J. 2019. More than 500 hours of content are now being uploaded to YouTube every minute. https://www.tubefilter.com/2019/05/07/number-hours-video-uploaded-to-youtube-per-minute/ (accessed May 7, 2019).

Hanfee, F. 2001. *Gentle Giants of the Sea: India's Whale Shark Fishery.* TRAFFIC-India, WWF-India.

Hazin, F. H. V., T. Vaske Júnior, P. G. Oliveira, B. C. L. Macena, and F. Carvalho. 2008. Occurrences of whale shark (*Rhincodon typus* Smith, 1828) in the Saint Peter and Saint Paul Archipelago, Brazil. *Brazilian Journal of Biology* 68:385–89.

Hearn, A. R., J. Green, M. H. Román, D. Acuña-Marrero, E. Espinoza, and A. P. Klimley. 2016. Adult female whale sharks make long-distance movements past Darwin Island (Galápagos, Ecuador) in the Eastern Tropical Pacific. *Marine Biology* 163:214.

Heupel, M. R., J. K. Carlson, and C. A. Simpfendorfer. 2007. Shark nursery areas: Concepts, definition, characterization and assumptions. *Marine Ecology Progress Series* 337:287–97.

Heupel, M. R., S. Kanno, A. P. B. Martins, and C. A. Simpfendorfer. 2019. Advances in understanding the roles and benefits of nursery areas for elasmobranch populations. *Marine and Freshwater Research* 70:897–907.

Hobbs, J. P. A., A. J. Frisch, T. Hamanaka, C. A. McDonald, J. J. Gilligan, and J. Neilson. 2009. Seasonal aggregation of juvenile whale sharks (*Rhincodon typus*) at Christmas Island, Indian Ocean. *Coral Reefs* 28:577.

Holmberg, J., B. Norman, and Z. Arzoumanian. 2008. Robust, comparable population metrics through collaborative photo-monitoring of whale sharks *Rhincodon typus*. *Ecological Applications* 18:222–33.

Holmberg, J., B. Norman, and Z. Arzoumanian. 2009. Estimating population size, structure, and residency time for whale sharks *Rhincodon typus* through collaborative photo-identification. *Endangered Species Research* 7:39–53.

Horton, T. W., N. Hauser, A. N. Zerbini, et al. 2017. Route fidelity during marine megafauna migration. *Frontiers in Marine Science* 4:422.

Hsu, H. H., C. Y. Lin, and S. J. Joung. 2014. The first record, tagging and release of a neonatal whale shark *Rhincodon typus* in Taiwan. *Journal of Fish Biology* 85:1753–56.

Hueter, R. E., J. P. Tyminski, and R. de la Parra. 2013. Horizontal movements, migration patterns, and population structure of whale sharks in the Gulf of Mexico and northwestern Caribbean Sea. *PloS One* 8:e71883.

Irigoien, X., T. A. Klevjer, A. Røstad, et al. 2014. Large mesopelagic fishes biomass and trophic efficiency in the open ocean. *Nature Communications* 5:1–10.

Joung, S.-J., C.-T. Chen, E. Clark, S. Uchida, and W. Y. P. Huang. 1996. The whale shark, *Rhincodon typus*, Is a livebearer: 300 embryos found in one 'megamamma' supreme. *Environmental Biology of Fishes* 46:219–23.

Ketchum, J. T., F. Galván-Magaña, and A. P. Klimley. 2013. Segregation and foraging ecology of whale sharks, *Rhincodon typus*, in the southwestern Gulf of California. *Environmental Biology of Fishes* 96:779–95.

Kukuyev, E. I. 1996. The new finds in recently born individuals of the whale shark *Rhiniodon typus* (Rhiniodontidae) in the Atlantic Ocean. *Journal of Ichthyology* 36:203.

Macena, B. C. L., and F. H. V. Hazin. 2016. Whale shark (*Rhincodon typus*) seasonal occurrence, abundance and demographic structure in the mid-equatorial Atlantic Ocean. *PloS One* 11:e0164440.

Marcus, L., P. Virtue, P. D. Nichols, L. C.Ferreira, H. Pethybridge, and M. Meekan. 2019. Stable isotope analysis of dermal tissue reveals foraging behavior of whale sharks at Ningaloo Reef, Western Australia. *Frontiers in Marine Science* 6:546.

Martin, R. A. 2007. A review of behavioural ecology of whale sharks (*Rhincodon typus*). *Fisheries Research* 84:10–16.

Matsumoto, R., Y. Matsumoto, K. Ueda, M. Suzuki, K. Asahina, and K. Sato. 2019. Sexual maturation in a male whale shark (*Rhincodon typus*) based on observations made over 20 years of captivity. *Fishery Bulletin* 117:78–86.

McCoy, E., R. Burce, D. David, et al. 2018. Long-term photo-identification reveals the population dynamics and strong site fidelity of adult whale sharks to the coastal waters of Donsol, Philippines. *Frontiers in Marine Science* 5:271.

McKinney, J. A., E. R. Hoffmayer, J. Holmberg, et al. 2017. Long-term assessment of whale shark population demography and connectivity using photo-identification in the western Atlantic Ocean. *PloS One* 12:e0180495.

Meekan, M. G., B. M. Taylor, E. Lester, et al. 2020. Asymptotic growth of whale sharks suggests sex-specific life-history strategies. *Frontiers in Marine Science* 7:774.

Meenakshisundaram, A., L. Thomas, W. J. Kennington, M. Thums, E. Lester, and M. Meekan. In press. DOI: https://doi.org/10.3354/meps13729 available at https://www.int-res.com/prepress/m13729.html Genetic markers validate photo-identification and uniqueness of spot patterns in whale sharks. *Marine Ecology Progress Series*.

Miranda, J. A., N. Yates, A. Agustines, et al. 2020. Donsol: An important reproductive habitat for the world's largest fish *Rhincodon typus*? *Journal of Fish Biology*. doi:10.1111/jfb.14610.

Nakamura, I., R. Matsumoto, and K. Sato. 2020. Body temperature stability in the whale shark, the world's largest fish. *The Journal of Experimental Biology* 223:jeb210286.

Nelson, J. D., and S. A. Eckert. 2007. Foraging ecology of whale sharks (*Rhincodon typus*) within Bahía de Los Angeles, Baja California Norte, México. *Fisheries Research* 84:47–64.

Newman, G., A. Wiggins, A. Crall, E. Graham, S. Newman, and K. Crowston. 2012. The future of citizen science: Emerging technologies and shifting paradigms. *Frontiers in Ecology and the Environment* 10:298–304.

Norman, B. M., J. A. Holmberg, Z. Arzoumanian, et al. 2017. Undersea constellations: The global biology of an endangered marine megavertebrate further informed through citizen science. *Bioscience* 67:1029–43.

Norman, B. M., and D. L. Morgan. 2016. The return of 'Stumpy' the whale shark: Two decades and counting. *Frontiers in Ecology and the Environment* 14:449–50.

Nosal, A. P., D. P. Cartamil, N. C. Wegner, C. H. Lam, and P. A. Hastings. 2019. Movement ecology of young-of-the-year blue sharks *Prionace glauca* and shortfin makos *Isurus oxyrinchus* within a putative binational nursery area. *Marine Ecology Progress Series* 623:99–115.

Ong, J. J. L., M. G. Meekan, H. H. Hsu, L. P. Fanning, and S. E. Campana. 2020. Annual bands in vertebrae validated by bomb radiocarbon assays provide estimates of age and growth of whale sharks. *Frontiers in Marine Science* 7:188.

Pajuelo, M., J. Alfaro-Shigueto, M. Romero, et al. 2018. Occurrence and bycatch of juvenile and neonate whale sharks (*Rhincodon typus*) in Peruvian waters. *Pacific Science* 7:463–73.

Parsons, G. R., E. R. Hoffmayer, J. M. Hendon, and W. V. Bet-Sayad. 2008. A review of shark reproductive ecology: Life history and evolutionary implications. In *Fish Reproduction*, eds. M. A. Rocha, A. Arukwe, B. G. Kapoor, 435–69. Enfield: Science Publishers.

Perry, C. T., E. Clingham, D. H. Webb, et al. 2020. St. Helena: An important reproductive habitat for whale sharks (*Rhincodon typus*) in the south central Atlantic. *Frontiers in Marine Science* 7:899.

Perry, C. T., J. Figueiredo, J. J. Vaudo, J. Hancock, R. Rees, and M. Shivji. 2018. Comparing length-measurement methods and estimating growth parameters of free-swimming whale sharks (*Rhincodon typus*) near the South Ari Atoll, Maldives. *Marine and Freshwater Research* 69:1487–95.

Pierce, S. J., J. Holmberg, A. A. Kock, and A. D. Marshall. 2019. Photographic identification of sharks. In *Shark Research: Emerging Technologies and Applications for the Field and Laboratory*, eds. J. C. Carrier, M. R. Heithaus, and C. A. Simpfendorfer, 1–37. Boca Raton, FL: CRC Press.

Pravin, P. 2020. Whale shark in the Indian coast - Need for conservation. *Current Science* 79(3):310–315.

Prebble, C. E. M., C. A. Rohner, S. J. Pierce, et al. 2018. Limited latitudinal ranging of juvenile whale sharks in the Western Indian Ocean suggests the existence of regional management units. *Marine Ecology Progress Series* 601:167–83.

Ramírez-Macías, D., N. Queiroz, S. J. Pierce, N. E. Humphries, D. W. Sims, and J. M. Brunnschweiler. 2017. Oceanic adults, coastal juveniles: Tracking the habitat use of whale sharks off the Pacific coast of Mexico. *PeerJ* 5:e3271.

Ramírez-Macías, D., A. Vázquez-Haikin, and R. Vázquez-Juárez. 2012. Whale shark *Rhincodon typus* populations along the west coast of the Gulf of California and implications for management. *Endangered Species Research* 18:115–28.

Reynolds, S. D., B. M. Norman, M. Beger, C. E. Franklin, and R. G. Dwyer. 2017. Movement, distribution and marine reserve use by an endangered migratory giant. *Diversity and Distributions* 23:1268–79.

Riley, M. J., M. S. Hale, A. Harman, and R. G. Rees. 2009. Analysis of whale shark *Rhincodon typus* aggregations near South Ari Atoll, Maldives archipelago. *Aquatic Biology* 8:145–50.

Robinson, D. P., M. Y. Jaidah, S. S. Bach, et al. 2016. Population structure, abundance and movement of whale sharks in the Arabian Gulf and the Gulf of Oman. *PloS One* 11:e0158593.

Robinson, D. P., M. Y. Jaidah, S. S. Bach, et al. 2017. Some like it hot: Repeat migration and residency of whale sharks within an extreme natural environment. *PloS One* 12:1–23.

Robinson, D. P., M. Y. Jaidah, R. W. Jabado, et al. 2013. Whale sharks, *Rhincodon typus*, aggregate around offshore platforms in Qatari waters of the Arabian Gulf to feed on fish spawn. *PloS One* 8:e58255.

Rohner, C. A., A. J. Armstrong, S. J. Pierce, et al. 2015a. Whale sharks target dense prey patches of sergestid shrimp off Tanzania. *Journal of Plankton Research* 37:352–62.

Rohner, C. A., J. E. M. Cochran, E. F. Cagua, et al. 2020. No place like home? High residency and predictable seasonal movement of whale sharks off Tanzania. *Frontiers in Marine Science* 7: 423.

Rohner, C. A., S. J. Pierce, A. D. Marshall, S. J. Weeks, M. B. Bennett, and A. J. Richardson. 2013. Trends in sightings and environmental influences on a coastal aggregation of manta rays and whale sharks" *Marine Ecology Progress Series* 482:153–68.

Rohner, C. A., A. J. Richardson, C. E. M. Prebble, et al. 2015b. Laser photogrammetry improves size and demographic estimates for whale sharks. *PeerJ* 3:e886.

Rowat, D., and K. S. Brooks. 2012. A review of the biology, fisheries and conservation of the whale shark *Rhincodon typus*. *Journal of Fish Biology* 80:1019–56.

Rowat, D., K. S. Brooks, A. March, et al. 2011. Long-term membership of whale sharks (*Rhincodon typus*) in coastal aggregations in Seychelles and Djibouti. *Marine and Freshwater Research* 62:621–27.

Rowat, D., M. A. Gore, B. B. Baloch, et al. 2008. New records of neonatal and juvenile whale sharks (*Rhincodon typus*) from the Indian Ocean. *Environmental Biology of Fishes* 82:215–19.

Rowat D., Meekan M.G. and U. Englehardt. 2007. Aggregations of juvenile whale sharks (*Rhincodon typus*) in the Gulf of Tadjoura, Djibouti. *Environmental Biology of Fishes* 80(4):465–472.

Rowat, D., C. W. Speed, M. G. Meekan, M. A. Gore, and C. J. A. Bradshaw. 2009. Population abundance and apparent survival of the vulnerable whale shark *Rhincodon typus* in the Seychelles aggregation. *Oryx* 43:591–98.

Ryan, J. P., J. R. Green, E. Espinoza, and A. R. Hearn. 2017. Association of whale sharks (*Rhincodon typus*) with thermo-biological frontal systems of the eastern tropical Pacific. *PloS One* 13:e0196443.

Schaefer, H. M., and N. Stobbe. 2006. Disruptive coloration provides camouflage independent of background matching. *Proceedings of the Royal Society B: Biological Sciences* 273:2427–32.

Silas, E. G. 1986. The whale shark (*Rhiniodon typus* Smith) in Indian coastal waters. Is the species endangered or vulnerable? *Marine Fisheries Information Service* 66: 1–21.

Sims, D. W. 2005. Differences in habitat selection and reproductive strategies of male and female sharks. In *Sexual Segregation in Vertebrates: Ecology of the Two Sexes*, ed. K. E. Ruckstuhl, and P. Neuhaus, 127–47. Cambridge: Cambridge University Press.

Sims, D. W., E. J. Southall, V. A. Quayle, and A. M. Fox. 2000. Annual social behaviour of basking sharks associated with coastal front areas. *Proceedings of the Royal Society B: Biological Sciences* 267:1897–904.

Sims, D. W., M. J. Witt, A. J. Richardson, E. J. Southall, and J. D. Metcalfe. 2006. Encounter success of free-ranging marine predator movements across a dynamic prey landscape. *Proceedings of the Royal Society B: Biological Sciences* 273:1195–201.

Speed, C. W., M. G. Meekan, D. Rowat, S. J. Pierce, A. D. Marshall, and C. J. A. Bradshaw. 2008. Scarring patterns and relative mortality rates of Indian Ocean whale sharks. *Journal of Fish Biology* 72:1488–503.

Springer, S. 1967. Social organisation of shark populations. In *Sharks, Skates and Rays*, eds. P. W. Gilbert, R. F. Mathewson, and D. P. Rall, 149–74. Baltimore, MD: Johns Hopkins Press.

Stevens, J. D. D. 2007. Whale shark (*Rhincodon typus*) biology and ecology: A review of the primary literature. *Fisheries Research* 84:4–9.

Taylor, G. 1994. *Whale Sharks. The Giants of Ningaloo Reef*. Sydney: Angus and Robertson.

Taylor, G. 2019. *Whale Sharks: The Giants of Ningaloo Reef*. 2nd Revised Edition. Perth: Scott Print.

Taylor, J. G. 1996. Seasonal occurrence, distribution and movements of the whale shark, *Rhincodon typus*, at Ningaloo Reef, Western Australia" *Marine and Freshwater Research* 47:637–42.

Taylor, J. G. 2007. Ram filter-feeding and nocturnal feeding of whale sharks (*Rhincodon typus*) at Ningaloo Reef, Western Australia. *Fisheries Research* 84:65–70.

Taylor, J. G., and G. Grigg. 1991. *Whale Sharks of Ningaloo Reef*. Canberra: Australian National Parks and Wildlife Service.

Tew Kai, E., V. Rossi, J. Sudre, et al. 2009. Top marine predators track Lagrangian coherent structures. *Proceedings of the National Academy of Sciences* 106:8245–50.

Theberge, M. M., and P. Dearden. 2006. Detecting a decline in whale shark *Rhincodon typus* sightings in the Andaman Sea, Thailand, using ecotourist operator-collected data. *Oryx* 40:337.

Thums, M., M. Meekan, J. Stevens, S. Wilson, and J. Polovina. 2013. Evidence for behavioural thermoregulation by the world's largest fish. *Journal of the Royal Society Interface* 10:20120477.

Vandeperre, F., A. Aires-da-Silva, J. Fontes, M. Santos, R. Serrão Santos, and P. Afonso. 2014. Movements of blue sharks (*Prionace glauca*) across their life history. *PloS One* 9:e0103538.

Wearmouth, V. J., and D. W. Sims. 2008. Sexual segregation in marine fish, reptiles, birds and mammals: Behaviour patterns, mechanisms and conservation implications. *Advances in Marine Biology* 54:107–70.

Whitehead, H. 2001. Analysis of animal movement using opportunistic individual identifications: Application to sperm whales. *Ecology* 82:1417–32.

Whitney, N. M., H. L. Pratt Jr., T. C. Pratt, and J. C. Carrier. 2010. Identifying shark mating behaviour using three-dimensional acceleration loggers. *Endangered Species Research* 10:71–82.

Williams, J. L., S. J. Pierce, M. M. P. B. Fuentes, and M. Hamann. 2015. Effectiveness of recreational divers for monitoring sea turtle populations. *Endangered Species Research* 26:209–19.

Wilson, S. G., and R. A. Martin. 2003. Body markings of the whale shark: Vestigial or functional? *Western Australian Naturalist* 124:118–34.

Wolfson, F. H. 1983. Records of seven juveniles of the whale shark, *Rhiniodon typus*. *Journal of Fish Biology* 22:647–55.

Wolfson, F. H. 1986. Occurrences of the whale shark, *Rhincodon typus*, Smith. In *Indo-Pacific Fish Biology: Proceedings of the Second International Conference on Indo-Pacific Fishes*, eds. T. Uyeno, R. Arai, T. Taniuchi, and K. Matsuura, 208–26. Tokyo: Ichthyological Society of Tokyo.

Worm, B., H. K. Lotze, and R. A. Myers. 2003. Predator diversity hotspots in the blue ocean. *Proceedings of the National Academy of Sciences of the United States of America* 100:9884–88.

Worm, B., M. Sandow, A. Oschlies, H. K. Lotze, and R. A. Myers. 2005. Global patterns of predator diversity in the open oceans. *Science* 309:1365–69.

Wyatt, A. S. J., R. Matsumoto, Y. Chikaraishi, et al. 2019. Enhancing insights into foraging specialization in the world's largest fish using a multi-tissue, multi-isotope approach." *Ecological Monographs* 89:e01339.

CHAPTER **8**

Whale Shark Foraging, Feeding, and Diet

Christoph A. Rohner and Clare E. M. Prebble
Marine Megafauna Foundation

CONTENTS

8.1 INTRODUCTION

While the whale shark is the world's largest fish, they feed on some of the ocean's smallest inhabitants. The filter-feeding mechanism of whale sharks makes their feeding ecology more similar to baleen whales than to most other sharks. Of the ~505 shark species in the world, only three have independently evolved a filter-feeding strategy (Sanderson & Wassersug 1993), and the mechanisms used by each differ. In contrast to most other large, filter-feeding animals, whale sharks only live in the warm waters of the tropics and sub-tropics. This area of the globe has the lowest plankton productivity, and the dense plankton patches that do form are highly variable in both space and time (Lalli & Parsons 1997; Sarmiento & Gruber 2006). The difficulty of "sieving a living" in such an unaccommodating environment is a reason why most large baleen whales mainly feed in the cold, plankton-rich waters around the poles (Zerbini et al. 2006), and why the second largest fish, the filter-feeding basking shark *Cetorhinus maximus*, lives in higher latitudes, where plankton are bigger and more abundant than in the tropics (San Martin et al. 2006; Sims 2008). Whale sharks fill a unique and challenging niche as warm-water (>21°C) specialists (Colman 1997; Duffy 2002; Afonso et al. 2014), and they employ a variety of strategies to find their prey in these generally unproductive waters. In this chapter, we describe the feeding ecology of whale sharks and explore how they sustain themselves on small, scarce prey that are patchily distributed in the sharks' wide geographic range.

8.2 FORAGING STRATEGIES

"Foraging" is the wide search for food or provisions. For whale sharks, this largely refers to the search for plankton patches in coastal waters and in the open ocean. Observed and inferred foraging behaviors from in-water and tracking studies (see Chapter 6) have revealed important insights, though this information is biased towards the habits of juvenile, predominantly male, whale sharks feeding in coastal constellations (see Chapter 7). How whale sharks forage while away from the coast is still poorly understood. The two main foraging strategies that have been observed in whale sharks are (a) the regular return of individual sharks to predictable feeding hotspots and (b) large-scale, three-dimensional movements, searching for cues and prey while away from these ephemeral feeding areas.

8.2.1 Returning to Predictable Feeding Hotspots

At least 30 whale shark hotspots have now been identified, with shark presence driven by high prey availability at most of these sites (Norman et al. 2017). Multiple whale sharks feed together in these hotspots (Figure 8.1a), with up to 400 individuals counted off Isla Mujeres in Mexico (de la Parra et al. 2011). Whale shark research output increased dramatically following the discovery of whale shark feeding aggregations, or constellations, thanks to the regular and predictable access to whale sharks that such areas provide (Colman 1997; Rowat & Brooks 2012; Norman et al. 2017). Predictable feeding hotspots are integral to our understanding of whale sharks, and their behavior in these constellations is the most well-understood aspect of whale shark foraging and feeding ecology. However, it is important to remember that the population structure at these feeding hotspots is biased towards male sharks, mostly large juveniles (Chapter 7). In addition to the segregation based on sex and size, only a small proportion of the larger juvenile sharks aggregate at these hotspots, while others are more transient. Newborn, small juvenile, mature, and female sharks, as well as transient large juveniles all need to forage and feed as well, but we lack knowledge of their ecology. This shows that, on the whole, we still know relatively little about the foraging ecology of whale sharks. For the moment, let us focus instead on the aspects that are comparatively well understood.

Feeding hotspots that predictably offer high densities of prey (Figure 8.1b) are understandably attractive to whale sharks, living as they do in an environment where good feeding opportunities will often be unreliable, few, and far between (Lalli & Parsons 1997; Sarmiento & Gruber 2006). Some feeding hotspots are characterized by a defined biological event, such as fish spawning (Heyman et al. 2001; de la Parra et al. 2011; Robinson et al. 2013), while others may be driven by multiple environmental factors that combine to promote the formation of dense zooplankton patches (Rohner et al. 2015b). Potential environmental drivers of high-density prey patches in feeding aggregations include upwelling, coastal run-off, and wind- or current-induced concentration

Figure 8.1   Whale shark feeding aggregations. (a) Multiple whale sharks feed together in constellations, as here off
Isla Mujeres in Mexico. Photo by Rafael de la Parra. (b) Whale sharks often feed in dense zooplankton
patches, as illustrated here off Mafia Island in Tanzania where they target sergestid shrimps Photo by
Christoph Rohner. (c) Day-time surface feeding in coastal constellations is most commonly observed,
although whale sharks likely also feed at depth, at night, and in oceanic waters. Photo by Christoph
Rohner. (d) Whale sharks also feed on small fishes. Photo by Clare Prebble.

of zooplankton. Whale sharks display high site fidelity at many of these aggregation sites, and a
portion of individuals are often sighted year after year, even for decades in some cases (Norman &
Morgan 2016). Individual sharks have rarely been identified at more than one aggregation site, and
there can be several years between return visits to a given feeding hotspot (Brooks et al. 2010;
McKinney et al. 2017; Norman et al. 2017). Outside periods of high prey abundance, though, these
same whale sharks routinely travel thousands of kilometers (e.g. Berumen et al. 2014; Robinson
et al. 2017; Rohner et al. 2018). Presumably, they return to these hotspots in anticipation of feeding opportunities, as it is unlikely that the sharks can pick up specific prey cues from hundreds
of kilometers away. Indeed, our field observations from places like Mafia Island in Tanzania suggest that whale sharks sometimes arrive at these hotspots weeks before actual feeding events are
observed. They display searching behavior, apparently waiting until dense zooplankton patches
form, triggering a switch to feeding behavior. Alternatively, it is also possible that they feed at night
or at depth during this perceived waiting time (Figure 8.1c). Assuming that they do arrive at some
hotspots in anticipation of good feeding conditions, this suggests a learnt behavior. While information on shark cognition is still scarce, other shark species do have the ability to retain spatial knowledge (Schluessel 2014). The impressive feats of navigation and spatial memory that are presumably
involved in whale sharks returning to feeding hotspots have not yet been researched.

The specifics of feeding aggregations vary from site to site, but they are generally seasonal and
involve surface feeding. Within a season, prey availability can fluctuate from day to day, and whale
sharks, although present, are not always seen feeding every day in some locales, such as Mafia
Island in Tanzania (Rohner et al. 2015b) or Tofo in Mozambique (Haskell et al. 2015). In other
places, such as Al Shaheen in Qatar (Robinson et al. 2013) and Isla Mujeres off Mexico (de la Parra
Venegas et al. 2011), there appears to be a continuous supply of prey (tuna spawn, in both cases) for
months at a time. These are also the sites that attract the most individual sharks (de la Parra Venegas

et al. 2011). The taxonomic composition of prey, and the specific feeding behaviors employed by the sharks, also varies among sites. We shall address that later in this chapter.

8.2.2 Large-Scale Movements away from Feeding Hotspots

Despite the observational bias, predictable constellations are unlikely to be the only places where whale sharks feed. Clearly, the small juveniles, adults, females, and transient whale sharks absent from these sites all need to feed as well. Even the large juveniles with a high site fidelity to feeding constellations are expected to continue foraging on their wider-scale movements, though this has rarely been directly observed. Similar to other large, free-ranging predators, whale sharks in oceanic waters have to make foraging decisions with little or no advance knowledge of prey distribution and availability, because the potential prey are located outside the sharks' sensory range. To overcome this, animals tend to use a Lévy-like search behavior. This behavior is characterized by clusters of short steps and high turning angles with longer, straight-line movement steps in between them (Sims et al. 2008). This movement pattern is thought to be the optimal search pattern while foraging for sparse, unpredictably distributed prey. Satellite tracks from tagged whale sharks reflect this strategy, with clusters of surface activity interspersed by long steps of no transmissions (e.g. Hearn et al. 2016, Rohner et al. 2018), suggesting surface feeding bouts between a broader search for prey or long-range movement. Even though feeding opportunities in the open ocean may not be associated with specific locations, there are more fluid features such as fronts and eddies that concentrate prey that whale sharks may orientate towards (Ryan et al. 2017).

This approach to foraging also means that there are likely to be periods when whale sharks are not feeding. Captive whale sharks sometimes intentionally forgo feeding opportunities (Wyatt et al. 2019), and it is possible that this is also the case in wild whale sharks, for example, when they prioritize mating or pupping over feeding. In other instances, particularly in juveniles, whale sharks might want to feed but cannot find their prey. It is difficult for us to assess intent and the physiological consequences of periods with no food intake. We therefore will call all scenarios in which ingestion is less than the basal metabolic rate "fasting" here, whether the lack of feeding is intentional or not, and also whether they are catabolic or not (Wyatt et al. 2019). In aquaria, healthy whale sharks routinely fasted for up to 8 days, while another shark did not feed for 151 days prior to death (Wyatt et al. 2019). Evidence from comparative blood samples of wild and captive whale sharks further suggests that long-term fasting during oceanic migrations in wild individuals is common (Wyatt et al. 2019). Ecologically similar elasmobranchs, such as manta and mobula rays (Mobulidae), also appear to undergo periods of fasting between feeding events while away from the coast (Rohner et al. 2017). Some shark species that feed irregularly will save energy by discontinuing acid secretion in the stomach, while others always produce acid to be in a ready state for digestion (Papastamatiou & Lowe 2005). Little is known about such physiological responses to fasting in whale sharks, but its effect on metabolism can now be monitored with biochemical analyses of blood samples (Wyatt et al. 2019). The health and growth of individuals likely varies based on their nutritional history, and further research in this area is needed.

There is further evidence that whale sharks continue to search for food while away from aggregation sites. Satellite-tagged whale sharks in Mozambique actively chose to swim in areas with higher chl-a levels – chlorophyll is the green pigment found in phytoplankton, often used by researchers as a proxy for zooplankton availability because it can be measured via satellite – compared to randomized computer-generated model sharks (Rohner et al. 2018). Basking sharks are also more successful at finding zooplankton than (hypothetical) sharks employing random movements (Sims et al. 2006). Large-scale zooplankton density data were available for that study, indicating that the foraging success of basking sharks may improve with age; larger individuals performed better than the juvenile they tagged (Sims et al. 2006). The fact that large whale sharks are rarely encountered at coastal aggregation sites suggests that they, too, might learn how to find sufficient food in the patchy prey field of the open ocean as they get older, thereby reducing their dependence on ephemeral coastal prey patches.

Whale sharks might also change their diet preferences and growth trajectories as they increase in size, influencing differing habitat choices between juveniles and adults (Meekan et al. 2020). Although the diet of adult whale sharks is poorly known, it is possible that they shift to targeting larger prey, such as the small mesopelagic fishes that are extremely abundant in the deeper layers of oceanic waters (Irigoien et al. 2014). Such a diet change would mean that adult whale sharks have a more reliable food source than juveniles targeting ephemeral zooplankton patches, potentially explaining their growth to a truly gigantic size. The large size of mature whale sharks means they naturally maintain a more stable body temperature during brief exposure to cold water (Nakamura et al. 2020), which allows them to dive deeper or longer than smaller juveniles. Larger sharks also have a larger gape and perhaps a stronger suction-feeding capacity. Their physiology and habitat choice indicate an ontogenetic diet change is likely, but further examination of their diet is needed to assess this hypothesis.

It is relatively easy to observe and study whale sharks while they are targeting dense patches of zooplankton near the surface. However, whale sharks are also likely to forage and feed at depth. Such behaviors are far more difficult to investigate and document from animal-borne sensor data. Whale shark deep (>500 m) diving behavior has, however, been loosely linked to the diel vertical migration of dense layers of zooplankton, similar to observations on other shark species (Graham et al. 2006). Their deep dives through the mesopelagic into the >1,000 m bathypelagic zone are thought to facilitate foraging, as the sharks traverse multiple water layers and are exposed to potential prey cues, or the prey themselves (Brunnschweiler et al. 2009; Rohner et al. 2013). Deeper gliding dives are also thought to be an energy-efficient means of traveling long distances (Gleiss et al. 2011). Drivers behind diving behaviors are still difficult to interpret, but with advances in sensor technology, specifically the increasing use of cameras, biologgers, accelerometers, and gyroscopes in animal-mounted tags, the nuances of behaviors that whale sharks use during their dives can now be revealed (Gleiss et al. 2013, Cade et al. 2020).

8.3 SENSES INVOLVED IN FORAGING AND FEEDING

Sensory capabilities are among the least-studied parts of whale shark biology, but there are some clues as to which senses are likely to be useful in foraging and feeding (see also Chapter 4). When whale sharks are in close proximity to a prey patch, olfaction and perhaps the electro-magnetic senses are likely to be key in guiding the shark to its prey and in determining when to switch from foraging to feeding. Captive whale sharks respond readily to chemosensory cues from krill homogenate and also, to a lesser extent, cues from solutions of dimethyl sulfide (DMS; Dove 2015). This chemical is released by phytoplankton when they are grazed on by zooplankton and can thus indirectly indicate patches of high zooplankton density. DMS, together with its precursor dimethyl sulfoniopropionate, is a major cue used by many seabirds (Nevitt 2008) and fishes (DeBose & Nevitt 2007) to find prey patches in the marine environment (Savoca & Nevitt 2014). Whale sharks may be able to sense the aqueous and the volatile forms of DMS in water and in air, as their large nares located on the leading edge of the upper jaw are exposed to air when they are feeding at the surface (Dove 2015). Chemical stimuli alone have elicited feeding behaviors in whale sharks, demonstrating that olfaction is a key sense for feeding (see Chapter 9).

Sharks can detect weak electric and magnetic fields with the electroreceptors, called Ampullae of Lorenzini, which are located on the skin and concentrated around their heads. Some species use this sense to capture hidden prey (Kalmijn 1972). While the electro-magnetic sense has not been investigated in whale sharks, they might be able to detect the minute bioelectric fields of their zooplankton prey, as has been shown for the Mississippi paddlefish (Wilkens & Hofmann 2007).

Experiments showed that *Daphnia* water fleas generated electric field potentials of up to 1 mV, meaning that zooplankton produce potentials equivalent to marine fishes and crustaceans that can be picked up by elasmobranchs (Kalmijn 1972; Wilkens & Hofmann 2007). However, the detection radius of the electromagnetic sense is short (~30 cm; Kajiura & Holland 2002). Whale sharks may thus use it mostly to assess the density of prey within a feeding patch and adjust their behavior accordingly, similar to the mechanism proposed for basking sharks (Kempster & Collin 2011).

Whale sharks also have other senses that could play a role in their foraging and feeding behaviors. They have the largest inner ear of any animal (Muller 1999) and are likely to respond mostly to long wavelength, low-frequency sounds. Whale sharks may, for example, be able to hear the splashing sounds of baitfishes at the surface being attacked by predators, such as tuna, and proceed to feed on these baitfishes themselves (Figure 8.1d; Fox et al. 2013). At depth, where there is little to no ambient light and sound is the optimal way of detecting prey, whale sharks may use their hearing abilities to locate mesopelagic fish prey (Baumann-Pickering 2016; McCauley & Cato 2016). The visual sense may also play a role in whale shark feeding behavior, although two lines of contrasting evidence exist. Their eyes are small and located on the side of the head, suggesting that they are set up to increase field of view rather than binocular vision (Clark & Nelson 1997). Whale shark brain organization also supports a minor role played by vision, with a relatively reduced mesencephalon (an area of the midbrain associated with vision, hearing, and motor control) compared to other pelagic predators (Yopak & Frank 2009; also see Chapter 4). However, whale sharks also have protective denticles on their eyes, indicating that vision may be more important than previously thought (Tomita et al. 2020). In relation to feeding, vision is likely to be used at short distances only and may be used particularly to locate relatively large baitfishes rather than small prey particles such as fish eggs.

8.4 FEEDING MECHANICS

The filter-feeding nature of whale sharks is made possible by their specialized anatomy. Their gill rakers have been modified into five lower and five upper filter pads, on each side of the pharynx, i.e. 20 filter pads in total (Figure 8.2). The lower filter pads are larger than the upper pads, and the posterior-most pads are the largest (Motta et al. 2010). Unlike the gill rakers of bony fishes, these are not filamentous and instead form a dorso-ventrally flattened reticulate mesh (Motta et al. 2010). This mesh makes their filtering pads different to the comb-like gill rakers of basking and megamouth sharks (Paig-Tran et al. 2013; Paig-Tran & Summers 2014). The mesh is supported by cartilaginous primary and secondary vanes and has a mesh diameter of ~1 mm (Motta et al. 2010). The individual filter pads are joined via a flexible ridge of connective tissue, meaning that the whole pharyngeal basket can contract and expand (Motta et al. 2010). This anatomical arrangement also means that water has to pass through the filter pads to exit the pharynx, and particles have to pass through the filter pads, or be swallowed, or be coughed back up and exit through the mouth. The filter pads of neonates have no denticles on their surface, but larger whale sharks do have denticles. These denticles are thought to protect the underlying mesh, similar to their function in megamouth sharks and some mobulids (Motta et al. 2010; Paig-Tran & Summers 2014). Water flows through the reticulated mesh on the filtering pads, passes between the primary and secondary vanes of the gill rakers and then flows over the gill tissue to exit through the gill slits (Figure 8.2).

During feeding, water enters the whale shark's mouth and pharynx and proceeds through the filter pads before passing over the gills for gas exchange, exiting through the external gill slits (Figure 8.2). But how exactly are the food particles separated from the water and ingested by the shark? Early observations of filter-feeding proposed two methods: *dead-end sieving*, in which particles larger than the mesh pores of the filter are trapped in the filter, and *hydrosol filtration*, in which a sticky mucus covering the filter traps particles that are even smaller than the mesh pores. Dead-end

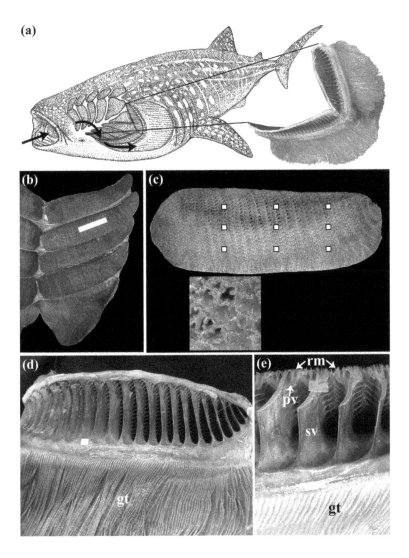

Figure 8.2 The anatomy of the filter-feeding apparatus of whale sharks from Motta et al. (2010) who dissected two male whale sharks with total lengths of 593 and 622 cm. (a) Schematic representation of a feeding whale shark, showing the approximate position of the filtering pads and the direction of waterflow through them. Detail drawing shows a lateral view of the vanes deep to the filtering mesh, as well as the primary gill filaments on the first branchial arch over which the water flows. (This illustration is, with permission, based upon a copyrighted illustration by Emily S. Damstra.) (b) Dorsal view of the lower filtering pads. The most posterior lower pad at the bottom is triangular in shape, and the lateral side of the pads is to the left. The lateral ridge between the lower and upper pads is visible towards the left. All other soft tissue has been removed. White ruler is 15 cm. (c) The upper second filtering pad where lateral is to the left and posterior towards the top. Upper pads are not as falcate on their medial margin as the lower pads. The squares (1 cm²) indicate areas sampled to measure mesh diameter, and the detail is a representative square area showing the irregularly shaped holes of the reticulated mesh. (d) External view of the first upper left pad with lateral margin towards the left. Note that the secondary vanes direct water laterally into the parabrachial chamber and over the gill tissue (gt) before it exits the pharyngeal slit (not shown). The white square is 1 cm². (e) Close-up of a section through the third left lower filtering pad showing the reticulated mesh (rm), primary vanes (pv), secondary vanes (sv), and gill tissue (gt).

sieving requires removal of the particles trapped in the filter, and hydrosol filtration requires a lot of mucus to concentrate the particles and transport them to the esophagus (Sanderson et al. 1996). Most of the filter-feeding elasmobranchs, including whale sharks, do not have mucus-producing goblet cells along the filter epithelium, nor epithelial cilia that could transport particles, but they do feed on prey smaller than the diameter of the mesh pores (Motta et al. 2010; Paig-Tran & Summers 2014). In the absence of these mechanisms, many fishes utilize a more recently discovered third option: *crossflow filtration* (Sanderson et al. 2001). Whale sharks, too, have been hypothesized to use biological crossflow filtration based on the morphology of their filter pads, and this is supported by field observations (Motta et al. 2010).

The premise of the crossflow filtration hypothesis is that water entering the pharynx of a whale shark will be directed parallel to the filter pads, not directly through them; hence, the name crossflow (Sanderson et al. 2001). It works in much the same way as a zooplankton net towed behind a boat, with the plankton accumulating in the end pocket of the net while the filtered water exits through the pores in the net. In whale sharks, particles, such as zooplankton, will get transported towards the back of the pharynx at high speeds, while most water exits through the gills via the pads. Tangential shearing clears the filter of particles and prevents clogging, as well as moving particles towards the esophagus, but this only works for particles larger than their filter pores (Sanderson et al. 2001; Paig-Tran & Summers 2014). The accumulated bolus of food particles in suspension at the back of the mouth can then be swallowed. The crossflow filtration hypothesis explains how whale sharks might avoid clogging or fouling of the filter pads, as most particles do not contact the filters themselves. This is especially important when feeding in high density prey patches, as whale sharks often do (Rohner et al. 2015b). When the filter pads do get clogged, whale sharks can "cough" to backflush the filter pads, removing any particles from the filter pad, and then expelling them through the mouth (Taylor 2007). Other challenges are not as clearly explained by crossflow filtration. For example, the mechanisms of keeping particles in suspension while in transit from the filter to the esophagus, where they can be swallowed, are not understood. Another major challenge for whale sharks is how to filter particles that are smaller than their mesh size. For example, they feed on fish spawn, which are smaller (<0.8 mm diameter) than the pores (1.2 mm diameter) of their filtering pads (Heyman et al. 2001; Motta et al. 2010). The mechanisms of how crossflow filtration could achieve this are not understood at present. It may be that whale sharks use a different mechanism altogether, such as the "ricochet separation" used by manta rays to separate particles smaller than the filter pores (Divi et al. 2018). It will take further investigation before the feeding mechanics of whale sharks are clarified, along with any ontogenetic changes that may affect the physical ability of whale sharks to target different prey as they grow.

8.5 FEEDING BEHAVIORS

Most filter-feeding elasmobranchs are passive ram filtration feeders, opening their mouth and swimming forwards to create water flow through the filtration device. Whale sharks differ from the others in their capacity to actively draw water into their mouth. This suction feeding has allowed whale sharks to develop a variety of novel behaviors. While slightly different names have been used by different authors (e.g. Nelson & Eckert 2007; Motta et al. 2010, Ketchum et al. 2013), the three main recognized feeding behaviors are (a) *passive feeding*, also called sub-surface passive or ram feeding; (b) *active feeding*, also called surface ram filter-feeding, surface active feeding or dynamic suction feeding; and (c) *stationary feeding*, also called vertical feeding or stationary/vertical suction feeding. A major benefit of active suction feeding is that whale sharks can capture more mobile prey and thus exploit a greater diversity of functional prey groups. However, it also means that they filter a smaller amount of water than the passive-filtering basking shark, and are thus more dependent on high prey densities (Compagno 1984). The megamouth shark is the only other filter-feeding

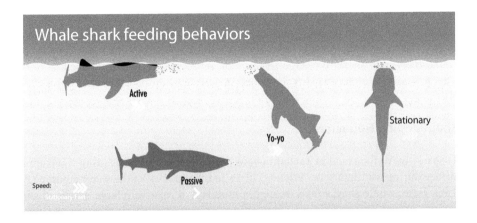

Figure 8.3 Whale shark feeding behaviors. The main feeding behaviors observed in surface waters, with speed indicated by arrows.

elasmobranch that has been suggested to use suction feeding (Compagno 1990). However, recent anatomical studies suggest that passive filter-feeding is more likely in that species (Nakaya et al. 2008; Tomita et al. 2011). The main feeding behaviors of whale sharks are illustrated in Figure 8.3 and discussed below.

8.5.1 Passive Feeding

Passive feeding involves the shark slowly swimming forwards with its mouth partially or fully open (Taylor 2007), and it can be used at depth or at the surface. Passive feeding is usually observed when prey densities are low (Nelson & Eckert 2007) or when prey items are small and immobile (e.g. fish eggs). Larger, more mobile prey can actively avoid the approaching gape of a whale shark. Other large filter-feeders that are restricted to passive filtering overcome this obstacle by swimming faster or concentrating mobile prey in different ways, such as collaborative cyclone feeding in manta rays (Law 2010).

8.5.2 Active Feeding

Active feeding involves the animal concurrently moving forward and rapidly expanding its mouth and gill basket to actively draw water into the pharynx. When active feeding happens at the surface, the shark's dorsal and upper caudal fins, parts of their dorsal surface, and the upper jaw are often visible above the water surface. The whale sharks take large gulps as they swim through the water, and this behavior is thus also sometimes called intermittent suction filter-feeding. Active feeding is usually observed when plankton densities are high or when highly mobile prey, such as small fishes or sergestids, are the target (Nelson & Eckert 2007; Rohner et al. 2015b).

8.5.3 Stationary Feeding

Whale sharks can also feed while remaining almost stationary by actively drawing water into their mouths. In this case, they take in big gulps of water, usually at the surface, without having to move forward at the same time. They often allow the tail to sink, assuming a vertical position in the water with their mouth near the surface and creating a vortex of water to suck prey in. This behavior is often seen in the afuera aggregation of the Mexican Caribbean, where these sharks are called "botella" or "botellita" as they resemble an empty bottle that is floating upright at the surface (R. de la Parra, pers. comm.). Stationary feeding is used when prey is concentrated in small patches (Heyman et al. 2001) and has been reported when whale sharks feed on small fishes (Duffy 2002). Sometimes, especially when small fishes are the target, prey appear to seek refuge around the whale shark to escape from other, more mobile predators such as tunas and seabirds, upon which the shark exploits the situation and sucks in the fishes themselves (see also Chapter 4).

8.5.4 Other Feeding Behaviors

Whale sharks at Mafia Island in Tanzania have been observed active feeding when approaching the surface during oscillatory or "bounce" dives in the top 10 m of the water column (Rohner et al. 2015b). Here, whale sharks feed on highly mobile and relatively large sergestid shrimps. This behavior, termed "yo-yo feeding", appears to concentrate fleeing prey against the water surface. Whale sharks can also deploy learned feeding behaviors in response to human activities. As part of a tourism industry, whale sharks in Oslob in the Philippines are fed "uyap" (a general term for a mixture of local shrimp species) from small boats where the sharks feed in a stationary vertical position, sometimes for hours at a time (Thomson et al. 2017). At Georgia Aquarium, USA, four sharks have each been trained using different colored ladles to follow a specific feeder boat to ensure correct daily feeding (Coco & Schreiber 2017). Whale sharks have also learned to target the lift-nets of fishers in Cenderawasih Bay, Indonesia, where the sharks feed on baitfish under the boats at night, get hand fed by the fishers, and sometimes suck small baitfish out of the holes in the nets (A. Dove pers. comm.). We have also observed whale sharks spitting out a large piece of plastic after inadvertently sucking it in during feeding, and seen a similar reaction to incidental jellyfish ingestion. Both infer an element of conscious selectivity to their feeding, at least for large items. However, small microplastics, which are similar to zooplankton in size, are not likely to be excluded during feeding, and can cause mechanical damage to the digestive tract and exposure to contamination and toxins (See Chapter 11; Germanov et al. 2018, 2019).

8.6 WHALE SHARK PREY SPECIES

Whale sharks are often seen feeding at the surface, and hence, there is no shortage of information on observed prey items in (coastal) habitats. As we will see in the next section, however, the focus on day-time surface observations has likely biased the early understanding of whale shark diet. We now know that they are likely to also feed on other prey at night and in different habitats (Figure 8.4). First though, let us focus on the direct observations of their prey.

8.6.1 Methods of Identifying Prey

When whale sharks are feeding, researchers can tow a plankton net through the water simultaneously to identify what the sharks are eating. To adequately capture what whale sharks are likely to ingest, a net with a mouth diameter of ~50 cm and a mesh size of ~200 μm is recommended (Motta et al. 2010; Rohner et al. 2015b). The net can be towed at the surface, or below it, depending on what part of the water column needs to be sampled. Alternatively, a dedicated neuston net with a square mouth can be used to quantify plankton at the water surface. To avoid the bias that can be introduced by predation among the collected zooplankton, samples are either immediately preserved in a ~5% formaldehyde solution or first stored on ice on the boat before preserving them. The *Zooplankton Methodology Manual* by Roger Harris et al. (2000) is a useful resource for further information on sampling protocols.

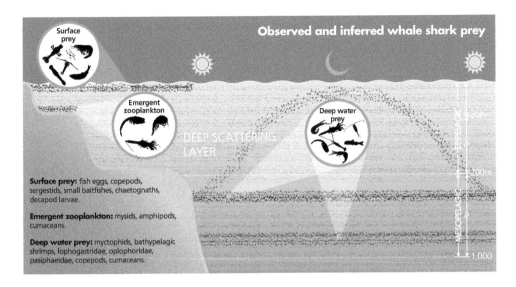

Figure 8.4   Observed and inferred whale shark prey. Whale shark prey items that have been observed or inferred grouped into functional units and illustrated in their habitat, with depth (m) on the right and a diel cycle indicated for the diel vertical migrators of the deep scattering layers. *Surface prey*: Fish eggs, copepods, sergestids, small baitfish, chaetognaths, decapod larvae. *Emergent zooplankton*: mysids, amphipods, cumaceans. *Deep water prey*: myctophids, bathypelagic shrimps and mysids, cumaceans, and copepods.

Plankton collection and analysis is a relatively simple and cost-effective method of studying whale shark diet. However, it is not possible to identify if whale sharks ingest all or only some of the available prey community, especially with regard to small or large mobile prey items. It is further largely limited to accessible areas, such as coastal surface waters. A more direct method of examining whale shark diet is stomach content analysis (Rohner et al. 2013). This method involves identifying prey items in a semi-digested state, often with only parts of a prey animal available, meaning that the ability to identify species is lower than in plankton tows. Alternatively, stomach contents can be genetically analyzed with a variety of primers to capture the sequences of prey items. This has not yet been done for whale sharks, but it could be useful: easily digestible prey items, such as ctenophores or jellyfishes, would be digested or broken down quickly in the stomach and thus avoid detection under the microscope. Genetic methods minimize such bias. The benefit of stomach content analysis is the certainty that all prey items identified were actually ingested. It also avoids the bias of only sampling day-time surface waters, as it is possible to find prey groups that were targeted in other habitats and at night (Rohner et al. 2013). The main drawback is that it is a lethal method. Opportunistic samples can be collected from infrequent natural strandings of whale sharks, which do not fulfil the requirement of a large sample size to obtain representative results (Pethybridge et al. 2011). Furthermore, stomach contents analysis can overestimate certain prey groups that digest slower than others (Richardson et al. 2000), and it only describes the most recent meals (Iverson et al. 2004). Last, it is possible to collect and genetically analyze fecal samples of whale sharks (Jarman & Wilson 2004; Meekan et al. 2009). Using these three direct methods, whale sharks have been shown to mostly feed on a variety of zooplankton.

8.6.2 Identified Prey Items

Whale sharks feed on different prey species in different locations. Some of the major prey groups include copepods, fish eggs, euphausiid krill, sergestid shrimps, and small schooling fishes (Table 8.1). Other prey are often also present in lower densities, such as amphipods or gastropods,

Table 8.1 Whale Shark Prey Identified in Stomach Contents, Plankton Tows Near Feeding Whale Sharks, and Direct Observations of Prey Ingestion. Bolded Text Denotes Major Prey Species.

Category	Prey Group	Species or Taxonomic Groups	Main Prey	Location	Rationale	Citations
Fishes	Fishes	*Trachurus picturatus* juveniles	Yes	Azores	Direct observation	Afonso et al. (2014)
		Sardines	Yes	Cuba, Seychelles	Direct observation, Stomach contents	Gudger (1941)
		Anchovies	Yes	Djibouti	Direct observation	Boldrocchi and Bettinetti (2019)
			Yes	Honduras, Indonesia, Madagascar, New Zealand	Direct observation	Duffy (2002), Fox et al. (2013), Rohner et al. (2013), Diamant et al. (2018)
		Encrasicholina devisi, Escualosa thoracata	Yes	India	Stomach contents	Rao (1986)
		Engraulis australis	Yes	New Zealand	Direct observation	Duffy (2002)
		Encrasicholina oligobranchus	Yes	Philippines (Palawan)	Direct observation	Araujo et al. (2019)
			No	Australia, Mozambique	Direct observation, Stomach contents	Rohner et al. (2013), Taylor (2007)
	Fish spawn	***Lutjanus cyanopterus, L. jocu***	Yes	Belize	Plankton tow	Heyman et al. (2001)
		Euthynnus alletteratus	Yes	Mexico (Yucatan), USA (Gulf of Mexico)	Plankton tow	Hoffmayer and Franks (2007), de la Parra Venegas et al. (2011)
		Euthynnus affinis	Yes	Qatar	Plankton tow	Robinson et al. (2013)
		Katsuwonus pelamis	Yes	St Helena	Plankton tow	Perry et al. (2020)
			No	Tanzania (Mafia Island)	Plankton tow	Rohner et al. (2015a)
		Caranx hippos	No	USA (Gulf of Mexico)	Plankton tow	Hoffmayer and Franks (2007)
	Other fishes	*Remora remora*	No	India	Stomach contents	Karbhari and Josekutty (1986)
		Saurida spp., Gobiidae	No	Indonesia	Stomach contents	van Kampen (1908)
		Fish larvae	No	Mexico (Yucatan)	Plankton tow	Motta et al. (2010)
Crustaceans	Amphipods		No	India	Stomach contents	Silas and Rajagopalan (1963)
			No	Mexico (Yucatan)	Plankton tow	Motta et al. (2010)
			No	Mozambique and South Africa	Stomach contents	Rohner et al. (2013)

(Continued)

Table 8.1 (Continued) Whale Shark Prey Identified in Stomach Contents, Plankton Tows Near Feeding Whale Sharks, and Direct Observations of Prey Ingestion. Bolded Text Denotes Major Prey Species.

Category	Prey Group	Species or Taxonomic Groups	Main Prey	Location	Rationale	Citations
		Hyperiidae	No	USA (Gulf of Mexico)	Plankton tow	Hoffmayer and Franks (2007)
	Cladocera		No	Tanzania (Mafia Island)	Plankton tow	Rohner et al. (2015a)
	Copepods	***Acartia* spp., *Centropages* spp., *Calanus* spp.**	Yes	Mexico (Baja California)	Plankton tow	Nelson and Eckert (2007), Clark and Nelson (1997)
		Calanoid, Cyclopoid	Yes	India	Stomach contents	Silas and Rajagopalan (1963), Rao (1986)
			No	Australia (Ningaloo)	Plankton tow	Taylor (2007)
		Calanoid	No	Mexico (Yucatan)	Plankton tow	Motta et al. (2010)
			No	Mozambique	Stomach contents	Rohner et al. (2013)
		Calanoid	No	Qatar	Plankton tow	Robinson et al. (2013)
		Acartia spp., *calanoid*, *Corycaeus* spp., *Oncea* spp., *Oithona* spp.	No	Tanzania (Mafia Island)	Plankton tow	Rohner et al. (2015a)
		Calanoid	No	USA (Gulf of Mexico)	Plankton tow	Hoffmayer and Franks (2007)
			No	Djibouti	Fecal microscopy	Rowat and Brooks (2012)
	Crab larvae	***Gecarcoidea natalis***	Yes	Australia (Christmas Island)	Fecal DNA	Meekan et al. (2009)
			Yes	Australia (Ningaloo)	Plankton tow	Taylor (2007)
			No	India	Stomach contents	Silas and Rajagopalan (1963)
			No	Mozambique and South Africa	Stomach contents	Rohner et al. (2013)
			No	Tanzania (Mafia Island)	Plankton tow	Rohner et al. (2015a)
			No	USA (Gulf of Mexico)	Plankton tow	Hoffmayer and Franks (2007)
			No	Djibouti	Fecal microscopy	Rowat and Brooks (2012)

(Continued)

Table 8.1 (Continued) Whale Shark Prey Identified in Stomach Contents, Plankton Tows Near Feeding Whale Sharks, and Direct Observations of Prey Ingestion. Bolded Text Denotes Major Prey Species.

Category	Prey Group	Species or Taxonomic Groups	Main Prey	Location	Rationale	Citations
	Euphausiid krill	***Pseudeuphausia latifrons***	Yes	Ningaloo Reef, Australia	Fecal DNA, direct observation	Jarman and Wilson (2004), Taylor (2007)
			No	Mexico (Baja California)	Plankton tow	Whitehead et al. (2020)
	Isopod		No	India	Stomach contents	Silas and Rajagopalan (1963)
			No	Mozambique and South Africa	Stomach contents	Rohner et al. (2013)
	Mysid shrimp		Yes	Mozambique and South Africa	Stomach contents	Rohner et al. (2013)
	Ostracod		No	India	Stomach contents	Silas and Rajagopalan (1963)
			No	Mozambique and South Africa	Stomach contents	Rohner et al. (2013)
	Sergestid shrimp	***Lucifer* spp.**	Yes	Mexico (Yucatan)	Plankton tow	Motta et al. (2010)
			Yes	Mozambique	Stomach contents	Rohner et al. (2013)
			Yes	Philippines (Leyte)	Direct observation	Araujo et al. (2017)
		Belzebub hanseni	Yes	Tanzania (Mafia Island)	Plankton tow	Rohner et al. (2015a)
		Lucifer spp.	No	India	Stomach contents	Silas and Rajagopalan (1963)
			No	USA (Gulf of Mexico)	Plankton tow	Hoffmayer and Franks (2007)
			N/A	Philippines (Oslob)	Provisioning	Schleimer et al. (2015)
	Stomatopod larvae		No	Australia (Ningaloo)	Direct observation	Taylor (2007)
			No	Djibouti	Direct observation	Rowat et al. (2007)
			No	Mozambique and South Africa	Stomach contents	Rohner et al. (2013)
Gelatinous plankton	Gelatinous plankton	Thimble jellyfish (*Linuche unguiculata*)	No	Belize	Direct observation	Heyman et al. (2001)
		Hydromedusae	No	Mexico (Baja California)	Plankton tow	Clark and Nelson (1997)
		Salps	No	Djibouti	Direct observation	Rowat et al. (2007)

(Continued)

Bolded Text Denotes Major Prey Species.

Category	Prey Group	Species or Taxonomic Groups	Main Prey	Location	Rationale	Citations
Other zooplankton	Appendicularian		No	Mexico (Baja California)	Plankton tow	Clark and Nelson (1997)
		Oikopleuridae	No	Tanzania (Mafia Island)	Plankton tow	Rohner et al. (2015a)
	Bivalves		No	India	Stomach contents	Silas and Rajagopalan (1963)
			No	Mozambique and South Africa	Stomach contents	Rohner et al. (2013)
	Chaetognaths		Yes	Djibouti	Direct observation, fecal microscopy	Rowat et al. (2007)
			No	Australia (Ningaloo)	Plankton tow	Taylor (2007)
		Sagitta spp.	No	India	Stomach contents	Silas and Rajagopalan (1963)
			No	Mexico (Yucatan)	Plankton tow	Motta et al. (2010)
			No	Mexico (Baja California)	Plankton tow	Clark and Nelson (1997), Nelson and Eckert (2007), Whitehead et al. (2020)
	Molluscs	Larvae	No	Mexico (Baja California)	Plankton tow	Nelson and Eckert (2007)
		Gastropod shells	No	Tanzania (Mafia Island)	Plankton tow	Rohner et al. (2015a)
		Pteropod	No	India	Stomach contents	Silas and Rajagopalan (1963)
		Pteropod	No	Mozambique and South Africa	Stomach contents	Rohner et al. (2013)
	Radiolaria		No	Qatar	Plankton tow	Robinson et al. (2013)
			No	India	Stomach contents	Silas and Rajagopalan (1963), Rao (1986); Karbhari and Josekutty (1986)
Marine algae			No	India	Stomach contents	McCann (1954)
			No	India	Stomach contents	Kaikini et al. (1959)
			No	Mozambique and South Africa	Stomach contents	Rohner et al. (2013)
			No	Seychelles	Stomach contents	Wright (1870)
			No	USA	Stomach contents	Gudger (1932)
	Cuttlefish	Sepia spp.	No	Indonesia	Stomach contents	van Kampen (1908)

In bold are the main prey present in high densities that appeared to be the target of feeding whale sharks, and other prey were present at low densities. Prey are grouped into larger categories, and groups within are ordered alphabetically, with additional detail given where available.

but these are unlikely to be the main target for feeding whale sharks (Table 8.1). One unifying theme among this long list of prey items is that the main prey often form dense patches. Individual fish can release millions of eggs during spawning events (Duarte and Alcaraz 1989), and larger zooplankton, such as sergestids and mysids, form dense feeding or spawning aggregations. In aquaria, whale sharks get fed a mix of euphausiids, fish, squid, and dietary supplements (Coco & Schreiber 2017).

8.6.3 Size and Biomass

Whale sharks clearly eat a wide range of prey. In some locations, whale sharks feed in plankton patches of large zooplankton species or even on larger baitfish, but in other locations, they feed on fish eggs which, individually, are very small. The specific prey species present in an area, and their individual size, are therefore not likely to be the main drivers of feeding behavior, so long as prey are big enough to be captured and small enough to be swallowed (Rohner et al. 2015b). Instead, the total amount of energy that can be gained, which is largely a function of biomass, likely has the biggest influence on whether whale sharks feed or not. In Mexico, plankton biomass was 2.5 times greater at whale shark feeding locations compared to areas nearby where they were not feeding (Motta et al. 2010), and in Tanzania, that difference was 10-fold (Rohner et al. 2015b).

Feeding is energetically costly for whale sharks because the large open mouth breaks their hydrodynamic profile and adds drag. The cost of feeding differs among the feeding behaviors, with active feeding on fast, mobile prey likely the most energetically expensive behavior. The cost of feeding also varies with depth; the difference in energy expenditure between swimming and feeding is over three times higher at the water surface than below the water (Cade et al. 2020). It makes sense then that whale sharks only feed in patches with a high enough prey biomass to ensure an energy net gain. Their prey biomass threshold to trigger feeding in Tanzania was estimated to be ~12 mg of dry plankton per cubic meter of filtered water (Rohner et al. 2015b), similar to the measured threshold for reef manta rays (11.2 mg m^{-3}) at a feeding area in eastern Australia (Armstrong et al. 2016) using the same study methods. Passive filter-feeding basking sharks had a mean feeding threshold of 620 mg of wet zooplankton per cubic meter of filtered water (Sims 1999), which was much lower than what Motta et al. (2010) recorded for feeding whale sharks in Mexico (4,500 mg m^{-3}). Comparatively, whale sharks are more dependent on high-density prey patches than are basking sharks, as predicted by Compagno (1984). This is largely explained by the differences in their anatomy and resulting feeding behaviors (active vs. passive feeding). Interestingly, manta rays had a similar feeding threshold to whale sharks, even though manta rays are limited to passive feeding. Anatomically, however, their filter structure is similar to that of whale sharks, with their robust, flattened filter pad, and quite different to the comb-like gill rakers of basking sharks (Paig-Tran & Summers 2014).

The caloric intake of two whale sharks was estimated in Mexico (Motta et al. 2010). A 6.6 m and a 4.4 m long whale shark were estimated to ingest 28,121 and 14,931 kJ per day, respectively. Using the length to mass conversion from Wyatt et al. (2019), these sharks weighed 1,716 and 571 kg, respectively, leading to a daily intake of 16.4 and 26.1 kJ kg^{-1}. Their respective base metabolic rate, as per Wyatt et al. (2019), is estimated at 16,900 and 6,708 kJ day^{-1}. The net gain in this back-of-the-envelope calculation is thus 11,221 kJ day^{-1} for the 6.6 m long whale shark and 8,223 kJ day^{-1} for the 4.4 m long individual, meaning that feeding in this aggregation is energetically highly beneficial. As discussed earlier, whale sharks need to make the most of food when available, as they are likely to undergo periods of fasting when away from feeding hotspots (Wyatt et al. 2019).

8.7 POTENTIAL PREY

Observations from feeding aggregations have shown that whale sharks feed on a variety of mostly zooplanktonic prey, but these are unlikely to be the only food sources for this species. First, whale sharks that form constellations are heavily biased towards juveniles – and mostly males – ranging from around 3–9 m in length (Chapter 7). Smaller juveniles (0.6–3 m long), females, and larger, mature sharks (>9 m in length; Chapter 2) rarely feed in the same areas as these juvenile males. Indeed, the evidence available to date suggests that large whale sharks are oceanic and seldom visit coastal areas (Hearn et al. 2016; Ramírez-Macías et al. 2017). Second, whale sharks may also feed at night, in deeper water, and away from the coast, where it is logistically more difficult to make observations. Researchers have thus developed indirect methods to investigate whale shark diet. These techniques show that prey other than coastal surface zooplankton are likely to be a significant component of their overall diet (Rohner et al. 2013).

8.7.1 Methods for Indirect Diet Analyses

The composition of an animal's tissue acts as a "biochemical tag" that keeps a record of their previous movements and feeding habits. In other words, you are literally both what you eat and where you eat it (De Niro & Epstein 1976). Analysis of these tissues reveals clues about the sharks' longer-term feeding habits, as opposed to the short-term nature of stomach content analysis and the bias towards day-time surface-feeding observations that we mentioned previously. The two main biochemical methods currently in use are stable isotope analysis and fatty acid analysis. Both methods allow for some reconstruction of retrospective longer-term diet and trophic ecology of whale sharks. The stable isotope ratios and the types and proportions of essential fatty acids in the whale sharks' tissues can be compared to potential prey, to other consumer species found in the same place, or to species from other areas and habitats in which we think whale sharks might be spending some of their time.

Stable isotope analysis is based on the fact that some elements naturally occur in the environment in multiple forms with different atomic weights (isotopes). Heavier isotopes are generally rarer than lighter ones. Heavy and light isotopes of an element behave differently in biochemical reactions, including during photosynthesis. In the marine environment, this means that there is a natural, and predictable, variation in the ratios of heavy to light isotopes in phytoplankton across each ocean, and also among different ecosystems (Hobson 1999). The isotopic composition of a particular habitat or area is determined by the primary producers (phytoplankton) that are present and is – somewhat – predictably transferred through the food web to the tissues of consumers (McMahon et al. 2013; Trueman et al. 2017). This creates a chemical record in the consumers' tissues of where and what the consumers ate (Graham et al. 2010). In marine systems, stable carbon isotopes (recorded as $\delta^{13}C$ values) are traditionally used to provide information on feeding location, as the ratios vary only slightly among organisms that share a habitat. Stable nitrogen isotopes (recorded as $\delta^{15}N$ values) increase predictably with trophic level, thereby allowing the relative position of the animal in the food web to be identified (Graham et al. 2010, Trueman et al. 2012).

Fatty acid signatures can be used to infer diet in a similar way to stable isotope ratios. Fatty acids are distinct types of lipids (fats), and their use as markers is based on phytoplankton having recognizable fatty acid profiles that are transferred to consumers with little alteration (Iverson 2009). The essential long-chain polyunsaturated fatty acids, in particular, are important as markers for diet studies of higher-level consumers, as higher trophic-level marine animals lack the ability to synthesize them. These fatty acids in the consumer must, therefore, originate from their diet (Iverson 2009), hence the term *essential*. The diversity and unique origins of these fatty acids can help reveal marine ecosystem structure and diet composition, although the evidence is not always clear-cut.

Something akin to a "detective approach" to scientific investigation is often required (Budge et al. 2006; Iverson, 2009).

Biochemical techniques have emerged as versatile and routine tools to complement traditional methods in food web ecology studies. While the seminal work underpinnning much of the research on biochemical tracers started ~60–70 years ago (Meyer et al. 2019), biochemical research in the marine environment, and on elasmobranchs in particular, is still in its infancy. This is especially true for whale sharks. As such, there are a few considerations and caveats that need to be addressed when interpreting biochemical data. We will address some of these in more detail later, as there are a lot of factors that may influence biochemical results in addition to dietary input, but be aware that we have only a rudimentary understanding of many important aspects.

8.7.2 Potential Prey Items

Carbon and nitrogen stable isotope analysis and fatty acid analysis techniques have both been used successfully to investigate different aspects of the trophic ecology of whale sharks (Borrell et al. 2011; Rohner et al. 2013; Couturier et al. 2013a; Marcus 2017; Marcus et al. 2019; Wyatt et al. 2019). Some initial ecological interpretations of biochemical data have been made. The first stable isotope study on whale sharks, from Gujarat on the northwest coast of India, calculated that whale sharks are secondary consumers, a result we might expect for a planktivore. Based on their comparable $\delta^{15}N$ values, whale sharks had a similar position in the food chain to other fishes that fed on surface plankton in the area. The study also suggested that larger sharks could be eating a larger proportion of small fishes and larger zooplankton, as larger sharks had more enriched $\delta^{13}C$ and $\delta^{15}N$ values than smaller individuals. More recently, a study conducted on both the Atlantic and Pacific coasts of Mexico indicated a similar ontogenetic change, suggesting that larger sharks have a different and more diverse diet than smaller individuals (Whitehead et al. 2020). Stable isotope studies at Ningaloo Reef in Western Australia (Marcus et al. 2019) and at Mafia Island in Tanzania (C. Prebble unpublished data) have also shown that the smaller, juvenile whale sharks commonly seen at coastal aggregation sites feed on similar prey and at a similar trophic level as other planktivores in their respective aggregation sites. Interestingly, most studies – excepting the Whitehead et al. (2020) research from Mexico – reported slight isotopic differences between male and female sharks. More enriched $\delta^{13}C$ and $\delta^{15}N$ values in males, and the stronger relationship between increasing $\delta^{15}N$ values as female sharks grow larger, suggest that the males and females feed in different places or on different prey. Few data are available from large, mature whale sharks of either sex, precluding a detailed examination of dietary changes from the juvenile to adult stage. On a broad spatial scale, stable isotope analyses have been used as a complementary method to confirm the long-term site fidelity of juvenile whale sharks to their respective coastal aggregation sites in the western Indian Ocean (Prebble et al. 2018). This suggested that, although whale sharks are highly mobile (Chapter 6), juveniles largely feed within a few 100 km from a known productive seasonal feeding site.

Compound-specific stable isotope analysis can provide a more detailed result than the bulk tissue analyses in the previous studies. This is where the nitrogen and carbon isotope ratios are measured for specific compounds, such as individual lipids or amino acids. In a recent study from Japan, the isotope ratios in specific amino acids from whale shark cartilage tissue found the lowest trophic position estimates presently published for the species (Wyatt et al. 2019). The authors emphasized that there is uncertainty in some of the assumed parameter values which influence this result. Assuming the relatively low trophic-level estimate derived from compound-specific stable isotope analysis is correct, the authors argued that a degree of herbivory in the diet of whale sharks may explain the result (Wyatt et al. 2019). While marine macroalgae have been found in the stomachs of stranded whale sharks (Rowat & Brooks 2012; Rohner et al. 2013), we suggest that this is likely a result of incidental ingestion. Field observations suggest that whale sharks do not target floating marine algae even when they are present in large quantities. It is also still unclear physiologically how marine algae could be filtered in large amounts (Paig-Tran et al. 2011), and to what degree digestion and assimilation could occur, though it is perhaps possible (Leigh et al. 2018).

Fatty acid analysis has been used as a tool to identify potential prey items for whale sharks in Mozambique, Australia, Mexico, and Tanzania (Rohner et al. 2013; Marcus et al. 2016; Cárdenas-Palomo et al. 2018, C. Prebble unpublished data). The fatty acid profiles of whale sharks differed among individuals, suggesting that each shark had a somewhat distinctive diet. The biggest surprise, though, was that all studies found the whale shark fatty acid profiles, though similar to each other, were mostly at odds with the local marine fauna, including known prey species, and most other chondrichthyan species (Meyer et al. 2019). Whale sharks differed from their observed prey in a similar way across these studies: the sharks had elevated levels of omega-6 essential fatty acids compared to their observed prey, day-time surface zooplankton, which were dominated by omega-3 fatty acids. Marine fishes generally have high levels of omega-3 fatty acids, which are well-known for their health benefits to humans (Calder 2017). Marine fishes typically do contain a small proportion of omaga-6 fatty acids, but high levels of these are more common in farmed terrestrial plants and terrestrial animal proteins (Simopoulos 2002).

Why, then, do whale sharks have more omega-6 than omega-3 essential fatty acids? The first inference was that whale sharks do not feed exclusively on day-time surface zooplankton (Couturier et al. 2013a). Initial evidence, based on the types and proportions of fatty acids in their tissues, suggested that their diet could include deep-sea zooplankton and "emergent" zooplankton, which live in the substrate during the day and only enter the water column at night (Rohner et al. 2013; Marcus 2017). Some potential prey in these two functional groups also had higher levels of omega-6 fatty acids in their profiles, though not in proportions as high as whale sharks. Behaviorally, though, this makes sense, and mysids found in stomach contents support that at least emergent zooplankton is part of their diet (Rohner et al. 2013). Whale sharks are also deep divers (see Chapter 6), so mesopelagic prey species are certainly within their reach. However, their fatty acid profiles differ from other deep-water sharks and fishes as well (Rohner et al. 2013), although whale sharks likely feed in a broader variety of habitats than most of these species.

Trophic guild has been linked to fatty acid profiles in other chondrichthyans (Meyer et al. 2019), so perhaps it's not surprising that so far, the only consumer with a similar fatty acid profile to whale sharks is the reef manta ray *Mobula alfredi* (Couturier et al. 2013a,b). They too have been observed feeding on day-time surface zooplankton, but are now also thought to routinely feed on other prey, similar to whale sharks. Whale sharks and manta rays are ecologically similar in that they are both large filter-feeders living in warm water, where food is scarce and patchily distributed. It thus makes sense that both species feed on different prey in a variety of habitats as and when they are available. Their wide range and different feeding habitats then make it possible that an important prey item that both are eating a lot of has so far been missed by researchers and is yet to be analyzed. However, none of the more obvious candidate groups, such as mesopelagic prey, emergent zooplankton, macroalgae, fishes, or gelatinous zooplankton matched the fatty acid signature of whale sharks (Rohner et al. 2013), making a non-dietary explanation more likely.

There may be a physiological explanation for the unusual fatty acid signature of whale sharks. The patchy distribution of their food in space and time means there are probably periods of fasting between feeding opportunities. In captive whale sharks, periods of fasting can affect the isotopic ratios in their tissues (Wyatt et al. 2019). Given the potentially more extreme "feast or famine" nature of whale shark feeding in the wild, this may also affect their fatty acid profiles. Evidence from other fishes supports this explanation; some fishes can actually change the allocation of fatty acids during times of environmental change, stress, and reproduction (Bell & Sargent 2003; Emata et al. 2003). Another potential explanation is that whale sharks preferentially store omega-6 fatty acids in their subdermal tissues, which is the tissue type samples by standard biopsy field techniques (i.e. Prebble et al. 2018). Elasmobranchs, in general, do store fatty acids differently in tissues depending on the tissue function (Remme et al. 2006). There are a number of different theories to explain the unique fatty acid profiles of whale shark subdermal tissue. It is difficult to evaluate these further at this point because our specific understanding of whale shark metabolism and bioprocessing of lipids is so limited. As a result, ecological interpretations of both stable isotope and fatty acid studies have necessarily been cautious and explicitly speculative so far.

8.7.3 Caveats of Diet Reconstruction from Whale Shark Tissue Samples

The retrospective reconstruction of whale shark diet from stable isotope and fatty acid analyses is complicated by the fact that many other variables can influence the observed data. As seen above, the feast and famine style feeding ecology affects biochemical signatures, and physiological pathways, such as preferential routing of specific compounds to specific tissues, also likely play a role. Additionally, there is a suite of biotic and abiotic factors, not related to feeding, that could also influence the observed data. These factors are poorly understood in whale sharks, and we thus have to be cautious with ecological interpretations of biochemical signature (Meyer et al. 2019). To complicate things even further, some of these factors, including metabolic rate, tissue turnover rate, and diet tissue discrimination factors, are tissue-specific, species-specific, and probably even individual-specific. These parameters are most accurately estimated through controlled feeding experiments, which require samples of multiple tissue types for analyses (Logan & Lutcavage 2010, Kim et al. 2012). Lethal studies are out of the question for whale sharks, which are an endangered species. Without solid assessments of these various metabolic factors, however, it is difficult to accurately interpret their isotopic values or fatty acid profiles. We thus have to estimate some of these factors through non-lethal methods or make inferences from other species. Researchers at the Okinawa Churaumi Aquarium in Japan have analyzed blood plasma and cartilage samples from live, captive whale sharks and determined isotopic tissue turnover rates of several months up to a few years (Wyatt et al. 2019). Muscle is normally used for biochemical diet studies in other species, but is difficult to collect in whale shark field studies due to their thick skin and underlying connective tissue. Subdermal connective tissue is more commonly sampled from whale sharks in the wild, as this is accessible via standard biopsy equipment. While not directly measured in the Okinawa study, the authors predicted muscle and subdermal tissue turnover times to be between those of blood and cartilage (i.e. from months to a couple of years). These slow turnover rates might be expected for such a large, slow-growing animal, but could perhaps be faster in wild whale sharks if they have greater metabolic needs. Body mass, body temperature, growth rate, and diet quality also affect turnover times, so these will slightly vary in each individual whale shark (Wyatt et al. 2019). No samples from the liver, where lipids are oxidized (Speers-Roesch 2006), have yet been examined for whale sharks. Liver samples are likely to be more indicative of diet than is subdermal connective tissue. Despite some unresolved complexities, these first tissue turnover rate estimates do give researchers an improved idea of the timeframe over which whale sharks' diet can be (provisionally) evaluated.

The subdermal tissue typically sampled in whale sharks is located on the shark's flank near the dorsal fin. The primary function of the tissue in this region is structural, and there are many layers of collagen fibers to support the dorsal fin within the tissue. The major amino acid in the type I collagens found in shark skin is glycine, a non-essential amino acid. This compound alone is typically ^{13}C-enriched compared to muscle tissue protein as a whole (McMahon et al. 2013). As in many other elasmobranchs, whale sharks are likely to store preferentially certain lipids or isotopes in different tissues, depending on tissue function and storage requirements (Beckmann et al. 2013; Gladyshev et al. 2018). Stable isotope analyses must take this probable variation into account when comparing values from other tissues to data derived from non-structural muscle tissue. There are similar considerations for fatty acid analysis of subdermal tissue, which has a much lower proportion of polyunsaturated fatty acids than is found in muscle and the liver (Meyer et al. 2019). Examining isotopic and lipid profile differences among tissue types, and gaining a deeper understanding of chondrichthyan metabolic processes in general, is important to build confidence in ecological inter-pretations of biochemical work on whale sharks. Further emphasis of the preliminary nature of the dietary inferences to date comes from a recent global meta-analysis on fatty acids in chondrichthy-ans (Meyer et al. 2019). This study highlighted trophic guild, phylogeny, habitat type, and water temperature as all being reliable predictors of lipid profiles (Meyer et al. 2019).

In this section, we have discussed some of the many considerations and complications that are present when interpreting biochemical data from whale sharks. Bulk stable isotope analysis of subdermal tissue, along with fatty acid analysis, is a useful tool for investigating broad ecological questions for whale sharks, and a valuable retrospective complement (i.e., projecting backwards in time) to other techniques such as photo-identification and tagging. Obtaining samples other than subdermal tissue has been the major challenge to date. Indeed, the few available stable isotope stud-ies on whale sharks all called for multi-tissue, multi-method projects to help resolve the trophic and movement behavior of whale sharks (Marcus et al. 2016, 2017, 2019; Prebble et al. 2018). However, opportunities for further studies do exist. Stranded, deceased whale sharks are rare, but present a realistic opportunity to collect muscle, liver, and subdermal tissue from the same animal. It is now also possible to collect blood samples from free-swimming whale sharks (Chapter 2), which are amenable to biochemical analyses. While the biochemical work to date on whale sharks may have raised more questions than it has answered, these novel tools present an exciting opportunity for minimally invasive studies of these enigmatic fish.

8.8 ECOLOGICAL ROLE OF WHALE SHARKS

Being largely zooplanktivorous, whale sharks inhabit a lower trophic position than large predatory sharks whose ecological role has been extensively studied (Hussey et al. 2014). Few species-specific data on the role of whale sharks in ocean ecosystems are available, and they are somewhat unique as a huge, ectothermic tropical filter-feeder, but some inferences can be made from ecologically comparable marine megafauna species such as baleen whales (Estes et al. 2016). Like these species, whale sharks are consumers, are a prey themselves, become detritus after death, and live sharks store and transport energy widely through marine systems.

Whale sharks consume large quantities of biomass (Motta et al. 2010; Rohner et al. 2015b) in warm-water ecosystems that are typically low in productivity (Lalli & Parsons 1997). This uptake of biomass may itself impact on trophic dynamics (Estes et al. 2016). Their large size and extensive vertical and horizontal movements (see Chapter 6) suggest that whale sharks are important vectors for transporting nutrients through ocean ecosystems (Estes et al. 2016). Deep-diving pelagic consumer species play a large role in cycling nutrients between the sur-face and the mesopelagic zone, between 200 and 1,000 m depth (Roman & McCarthy 2010; Saba & Steinberg 2012). As more species are tracked with depth sensors, many are found to travel between the surface and the mesopelagic zone (Ponganis 2015). Whale sharks dive extensively and likely play a role in opposing the downward flux of carbon to the deep ocean, while transferring energy and materials, including key limiting nutrients such as nitrogen, from the mesopelagic into the euphotic zone (0–200 m). As a result, phytoplankton growth is encouraged in areas that are resource-limited, perpetuating up the food chain to create a positive feedback system (Estes et al. 2016).

Whale sharks are also prey to large marine predators, especially while they are still relatively small (<3 m long) (Chapter 7). Larger whale sharks are unlikely to be predated often (Fertl et al. 1996), but natural deaths or those caused by non-extractive human activities both have the same ecological consequences. Whale shark carcasses sink to the seafloor, where they provide food and habitat for deep-sea organisms (Higgs et al. 2014; Estes et al. 2016). While whale shark car-casses have seldom been reported (Higgs et al. 2014), inferences can be made from better-studied whale carcasses or "whale falls". The progressive changes that whale carcasses go through during decomposition can continue for decades, during which time hundreds of different species are supported by this nutritionally rich, large magnitude, and infrequent resource (Goffredi et al. 2004; Fujiwara et al. 2007; Estes et al. 2016). As well as providing nutrition, whale falls are essential habitat for over 60 species of deep-sea macrofauna known only from whale falls, and numerous other species associated with cold seeps and hydrothermal vents have also been found on whale falls (Estes et al. 2016). Given the inherently short-lived nature of vent systems, whale – and whale shark – falls are likely to be important stepping-stones in the spatial ecology of the deep sea (Estes et al. 2016).

8.9 SUMMARY

Juvenile whale sharks often return to predictable feeding hotspots and will almost certainly exploit other ephemeral food sources while ranging more widely. Remembering the timing and location of seasonal, but reliable, feeding areas appears to play an important role in foraging, particularly in juvenile male whale sharks, coupled with the employment of a probabilistic movement strategy to find patchy prey fields in the open ocean. Whale sharks are thought to pick up chemosensory and auditory cues to locate their prey, and they might use their electromagnetic sense to determine prey density at short distances within a feeding patch. Whale sharks can use a variety of feeding modes to capture prey and select the appropriate method depending on the mobility and density of their prey. How exactly whale sharks filter and ingest their prey is not yet clear, but we know that they can feed on prey smaller than the pore diameter of their filters. Whale sharks feed on a variety of zooplankton and small fishes, as has been observed at the surface in coastal waters. Prey biomass is the main factor that drives the switch from foraging to feeding. Biochemical methods suggest that whale sharks are likely to also feed on other prey types, in other parts of their habitat, away from coastal surface waters. Whale sharks have an important ecological role to play as consumers, as prey, as detritus, and by storing and transporting energy from high- to low-productivity marine habitats. While some aspects of whale shark feeding and foraging have been studied in detail, such as day-time surface feeding events, many aspects of their feeding ecology remain poorly known. With advances in methodologies, such as novel animal-borne sensors, and compound-specific stable isotope analysis, our understanding of their feeding and foraging behaviors will no doubt expand in the near future.

REFERENCES

Afonso, P., N. McGinty, and M. Machete. 2014. Dynamics of whale shark occurrence at their fringe oceanic habitat. *PLoS One* 9:e102060.

Araujo, G., A. Agustines, B. Tracey, S. Snow, J. Labaja, and A. Ponzo. 2019. Photo-ID and telemetry highlight a global whale shark hotspot in Palawan, Philippines. *Scientific Reports* 9(1):17209.

Araujo, G., S. Snow, C. Lee So, J. Labaja, R. Murray, A. Colucci, et al. 2017. Population structure, residency patterns and movements of whale sharks in Southern Leyte, Philippines: Results from dedicated photo-ID and citizen science. *Aquatic Conservation: Marine and Freshwater Ecosystems* 27(1): 237–52.

Armstrong, A. O., A. J. Armstrong, F. R. A. Jaine, et al. 2016. Prey density threshold and tidal influence on reef manta ray foraging at an aggregation site on the Great Barrier Reef. *PLoS One* 11(5):e0153393.

Baumann-Pickering, S., D.M. Checkley, and D. A. Demer. 2016. Do top predators cue on sound production by mesopelagic prey? Ocean Sciences Meeting, 22 February 2016, New Orleans, Louisiana, USA.

Beckmann, C. L., J. G. Mitchell, L. Seuront, D. A. J. Stone, and C. Huveneers. 2013. Experimental evaluation of fatty acid profiles as a technique to determine dietary composition in benthic elasmobranchs. *Physiological and Biochemical Zoology* 86(2):266–78.

Bell, J. G., and J. R. Sargent. 2003. Arachidonic acid in aquaculture feeds: Current status and future opportunities. *Aquaculture* 218(1–4):491–99.

Berumen, M. L., C. D. Braun, J. E. M. Cochran, G. B. Skomal, and S. R. Thorrold. 2014. Movement patterns of juvenile whale sharks tagged at an aggregation site in the Red Sea. *PLoS One* 9(7):e103536.

Boldrocchi, G. and R. Bettinetti. 2019. Whale shark foraging on baitfish off Djibouti. *Marine Biodiversity* 49:2013–2016.

Borrell, A., L. Cardona, R. P. Kumarran, and A. Aguilar. 2011. Trophic ecology of elasmobranchs caught off Gujarat, India, as inferred from stable isotopes. *ICES Journal of Marine Science* 68(3):547–54.

Brooks, K., D. Rowat, S. J. Pierce, D. Jouannet, and M. Vely. 2010. Seeing spots: Photo-identification as a regional tool for whale shark identification. *Western Indian Ocean Journal of Marine Science* 9(2):185–94.

Brunnschweiler, J. M., H. Baensch, S. J. Pierce, and D. W. Sims. 2009. Deep-diving behaviour of a whale shark *Rhincodon typus* during long-distance movement in the Western Indian Ocean. *Journal of Fish Biology* 74(3):706–14.

Budge, S. M., S. J. Iverson, and H. N. Koopman. 2006. Studying trophic ecology in marine ecosystems using fatty acids: A primer on analysis and interpretation. *Marine Mammal Science* 22(4):759–801.

Cade, D.E., J. J. Levenson, R. Cooper, R. de la Parra, D. H. Webb, and A. D. Dove. 2020. Whale sharks increase swimming effort while filter feeding, but appear to maintain high foraging efficiencies. *Journal of Experimental Biology* 223(11):1–13.

Calder, P. C. 2017. New evidence that omega-3 fatty acids have a role in primary prevention of coronary heart disease. *Journal of Public Health and Emergency* 1(6):35–35.

Cárdenas-Palomo, N., E. Noreña-Barroso, J. Herrera-Silveira, F. Galván-Magaña, and A. Hacohen-Domené. 2018. Feeding habits of the whale shark (*Rhincodon typus*) inferred by fatty acid profiles in the northern Mexican Caribbean. *Environmental Biology of Fishes* 101:1599–1612.

Clark, E., and D. R. Nelson. 1997. Young whale sharks, *Rhincodon typus*, feeding on a copepod bloom near La Paz, Mexico. *Environmental Biology of Fishes* 50(1):63–73.

Coco, C., and C. Schreiber. 2017. Husbandry of whale sharks. In *The Elasmobranch Husbandry Manual II*, edited by M. Smith, D. Warmolts, D. Thoney, R. Hueter, M. Murray, and J. Ezcurra, 87–98.

Colman, J. G. 1997. A review of the biology and ecology of the whale shark. *Journal of Fish Biology* 51(6):1219–34.

Compagno, L. J. V. 1984. FAO species catalogue. Vol. 4. Sharks of the world. An annotated and illustrated catalogue of shark species known to date. Part 1. Hexanchiformes to Lamniformes. *FAO Fisheries Synopsis (125)* 4(1):249 p.

Compagno, L. J. V. 1990. Relationships of the megamouth shark, *Megachasma pelagios* (Lamniformes: Megachasmidae), with comments on its feeding habits. In *Elasmobranchs as Living Resources: Advances in the Biology, Ecology, Systematics, and the Status of the Fisheries, NOAA Tech. Rep. NMFS 90*, edited by H. L. Pratt Jr., S. H. Gruber, and T. Taniuchi, 357–79.

Couturier, L. I. E., C. A. Rohner, A. J. Richardson, et al. 2013a. Unusually high levels of Ω-6 polyunsaturated fatty acids in whale sharks and reef manta rays. *Lipids* 48(10):1029–34.

Couturier, L. I. E., C. A. Rohner, A. J. Richardson, et al. 2013b. Stable isotope and signature fatty acid analyses suggest reef manta rays feed on demersal zooplankton. *PLoS One* 8(10):e77152.

de la Parra Venegas, R., R. Hueter, J. González Cano, et al. 2011. An unprecedented aggregation of whale sharks, *Rhincodon typus*, in Mexican coastal waters of the Caribbean Sea. *PloS One* 6(4):e18994.

De Bose, J. L., and G. A. Nevitt. 2007. Investigating the association between pelagic fish and dimethylsulfoniopropionate in a natural coral reef system. *Marine and Freshwater Research* 58(8):720.

De Niro, M. J., and S. Epstein. 1976. You are what you eat (plus a few per mil): The carbon isotope cycle in food chains. *Geological Society of America Abstracts with Programs* 8:834–35.

Diamant, S., C. A. Rohner, J. J. Kiszka, et al. 2018. Movements and habitat use of satellite-tagged whale sharks off Western Madagascar. *Endangered Species Research* 36:49–58.

Divi, R. V., J. A. Strother, and E. W. M. Paig-Tran. 2018. Manta rays feed using ricochet separation, a novel nonclogging filtration mechanism. *Science Advances* 4(9):eaat9533.

Dove, A. D. M. 2015. Foraging and ingestive behaviors of whale sharks, *Rhincodon typus*, in response to chemical stimulus cues. *Biological Bulletin* 228:65–74.

Duarte, C. M., and M. Alcaraz. 1989. To produce many small or few large eggs: A size-independent reproductive tactic of fish. *Oecologia* 80(3):401–4.

Duffy, C. A. J. 2002. Distribution, seasonality, lengths, and feeding behaviour of whale sharks (*Rhincodon typus*) observed in New Zealand waters. *New Zealand Journal of Marine and Freshwater Research* 36:565–70.

Emata, A. C., H. Y. Ogata, E. S. Garibay, and H. Furuita. 2003. Advanced broodstock diets for the mangrove red snapper and a potential importance of arachidonic acid in eggs and fry. *Fish Physiology and Biochemistry* 28(1–4):489–91.

Estes, J. A., M. Heithaus, D. J. McCauley, D. B. Rasher, and B. Worm. 2016. Megafaunal impacts on structure and function of ocean ecosystems. *Annual Review of Environment and Resources* 41(1):83–116.

Fertl, D., A. Acevedo-Gutierrez, and F. L. Darby. 1996. A report of killer whales (*Orcinus orca*) feeding on a carcharhinid shark in Costa Rica. *Marine Mammal Science* 12(October):606–11.

Fox, S., I. Foisy, R. De La Parra Venegas, et al. 2013. Population structure and residency of whale sharks *Rhincodon typus* at Utila, Bay Islands, Honduras. *Journal of Fish Biology* 83(3):574–87.

Fujiwara, Y., M. Kawato, T. Yamamoto, et al. 2007. Three-year investigations into sperm whale-fall ecosystems in Japan. *Marine Ecology* 28(1):219–32.

Germanov, E. S., A. D. Marshall, L. Bejder, M. C. Fossi, and N. R. Loneragan. 2018. Microplastics: No small problem for filter-feeding megafauna. *Trends in Ecology & Evolution* 33(4):227–32.

Germanov, E. S., A. D. Marshall, I. G. Hendrawan, et al. 2019. Microplastics on the menu: Plastics pollute Indonesian manta ray and whale shark feeding grounds. *Frontiers in Marine Science* 6:679.

Gladyshev, M. I., N. N. Sushchik, A. P. Tolomeev, and Y. Y. Dgebuadze. 2018. Meta-analysis of factors associated with omega-3 fatty acid contents of wild fish. *Reviews in Fish Biology and Fisheries* 28(2):277–99.

Gleiss, A. C., B. Norman, and R. P. Wilson. 2011. Moved by that sinking feeling: Variable diving geometry underlies movement strategies in whale sharks. *Functional Ecology* 25(3):595–607.

Gleiss, A. C., S. Wright, N. Liebsch, R. P. Wilson, and B. Norman. 2013. Contrasting diel patterns in vertical movement and locomotor activity of whale sharks at Ningaloo Reef. *Marine Biology* 160(11):2981–92.

Goffredi, S. K., C. K. Paull, K. Fulton-Bennett, L. A. Hurtado, and R. C. Vrijenhoek. 2004. Unusual benthic fauna associated with a whale fall in Monterey Canyon, California. *Deep-Sea Research Part I: Oceanographic Research Papers* 51(10):1295–1306.

Graham, B. S., P. L. Koch, S. D. Newsome, K. W. Mcmahon, and D. Aurioles. 2010. Using isoscapes to trace the movements and foraging behavior of top predators in oceanic ecosystems. In *Isoscapes: Understanding Movement, Pattern, and Process on Earth Through Isotope Mapping*, edited by J. B. West, G. J. Bowen, T. E. Dawson, and K. P. Tu, 299–318. London, UK.

Graham, R. T., C. M. Roberts, and J. C. Smart. 2006. Diving behaviour of whale sharks in relation to a predictable food pulse. *Journal of the Royal Society Interface* 3(6):109–116.

Gudger, E. W. 1932. The fifth Florida whale shark. *Science* 75:412–413.

Gudger, E. W. 1941. The whale shark unafraid the greatest of the sharks, *Rhineodon typus*, fears not shark, man nor ship. *The American Naturalist* 75(761):550–68.

Harris, R., P. H. Wiebe, J. Lenz, H.-R. Skjoldal, and M. Huntley. 2000. *ICES Zooplankton Methodology Manual*. Academic Press, Cambridge, MA, 684 pp.

Haskell, P. J., A. McGowan, A. Westling, et al. 2015. Monitoring the effects of tourism on whale shark *Rhincodon typus* behaviour in Mozambique. *Oryx* 49(3):492–99.

Hearn, A. R., J. Green, M. H. Román, D. Acuña-Marrero, E. Espinoza, and A. P. Klimley. 2016. Adult female whale sharks make long-distance movements past Darwin Island (Galapagos, Ecuador) in the Eastern Tropical Pacific. *Marine Biology* 163(10):214.

Heyman, W. D., R. T. Graham, B. Kjerfve, and R. E. Johannes. 2001. Whale sharks *Rhincodon typus* aggregate to feed on fish spawn in Belize. *Marine Ecology Progress Series* 215:275–82.

Higgs, N. D., A. R. Gates, and D. O. B. Jones. 2014. Fish food in the deep sea: Revisiting the role of large food-falls. *PLoS One* 9(5):e96016.

Hobson, K. A. 1999. Tracing origins and migration of wildlife using stable isotopes: A review. *Oecologia* 120(3):314–26.

Hoffmayer, E. R., and J. S. Franks. 2007. Observations of a feeding aggregation of whale sharks, *Rhincodon typus*, in the North Central Gulf of Mexico. *Gulf and Caribbean Research* 19(2):69–73.

Hussey, N. E., M. A. Macneil, B. C. Mcmeans, et al. 2014. Rescaling the trophic structure of marine food webs. *Ecology Letters* 17:239–50.

Irigoien, X., T. A. Klevjer, A. Røstad, et al. 2014. Large mesopelagic fishes biomass and trophic efficiency in the open ocean. *Nature Communications* 5:3271.

Iverson, S. J. 2009. Tracing aquatic food webs using fatty acids: from qualitative indicators to quantitative determination. In *Lipids in Aquatic Ecosystems*, eds. M. Kainz, M. T. Brett, and M. T. Arts. New York, NY: Springer, 281–308.

Iverson, S. J., C. Field, W. D. Bowen, and W. Blanchard. 2004. Quantitative fatty acid signature analysis: A new method of estimating predator diets. *Ecological Monographs* 74(2):211–35.

Jarman, S. N., and S. G. Wilson. 2004. DNA-based species identification of krill consumed by whale sharks. *Journal of Fish Biology* 65(2):586–91.

Kaikini A. S., V. R. Rao, and M. H. Dhulkhed. 1959. A note on the whale shark *Rhincodon typus* Smith, stranded off Man-galore. *J Mar Biol Assoc India* 4:92–93.

Kajiura, S. M., and K. N. Holland. 2002. Electroreception in juvenile scalloped hammerhead and sandbar sharks. *The Journal of Experimental Biology* 205:3609–21.

Kalmijn, A. J. 1972. Bioelectric fields in sea water and the function of the ampullae of lorenzini in elasmo-branch fishes. La Jolla, CA, 23pp.

Karbhari, J. P., and C. J. Josekutty. 1986. On the largest whale shark *Rhincodon typus* Smith landed alive at Cuffe Parade, Bombay. *Marine Fisheries Information Service Technical and Extension Series* 66: 31–35.

Kempster, R. M., and S. P. Collin. 2011. Electrosensory pore distribution and feeding in the basking shark *Cetorhinus maximus* (Lamniformes: Cetorhinidae). *Aquatic Biology* 12(1):33–36.

Ketchum, J. T., F. Galván-Magaña, and A. P. Klimley. 2013. Segregation and foraging ecology of whale sharks, *Rhincodon typus*, in the southwestern Gulf of California. *Environmental Biology of Fishes* 96(6):779–95.

Kim, S. L., C. M. del Rio, D. Casper, and P. L. Koch. 2012. Isotopic incorporation rates for shark tissues from a long-term captive feeding study. *Journal of Experimental Biology* 215(14): 2495–500.

Lalli, C. M., and T. R. Parsons. 1997. *Biological Oceanography - An Introduction*. Second Edition. Elsevier, Amsterdam, 320pp.

Law, M. 2010. The twister of mantas. *Ocean Geographic* 11:58–69.

Leigh, S. C., Y. P. Papastamatiou, and D. P. German. 2018. Seagrass digestion by a notorious 'carnivore'. *Proceedings of the Royal Society B* 285:20181583.

Logan, J.M., and M. E. Lutcavage. 2010. Stable isotope dynamics in elasmobranch fishes. *Hydrobiologia* 644(1):231–44.

Marcus, L. 2017. Feeding ecology of whale sharks at Ningaloo Reef, Western Australia. PhD Thesis, University of Tasmania, 96pp.

Marcus, L., P. Virtue, H. R. Pethybridge, M. G. Meekan, M. Thums, and P. D. Nichols. 2016. Intraspecific variability in diet and implied foraging ranges of whale sharks at Ningaloo Reef, Western Australia, from signature fatty acid analysis. *Marine Ecology Progress Series* 554:115–28.

Marcus, L., P. Virtue, P. D. Nichols, M. G. Meekan, and H. Pethybridge. 2017. Effects of sample treatment on the analysis of stable isotopes of carbon and nitrogen in zooplankton, micronekton and a filter-feeding shark. *Marine Biology* 164(6):124.

Marcus L., P. Virtue, P. D. Nichols, L. C. Ferreira, H. Pethybridge, M. G. Meekan. 2019. Stable isotope analy-sis of dermis and the foraging behavior of whale sharks at Ningaloo Reef, Western Australia. *Frontiers in Marine Science* 6:546.

McCann, C. 1954. The whale shark *Rhineodon typus* Smith. *J Bombay Nat Hist Soc* 52:326–333.

McCauley, R. D. and D. H. Cato 2016. Evening choruses in the Perth Canyon and their potential link with Myctophidae fishes. *The Journal of the Acoustical Society of America* 140:2384.

McKinney, J. A., E. R. Hoffmayer, J. Holmberg, et al. 2017. Long-term assessment of whale shark popula-tion demography and connectivity using photo-identification in the Western Atlantic Ocean. *PLoS One* 12(8):e0180495.

McMahon, K. W., L. L. Hamady, and S. R. Thorrold. 2013. A review of ecogeochemistry approaches to esti-mating movements of marine animals. *Limnology and Oceanography* 58(2):697–714.

Meekan, M. G., S. N. Jarman, C. McLean, and M. B. Schultz. 2009. DNA evidence of whale sharks (*Rhincodon typus*) feeding on red crab (*Gecarcoidea natalis*) larvae at Christmas Island, Australia. *Marine and Freshwater Research* 60(6):607–9.

Meekan, M. G., B. M. Taylor, E. Lester, et al. 2020. Asymptotic growth of whale sharks suggests sex-specific life-history strategies. *Frontiers in Marine Science* 7:774.

Meyer L., H. Pethybridge, P. D. Nichols, C. Beckmann, and C. Huveneers. 2019. Abiotic and biotic driv-ers of fatty acid tracers in ecology: A global analysis of chondrichthyan profiles. *Functional Ecology* 33:1243–55.

Motta, P. J., M. Maslanka, R. E. Hueter, et al. 2010. Feeding anatomy, filter-feeding rate, and diet of whale sharks *Rhincodon typus* during surface ram filter feeding off the Yucatan Peninsula, Mexico. *Zoology* 113(4):199–212.

Muller, M. 1999. Size limitations in semicircular duct systems. *Journal of Theoretical Biology* 198:405–37.

Nakamura, I., R. Matsumoto, and K. Sato. 2020. Body temperature stability observed in the whale sharks, the world's largest fish. *The Journal of Experimental Biology*, 223:jeb210286.

Nakaya, K., R. Matsumoto, and K. Suda. 2008. Feeding strategy of the megamouth shark *Megachasma pelagios* (Lamniformes: Megachasmidae). *Journal of Fish Biology* 73(1):17–34.

Nelson, J. D., and S. A. Eckert. 2007. Foraging ecology of whale sharks (*Rhincodon typus*) within Bahía de Los Angeles, Baja California Norte, México. *Fisheries Research* 84(1):47–64.

Nevitt, G. A. 2008. Sensory ecology on the high seas: The odor world of the procellariiform seabirds. *Journal of Experimental Biology* 211(11):1706–13.

Norman, B. M., J. A. Holmberg, Z. Arzoumanian, et al. 2017. Undersea constellations: The global biology of an endangered marine megavertebrate further informed through citizen science. *BioScience* 67(12):1029–43.

Norman, B. M., and D. L. Morgan. 2016. The return of 'Stumpy' the whale shark: Two decades and counting. *Frontiers in Ecology and the Environment* 14:449–50.

Paig-Tran, E. W. M., J. J. Bizzarro, J. A. Strother, and A. P. Summers. 2011. Bottles as models: Predicting the effects of varying swimming speed and morphology on size selectivity and filtering efficiency in fishes. *The Journal of Experimental Biology* 214(10):1643–54.

Paig-Tran, E. W. M., T. Kleinteich, and A. P. Summers. 2013. The filter pads and filtration mechanisms of the devil rays: Variation at macro and microscopic scales. *Journal of Morphology* 274(9):1026–43.

Paig-Tran, E. W. M., and A. P. Summers. 2014. Comparison of the structure and composition of the branchial filters in suspension feeding elasmobranchs. *Anatomical Record* 297(4):701–15.

Papastamatiou, Y. P., and C. G. Lowe 2005. Variations in gastric acid secretion during periods of fasting between two species of shark. *Comparative Biochemistry and Physiology* 141(2):210–14.

Perry, C. T., E. Clingham, D. H. Webb, R. de la Parra, S. J. Pierce, A. Beard, et al. 2020. St. Helena: An important reproductive habitat for whale sharks (*Rhincodon typus*) in the South Central Atlantic. *Frontiers in Marine Science* 7:899.

Pethybridge, H., R. K. Daley, and P. D. Nichols. 2011. Diet of demersal sharks and chimaeras inferred by fatty acid profiles and stomach content analysis. *Journal of Experimental Marine Biology and Ecology* 409(1–2):290–99.

Ponganis, P. J. 2015. Diving behavior. In *Diving Physiology of Marine Mammals and Aeabirds*, ed. Ponganis, P. J., Cambridge University Press, Cambridge, UK, 321–27.

Prebble, C. E. M., C. A. Rohner, S. J. Pierce, et al. 2018. Limited latitudinal ranging of juvenile whale sharks in the Western Indian Ocean suggests the existence of regional management units. *Marine Ecology Progress Series* 601:167–83.

Ramírez-Macías, D., N. Queiroz, S. J. Pierce, N. E. Humphries, D. W. Sims, and J. M. Brunnschweiler. 2017. Oceanic adults, coastal juveniles: Tracking the habitat use of whale sharks off the Pacific Coast of Mexico. *PeerJ* 5:e3271.

Rao, K. S. 1986. On the capture of whale sharks off Dakshina Kannada Coast. *Marine Fisheries Information Service* 66:22–29.

Remme J. F., W. E. Larssen, I. Bruheim, P. C. Sæbø, A. Sæbø, and I. S. Stoknes. 2006. Lipid content and fatty acid distribution in tissues from Portuguese dogfish, leafscale gulper shark and black dogfish. *Comparative Biochemistry and Physiology - B Biochemistry and Molecular Biology* 143:459–64.

Richardson, A. J., C. Lamberts, G. Isaacs, C. L. Moloney, and M. J. Gibbons. 2000. Length-weight relationships of some important forage crustaceans from South Africa. *The ICLARM Quarterly* 23(2):29–33.

Robinson, D. P., M. Y. Jaidah, S. S. Bach, et al. 2017. Some like it hot: Repeat migration and residency of whale sharks within an extreme natural environment. *PLoS One* 12(9):1–23.

Robinson, D. P., M. Y. Jaidah, R. W. Jabado, et al. 2013. Whale sharks, *Rhincodon typus*, aggregate around offshore platforms in Qatari waters of the Arabian Gulf to feed on fish spawn. *PLoS One* 8(3):e58255.

Rohner, C. A., A. J. Armstrong, S. J. Pierce, et al. 2015a. Whale sharks target dense prey patches of sergestid shrimp off Tanzania. *Journal of Plankton Research* 37(2):352–62.

Rohner, C. A., K. B. Burgess, J. M. Rambahiniarison, J. D. Stewart, A. Ponzo, and A. J. Richardson. 2017. Mobulid rays feed on euphausiids in the Bohol Sea. *Royal Society Open Science* 4:161060.

Rohner, C. A., L. I. E. Couturier, A. J. Richardson, et al. 2013. Diet of whale sharks *Rhincodon typus* inferred from stomach content and signature fatty acid analyses. *Marine Ecology Progress Series* 493:219–35.

Rohner, C. A., A. J. Richardson, F. R.A. Jaine, et al. 2018. Satellite tagging highlights the importance of productive Mozambican coastal waters to the ecology and conservation of whale sharks. *PeerJ* 6:e4161.

Rohner, C. A., A. J. Richardson, C. E. M. Prebble, et al. 2015b. Laser photogrammetry improves size and demographic estimates for whale sharks. *PeerJ* 3:e886.

Roman, J., and J. J. McCarthy. 2010. The whale pump: Marine mammals enhance primary productivity in a coastal basin. *PLoS One* 5(10):e13255.

Rowat, D., M. G. Meekan, U. Engelhardt, B. Pardigon, and M. Vely. 2007. Aggregations of juvenile whale sharks (*Rhincodon typus*) in the Gulf of Tadjoura, Djibouti. *Environmental Biology of Fishes* 80(4):465–72.

Rowat, D., and K. S. Brooks. 2012. A review of the biology, fisheries and conservation of the whale shark *Rhincodon typus*. *Journal of Fish Biology* 80(5):1019–56.

Ryan, J. P., J. R. Green, E. Espinoza, and A. R. Hearn. 2017. Association of whale sharks (*Rhincodon typus*) with thermo-biological frontal systems of the eastern tropical Pacific. *PloS One*, 13(4):e0196443.

Saba, G. K., and D. K. Steinberg. 2012. Abundance, composition, and sinking rates of fish fecal pellets in the Santa Barbara Channel. *Scientific Reports* 2:1–6.

San Martin, E., R. P. Harris, and X. Irigoien. 2006. Latitudinal variation in plankton size spectra in the Atlantic Ocean. *Deep-Sea Research Part II: Topical Studies in Oceanography* 53(14–16):1560–72.

Sanderson, S. L., A. Y. Cheer, J. S. Goodrich, J. D. Graziano, and W. T. Callan. 2001. Crossflow filtration in suspension-feeding fishes. *Nature* 412:439–441.

Sanderson, S. L., M. Stebar, K. Ackermann, S. Jones, I. Batjakas, and L. Kaufman. 1996. Mucus entrapment of particles by a suspension-feeding Tilapia (Pisces: Cichlidae). *The Journal of Experimental Biology* 199:1743–56.

Sanderson, S. L., and R. Wassersug. 1993. Convergent and alternative designs for vertebrate suspension feeding. In *The Skull*, eds. J. Hanken and B. Hall, Vol. 3, 37–112. University of Chicago Press, Chicago, IL.

Sarmiento, J. L., and N. Gruber. 2006. *Ocean Biogeochemical Dynamics*, Vol. 1, 503pp., Princetown University Press, Princetown, NJ.

Savoca, M. S., and G. A. Nevitt. 2014. Evidence that dimethyl sulfide facilitates a tritrophic mutualism between marine primary producers and top predators. *Proceedings of the National Academy of Sciences* 111(11):4157–61.

Schleimer, A., G. Araujo, L. Penketh, A. Heath, E. McCoy, J. Labaja. 2015. Learning from a provisioning site: Code of conduct compliance and behaviour of whale sharks in Oslob, Cebu, Philippines." *PeerJ* 3:e1452.

Schluessel, V. 2014. Who would have thought that 'Jaws' also has brains? Cognitive functions in elasmobranchs. *Animal Cognition* 18(1):19–37.

Silas, E. G., and M. S. Rajagopalan. 1963. On a recent capture of a whale shark (*Rhincodon typus* Smith) at Tuticorin, with a note on information to be obtained on whale sharks from Indian Waters. *Journal of the Marine Biological Association of India* 5(1):153–57.

Simopoulos, A. P. 2002. The importance of the ratio of omega-6/omega-3 essential fatty acids. *Biomedicine and Pharmacotherapy* 56(8):365–79.

Sims, D. W. 1999. Threshold foraging behaviour of basking sharks on zooplankton: Life on an energetic knife-edge? *Proceedings of the Royal Society B: Biological Sciences* 266(1427):1437–43.

Sims, D. W., E. J. Southall, N. E. Humphries, et al. 2008. Scaling laws of marine predator search behaviour. *Nature* 451(7182):1098–1102.

Sims, D. W., M. J. Witt, A. J. Richardson, E. J. Southall, and J. D. Metcalfe. 2006. Encounter success of free-ranging marine predator movements across a dynamic prey landscape. *Proceedings of the Royal Society B: Biological Sciences* 273(1591):1195–1201.

Sims, D. W. 2008. Sieving a living: A review of the biology, ecology and conservation status of the plankton-feeding basking shark *Cetorhinus maximus*. *Advances in Marine Biology*, 54:171–220.

Speers-Roesch, B. 2006. Metabolic organization of freshwater, euryhaline, and marine elasmobranchs: Implications for the evolution of energy metabolism in sharks and rays. *Journal of Experimental Biology* 209(13):2495–2508.

Taylor, J. G. 2007. Ram filter-feeding and nocturnal feeding of whale sharks (*Rhincodon typus*) at Ningaloo Reef, Western Australia. *Fisheries Research* 84(1):65–70.

Thomson, J. A., G. Araujo, J. Labaja, E. Mccoy, R. Murray, and A. Ponzo. 2017. Feeding the world's largest fish: Highly variable whale shark residency patterns at a provisioning site in the Philippines. *Royal Society Open Science*, 4:170394.

Tomita, T., K. Sato, K. Suda, J. Kawauchi, and K. Nakaya. 2011. Feeding of the megamouth shark (Pisces: Lamniformes: Megachasmidae) predicted by its hyoid arch: A biomechanical approach. *Journal of Morphology* 272(5):513–24.

Tomita, T., K. Murakumo, S. Komoto, et al. 2020. Armored eyes of the whale shark. *PloS One* 15(6):e0235342.

Trueman, C. N., K. M. MacKenzie, and M. R. Palmer. 2012. Identifying migrations in marine fishes through stable-isotope analysis. *Journal of Fish Biology* 81(2):826–47.

Trueman, C. N., K. M. MacKenzie, and K. St John Glew. 2017. Stable isotope-based location in a shelf sea setting: Accuracy and precision are comparable to light-based location methods. *Methods in Ecology and Evolution* 8(2): 232–40.

van Kampen, P. N. 1908. Kurze Notizen über Fische des Java-Meeres. *Natuurkundig tijdschrift voor Nederlandsch Indië* 67:124.

Whitehead, D. A., D. Murillo-Cisneros, F. R. Elorriaga-Verplancken, et al. 2020. Stable isotope assessment of whale sharks across two ocean basins: Gulf of California and the Mexican Caribbean. *Journal of Experimental Marine Biology and Ecology* 527: 151359.

Wilkens, L. A., and M. H. Hofmann. 2007. The paddlefish rostrum as an electrosensory organ: A novel adaptation for plankton feeding. *BioScience* 57(5):399–407.

Wright, E. P. 1870. Six months at Seychelles. *Spicilegia Zoologica*, 1:64–65.

Wyatt, A. S. J., R. Matsumoto, Y. Chikaraishi, et al. 2019. Enhancing insights into foraging specialization in the world's largest fish using a multi-tissue, multi-isotope approach. *Ecological Monographs* 89(1):e01339.

Yopak, K. E., and L. R. Frank. 2009. Brain size and brain organization of the whale shark, *Rhincodon typus*, using magnetic resonance imaging. *Brain, Behavior and Evolution* 74(2):121–42.

Zerbini, A. N., A. Andriolo, M. P. Heide-Jørgensen, et al. 2006. Satellite-monitored movements of humpback whales *Megaptera novaeangliae* in the Southwest Atlantic Ocean. *Marine Ecology Progress Series* 313: 295–304.

CHAPTER 9

Lessons from Care of Whale Sharks in Public Aquariums

Alistair D. M. Dove
Georgia Aquarium

Rui Matsumoto
Okinawa Churaumi Aquarium

Christian Schreiber
Georgia Aquarium

Kiyomi Murakumo
Okinawa Churaumi Aquarium

Christopher Coco
Georgia Aquarium

Makio Yanagisawa
Marine Palace Co. Ltd.

Tonya Clauss and Lisa Hoopes
Georgia Aquarium

Keiichi Sato
Okinawa Churshima Research Foundation

CONTENTS

9.1 INTRODUCTION

The majority of field research with whale sharks is conducted during frustratingly brief encounters in open water. Effective whale shark field research is predicated on the reliable presence of animals in accessible places at known times, so that these rendezvous can be planned for and their benefits maximized. Researchers typically snorkel or SCUBA dive alongside free-ranging animals, and all necessary data collection, tagging and sampling has to happen during those fleeting moments. This can and has significantly limited the amount and type of science that can be directed to this species in the field. For example, the difficulty of observing whale sharks feeding at depth and at night forces researchers to rely on indirect methods such as stable isotope ratios and fatty acid analysis in tissue biopsies to determine when and where whale sharks feed, and on which prey (Borrell et al. 2011, Rohner et al. 2013, Marcus et al. 2016, Prebble et al. 2018, Marcus et al. 2019). Even simple questions, such as how long is a whale shark, can be hard to answer in the field. Techniques such as laser photogrammetry have been employed with some success (Rohner et al. 2011, Jeffreys et al. 2013), but many researchers still rely on visual estimates, which are notoriously inaccurate and imprecise, and subject to systemic biases in which human observers tend to overestimate the size of small animals and underestimate the size of large ones (Perry et al. 2018).

There are, however, a handful of large public aquariums in the world that have the capacity to safely and successfully display and study whale sharks (Dove et al. 2010b, Coco and Schreiber 2017, Ito et al. 2017, Matsumoto et al. 2017). These facilities provide extraordinary opportunities for the general public to experience, learn about and connect to whale sharks, which is a species that the overwhelming majority of people will never experience in its natural setting. In addition to the educational potential, the maintenance of whale sharks in aquariums provides opportunities for scientists to conduct research of a sort that is simply not possible in the field. The types of scientific questions that can be explored this way are different from, and complementary to, those addressed in open water. Rather than focusing on temporal snapshots and population-level studies such as movement ecology, demographics or conservation genetics, studies in aquariums can focus more on the individual animal, following behavior, growth, nutrition, maturation, health and development over extended periods of time.

There are definitely limitations and caveats on research with whale sharks in public aquariums. First, just like a laboratory, the aquarium setting is artificial and obviously different from the natural setting, a fact which must be kept in mind when interpreting the results of these sorts of studies. Second is low sample size. At the time of writing, Okinawa Churaumi Aquarium (Japan) houses two whale sharks and Georgia Aquarium (USA) houses four, and the number of whale sharks in other facilities worldwide is probably less than a dozen. Population-level studies are simply not possible in aquariums because the "population" in question is relatively tiny. Again, this leads more to focal animal studies involving longer time series, which is very difficult to do in the field. Third among the limitations of research in public aquariums is interference from other species. An aquarium is not a controlled environment in the laboratory sense, and the impact of cohabiting animals and human activities in and around the habitat must be considered when interpreting findings, particularly with behavioral studies. By design, public aquariums also maintain animal densities that are higher than those found in nature, on a biomass per unit volume basis, which may have an effect as well. Finally, there are ethical and operational limits on the sort of manipulations that can be carried out for research using whale sharks in a public aquarium setting. For the most part, the science must be conducted in a hands-off fashion or as part of required animal husbandry or veterinary care procedures. Invasive procedures and other noxious stimuli for the sole purposes of research are generally not desirable, since animal welfare has to remain the highest priority. Usually, these limitations are enshrined in the type of government permits under which the animals are being displayed.

Despite the limitations described above, the opportunities afforded by display of whale sharks have provided for some unique scientific studies. In this chapter, we present some of these scientific lessons learned at two major non-profit aquariums that have extensive track records successfully caring for, studying and displaying whale sharks: the Okinawa Churaumi Aquarium in Okinawa, Japan (OCA), and Georgia Aquarium in Atlanta, USA (GAI). Specifically, we address techniques used in the acquisition and disposition of whale sharks, husbandry requirements, behavior in the aquarium and its management, veterinary care, nutrition and feeding, and growth and maturation/reproduction.

9.2 A BRIEF HISTORY OF WHALE SHARKS IN PUBLIC AQUARIUMS

Japanese public aquariums were the pioneers of the captive care and public display of whale sharks. The first captive whale shark was maintained in a net-pen in Japan in 1934 by the Mito Aquarium and lived for 4 months (Nishida 2001). Okinawa Aquarium has housed whale sharks intermittently since 1980 (as Ocean Expo Park) and continuously since the aquarium renovated and reopened as the Okinawa Churaumi Aquarium in 2002. Since 1993, Osaka Kaiyukan Aquarium has housed whale sharks, as have a number of other facilities in Japan including Notojima Aquarium, Kagoshima City Aquarium, and Yokohama Hakkeijima Aquarium. The Kenting National Museum of Marine Biology and Aquarium in Taiwan has held whale sharks in the past but not at the time of writing. Georgia Aquarium is the only aquarium in the Western Hemisphere to house this species and has done so since 2005. Several newer aquariums in China also have whale sharks, including the Chimelong Ocean Kingdom in Hengqin, Cube Aquarium in Chengdu, and Yantai Haicheng Aquarium; the Fenjiezhou Island Ocean Park displays whale sharks in a sea pen setting. All of the nine largest public aquariums in the world have housed whale sharks at some point; they are (in decreasing order of size): Chimelong Ocean Kingdom (China), Georgia Aquarium (USA), S.E.A. Aquarium (Singapore), Okinawa Churaumi Aquarium (Japan), Cube Aquarium (China), Osaka Kaiyukan Aquarium (Japan), Kenting NMMB Aquarium (Taiwan), Dubai Aquarium and Underwater Zoo (UAE), and Aqua Planet Jeju Aquarium (South Korea).

9.3 HABITAT DESIGN

It is not a surprise that the successful care and public display of whale sharks requires the largest aquariums in existence. OCA was the largest aquarium in the world by water volume until it was surpassed by GAI in 2005, and both of these have since been surpassed by Chimelong Ocean Kingdom in China, which is currently home to the largest single aquarium exhibit in the world, which also houses several whale sharks. But size alone is not sufficient to provide for the long-term care of whale sharks; other key considerations include the cohabiting animals, habitat shape, depth, décor and substrate, and water quality parameters, as well as circulation patterns, lighting, filtration, and the availability of expertly trained husbandry and veterinary staff (Dove et al. 2010b, Matsumoto et al. 2017, Schreiber and Coco 2017).

The Ocean Voyager exhibit at GAI holds approximately 24 million liters (6.3 million US gallons) and is illustrative of the sort of requirements demanded by this species. The main display habitat is approximately dumbbell-shaped, 80 m long by 24–37 m wide and with depth ranging from 9 to 11 m. The dumbbell shape of the habitat is important for maintaining appropriate swimming patterns, not just in whale sharks but among all elasmobranchs; the presence of a narrow point encourages animals to turn both left and right as they follow the perimeter of the habitat, which ensures healthy use of the musculature on both sides of the body (Figure 9.1).

The substrate in the Ocean Voyager habitat is a mix of calcareous and siliceous sand, punctuated by formed concrete rockwork that comprises elements of visual interest and for the strategic

Figure 9.1   The Ocean Voyager habitat at Georgia Aquarium. (© Georgia Aquarium.)

obscuring of viewing windows, such than any given window cannot be seen from any other window. The walls are vertical, epoxy-coated, reinforced concrete. Public viewing occurs through nine poly-laminated acrylic windows including several domes, a tunnel, and a single large 23 m wide by 10 m high viewing window at the deep end. Lighting is provided through a combination of natural light from an overhead skylight and high-intensity metal halide light fixtures in a range of color temperatures.

The whole system, including the life support components and a 300,000 L holding pool, contains approximately 24 million L of synthetic seawater and operates as a closed recirculating system. Filtration begins with removal of water from the exhibit by surface skimmers along the perimeter and from grated drains scattered across the bottom. Water from these sources is combined in a hydrodynamic sump and then divided evenly between two process treatment paths, one of which leads to 34 quad-cone foam fractionating protein skimmers, and the other to 72 high-rate sand filters, collectively impelled by 70 centrifugal 25 hp pumps. After primary treatment by protein skimming or sand filtration, depending on the path taken, a 30% side-stream is subjected to ozone disinfection and clarification in 12 telescoped contact towers, within which ozone dosing is controlled by an array of ORP (oxidation–reduction potential) probes and rotameters that dose ozone through Venturi injectors. A manifold of six sulfur-based autotrophic denitrification bioreactors is used on a small side-stream (<1%) to gradually reduce dissolved nitrate to gaseous nitrogen (Burns et al. 2018, Dove and Hall 2018). Appropriate gas balance is maintained by a large nitrification/de-aeration tower (1.5 million L water capacity) containing a 2.5 m thick layer of convoluted plastic media with a total surface area of 107,000 m^2, over which the water cascades, facilitating gas exchange. After dissolved gas correction, water is gravity-fed back into the exhibit through a range of subsurface return nozzles concealed within the rockwork décor, some of which have rotating grates to vary haphazardly the circulation patterns. Total system flow is 484,000 L/min, which results in a system turnover of around 1 hour. A recovery system captures backwash water from the sand filters, treats it and then returns it to the system, reducing the need for addition of new water. The habitat requires the addition of less than 1% of new water per month, compared to 20%–30% of new water per *day* for flow through systems.

For comparison, the Kuroshio Current habitat at OCA measures 35 m long by 27 m wide, by 10 m deep, with a volume of 7.5 million L of natural seawater, piped at 0.6–1.2 million L h^{-1} from a location 300 m off the coast of Okinawa island at a depth of 20 m. Incoming seawater is filtered through industrial fiberglass vessels containing balls of fibrous acrylic material while exhibit water filtered through high-rate sand filters similar to those at GAI. Filtration flow is approximately 3.8 million L h^{-1}, giving the habitat a turnover rate of about 1.5–2 hours.

Water quality requirements for whale sharks are consistent with those of other tropical and subtropical elasmobranchs. In the Ocean Voyager habitat at GAI, temperature is maintained at a constant 24.6°C, with a pH of 8.1 and a salinity of 32 ppt. In contrast, the Kuroshio Current habitat at OCA receives a constant supply of new seawater and is therefore subject to seasonal fluctuations in temperature between 21°C and 29°C, and natural fluctuations in other water chemistry parameters.

9.4 ACQUISITION/DISPOSITION TECHNIQUES

Since 1980, 33 whale sharks have been acquired by OCA. Of these animals, 31 were accidentally caught in set fishing nets along the coast of Okinawa, and the other two were stranded animals rescued by OCA staff (Matsumoto et al. 2017). Most of the whale sharks acquired by OCA were released after a period of several months to several years in a sea pen or on public display, or some combination of both. By contrast, animals acquired by Georgia Aquarium are intended to be displayed permanently. GAI has acquired a total of six whale sharks since 2005, all of which came from a quota pelagic trap fishery in Taiwan; that quota has since been reduced to zero (Pierce and Norman 2016). Because of the logistic challenges of large animal transports, generally only animals of less than 5 m in total length are suitable for acquisition.

For both GAI and OCA, net pens in the ocean play an important role in the acquisition process. After capture, whale sharks are staged in the net pen for a period of acclimation and behavioral modification lasting from several weeks to several months, which is necessary to ensure successful transition to the aquarium environment. During this period, aquarists work intensively with the animals to desensitize them to the presence of people and to acclimatize them to feeding on a diet provided by animal care staff. Animals must be feeding consistently from aquarists before transport can be attempted (Figure 9.2).

Another commonality between the two aquariums concerns the use of customized fiberglass shipping boxes for transporting animals to or from the aquarium. These shipping boxes are reinforced rectangular tanks approximately 6.1 m long, 2.1 m deep and 2.3 m wide, with life-support systems built into the walls and lids, featuring paper cartridge filters, activated

Figure 9.2 A whale shark transport container at Georgia Aquarium. (Picture reproduced with permission, Georgia Aquarium.)

carbon to absorb organic wastes, and compressed oxygen supplementation. Because whale sharks can aspirate their own gills by buccal pumping, they tolerate shipping well and certainly much better than many smaller shark species that have rigid pharyngeal baskets and are obligate ram ventilators.

At OCA, transport takes place over relatively small distances. The sea pen ($20 \times 30 \times 10$ m deep) is installed ca. 1 km offshore from the port closest to the aquarium. For transport, the animal is moved to a shipping container (inside dimensions: 2.3 W m $\times$ 7.4 L $\times$ 1.7 H), which is towed to the port, after which the animal is lifted, fully immersed in a bladder stretcher, into the transport container by crane, and finally driven by truck a few kilometers to the aquarium from the port. In the more complicated case of GAI, intercontinental transport took place by truck from the fishing site to Hualien Airport for a short flight to Taipei on a Belfast heavy lift aircraft, followed by transfer to a Boeing 747 freight aircraft for the longer leg from Taipei to Atlanta via Anchorage Alaska, a distance of approximately 13,000 km. After gradual water changes and chemical additions so that transport water closely matched exhibit water in temperature, pH and ammonia concentration (NH_3-N), the animals were released into the habitat. From early experiences, it was learned that proactive pH management early in transport provided the least amount of pH fluctuation and minimized the time needed for acclimation at the aquarium end. Total transport time from water to water ranged from 28 to 36 hours.

9.5 BEHAVIOR AND ITS MANAGEMENT

Despite challenges associated with their sheer size, whale sharks have proven very amenable to long-term care in the aquarium setting. One animal at OCA has been on display for more than 25 years at the time of writing. They are a docile species, and their behavior can be modified using operant conditioning methods (i.e. training), although not with as much facility as some other species. Both traits help to make the whale shark more manageable than many other pelagic species kept in aquariums, despite its size.

Training strategies have been employed at GAI to manage whale shark behavior. For example, one specimen began to exhibit a stereotypical swimming pattern, whereby the shark did not use the entire available habitat and instead repeatedly swam in clockwise circles. To interrupt this behavior pattern, divers on SCUBA fed the shark underwater using a squeeze bottle. Over several months, the animal was conditioned to swim in increasingly larger circles and then to do so in the opposite direction such that, over time, the stereotypical swimming behavior was successfully extinguished (Schreiber and Coco 2017) (Figure 9.3).

Figure 9.3 Squeeze bottle feeding of whale sharks. (© Chris Miller / Georgia Aquarium.)

Reproductive behaviors are beginning to be observed in both OCA and GAI at the time of writing. The male at OCA is mature, based on clasper differentiation and testosterone levels (Matsumoto et al. 2019), and it is possible that this individual will mate with the female with which it currently shares the Kuroshio Current habitat. If successful, this would be the first case of captive reproduction in whale sharks and would provide a treasure trove of basic life-history information that is sorely lacking for this species (see Chapter 2). At least one of the males at GAI is also maturing, based on clasper morphology. Both males have been observed rolling over onto their side or back and then externally rotating their claspers, which is interpreted to be a pre-copulatory behavior (Chapter 2).

Maintenance of whale sharks at GAI has facilitated a long-term behavioral study (Black et al. 2013). This collaboration of aquarium staff, university researchers, and volunteers has resulted in the establishment of ethograms of whale shark behavioral diversity, time budgets for different behaviors, and a very large data set of repeated observations over many years. These data should provide a rich tool to analyze the behavior of this species in human care.

Only one manipulative experiment has been conducted with the whale sharks at GAI (Dove 2015); the behavioral response of whale sharks to chemical odorants was tested, as these are often associated with foraging in epipelagic species (Atema 1995, Nevitt 2000, DeBose and Nevitt 2008). This experiment was only possible because it did not involve any physical contact with the animals; rather their behavioral response to the presence of odor plumes created at the periphery of the exhibit was observed. These plumes, which contained either filtered krill homogenate or an aqueous solution of dimethylsulfoxide (DMSO), both induced pre-feeding behaviors (stopping, inclining the body head-up, and buccal pumping), with the krill homogenate inducing a stronger response than did the DMSO. The fact that the DMSO-induced behavioral responses stronger than those of the sham plumes suggested that this compound is a chemical cue that can be detected by whale sharks in the field, but there are likely to be many other cues, and it is often complex mixtures, such as the krill homogenate, that trigger the strongest feeding responses in marine animals (Atema 1995). This remains a subject that desperately needs further research if we are to properly understand whale shark foraging behavior (Chapter 8).

9.6 VETERINARY CARE

In the course of caring for whale sharks in public aquariums, a great deal of creativity has been required in the development of techniques to facilitate veterinary exams, diagnostics, and treatment procedures on such large and water-breathing animals. These include restraint, anesthesia, phlebotomy (blood draw), diagnostic imaging (ultrasound, endoscopy), injections, gavage-feeding, and biopsy. Restraint is labor-intensive and difficult, and requires the use of divers, snorkelers, and dry-side staff to corral the animal into a vinyl stretcher suspended in the habitat. Anesthesia was used initially at GAI for veterinary exams, employing approximately 75 ppm MS-222 (Tricaine methanesulfonate, Finquel® Argent Laboratories, Redmond WA) in solution to successfully and reversibly anesthetize whale sharks. Later, sedation through elevated oxygen saturation was used. At present, restraint procedures can be avoided altogether because methods have been developed to take blood samples while swimming on SCUBA beside the animal (see below).

Studies at GAI showed at least three places on the body where blood samples could be reliably taken (Dove et al. 2010a): the ventral caudal vein and the dorsal caudal vein, both of which can be accessed via the pre-caudal pits at the caudal peduncle, and a medial sinus that can be accessed from the posterior insertion of the second dorsal fin (Figure 9.4). Care must be taken at interpreting the blood values from the second dorsal sinus because it appears to be a mixed venous/lymphatic sinus; hematocrit (packed red cell volume) and some chemical parameters are lower at this site than the caudal vein sites (Dove et al. 2010a). Hematology values for whale sharks are in line

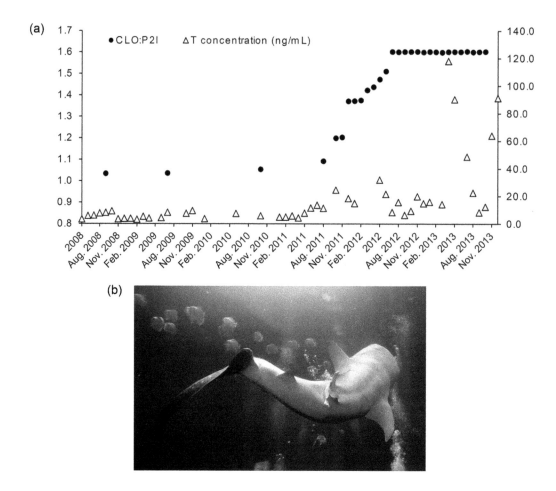

Figure 9.4   (a) Maturation of whale sharks demonstrated by a ratio of inner and outer margin clasper measurements, and plasma testosterone concentration. (Reproduced with permission from Matsumoto
et al. (2019). Figure in public domain.) (b) External rotation of the claspers in a maturing male whale
shark at Georgia Aquarium, accompanied by body rotation to the left, which can reach almost
180°. (Picture reproduced with permission, Georgia Aquarium.)

with most other sharks (Tables 9.1 and 9.2); their cell sizes, while on the large side for sharks in
general, are similar in size to those of other orectolobiforms and are not markedly different in morphology. Their red cells (erythrocytes) are ellipsoid, with a central ellipsoid nucleus. Their white
cells (leukocytes) consist of lymphocytes, monocytes, fine eosinophilic granulocytes (heterophils),
coarse eosinophilic granulocytes (eosinophils), neutrophils, non-granulated thrombocytes, and,
rarely, basophils (Figure 9.5). Metabolomics suggests that standard clinical chemistry analytes
derived from mammals are not good biomarkers of health in this species (Dove et al. 2012; Dove
2013) (Figure 9.6).

The subsequent development at OCA of techniques to take blood from unrestrained whale
sharks has reduced further the need for many veterinary exams. Interestingly, the site of choice for
phlebotomy on unrestrained whale sharks developed at OCA is the anterior dorsal surface of the
pectoral fin, which is not one of the sites previously used at GAI. Studies at OCA have shown that
blood can also be successfully collected from the lateral base of the pelvic fins. This illustrates the
rapid development of our knowledge of this species resulting from captive care. In the early days
of GAI, it required more than 40 people and 3–4 hours to get a blood sample from a whale shark,

Table 9.1 Average Hematological Parameters of the Whale Shark

Parameter	Relative (%)	Absolute (cells mL)
White blood cells		5090
Lymphocyte	46.0	2369
Heterophil	39.5	2009
Eosinophil	6.5	331
Monocyte	6.3	321
Basophil	1.8	90
Neutrophil	0.9	45

From Dove et al. (2011).

Table 9.2 Blood Chemistry of Whale Sharks

Analyte	Average Concentration at GAI	Average Concentration at OCA
Alkaline phosphatase	20.5 ± 5.3 mg dl^{-1}	38.3 ± 2.4 U l^{-1}
Alanine aminotransferase	9.0 ± 3.7 mg dl^{-1}	9.1 ± 0.8 U l^{-1}
Aspartate aminotransferase	13.6 ± 5.0 mg dl^{-1}	15.5 ± 0.6 U l^{-1}
Creatine kinase	88.1 ± 53.3 mg dl-1	241.0 ± 25.9 U l^{-1}
Albumin	Not detected	1.39 ± 0.12 g dl^{-1}
Amylase	Not detected	NA
Uric acid	0.9 ± 0.1 mg dl^{-1}	0.12 ± 0.01 mg dl^{-1}
Total bilirubin	0.22 ± 0.08 mg dl^{-1}	0.38 ± 0.03 mg dl^{-1}
Calcium	13.6 ± 0.4 mg dl^{-1}	16.1 ± 0.1 mg dl^{-1}
Ionized calcium	2.25 ± 0.21 mg dl-1	NA
Phosphorus	4.1 ± 0.3 mg dl^{-1}	2.5 ± 0.1 mg dl^{-1}
Creatinine	0.35 ± 0.07 mg dl^{-1}	1.69 ± 0.99 mg dl^{-1}
Glucose	64.0 ± 3.6 mg dl^{-1}	62.8 ± 3.5 mg dl^{-1}
Sodium	284.0 ± 19.5 mEq l^{-1}	296.8 ± 0.6 mmo l^{-1}
Potassium	5.54 ± 0.36 mEq l^{-1}	4.39 ± 0.07 mmol l^{-1}
Chloride	257.9 ± 19.9 mEq l^{-1}	249.7 ± 0.7 mmol l^{-1}
Total protein	2.4 ± 0.3 g dl^{-1}	3.1 ± 1.1 g dl^{-1}
pCO$_2$	7.2 ± 2.6 mm Hg	6.0 ± 0.1 mm Hg
TCO$_2$	<5.0 mmol l^{-1}	NA
pO$_2$	36.2 ± 21.9 mm Hg	28.8 ± 1.1 mm Hg
sO$_2$	$71.7 \pm 31.3\%$ saturation	NA
pH	7.37 ± 0.22	7.59 ± 0.01
Lactate	2.35 ± 1.32 mmol l^{-1}	0.44 ± 0.04 mmol l^{-1}
HCO$_3$	3.7 ± 0.6 mmol l^{-1}	NA
Base excess	-23.2 ± 1.2 mmol l^{-1}	NA
Blood urea nitrogen	NA	865.8 ± 2.5 mg dl^{-1}
Ammonia	NA	120.7 ± 15.1 µg dl^{-1}
Hemoglobin	NA	5.1 ± 0.1 g dl^{-1}
Total cholesterol	NA	48.0 ± 1.1 mg dl^{-1}
Triglyceride	NA	28.0 ± 0.8 g dl^{-1}
Lactate dehydrogenase	NA	53.0 ± 6.7 U l^{-1}
Magnesium	NA	5.0 ± 0.3 mg dl^{-1}

From Dove et al. (2011).

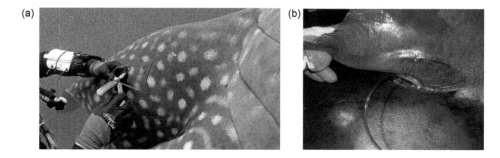

Figure 9.5   (a) Blood draw from left pectoral fin of a juvenile male whale shark in a bagan net in Cenderawasih Bay, Indonesia, showing phlebotomy apparatus for underwater blood draw from whale sharks. The extension set allows independent movement of shark and researcher, but it floods with seawater once submerged. The perpendicular syringe is therefore used to aspirate the seawater from the extension set until pure blood is being drawn up, after which the researcher can switch the three-way stopcock and draw an uncontaminated blood sample into the main syringe. Once full, the stopcock can be closed off so that the blood sample remains sealed within the syringe. The precise location for inserting the needle is not marked by any palpable surface feature (a "blind stick"), and it may be necessary to try several sites, but the area shown is a good starting point; (b) blood draw from the posterior sinus of the second dorsal fin. The skin is very thin at this site, and blood values may vary (see text). Images Alistair Dove/Georgia Aquarium.

a procedure that today can be conducted by a single diver during the course of one feeding session. Staff at OCA have also developed a waterproofed housing to conduct ultrasound exams on unrestrained whale sharks, which has been used successfully to confirm pregnancy in reef manta rays *Mobula alfredi* and other elasmobranch species at OCA (Murakumo et al. 2020). This technique has also been applied to examine digestive organs, measure heart rate and test for pregnancy whale sharks in the field, which is a remarkable technological achievement (more in Chapter 2) (Figure 9.7).

Giving subcutaneous or intramuscular injections to whale sharks is possible but complicated by the great thickness and toughness of the skin. Large gauge (~18 gauge) needles of lengths typically reserved for spinal injections (2.5–3.5 inches) are required, as are stylets, which are wire inserts to prevent plugging the lumen of the needle by the dense connective tissue of the skin. Intravenous infusions can be given at the second dorsal sinus. For both intramuscular and intravenous injections, the large body weight often requires formulation of highly concentrated solutions, or the infusion of very large volumes of fluid. A formulary of whale shark medicine is included in the second edition of the Elasmobranch Husbandry Manual (Schreiber and Coco 2017, appendix 2, available online at https://bit.ly/2E3XQuz).

9.7 DIET, FEEDING, AND NUTRITION

Whale sharks in aquariums are offered diet items consistent with wild diets and the anatomical features of their filter-feeding mechanism, presented in a manner consistent with the natural feeding behaviors of this species (described by Motta et al. 2010). Successful feeding of whale sharks has been achieved using several approaches. In a static station approach, whale sharks feed in the vertical modality of Motta et al. (2010) and food can be poured directly into the mouth of the shark using a ladle on a pole from the edge of the habitat or a gantry over the water. For larger animals, this approach tends to be impractical, and a different method has been developed at GAI, wherein an aquarist in a small inflatable raft maneuvers along a fixed tagline, trailing food in the water for the animal to eat. In this approach, the whale shark swims alongside the raft and uses the active surface suction feeding modality of Motta et al. (2010) (Figure 9.8).

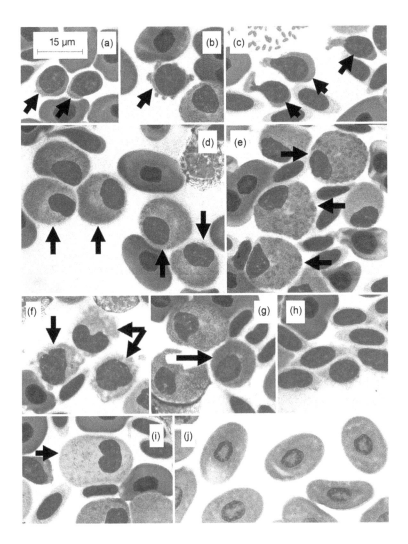

Figure 9.6 Differential hematology of the whale shark. (a–c) lymphocytes (arrows); (d) heterophils (arrows); (e) eosinophils (arrows); (f) monocytes (arrows); (g) basophil (arrow); (h) thrombocytes; (i) neutrophil (arrow); and (j) erythrocytes. All photographs to same scale. (Reproduced with permission, Interresearch (confirm).)

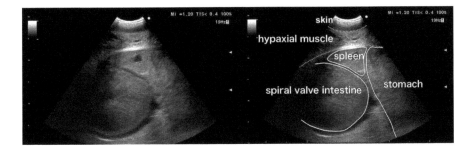

Figure 9.7 Underwater ultrasonography on female whale shark at OCA to check the health condition of digestive organs including stomach, gall bladder and spiral intestine. (Picture credit: Rui Matsumoto/ Okinawa Churaumi Aquarium.)

Figure 9.8 Feeding of whale sharks at Georgia Aquarium, using boats to maintain the active surface suction feeding modality. (© Georgia Aquarium.)

It is difficult to estimate the ration as a percentage of body weight due to the challenge of accurately weighing specimens, but a conservative figure of 0.3%–0.5% of body mass per day is likely at GAI. This is modest for a growth ration in such a large species in warm water, which suggests that they are either very efficient at feed conversion or very efficient swimmers, or both. Diet at GAI consists primarily of commercially harvested krill *Euphausia superba* and *E. pacifica*, supplemented with lance fish (*Ammodytes americanus*), juvenile silversides (*Menidia menidia*) and vitamin-supplemented Mazuri® aquatic gel diet (Purina Mills LLC, St Louis MO). OCA also relies heavily on superba and pacifica krill, supplemented with seasonal local foods such as sakura shrimp (*Lucensosergia lucens*) and Japanese anchovy (*Engraulis japonica*). Whale sharks at OCA are fed ~1% of their body mass per day; an 8.5 m male with an estimated mass of 5,100 kg, consumed around 100,400 kJ per day; and a 7.1 m female with an estimated mass of 2,800 kg, consumed around 60,500 kJ per day (Matsumoto et al. 2017). The ratios of caloric intake to body mass of these two individuals were 19.7 and 21.6 kJ/kg, for the male and female, respectively. Motta et al. (2010) estimated that the energy intakes of two wild whale sharks in Yucatan Mexico, approximately 6.6 and 4.4 m in length, were 28,121 and 14,931 kJ, respectively. Application of a length-weight equation (see next section) to the whale sharks in Motta's study suggests that the 6.6 m shark weighed 1,811 kg and the 4.4 m shark weighed 595 kg, and that the estimated daily energy intake for these animals was therefore 12.7 and 25.1 kJ/kg, respectively, which is in line with the captive ration. Gut transit time is another simple indicator of digestive function, and tracer studies at GAI have shown that gut transit time is around 3–4 days for healthy whale sharks (GAI unpublished data). Staff at OCA measure the girth (G) and TL of whale sharks monthly and monitor the effects of energy intake and water temperature on the body condition, expressed as the ratio of G/TL (Matsumoto et al. 2018). Body condition in the female decreased (−0.013) from June (G/TL: 0.422; 26.4°C) to August (G/TL: 0.409; 29.2°C), even though the animal was consuming 56,831 kJ/day. In contrast, this value increased slightly (+0.004) between January (G/TL: 0.361; 22.0°C) and February (G/TL: 0.365; 21.3°C), even at a lower energy intake of around 34,526 kJ/day. Similar results were obtained from the male whale shark in OCA. This monitoring suggests that changes in whale shark body shape (G/TL) were caused by changes in the metabolic rates of these ectothermic animals, which were in turn determined by the relationship between the water temperature and energy intake.

Wyatt et al. (2019) published a large study about diet in whale sharks, comparing collection animals at OCA with sea pen animals that were rescued by OCA staff but never moved to the aquarium. Results showed evidence that several of the sea pen animals had undergone prolonged periods of fasting prior to their capture, while others of this group showed evidence that plant material made up a significant portion of their diet (although see Chapter 8 for further discussion). This study also presented tissue turnover rates for stable isotopes, derived from studies of the collection animals at OCA. Fin cartilage turnover was approximately 3 years, whereas plasma turnover was around 9 months. These figures should be a valuable tool for field researchers attempting to understand the dynamics of diet in wild animals based on stable isotope ratios.

Blood nutrient values, which are presently unavailable for wild whale sharks, are helpful in determining the nutritional status of individual animals. Blood vitamin, trace mineral and fatty acid profiles from healthy whale sharks at GAI are shown in Tables 9.3 and 9.4 (Hoopes 2017; GAI unpublished data). Plasma fatty acid profiles of two whale sharks at GAI that had stopped eating (Dove et al. 2012) showed increased relative concentrations of the fatty acids 22:5n3, 18:1n11c, 18:2n6c, and 20:4n6, and decreased relative concentration of 20:5n3 (GAI unpublished data). Similar to stable isotopes, fatty acid profiles are useful biotracers of diet since long-chain fatty acids are conserved in the tissues of consumers relatively unchanged, allowing for the reflection of the fatty acid profile of prey to be represented in the tissues of the predator (see Chapter 8). With knowledge of the fatty acid profiles of available prey, it is possible to determine likely diet composition. This is another area of feeding ecology where aquarium animals that are feeding on known diets can help ground-truth field results, although those studies have not yet been conducted.

9.8 GROWTH AND REPRODUCTION

Precise length and mass measurements are needed to determine growth rates and, in practical husbandry terms, to provide appropriate transportation and the right quantities of food and medicine. However, the sheer size of whale sharks and their constant swimming behavior makes the measurement of live unrestrained specimens difficult. In the field, researchers have estimated total length (TL) by comparing the animal to the length of a boat (Heyman et al. 2001, Hobbs et al. 2009), or to snorkelers photographed with the shark (Graham and Roberts 2007, Norman and Stevens 2007, Bradshaw et al. 2008). Rohner et al. (2011) refined a method of estimating length using laser

Table 9.3   Average Vitamin and Trace Mineral Values (n = 10 Samples) from Four Healthy Whale Sharks at Georgia Aquarium

Analyte	Plasma	Whole Blood
Vitamin A	115.5 ng ml^{-1}	
Vitamin E	3.4 ug ml^{-1}	
Cobalt	221.1 ng ml^{-1}	
Copper	0.7 ug ml^{-1}	
Iron	14.4 ug ml^{-1}	
Iodine	278.3 ng ml^{-1}	
Manganese	57.3 ng ml^{-1}	
Molybdenum	3.0 ng ml^{-1}	
Zinc	0.3 ug ml^{-1}	
Selenium	312.1 ng ml^{-1}	928.3 ng ml^{-1}
Arsenic		6,053 ppb[a]
Lead		55.5 ppb
Mercury		60.7 ppb

[a] This seemingly high value reflects a predominance of organic arsenic complexes, especially arsenobetaine, in the total body load of arsenic in sharks. These are not toxic, so total arsenic may be deceptive.

Table 9.4 Average Plasma Fatty Acid Profiles for Four Healthy Whale Sharks at Georgia Aquarium

Fatty Acid Formula	Relative %
C6:0	<0.001
C7:0	<0.001
C8:0	<0.001
C9:0	<0.001
C10:0	<0.001
C11:0	<0.001
C12:0	<0.001
C13:0	<0.001
C14:0	2.3
C14:1n5	<0.001
C15:0	0.3
C15:1n5	<0.001
C16:0	11.9
C16:1n7	4.1
C17:0	0.3
C17:1n7	<0.001
C16:2	0.2
C18:0	4.5
C18:1n9T	0.4
C16:3n4	<0.001
C18:1n9C	6.8
C18:1n7C	3.7
C18:1	0.1
C19:0	0.2
C18:2	0.2
C16:4n1	<0.001
C18:2n6	2.5
C20:0	<0.001
C18:3	<0.001
C18:3n6	0.1
C18:3	<0.001
C20:1n11	0.6
C18:3n3	0.6
C21:0	<0.001
C18:4n3	0.8
C20:2n6	0.2
C22:0	0.1
C20:3n6	0.2
C22:1n11	0.1
C22:1n9	0.2
C20:3n3	0.1
C20:4n6	3.1
C23:0	<0.001
C20:4n3	0.4
C22:2n6	<0.001
C24:0	0.3
C20:5n3	12.8
C24:1n9	0.2
C21:5	0.2
C22:5n6	0.4
C22:5n3	1.7
C22:6n3	11.3

photogrammetry and measuring the distance from the fifth gill opening to the first dorsal fin as a correlate of TL. More recently, stereovideogrammetry (Spitz et al. 2000) has become a popular method of measuring large animals underwater. Whale sharks in aquariums offer the advantage of being easily observed at any time, allowing for direct measurement of TL or of any specific body parts.

A range of truss measurements have been taken at OCA for the 33 whale sharks collected for display there from 1979 to 2010 (Matsumoto et al. 2017). For eight of the truss measurements, there was a high correlation coefficient ($r^2 > 0.9$, $p < 0.01$) with total length: length from snout to origin of first dorsal fin, length from snout to origin of second dorsal fin, length from snout to dorsal pre-caudal pit (often called PCL in the literature), length from snout to origin of pelvic fins, length from snout to anus, length of first dorsal fin caudal margin, mouth width, and interorbital distance. The snout tip to first dorsal fin origin truss is on the midline of the body, on the dorsal surface, and is not affected by the serpentine motion of the back half of the body through the water, and for these reasons, it is another appropriate measurement to use as a practical correlate of TL, based on the equation:

$$\log TL = 0.964 \log PDL + 0.443$$

where PDL is the length from the snout to the origin of the dorsal fin in cm and TL is the total length in cm.

Laser photogrammetry was critically evaluated at GAI by repeating measurements on the same animal over and over, so that staff could understand the precision of this method (GAI unpublished data). Unfortunately, precision is not really satisfactory at roughly ± 0.5 m, such that any individual laser photogrammetric measurement of an animal must be presented with acknowledgement of this uncertainty. It was possible to obtain a more accurate estimate by taking the central tendency of many repeated measurements, which allows the presentation of confidence intervals of $\pm$ just a few centimeters (GAI unpublished data), but the method is tedious and time-consuming.

Body mass (BM) estimation is extremely helpful in order to determine growth rates and to set the correct food ration and medication dosage rates. Staff from OCA analyzed morphometrics and weight data from eight whale sharks that had stranded or were fishery bycatch (seven males and one female) at Okinawa island, all of which died within a few weeks of rescue (Matsumoto et al. 2017). Using these data, they developed a length–weight relationship equation:

$$BM = 4.51 * TL^{3.28}$$

where BM is body mass in kg and TL is total length in m.

Based on this length–weight relationship, the largest whale sharks observed at Darwin's Arch in the Galapagos (roughly 13.5 m), which are among the largest individuals of this species regularly encountered, probably weigh in excess of 21 metric tons. A more "average" 8.0 m individual would weigh approximately 5 metric tons.

9.9 CONCLUSIONS

The care of whale sharks in modern public aquariums is as recent as the study of their coastal aggregations, having started in 1980 in Japan and accelerated in 2002 with the reopening of Expo Park as the Okinawa Churaumi Aquarium, and again in 2005 with the opening of Georgia Aquarium in Atlanta. While field studies have focused on population-level subjects like movement ecology, demographics, and direct and indirect diet studies, research in the aquarium setting has focused more on longitudinal studies of behavior, growth, and maturation. Due to the limitations on the type of science that is possible in an aquarium, much of this research is retrospective and observational; it results simply from compiling and interpreting data collected during the course of routine husbandry and veterinary care. By presenting some of these data in summary form here, we hope to demonstrate the complementary nature of research in the public aquarium setting and to help fill some gaps in biological knowledge obtained in the field.

Aquarium collections hold continuing promise for answering some of the questions about whale sharks that have proven hard to investigate in the field. For example, we know very little about the sensory ecology of whale sharks, but the one study on olfaction at GAI (Dove 2015) is an example of what can be done in the aquarium setting. Future research might include, for instance, evoked auditory potential studies to determine the frequency-specific hearing ability of whale sharks, and whether it might play a role in foraging behavior. Continued study of whale shark behavior will be important too, and should breeding efforts at OCA or elsewhere prove successful, we should be able finally to answer some of the most vexing questions about reproduction in this species.

ACKNOWLEDGMENTS

AD appreciates the guidance and assistance of current and former Georgia Aquarium staff, especially Dr. Bruce Carlson, Ray Davis, Tim Binder, Mike Maslanka and Dr. Tim Mullican.

REFERENCES

Atema, J. 1995. Chemical signals in the marine environment: Dispersal, detection, and temporal signal analysis. *Proceedings of the National Academy of Sciences* 92:62–66.

Black, M. P., Grober, M., Schreiber, C., et al. 2013. Whale shark (*Rhincodon typus*) behavior: A multi-year analysis of individuals at Georgia Aquarium. *PeerJ PrePrints* 1:e88v81.

Borrell, A., Aguilar, A., Gazo, M., et al. 2011. Stable isotope profiles in whale shark (*Rhincodon typus*) suggest segregation and dissimilarities in the diet depending on sex and size. *Environmental Biology of Fishes* 92(4):559–67.

Borrell, A., Cardona, L., Kumarran, R. P., and A. Aguilar. 2011. Trophic ecology of elasmobranchs caught off Gujarat, India, as inferred from stable isotopes. *ICES Journal of Marine Science* 68(3):547–54.

Bradshaw, C., Fitzpatrick, B., Steinberg, C., et al. 2008. Decline in whale shark size and abundance at Ningaloo Reef over the past decade: The world's largest fish is getting smaller. *Biological Conservation* 141(7):1894–905.

Burns, A. S., Padilla, C. C., Pratte, Z. A., et al. 2018. Broad phylogenetic diversity associated with nitrogen loss through sulfur oxidation in a large public marine aquarium. *Applied and Environmental Microbiology* 4:e01250–18. doi:10.1128/AEM.01250-18.

Schreiber, C., and C. Coco. 2017. Husbandry of whale sharks. In *The Elasmobranch Husbandry Manual II: Recent Advances in the Care of Sharks, Rays and Their Relatives*, ed. M. Smith et al., 87–98. Columbus: Ohio Biological Survey.

De Bose, J. L. and G. A. Nevitt. 2008. The use of odors at different spatial scales: Comparing birds with fish. *Journal of Chemical Ecology* 34:867–81.

Dove, A. D. M. 2013. Metabolomics has great potential for clinical and nutritional care and research with exotic animals. *Zoo Biology* 32(3):246–50.

Dove, A. D. M. 2015. Foraging and ingestive behaviors of whale sharks, *Rhincodon typus*, in response to chemical stimulus cues. *The Biological Bulletin* 228(1):65–74.

Dove, A. D. M., Arnold, J., and T. M. Clauss. 2010a. Blood cells and serum chemistry in the world's largest fish: The whale shark *Rhincodon typus*. *Aquatic Biology* 9(2): 177–83.

Dove, A. D. M., Coco, C., Binder, T., et al. 2010b. Care of whale sharks in a public aquarium setting. *Proceedings of the 2nd US/Russian Bilateral Conference on Aquatic Animal Health*. East Lansing: Michigan State University Press.

Dove, A. D. M. and E. Hall. 2018. Removing nitrate from water. *US Patent No. 9994466.* Published August 6, 2018.

Dove, A. D. M., Leisen, J., Zhou, M., et al. 2012. Biomarkers of whale shark health: A metabolomic approach. *PLoS One* 7(11):e49379.

Graham, R. T., and C. M. Roberts. 2007. Assessing the size, growth rate and structure of a seasonal population of whale sharks (*Rhincodon typus* Smith 1828) using conventional tagging and photo identification. *Fisheries Research* 84(1):71–80.

Heyman, W. D., Graham, R. T., Kjerfve, B., and R. E. Johannes. 2001. Whale sharks *Rhincodon typus* aggregate to feed on fish spawn in Belize. *Marine Ecology Progress Series* 215:275–82.

Hobbs, J. P. A., Frisch, A. J., Hamanaka, T., et al. 2009. Seasonal aggregation of juvenile whale sharks (*Rhincodon typus*) at Christmas Island, Indian Ocean. *Coral Reefs* 28(3):577.

Hoopes, L. A. 2017. Elasmobranch mineral and vitamin requirements. In *The Elasmobranch Husbandry Manual II: Recent Advances in the Care of Sharks, Rays and Their Relatives*, ed. M. Smith et al., 135–46. Columbus: Ohio Biological Survey.

Ito, T., Onda, K., and K. Nishida. 2017. Effects of noise and vibration on the behavior and feeding activity of whale sharks, *Rhincodon typus* (Smith, 1828), in Osaka Aquarium Kaiyukan. In *The Elasmobranch Husbandry Manual II: Recent Advances in the Care of Sharks, Rays and Their Relatives*, ed. M. Smith et al., 159–68. Columbus: Ohio Biological Survey.

Jeffreys, G. L., Rowat, D., Marshall, H., and K. Brooks. 2013. The development of robust morphometric indices from accurate and precise measurements of free-swimming whale sharks using laser photogrammetry. *Journal of the Marine Biological Association of the United Kingdom* 93(2):309–20.

Marcus, L., Virtue, P., Pethybridge, H. R., et al. 2016. Intraspecific variability in diet and implied foraging ranges of whale sharks at Ningaloo Reef, Western Australia, from signature fatty acid analysis. *Marine Ecology Progress Series* 554:115–28.

Marcus, L., Virtue, P., Nichols, P. D., et al. 2019. Stable isotope analysis of dermis and the foraging behavior of whale sharks at Ningaloo Reef, Western Australia. *Frontiers in Marine Science* 6:546.

Matsumoto, R., Matsumoto, Y., Ueda, K., et al. 2019. Sexual maturation in a male whale shark (*Rhincodon typus*) based on observations made over 20 years of captivity. *Fishery Bulletin* 117(1):78–86.

Matsumoto, R., Murakumo, K., Furuyama, R., and S. Matsuzaki. 2018. Effects of energy intake and water temperature on the body shape of whale sharks at the Okinawa Churaumi Aquarium. *Proceedings of 10th International Aquarium Congress,* Fukushima, Japan, pp. 80–84.

Matsumoto, R., Toda, M., Matsumoto, Y., et al. 2017. Notes on husbandry of whale sharks, *Rhincodon typus*, in aquaria. In *The Elasmobranch Husbandry Manual II: Recent Advances in the Care of Sharks, Rays and Their Relatives*, ed. M. Smith et al., 87–98. Columbus: Ohio Biological Survey.

Motta, P. J., Maslanka M., Hueter R. E., et al. 2010. Feeding anatomy, filter-feeding rate, and diet of whale sharks *Rhincodon typus* during surface ram filter feeding off the Yucatan Peninsula, Mexico. *Zoology* 113(4):199–212.

Murakumo, K., Matsumoto, R., Tomita, T., et al. 2020. The power of ultrasound: Observation of nearly the entire gestation and embryonic developmental process of captive reef manta rays (*Mobula alfredi*). *Fishery Bulletin* 118:1–7.

Nevitt, G. A. 2000. Olfactory foraging by Antarctic procellariiform seabirds: Life at high Reynolds numbers. *Biological Bulletin* 198:145–253.

Nishida, K. 2001. Whale shark - the world's largest fish. In *Fishes of the Kuroshio Current, Japan*, eds. T. Kakabo, Y. Machida, K. Yamaoka, and K. Nishida, 20–26. Osaka: Osaka Aquarium Kaiyukan.

Norman, B. M. and J. D. Stevens. 2007. Size and maturity status of the whale shark (*Rhincodon typus*) at Ningaloo Reef in Western Australia. *Fisheries Research* 84(1):81–86.

Perry, C. T., Figueiredo, J., Vaudo, J. J., et al. 2018. Comparing length-measurement methods and estimating growth parameters of free-swimming whale sharks (*Rhincodon typus*) near the South Ari Atoll, Maldives. *Marine and Freshwater Research.* doi:10.1071/MF17393.

Pierce, S. J., and B. Norman. 2016. *Rhincodon typus.* The IUCN Red List of Threatened Species: e.T19488A2365291. Available from: https://dx.doi.org/10.2305/IUCN.UK.2016-1.RLTS.T19488A2365291.en.

Prebble, C., Rohner, C. A., Pierce, S. J., et al. 2018. Limited latitudinal ranging of juvenile whale sharks in the Western Indian Ocean suggests the existence of regional management units. *Marine Ecology Progress Series* 601:167–83.

Rohner, C. A., Richardson, A. J., Marshall, A. D., et al. 2011. How large is the world's largest fish? Measuring whale sharks *Rhincodon typus* with laser photogrammetry. *Journal of Fish Biology* 78(1):378–85.

Rohner, C. A., Couturier, L. I. E., Richardson, A. J., et al. 2013. Diet of whale sharks *Rhincodon typus* inferred from stomach content and signature fatty acid analyses. *Marine Ecology Progress Series* 493:219–35.

Schreiber, C. and C. Coco. 2017. Husbandry of whale sharks. In *The Elasmobranch Husbandry Manual II: Recent Advances in the Care of Sharks, Rays and their Relatives*, ed. M. Smith, et al., 87–98. Columbus: Ohio Biological Survey.

Spitz, S. S., Herman, L. M., and A. A. Pack. 2000. Measuring sizes of humpback whales (*Megaptera novaeangliae*) by underwater videogrammetry. *Marine Mammal Science* 16:664–76.

Wyatt, A. S. J., R. Matsumoto, Y. Chikaraishi, et al. 2019. Enhancing insights into foraging specialization in the world's largest fish using a multi-tissue, multi-isotope approach. *Ecological Monographs* 89(1):e01339.

CHAPTER 10

Whale Shark Tourism as an Incentive-Based Conservation Approach

Jackie Ziegler and Philip Dearden
University of Victoria

CONTENTS

10.1 INTRODUCTION

Whale shark tourism emerged in the late 1980s in Australia and the Maldives, along with an increasing interest in shark diving generally (Topelko and Dearden 2005). It has since exploded to become the most popular shark-watching activity in the world. Tourism activities range from captive aquariums and seapen experiences to non-captive provisioned and wild encounters. Whale sharks' large size, docile nature, predictable presence, accessibility, and growing popularity as charismatic megafauna has made "swim-with" non-captive whale shark tourism activities one of the fastest-growing sectors of the marine wildlife tourism sector overall (Dearden et al. 2008; Gallagher and Hammerschlag 2011). The activity now takes place in at least 46 sites worldwide, involving an estimated 25.5 million participants, in countries ranging from highly developed to chronically undeveloped.[1]

Non-captive tourism activities can be an important incentive for protecting whale sharks. Incentive-based conservation (IBC) focuses on providing incentives (e.g., employment, ecological services, compensation payments, health care, education, agroforestry, tourism development/promotion) as a means of gaining local support for conservation (Spiteri and Nepal 2006). IBC strives to integrate environmental protection with poverty reduction and community participation. The assumption is that economic benefits derived directly from natural resources will lead to the conservation and sustainable use of those resources (Brockelman and Dearden 1990; Eshoo et al. 2018). Social conservation outcomes of IBC projects may also include improved conservation awareness and attitudes and an increase in pro-conservation behaviors (Archabald and Naughton-Treves 2001; Lee et al. 2009; de Vasconcellos Pegas et al. 2013; Chaigneau and Daw 2015). The reduction in negative behaviors (e.g., poaching) and increase in positive behaviors (e.g., participation in conservation projects) can lead to positive ecological conservation outcomes (Holmes 2003; Mbaiwa 2011; de Vasconcellos Pegas et al. 2013).

However, whether a tourism activity results in overall positive conservation outcomes varies according to circumstance. The ideal situation occurs when a community that once hunted sharks can transition to working in shark tourism activities, as described by Bentz et al. (2014) in relation to shark fishing in the Azores and the subsequent rise in shark diving tourism. However, whale shark fishing is now banned in many jurisdictions and, where extraction still occurs (e.g., China), there are no developed tourism industries with economic benefits that can surpass extractive values. In addition, the people who gain the financial benefits from tourism are often not the ones who are undertaking the extractive activity. Hence, the conservation benefits of whale shark tourism are often somewhat indirect and intangible with management focusing on mitigation of possible negative environmental impacts. Nevertheless, working in community-based whale shark tourism can lead to improved pro-conservation attitudes and behaviors even if the locals involved did not hunt the sharks prior to the start of tourism activities (Ziegler et al. 2020). This chapter provides an overview of the global whale shark tourism industry and the main challenges it faces, as well as discusses management best practices to ensure this activity is contributing to the sharks' overall protection.

10.2 GLOBAL ASSESSMENT OF WHALE SHARK TOURISM

Whale shark tourism started in the late 1980s at several sites around the world where sightings were fairly predictable. It is possible that this interest was spurred by the rapid expansion of whale-watching tours and the fact that whale sharks could provide similar experiences (e.g., Duffus and Dearden 1993). These early sites include Ningaloo Reef in Australia, South Ari atoll in the Maldives, Bahia de Los Angeles and Bahia de La Paz in Mexico, and Phuket, Thailand. These sites are still among the premium sites for whale shark watching, but have been joined by many other sites worldwide.

The whale shark tourism industry has grown rapidly in the last 10 years (Figure 10.1). Tourism now occurs at 46 sites (ten aquariums, one seapen site, eight provisioned sites, 27 wild sites) in 23 countries, attracting over 25.5 million people (captive – 24.5 million, non-captive – 981,083), and worth an estimated US$1.9 billion in 2019 (captive – $1.8 billion, non-captive – $139.2 million; Table 10.1). The actual value of the non-captive industry is much greater, as reliable economic information is not available for many sites, especially in Asia. The economic value of captive whale shark tourism may be overvalued, on the other hand, as the approach used (i.e., using

---

[1] These statistics were generated, in part, from responses to a global survey that was distributed to whale shark operators and researchers at various sites around the world. We are very grateful to all those who contributed information.

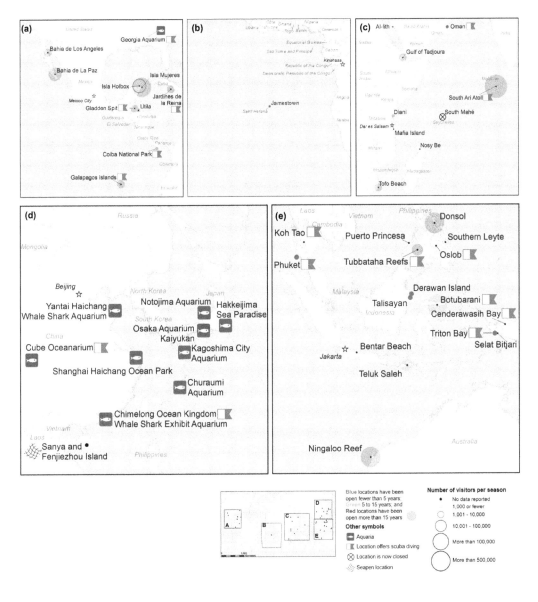

Figure 10.1 Map of captive and non-captive whale shark tourism sites around the world. (a) Americas. (b) South Atlantic. (c) Indian Ocean. (d) Asia. (e) Southeast Asia and Australia

the gross annual general admission ticket sales) assumes that whale sharks are a major draw for visitors to these facilities and the full value of the aquarium can be attributed to the presence of these sharks. However, no such research has been conducted, to our knowledge. Although whale shark-watching sites are spread throughout the tropics, the largest concentration of sites is in Southeast Asia; this region has also seen the most growth in tourism activities over the last decade (Figure 10.1).

Whale shark tourism differs from most other forms of shark tourism in that there are several different modes of interaction. The sharks can be observed at the surface from boats, by snorkel and SCUBA gear, or in captivity in large exhibits. Originally focused at sites where sightings were very predictable, other sites have developed recently where the normally transient whale sharks

Table 10.1 Wild Whale Shark Tourism Sites: Characteristics and Economics (US$)

Tourism Sites	No. Tour Operators	No. boats	No. Permits	Year Established	Visitation	Tour Cost	Annual Economic Returns	Regulatory Environment	Length of Season	Issues
Caribbean & Atlantic										
Georgia Aquarium, Atlanta, USA	n/a	n/a	n/a	2005	2.4 million	$39.95 admission $235.95 snorkel $335.95 scuba $469.95 rebreather	$315 million	Code of conduct	n/a	n/a
Isla Holbox, Mexico	30	25	25	2003	17,000	$150	$2.5 million	Legally enforceable regulations	4 months	Visitation has declined; crowding; lack of enforcement
Isla Mujeres, Mexico	110	200	243	2002	93,000	$150	$14.0 million	Legally enforceable regulations	4 months	Crowding; need better enforcement
Gladden Spit, Belize	>30	30+	–	1997	2,032	$195 (snorkel), $265 (dive)	$3.7 million	Legally enforceable regulations	6 weeks	Declining sightings; crowding
Utila, Honduras	15	25	none	1998	1,000	$60	$60,000	Voluntary code of conduct	year-round	Variable sightings, high non-compliance; lack of enforcement
Jardines de la Reina, Cuba	2	5–6	–	2002	1,250	$348–571 per day (liveaboard)	$6.5 million	Legally enforceable regulations	5 months	Variable whale shark sightings
Jamestown, St. Helena	4	4	–	2014	–	$112.50–250 for boat	–	Voluntary guidelines	3 months	–
Asia										
Churaumi Aquarium, Japan	n/a	n/a	n/a	1980	3.6 million	$16.25	$586.7 million	n/a	n/a	–
Hakkeijima Sea Paradise, Yokohama, Japan	n/a	n/a	n/a	–	4.8 million	$28	$134.4 million	n/a	n/a	–

(Continued)

Table 10.1 (Continued) Wild Whale Shark Tourism Sites: Characteristics and Economics (US$)

Tourism Sites	No. Tour Operators	No. boats	No. Permits	Year Established	Visitation	Tour Cost	Annual Economic Returns	Regulatory Environment	Length of Season	Issues
Kagoshima City Aquarium, Kagoshima, Japan	n/a	n/a	n/a	2000	691, 186	$14	$9.7 million	n/a	n/a	–
Notojima Aquarium, Japan	n/a	n/a	n/a	2010	–	$18	–	n/a	n/a	–
Osaka Aquarium Kaiyukan, Japan	n/a	n/a	n/a	1990	2.2 million	$22	$48.4 million	n/a	n/a	–
Chimelong Ocean Kingdom Whale Shark Exhibit Aquarium, Hengqin, China	n/a	n/a	n/a	2014	10.83 million	$56 viewing $254 scuba	$607 million	n/a	n/a	–
Cube Oceanarium, Chengdu, China	n/a	n/a	n/a	2015	–	$21	–	n/a	n/a	–
Shanghai Haichang Ocean Park	n/a	n/a	n/a	2018	~2 million	$45–55	$100 million	n/a	n/a	–
Sanya and Fenjiezhou Island, China	–	n/a	–	~2011	–	$130	–	–	–	Seapen tourism; sharks provisioned
Yantai Haichang Whale Shark Aquarium, China	n/a	n/a	n/a	2011	360,000	$19	$6.8 million	n/a	n/a	–
Donsol, Philippines	41 BIOs	56	–	1998	14,191	$75 for boat with up to 6 tourists	$800,000	Legally enforceable regulations	8 months	Sharks disappeared 2013, 2014 and 2018; crowding; visitation down
Southern Leyte, Philippines	10 dive shops, 50 KASAKA member	–	–	2005	800+	$40–100 pp	$56,000	Voluntary code of conduct	9 months	Highly variable sightings and tourist numbers

(Continued)

Table 10.1 (Continued) Wild Whale Shark Tourism Sites: Characteristics and Economics (US$)

Tourism Sites	No. Tour Operators	No. boats	No. Permits	Year Established	Visitation	Tour Cost	Annual Economic Returns	Regulatory Environment	Length of Season	Issues
Oslob, Philippines	1	>40	–	2011	508,000	$10 viewing $20 snorkel $30 diving	$31 million	Legally enforceable regulations	year-round	Overcrowding; shark provisioning causing changes in shark behaviors and ecosystem health; high physical contact rates between tourists and sharks
Puerto Princesa, Philippines	1	2	none	2007	200–300	$35 pp	$8,750	None present	7 months	Unregulated tourism; no guidelines
Tubbataha Reefs Natural Park, Philippines	–	15	–	1988	2,600	$2,000–4,000	$7.8 million	Legally enforceable regulations	4 months	Well managed, remote
Koh Tao, Thailand	70 dive shops	–	none	2010	100,000	$10 (snorkel); $25 (diving)	$1.8 million	None present	year round	Sightings recently increased in last year
Phuket, Thailand	80+	–	none	1990	–	–	–	–	8 months	Sightings recently increased
Bentar Beach, Probolinggo, Indonesia	–	12	none	2010	127	$0.75	–	–	4 months	Activities restricted to viewing from boat due to turbid waters
Botubarani Village, Gorontalo, Indonesia				~2016	~33,000	~$6	$17.8 million	–	~4 months	Sharks provisioned; crowding

(Continued)

Table 10.1 (Continued) Wild Whale Shark Tourism Sites: Characteristics and Economics (US$)

Tourism Sites	No. Tour Operators	No. boats	No. Permits	Year Established	Visitation	Tour Cost	Annual Economic Returns	Regulatory Environment	Length of Season	Issues
Cenderawasih Bay, Indonesia	4 liveaboards, 1 resort	–	–	~2009	~6,000	$225 (fast boats) $1,500–6,000 (liveaboard)	~$11.8 million	Voluntary code of conduct	year-round	No enforcement of code of conduct; whale sharks provisioned (tourists pay ~$20 for puri fish to feed the sharks at *bagans*); crowding
Derawan Island, Indonesia	–	–	–	2017	–	$50–100	–	Voluntary code of conduct	–	Lack of enforcement, overfeeding of sharks, high physical contact rates between tourists and sharks
Selat Bitjari, Kaimana, Indonesia	–	–	–	~2016	–	–	–	Voluntary code of conduct	year-round	Associated with *bagans*
Talisayan, Indonesia	1 liveaboard, unknown no. local operators offering day trips	–	–	2015	–	$3,000 (liveaboard)	$288,000 (liveaboard only)	Voluntary code of conduct	–	Lack of enforcement, overfeeding of sharks, high physical contact rates between tourists and sharks
Teluk Saleh, Sumbawa, Indonesia	6 operators, 1 liveaboard	7	–	2018	~600	$35	~$54,000	Voluntary code of conduct	year-round	Whale sharks associated with *bagans*
Triton Bay, Indonesia	1 dive shop, 4 liveaboards	–	–	2013	256 (liveaboard only)	$40–50 (snorkel, dive), $5,000 (liveaboard)	$1.2 million (liveaboard only)	Voluntary code of conduct	–	Whale sharks provisioned

(Continued)

Table 10.1 (*Continued*) Wild Whale Shark Tourism Sites: Characteristics and Economics (US$)

Tourism Sites	No. Tour Operators	No. boats	No. Permits	Year Established	Visitation	Tour Cost	Annual Economic Returns	Regulatory Environment	Length of Season	Issues
Australia										
Ningaloo Reef	11	15	15	1989	29,197	$325	$19 million	Legally enforceable regulations	6 months	Well managed; some issues with sharing sharks between operators
Indian Ocean										
South Mahé, Seychelles	1	2–3	none	1997–2014	560 in 2007, 60 in 2014, none since then	$155 (snorkel), $310 (ultralight)	$2.0 million	Voluntary code of conduct	4 months	Whale sharks disappeared in 2014, tourism no longer offered
South Ari Atoll, Maldives	150+	50+	none	1988	>100, 000	$97 (snorkel), $102 (dive), $247 (liveaboard)	$9.4 million	Voluntary code of conduct	year round	Low compliance with code of conduct; crowding, high physical contact rates; lack of enforcement; high whale shark injury rates due to tourism activities
Diani, Kenya				2005					4 months	
Tofo Beach, Mozambique	4–7	10	4–7	late 1990s	4,608	$45	$317,952	Voluntary code of conduct	year round	Declining sightings
Mafia Island, Tanzania	6	10	none	2005	2,002	$50–100	$100,000	Voluntary code of conduct	6 months	Increasing issues with crowding, lack of enforcement
Nosy Be, Madagascar	2	25	none	2011	5,000–8,000	$70 (snorkel), $110 (dive), $745 (liveaboard)	$1.5 million	Voluntary code of conduct	4 months	Some issues with crowding, compliance

(Continued)

Tourism Sites	No. Tour Operators	No. boats	No. Permits	Year Established	Visitation	Tour Cost	Annual Economic Returns	Regulatory Environment	Length of Season	Issues
Red Sea/Arabian Gulf/Horn of Africa										
Gulf of Tadjoura, Djibouti	2 or 3	21	none	2003	3,000+	$30–50 (skiffs), $150–200 (liveaboard)	$120,000	Voluntary code of conduct	5 months	Crowding serious problem at site, high physical contact rates, lack of enforcement, high whale shark injury rates
Oman	11	–	none	2017	–	$110 (dive)	–	Voluntary code of conduct	year round	Lack of enforcement
Al-lith, Saudi Arabia	1	2–5	none	2011	200–400	$400–700	$165,000	Voluntary code of conduct	3 months	Some issues with crowding, lack of enforcement
Eastern Pacific										
Galapagos Islands	7	8	8	2000	1,700	$5,000	$8.5 million	Legally enforceable regulations	5 months	Well managed
Bahia de Los Angeles, Mexico	28	28	28	1990	7,020	$11–44	$193,050	Legally enforceable regulations	6 months	Variable sightings; low compliance
Bahia de La Paz, Mexico	56	126	83	1990	45,000	$11–120 (snorkel), $1,700–3,000 (liveaboard)	$1.3 million	Legally enforceable regulations	7 months	Issues with compliance and high shark injury rates, trialing GPS chip in 2018 to improve compliance (see **Box 10.3**)
Coiba National Park, Panama	3–4	–	–	2005	–	$120–160 (day trip diving); $460–699 (multi-day diving trips)	–	–	4 months	Issues with boat strike injuries to the sharks; lack of regulations; crowding; sharks showing signs of avoidance

Sources: Allen (2018); Anna and Saputra (2017); Araujo et al. (2014); Araujo et al. (2017); Cagua et al. (2014); Cardenas-Torres et al. (2007); Carne (2005); Catlin and Jones (2010); S. Diamant (pers. comm., 2020); Djudnaidi et al. (2019); A. Dove (pers. comm. 2020); Graham (2004); Graham and Roberts (2007); Haiid and Abdul (2018); Hani (2019); Hanodoko et al. (2017); Hargreaves-Allen (2009); Huveneers et al. (2017); Imam et al. (2019); Kallsen (2018); Lupton (2008); Pierce et al. (2010); Price (2019); Rowat and Engelhardt (2007); Samodro (2016); K. Sato (pers. comm., 2018); Schleimer et al. (2015); Sea Stroll (2018); Theberge and Dearden (2006); Thomson et al. (2017); Wong et al. (2019); Yohana (2014); Ziegler et al. (2019a,b); Ziegler et al. (2012, 2016); and the expert surveys.

are attracted to stay via provisioning. Other sites have addressed the lack of natural aggregations by keeping whale sharks caught by fishers in seapens for several weeks. Most tourists interact with whale sharks at aquariums, followed by provisioned sites, wild sites, and then seapen tourism. Although there are significant gaps in the participation data (Table 10.1), the majority of sites where tourism activities occur are wild sites (Figure 10.1). This next section provides an overview of tourist activities at wild sites, provisioned sites, temporary seapen enclosures, and aquariums.

10.2.1 Wild Sites

Whale shark tourism at wild sites is possible because of the occurrence of predictable, seasonal aggregations of whale sharks. The industry has rapidly expanded in the last decade to encompass 27 sites attracting over 433,227 visitors in 2018 and worth approximately $77.2 million (Figure 10.1 and Table 10.1). Visitation among these sites varies wildly. Most sites have less than 5,000 participants per year, but there are some sites where numbers are estimated to be many times that amount (e.g., South Ari atoll, Maldives: > 100,000; Isla Mujeres, Mexico: 93,000; and Koh Tao, Thailand: 100,000). However, it should be emphasized that in most cases, the numbers provided are gross estimates; rarely are accurate data available. In Koh Tao, for example, the total number of divers per year is about 500,000. While the possibility to see whale sharks has been excellent in recent years, there are no data currently available on how many of the divers come specifically to see whale sharks. Hence, we have attempted to estimate that ratio, based on personal experience and conversations with other divers and dive operators on Koh Tao, at one in five divers, but obviously there is considerable room for error until further research is undertaken.

The length of the season can vary dramatically among sites, with some lucky to have 6 weeks (e.g., Gladden Spit, Belize), whereas others have virtually a year-round attraction (e.g., Mozambique, Honduras, Thailand, Maldives, Oman). Cardenas-Torres et al. (2007) noted high variability in whale shark seasonality in Bahia de Los Angeles, Mexico, with seasons starting as early as May in 2001 and as late as September in 2003 and ranging in length from 3 to 8 months. This variability makes tourism an unreliable source of additional income for locals (Cardenas-Torres et al. 2007).

There is also great variability in price of the tour from site to site depending on the mode of interaction and development level of the country (Table 10.1). The median price for such tours is $100. However, visitors can snorkel with whale sharks for as little as $0.03 at Botubarani, Indonesia, or SCUBA dive for as little as $25 at Koh Tao, Thailand. Alternatively, snorkeling at Gladden Spit in Belize costs $195, and SCUBA diving is $265, while a live-aboard SCUBA diving trip to the Galapagos may cost more than $5,000. In general, it is better in conservation terms to have fewer tourists paying more money per person to minimize the impact of ecotourism activities (Dearden et al. 2006).

10.2.2 Provisioned Sites

Feeding fish for tourist viewing is a widespread and growing activity that affects the target animals, the ecosystem, and the nature of the tourist experience (see Patroni et al. 2018 for a review). Much of the research has focused on shark provisioning, but relatively little attention has been directed towards whale sharks, perhaps due to the relatively recent development of the activity. However, whale shark provisioning activities have grown rapidly since first appearing in 2009 in Cenderawasih Bay, Indonesia, where whale sharks are attracted to large fishing platforms or "bagans," which have large nets full of small baitfish below them (Figure 10.2). There are now eight known sites where whale sharks are provisioned, attracting over 547,856 people and worth over $62.1 million in 2019.

Provisioning activities vary by site. For example, whale shark tourism is an incidental activity at the bagans in Cenderawasih Bay, Indonesia. The primary activity at these platforms is "puri" fishing, in which fishers use lights to attract baitfish to the bagans during the night. The whale

Figure 10.2   Whale shark at a fishing bagan in Cenderawasih Bay, Indonesia. (Photo: Alistair Dove / Georgia Aquarium.)

sharks come to feed on these aggregations of baitfish in between net hauls. During the day, the fishers handline for larger fish and, at the same time, feed the sharks small amounts of puri from the previous night's catch. They do this for good luck, to prevent interference with their line fishing efforts, and in case any, tourist boats show up that may be interested in paying to swim or dive with the sharks (A. Dove, pers. comm.).

In Oslob, Philippines, on the other hand, tourism is the primary goal of provisioning the sharks. It is the largest non-captive whale shark tourism site in the world in terms of number of visitors each year. The whale sharks are hand-fed in a small viewing area. Tour operators paddle small outrigger canoes 50–100 m out from shore, and tourists can view the sharks from the boat or enter the water (Figure 10.3). The site is community-run, and tour operators are all from Oslob, or related to someone who is. Crowding is a serious concern at this site, with over 96% of respondents to a 2016 survey reporting feeling crowded due to the number of boats in the viewing area, along with the almost complete lack of compliance to, or enforcement of, the whale shark encounter guidelines and changes to the sharks' natural feeding and migratory behaviors (Araujo et al. 2014, 2020; Schleimer et al. 2015; Thomson et al. 2017; Ziegler et al. 2018, 2019a, b; Legaspi et al. 2020; Penketh et al. 2020).

The ethics of provisioning a wild, highly mobile, endangered animal needs to be further assessed (Ziegler et al. 2018). In the case of Oslob, while there are significant economic benefits for the community (e.g., extrapolations from previous assessments suggest that the industry was worth approximately $31 million in 2018; Table 10.1), it is unclear if these benefits outweigh the costs to the sharks, the local community, and greater marine environment (Ziegler et al. 2018, 2019a, b). Studies have attempted to determine the impacts of provisioning activities on the sharks at Oslob in the Philippines. Araujo et al. (2014) concluded that differences in both residency and probability of being resighted of provisioned and non-provisioned whale sharks suggest that provisioning is leading to behavioral change in Oslob. Schleimer et al. (2015) found that sharks at the site demonstrated both associative learning (i.e., associating the site with food and adapting their feeding behavior) and habituation (i.e., increased tolerance levels to proximity to people and other sharks), which suggests that a combination of habituation and conditioning is changing whale shark behavior while at the site. Whale sharks with greater experience at the site were also more likely to display vertical feeding behavior, which indicates that sharks are capable of learned behavior as a result of conditioning from positive reinforcement (Schleimer et al. 2015). Furthermore, experienced whale sharks

Figure 10.3 Tourists swimming and diving with whale sharks in Oslob, Philippines. (Photo: Simon Pierce.)

were less likely to exhibit avoidance behaviors in response to either active or passive contact from humans, and contact with other sharks, which suggests that a combination of habituation and conditioning is changing whale shark behavior while at the site (Schleimer et al. 2015). These results were further corroborated by a similar study by Legaspi et al. (2020), who found that provisioned whale sharks with a long visitation history at the site were more likely to feed steadily either by horizontal or vertical feeding and did not react to external anthropogenic stimuli, indicating habituation at the site. This study also found that 93% of tourists were non-compliant to the minimum distance from the shark, showing high tourism pressure on the sharks. Thomson et al. (2017) found that some whale sharks exhibited prolonged residency at the site and concluded that provisioning could influence foraging success, alter whale shark distributions, and lead to dependency as the sharks get older. Araujo et al. (2020) estimated a 7.2%±3.7% higher metabolic rate when sharks frequent the provisioning site, which can affect their fitness in the long term. Furthermore, Penketh et al. (2020) assessed the scarring patterns of whale sharks at Oslob as a measure of the physical impacts of the tourism activities on individual sharks and compared these values to other whale shark sites in Australia, Mozambique, and Seychelles. They found that 94.7% of sharks identified ($n=144$) had at least one scar, and 90.8% ($n=138$) had more than one scar, with the amount of scarring significantly higher in those sharks that visited the provisioning site more frequently (Figure 10.4). Furthermore, sharks at Oslob had much higher rates of scarring than other, non-provisioned whale shark tourism sites globally. These findings suggest that provisioning activities are consumptive in nature, as they are affecting whale shark fitness and possibly survival. A recent study also found that the provisioning activities in Oslob resulted in the degradation of the local coral reefs, including higher microalgal cover, lower coral cover, and dominance of stress-tolerant coral genera compared to a control site (Wong et al. 2019).

Many fishers working in tourism at this site reported a significant decrease in the time they spent fishing since they started working in tourism reflecting a potential decline in fishing effort at the site (Ziegler et al. 2020). However, the large number of tourists visiting this site relative the local population (508,000 tourists in 2018 vs. 27,893 municipal population of Oslob) will increase demand for food in Oslob, which will increase pressure on the already depleted local marine resources and likely wipe out any positive benefits of reduced fishing pressure from tourism activities in local waters (Ziegler et al. 2020). Furthermore, these benefits do not extend to neighboring barangay waters. For example, fishers from a neighboring barangay within the municipality of Oslob reported

Figure 10.4   Whale shark with injuries caused by boat propellers at Oslob, Philippines. (Photos: Large Marine Vertebrates Research Institute, Philippines.)

problems with reduced catch, the use of illegal fishing methods, and commercial vessels fishing in municipal waters. Biological studies are needed to assess the health of local coral reefs and fish stocks at Oslob before conclusions can be drawn regarding the positive impact of whale shark tourism on ocean conservation (Ziegler et al. 2020), while socio-economic studies are needed to determine if the economic benefits to the community outweigh the costs to the sharks, community, and wider marine environment before a final determination regarding the ethics of the activity can be made (Ziegler et al. 2018).

10.2.3 Seapen Tourism

A recent trend in Asia, particularly, is to keep whale sharks in "temporary exhibits," which occur when local fishers opportunistically catch whale sharks and keep them in seapens similar to those used in fish farming. These seapens turn into temporary whale shark exhibits, where tourists pay to view the sharks and interact with them in the pens. The sharks are kept on the scale of weeks before being released and potentially replaced with a freshly caught shark. Such exhibits are found off Sanya and Fenjiezhou Island in southern China, although additional locations are likely (Obura et al. 2013). For example, a company applied for a license to keep two whale sharks in a 2,000 m by 600 m enclosure (termed a "community-based marine sanctuary") off the coast of Diani, Kenya, for tourism, research, and conservation purposes in 2012 (Reinl 2013). They argued that the ticket price of $134 to snorkel with the sharks would help fund conservation projects to protect the sharks, as well as fund a research center and potential breeding program (Reinl 2013). Although their application was rejected, it does highlight the interest in monetizing the artificial presence of whale sharks in such enclosures when feeding aggregations do not naturally occur at the location.

10.2.4 Aquarium Tourism

Whale sharks are kept in captivity at ten known aquariums, all but one of them in Asia. Valuing this sector is difficult because keeping whale sharks in captivity is a controversial topic and many aquariums are reticent to provide information. Economic information was only available for two aquariums – the Georgia Aquarium, USA ($315 million; A. Dove, pers. comm., 2020), and Okinawa Churaumi Aquarium, Japan ($586.7 million; K. Sato, pers. comm., 2018). Along with visitation and ticket prices for six of the remaining eight aquariums, over 24.5 million people spent over $1.86 billion to see whale sharks in 2019. This approach assumes that whale sharks are one of the major draws of the aquarium and therefore the revenue generated by the aquarium can be attributed to the presence of whale sharks on exhibit. While the sharks are likely an important factor in attracting visitors – many of these aquariums showcase whale sharks in their advertising materials and have whale shark statues and other multimedia throughout their facility – further research is needed to determine how important whale sharks are in attracting tourists to these aquariums.

Visiting aquariums can allow people to form an emotional connection with animals they may never otherwise encounter; this emotional connection can, potentially, improve social and ecological conservation outcomes for the sharks through an increase in pro-conservation attitudes, beliefs, and behaviors (Skibins et al. 2013; Jensen et al. 2017; Ballantyne et al. 2018; Godinez and Fernandez 2019). At least two of these aquariums, namely, the Georgia Aquarium and Churaumi Aquarium, are non-profits that also conduct field research or fund projects that provide tangible conservation benefits for wild whale shark populations, including whale shark tagging (de la Parra Venegas et al. 2011) and DNA analysis (Alam et al. 2014; Read et al. 2017).

Most aquariums only allow the viewing of the sharks, but the Georgia Aquarium (USA), Chimelong Ocean Kingdom Whale Shark Exhibit Aquarium (China), and the Cube Oceanarium (China) also allow visitors to snorkel and scuba dive in the main habitat with the whale sharks. These activities, worth an estimated $4.1 million in 2018, could conceivably introduce an element of competition with wild sites, although it is also plausible that those who see a whale shark in an aquarium may be more likely to visit a non-captive whale shark site. Aquariums may therefore act as a gateway for some people who may not have otherwise chosen to see the animals in the wild (Duffus and Dearden 1993).

There are, however, concerns with keeping a highly mobile, long-lived, endangered shark within an enclosed exhibit. It is unclear whether all aquariums have facilities of sufficient size or sufficient husbandry knowledge for long-term care of whale sharks. Mortality rates in aquariums are unknown, but high rates could result in continued pressure on already dwindling wild populations to supply new sharks for exhibit. The popularity and significant economic returns from having whale sharks on exhibit (i.e., 24.5 million visitors generating $1.8 billion in 2019) may incentivize result the black-market capture and sale of wild whale sharks (Master 2018; The Strait Times 2018). For example, in 2016, a major supplier of large marine megafauna to the illegal international wildlife trade was found in possession of two juvenile whale sharks in a seapen in Indonesia, awaiting transport to China (Hilton 2016). The poachers were selling the sharks for $1,476. The scope of the problem is assumed to be small, given the few facilities that could consider holding whale sharks, but is hard to assess given the difficulty in tracking the illegal trade in endangered species and the fact it is often linked with other criminal activities and international gangs (McConnell 2017; Winter 2018). However, given the fact that whale sharks are a protected species in most countries (including China), and the growing popularity of marine parks in China – 60 marine parks are already operating, and a further 36 large-scale projects are planned by 2020, many of which feature threatened species like whale sharks, dolphins, and manta rays (Master 2018; The Strait Times 2018) – it is possible that at least some of the sharks on exhibit are from the black market. Further research is obviously required on the sourcing of whale sharks for aquaria, along with investigations on whether tourists who visit aquariums with whale sharks are more likely to become involved in conservation activities, are stimulated to visit whale sharks in the wild, have improved pro-conservation attitudes or awareness towards whale sharks, and/or are more likely to donate money towards conservation.

10.3 ECONOMIC VALUATION OF WHALE SHARK TOURISM

Direct use value refers to the economic benefit from the direct use of an animal. This can either be consumptive, through the capture and sale of the animal, or non-consumptive through wildlife tourism (Catlin et al. 2013). Many whale shark fisheries are illegal, with the sharks caught and sold for meat and fins for restaurants (mainly for shark fin soup), oil for cosmetic and health products (e.g., lipstick, skin care, nutritional supplements), and skins for bags (Hilton and Hofford 2014). The total number of whale sharks caught and sold at market annually is unknown. China is the largest market globally for whale sharks and their by-products. Li et al. (2012) found that the recorded catch of whale sharks in China was grossly underreported: approximately 1,000 whale sharks were landed each year in coastal Chinese waters, although only 186 whale shark landings were reported between 1983 and 2011. In 2014, a factory in Zhejiang Province, China, was found to slaughter and process more than 600 whale sharks per year, despite the sharks being a protected species in China (Hilton and Hofford 2014). It is believed that there were many such factories in the region, and the number of whale sharks processed at these plants was therefore presumably much higher; however, see Chapter 12 for an updated assessment of Chinese whale shark fisheries.

The price at which whale sharks have been sold has varied widely, depending on location and shark size. The average market price for whale sharks in China was estimated to be $30,000 for a whole shark (Lee 2014), although in Taiwan, prices ranged from $14,000 for a 2 t shark to $70,000 for a 14 t shark during the peak demand period (Chen and Phipps 2002). Whale shark meat sold for up to $12–17/kg in Taiwan, while individual fins could reach prices of $57,000 (Clarke 2004). Chen and Phipps (2002) reported market prices of individual whale sharks in Taiwan from 1999 to 2002 of $7,116 to $23,543, with wholesale prices ranging from $2.10 to $6.92/kg and retail prices ranging from $4.91 to $17.16/kg. In Hong Kong, whale shark fins could sell for $247 in 2010 (Paddenburg 2010). Although they were previously considered to be of relatively poor quality and mainly used in restaurant displays to advertise the availability of shark fin soup (Chen and Phipps 2002), Li et al. (2012) reported that, by then, whale sharks were considered a high value shark and emphasized the potential for their meat and fins to increase in price with rising demand.

Norman and Catlin (2007), however, point out that such high prices do not reach the fishers who land these sharks. For example, a 370 kg whale shark caught in China in 2016 was sold for only $0.40/kg (Griffiths 2016; PDO 2016). Another Chinese fisherman planned to sell a 2 t whale shark caught in 2014 for $1,500–3,500, or $0.75–1.75/kg, at a market before the police intervened (Newport 2014; Withnall 2014). A 3–4 t whale shark caught in 2014 in Pakistan sold for about $250 ($0.07/kg; Ilyas 2014), while a 7.7 t whale shark sold for $2,200 ($0.29/kg) in 2012 (Vafeiadis 2012). In 2008, a seafood hawker in Hong Kong paid a fisherman the equivalent of $2,550 to release a live whale shark that he was planning to sell at the market (Chiu and Lo 2008). Similarly, a Russian tourist paid Indonesian fishers $75 to release a live whale shark they had brought to the beach (Pearl 2016). Norman and Catlin (2007) concluded that this discrepancy between the market value and the price paid to local fishers "clearly highlights that there is minimal economic benefit on a regional/ local scale from whale shark fishing, as the high prices are realized elsewhere."

The direct use value from tourism can bring significant economic returns to local communities. The valuation of whale sharks from tourism activities can occur at different scales, from individual animals to tourism sites to the national or global scale. For example, Graham (2004) estimated the individual value of a whale shark at Gladden Spit, Belize, at $34,906 annually and $2,094,340 over its lifetime (assuming life expectancy of 60 years), or $104,718 per shark, per year, if it visits multiple aggregation sites in the region (Table 10.2). Norman and Catlin (2007) valued an individual whale shark at Ningaloo, Australia, at $9,000 per year and $216,000 over its lifetime (assuming a 24-year generation gap). Although attaching a value to an individual shark is appealing from a conservation campaign standpoint and makes for a persuasive argument (i.e., the animal is worth more alive than dead), Catlin et al. (2013) argue that the individual valuation approach is not scientifically rigorous. They argue that such numbers are based on several assumptions, including a direct and positive relationship between the presence of wildlife and tourism value, a known population size of the targeted species at a specific location, the period over which the species is valued (e.g., life expectancy vs. generation gap), that all animals in a given population are equally exposed to tourism activities (which disregards demonstrated cases of habituation, attraction, and avoidance), and that estimated values of individual whale sharks for one location can be extrapolated to other locations (e.g., assuming that a whale shark at a smaller tourism site would have the same value as one from a more established site like Ningaloo).

Table 10.2   Tourism Value of Whale Sharks

Location	Economic Value (USD)	Number of Sharks	Individual Valuation	References
Ningaloo	$19 million	440[a]	$43,182 (individual); $1.0 million[b] (lifetime)	Huveneers et al. (2017)
Gladden Spit	$3.7 million	106	$34,906 (individual); $2.1 million[c] (lifetime)	Graham (2004)
Isla Mujeres	$14 million	1101[a]	$12,716 (individual); $305,177 (lifetime)[b]	Table 10.1
Oslob	$31 million	208	$149,462 (individual); $3.6 million[b] (lifetime)	Thomson et al. (2017); Table 10.1
South Ari Atoll, Maldives	$9.4 million	101[a]	$93,069 (individual); $2.2 million[b] (lifetime)	Cagua et al. (2014)

[a] used the total number of whale sharks identified over calendar identified at each location from Norman et al. (2017) to approximate population size in given season.
[b] used 24-year generation like Norman and Catlin (2007) to account for fact that aggregation is dominated by juvenile males.
[c] used 60-year life expectancy.

It is, therefore, more accurate to discuss the tourism value of whale sharks at the local or national level. The value of non-captive whale shark tourism at the local level ranges from a few thousand dollars at smaller sites like Puerto Princesa (Philippines) and Shib Habil (Saudi Arabia), to multi-million dollar industries at more established locations like Ningaloo (Australia) and Isla Mujeres (Mexico; Table 10.1). Graham (2003) estimated the national value of the 6-week whale shark tourism season to be $3.7 million, with a local value of $1.35 million, for the five communities involved in the management of the Gladden Spit Marine Reserve. Huveneers et al. (2017) valued the whale shark tourism industry at Ningaloo, Australia, at $8.8 million for the local economy and an additional $9.6 million to the regional economy for dedicated shark divers who specifically came to Ningaloo to swim with whale sharks. Using a different approach, Davis and Tisdell (1999) assessed visitors' willingness to pay for the whale shark encounter at Ningaloo and found that respondents were willing to pay a maximum of $230 (range $40–1,150), while Ziegler (2010) found that whale shark tour participants at Holbox, Mexico, were willing to pay a maximum of $151–200 for the experience.

Regardless of any issues with the validity of methodologies used, it is evident that whale shark watching generates a significant amount of money. At the global scale, the non-captive whale shark tourism industry is worth over $139 million annually (Table 10.1), and a significant proportion of this goes to developing nations.

10.4 WHALE SHARKS AND ECOTOURISM

Activities such as whale and shark watching have often been promoted by conservationists as a way to attach a higher economic value to protecting species for viewing rather than extracting them for consumption (Duffus and Dearden 1993). There are three key components to ecotourism. (a) The environment: ecotourism is a low-impact activity that occurs in natural environments and should contribute to the conservation of local flora and fauna; (b) local community: ecotourism should encourage local participation and management of socio-economically sustainable tourism activities for the local community; and (c) education: ecotourism should provide meaningful learning opportunities for tour participants (Sharpley 2006). In addition, for any tourist endeavor to be successful, it must continue to attract tourists, implying that they must continue to find the attraction satisfactory. This section looks at some of the challenges whale shark viewing faces in meeting these criteria, namely, through environmental impacts, community involvement, educational awareness, and tourist satisfaction, and some of the mitigating approaches that managers can take. We finish with an overall evaluation of whale shark tourism as ecotourism and a potential IBC mechanism.

10.4.1 Impacts of Tourism on Whale Sharks

There are numerous actual and potential impacts of whale shark tourism activities (Table 10.3). *Actual* impacts are defined as those that are measurable by current scientific techniques, while *potential* impacts are those that are possible based on the wildlife tourism literature and knowledge of whale shark biology/ecology. Interpreting the results of impacts is difficult. For example, one generally cannot tell if sharks with boat propeller scars are due to collisions with whale shark

Table 10.3   Real and Potential Negative Impacts of Whale Shark Tourism Activities on the Sharks

Tourism Impacts	Studied	Issues	References
		Actual	
Injuries (e.g., boat strikes)	Yes	Difficult to determine if observed injuries due to tourist boats or other private or commercial boats	Cardenas-Torres et al. (2007); Rowat et al. (2007); Lester et al. (2020); Penketh et al. (2020);
Short-term behavioral changes	Yes	Unclear if short-term behavioral changes affect shark fitness over longer term No control available for "natural" whale shark behaviors in absence of tourists	Araujo et al. (2017, 2020); Haskell et al. (2015); Legaspi et al. (2020); Lester et al. (2019); Norman (1999); Pierce et al. (2010); Quiros (2007); Schleimer et al. (2015); Thomson et al. (2017)
Changes in whale shark visitation/residency	Yes	Sharks may return to a given site even if experience disturbance; does not mean they are not experiencing stress May still experience reduced fitness	Araujo et al. (2014); Sanzogni et al. (2015); Thomson et al. (2017)
Disruption of normal feeding activities	Yes	Unknown if short-term disruptions to normal foraging behavior translates into longer term impacts on overall animal fitness	Quiros (2007); Thomson et al. (2017)
Physiological stress	No	Difficult to obtain blood samples without affecting results Difficult to obtain baseline value from sharks not affected by tourism	
Disease transfer	No		
Parasite loads	No		
		Potential	
Longer-term behavioral changes	Yes	Longest period assessed is 4 years Only completed at a few sites	Araujo et al. (2017); Haskell et al. (2015); Thomson et al. (2017)
Reduced health/fitness	No	Difficult to assess in long-lived, highly migratory marine species	
Reduced reproductive success	No	Difficult to assess in species for which mating behavior has rarely been observed in the wild, do not know where breeding/pupping occurs, and no parental care	
Loss of individuals	No	Difficult to assess in long-lived, highly migratory marine species	

tour boats or other private or commercial vessels at the site, or have occurred elsewhere, due to the highly migratory nature of whale sharks. Furthermore, only two preliminary studies (Rowat and Brooks 2012; Raudino et al. 2016) have included a control group of sharks without the presence of tourism, so there is no baseline benchmark of "normal" whale shark behaviors in the absence of tourism activities. Preliminary focal animal monitoring in the presence and absence of swimmers and boats has been completed in the Seychelles and Australia; however, initial findings are contradictory. Rowat and Brooks (2012) found that whale sharks spend significantly less time at the surface when tourists or boats are present, while Raudino et al. (2016) found that whale sharks at Ningaloo maintained neutral behaviors (e.g., surface swimming, swimming at depth, resurfacing, no reaction) while in the presence of tourist boats. The authors did, however, find that whale sharks were significantly more likely to make directional changes while in the presence of tourist boats, suggesting that tourism may be affecting the sharks (Raudino et al. 2016). More research is needed to better understand natural whale shark behavior in the absence of tourists.

Attempts to assess the impacts of whale shark tourism on the sharks are relatively new and have focused primarily on short-term behavioral studies (Table 10.4). These techniques involve creating an ethogram of a given shark's observed behaviors (e.g., feeding, diving, swim speed and direction) when interacting with humans in order to determine if there is a marked change in the shark's behavior in response to given human behaviors (Table 10.5). Araujo et al. (2017) found that over 50% of interactions in Southern Leyte ended within 2 minutes, mostly due to the sharks diving away from the surface, suggesting short-term disturbance was occurring. Norman (1999) found an increase in the frequency of avoidance behaviors at Ningaloo over a 3-year period, from 56% in 1995 to 71% in 1997, suggesting that the sharks' tolerance to tourism activities may have decreased over that time period. A similar pattern was observed in Mozambique, where Pierce et al. (2010) found that 34.6% of observed encounters between an individual shark and swimmers resulted in avoidance responses at Tofo Beach, Mozambique, while Haskell et al. (2015) found that 64.7% of observed interactions resulted in avoidance responses at the same site, suggesting that the sharks' tolerance for tourism activities also decreased over the study period. Furthermore, Pierce et al. (2010) and Haskell et al. (2015) found that sharks exhibiting avoidance behaviors resulted in shorter encounters, indicating that tourist presence did affect the shark's short-term behavior. However, Pierce et al. (2010) noted that encounter length was also dependent on the shark's prior reaction to the boat, such that in-water

Table 10.4 Whale Shark Tourism Impact Studies

Method Used	Location	Length of Study	References
Behavioral	Tofo Beach, Mozambique	20 months	Pierce et al. (2010)
		30 months	Haskell et al. (2015)
	Oslob, Philippines	2 years	Schleimer et al. (2015)
		1 year	Araujo et al. (2020)
		2 years	Legaspi et al. (2020)
	Donsol, Philippines	2 years	Quiros (2007)
	Leyte, Philippines	4 years	Araujo et al. (2017)
	Ningaloo, Australia	3 years	Norman (1999)
	Bahia Los Angeles, Mexico	2 months	Montero-Quintana et al. (2020)
Photo-ID and/or logbook data	Ningaloo, Australia	5 years	Sanzogni et al. (2015)
		1 year	Lester et al. (2019)
		6 years	Lester et al. (2020)
	Oslob, Philippines	21 months	Araujo et al. (2014)
		3 years	Thomson et al. (2017)
		3 years	Penketh et al. (2020)

Table 10.5   Known Whale Shark Avoidance Behaviors and the Tourist Behaviors that Elicit Such Responses

Shark Avoidance Behaviors	Definition
Banking	Shark rolls laterally and presents dorsal side to perceived threat
Eye-rolling	Shark retracts its eye as a defensive measure
Fast swimming	Shark increases swim speed in response to perceived threat
Diving	Shark dives away from surface
Change in direction	Shark alters its direction of swimming in response to swimmers
Violent shudder	Shark shakes its body violently along lateral axis
Tourist Behaviors	**Definition**
Free diving	Person swims below surface to get closer to shark
Touching	Person makes physical contact with shark
Flash photography	Photography with flash, strobes or lights
Splash entry	Person forcefully enters water
Blocking path of shark	Person places body in path of oncoming shark
Swimmer proximity	Person too close to shark based on encounter guidelines

interactions with sharks displaying boat avoidance behaviors were significantly shorter than interactions with sharks that did not display such avoidance behaviors.

There are, however, conflicting results regarding the effect of swimmer presence on whale sharks. For example, Araujo et al. (2017) and Haskell et al. (2015) found that the number of previous encounters with tourists a shark experienced did not affect interaction duration in Leyte and Mozambique. Quiros (2007), meanwhile, found that sharks that had been swum with repeatedly were more likely to exhibit a dive response compared to those that had been sighted for the first time, which suggests that sharks learn tourist avoidance over time. Araujo et al. (2017), Haskell et al. (2015) and Quiros (2007) found that encounter duration was longer with scarred individuals, while Norman (1999) found the opposite to be true at Ningaloo. However, Norman (1999) reported that sharks with recent propeller wounds appeared calm with swimmers but would dive upon hearing the sound of an approaching vessel; vessel avoidance is therefore the likely cause of the avoidance behavior and not swimmer presence. Pierce et al. (2010), meanwhile, found that scarring had no effect on likelihood of exhibiting avoidance behaviors and therefore interaction length, although the authors noted that they did not specifically examine boat injuries in their analysis. Both Haskell et al. (2015) and Araujo et al. (2017) found that there was no relationship between shark size and interaction length, suggesting that these sharks do not display avoidance learning (i.e., larger, presumably older sharks avoiding tourist interactions due to previous experiences with tourists over past seasons). Araujo et al. (2017) concluded that the variability in observed interaction durations between sharks and tourists was likely due to differing personalities among the sharks, with some sharks more sensitive to tourist behaviors than others.

It is difficult to interpret the results of behavioral studies because we do not know if the observed short-term behavioral changes translate into longer-term impacts on shark health and fitness. Haskell et al. (2015) and Araujo et al. (2017) did not find any long-term behavioral changes in individual sharks, despite sharks displaying avoidance behaviors in the short term. For example, individual whale sharks were not more likely to display avoidance behaviors if they had done so in the past. Norman (1999) concluded that whale sharks appeared to be tolerant of any tourism pressure present at Ningaloo because they did not actively avoid swimmer interactions. Lester et al. (2019) used photo identification and electronic monitoring system data to assess the impacts of tourism activities on sharks at Ningaloo and concluded that there was no evidence of long-term tourism impacts, since individual sharks did not demonstrate declines in encounter length after repeated tourist interactions, even though individual sharks had an average of 2.4 interactions with tourists per day and up to 16 times over the course of the 2-year study. However, the studies varied in length from 1 to 6 years (Table 10.4); it is unclear if this is a sufficient timeline to identify so-called "long-term" impacts. Furthermore, none of the impact studies used baseline values or a control population of sharks without tourism present, nor do they assess the sharks' physiological state in the presence of tourism activities. Assessing the effects of tourism on the sharks would best be accomplished using physiological indicators (e.g., changes in hormone levels, heart rate, metabolic rate, growth rate; Bateman and Fleming 2017). For example, an increased heart rate in the presence of tourists would result in increased energy expenditure even if there was no observed behavioral response to a tourist disturbance (Bateman and Fleming 2017). Stressed animals may divert energy reserves away from the immune system, thereby becoming immunocompromised; immune suppression can result in animals carrying higher parasite loads (Bateman and Fleming 2017). Other potential impacts of tourism activities can include increased metabolic rates, reduced growth rates, increased rate of injuries or slower wound healing, and reduced reproductive success (Bateman and Fleming 2017).

Semeniuk et al. (2009) highlighted the main challenges in determining the impacts of tourism activities on marine animals, especially non-mammals, namely, (a) the difficulty in accessing and/or observing animals that do not need to breathe at the surface, (b) the difficulty in measuring reproductive success in animals that do not have communal nursing grounds, or do not have any parental care, and (c) their long-lived nature precluding the direct measurement of mortality. Due to these difficulties, the authors suggest the use of physiological indicators to measure tourism impacts on non-marine mammal species, since these indicators are able to reflect the health state and predict survival and reproduction of animals exposed to tourism activities. Possible indicators include hematocrit, total serum protein concentration, blood pH, cortisol, lactate, glucose, differential white blood cell counts, and antioxidant capacity and oxidative status (Semeniuk et al. 2009; A. Dove, pers. comm.). Semeniuk et al. (2009) used physiological indicators as evidence that provisioned stingrays at Stingray City Sandbar, Cayman Islands, were exhibiting symptoms of immunosuppression when compared to stingrays at non-tourist sites. The authors concluded that while there was no direct evidence of reduced survival at the site, the physiological parameters measured suggest that it is quite likely. The authors further noted that the use of physiological stress indicators to determine tourism impacts is preferred over behavioral studies, as it provides a better understanding of the actual costs to the animals, especially when long-term population data are not available.

In the absence of any physiological data from whale shark tourism impact studies, researchers attempting to ascertain the longer-term impacts of short-term behavioral changes can use whale shark aggregation demographics. For example, most aggregations across the globe are dominated by juvenile males, suggesting that tourism activities do not disrupt reproduction of the species at these sites, as reproduction is not likely to occur in any case (e.g., Philippines, Mozambique, Tanzania, Djibouti, Australia, Mexico; Araujo et al. 2017; Haskell et al. 2015; Sanzogni et al. 2015). However, Miranda et al. (2020) note the potential of Donsol, Philippines, to be an important reproductive habitat for whale sharks due to anecdotal reports of mating and precopulatory behavior and the sighting of neonatal sharks (i.e., 0.4–0.6 m in total length). It is important to note that this aggregation is fairly unique, in that it also has a number of adult males and females present (McCoy et al. 2018), unlike most other (juvenile-dominated) aggregation sites. Feeding habits are another characteristic that can be used to assess longer-term tourism impacts. At Ningaloo, for example, whale sharks feed at night and not during the day, when tours occur, suggesting that tourism activities do not disrupt whale shark feeding (Sanzogni et al. 2015). In Donsol, whale sharks do feed during tour hours and feeding sharks were more likely to exhibit avoidance behaviors including violent shudders, banking, and abrupt changes in direction, suggesting that sharks were diverting their energies from feeding to avoidance behaviors, which can affect their long-term survival (Quiros 2007).

Shark residency or site fidelity is another criterion used to assess tourism impacts, as those aggregations with more resident sharks (e.g., Southern Leyte in the Philippines, Mafia Island in Tanzania) will likely experience greater tourism impacts than those consisting of mostly transient sharks (e.g.,

Honduras, Mozambique; Araujo et al. 2017; Haskell et al. 2015). Furthermore, tourism disturbance could affect site residency or fidelity, as individual sharks may no longer return to the aggregation site in future seasons due to disturbance from tourism activities at the site. Consequently, if sharks continue to return to the site, it suggests that they are not experiencing significant disturbance from tourism activities. For example, Sanzogni et al. (2015) assumed that if tourism activities were having negative impacts on the sharks at Ningaloo, that "total encounters per shark would decline, the day of first encounter would occur later each season, the days between encounters within a season would lengthen, residency time would decrease, and the rate of departure should increase" (Sanzogni et al. 2015). The authors concluded that since sharks continued to return to key feeding habitat, they are not experiencing sublethal impacts from tourism activities at this site.

However, apparent wildlife tolerance of human presence is not necessarily a sign that tourism activities are harmless to the focal species (Higham and Shelton 2011). Bejder et al. (2009) outlined three factors that can explain this apparent tolerance: (a) displacement – less tolerant individuals are displaced leaving more tolerant individuals at the site; (b) physiology – physiological impairment prevents an individual from reacting to human presence; and (c) ecology – there is no suitable habitat to which the animals may relocate. Shark residency or fidelity, therefore, may not be a good predictor of tourism disturbance (Araujo et al. 2017).

More research is needed to understand the impacts of whale shark tourism activities. Without data regarding the physiological impacts of these activities on the sharks, it is difficult to state with any scientific certainty that these activities do not negatively affect the health/well-being of the sharks in the long term. When scientific data are unavailable, especially when dealing with an endangered species in critical habitat, it is important to use a precautionary approach that minimizes potential negative impacts (Sorice et al. 2003).

10.4.2 Community Involvement

The success of IBC projects in meeting social and ecological conservation goals is highly dependent on local participation in these programs (Salafsky et al. 2001; Brooks et al. 2013). At some sites, tourism activities were developed at the impetus of the local community, and they are directly involved in their management (e.g., Donsol, Oslob, Holbox, Isla Mujeres, Bahia La Paz). In Donsol, whale shark tourism developed after locals contacted the government regarding the tourism potential of a large aggregation of whale sharks in local waters. After news coverage resulted in poachers killing seven of the sharks, the waters were declared a whale shark sanctuary and the sharks were protected at the national level (Quiros 2005). WWF-Philippines worked with the Donsol Municipal Tourism Council, local government, and local fishers to develop a community-based ecotourism program, including the establishment of the marine sanctuary, the setting of fees and regulations, and the training of guides and boatmen (Yaptinchay 1999, as cited by Quiros 2005). Boatmen and guides are selected using an alphabetic list such that everyone has an equal opportunity to work, and economic benefits are equitably distributed within the group (Ziegler et al. 2012). At other sites, for example, in Holbox, Mexico, local fishers also work as tour guides, but there are several competing tour companies ranging in size from a single boat to a fleet of 10–15 boats. Subsequently, three or four of the larger tour operators own at least 60% of the market, leaving the 40 other operators to fight over the remaining 40% of the market (Zenteno 2007). That leads to a high turnover rate of permit holders because the operating costs for the 3-month season are often greater than the money generated from selling the tours (Zenteno 2007).

In Oslob, only locals from the barangay of Tan-awan or their relatives are allowed to participate in whale shark tourism. This has created conflict within the community as neighboring barangays also want to have access to this economic benefit. Potential solutions suggested included rotating the barangay responsible for running the tourism operations on a monthly basis or allowing multiple barangays within the municipality of Oslob to open their own whale shark tourism operations.

The mayor at the time, however, decided against these options, only allowing tourism activities in Tan-awan, although economic benefits are shared – the municipality of Oslob receives 30% of money generated from ticket sales, while the tour operator association (TOSWFA) keeps 60% and Tan-awan 10%. The mayor did give the neighboring barangay of Bangcogon a 50% share of the profits from the tourist entrance fees to Sumilon Island, across the bay from Oslob, which can translate into monthly payments of $20–30 or more to each household – a significant amount considering most households were impoverished prior to tourism activities.

At most other sites, the majority of tourism activities are managed and run by foreign-owned companies (e.g., Madagascar, Galapagos, Mozambique, Saudi Arabia, Oman). For example, in Southern Leyte (Philippines), although there is a local whale shark tourism group (KASAKA) consisting of fishers who work as guides and spotters in whale shark tourism activities, the activity itself is reliant on foreign-owned dive shops to bring them tourists. The dive shops made an agreement with the municipal government that they would hire members of KASAKA to work as spotters and guides for a set amount of money. Spotters are paid approximately $5 per day but have to pay a $1 rental fee for the boat they use if they do not own one, while guides make $6/day and do not have to pay any additional fees. On average, spotters and guides work 2–3 days per week (Ziegler et al. 2020). The dive shops charge upwards of $100 per person with an average of eight tourists per boat.

One of the primary justifications for whale shark tourism is that it helps protect the sharks by providing economic benefits to local communities. Although the presence of whale shark tourism has led to the establishment of marine protected areas (e.g., Whale Shark Biosphere Reserve in the Mexican Caribbean), and is used to argue for the further protection of the species in national waters (e.g., Madagascar), no studies, to date, have actually made a direct link between whale shark tourism and improved ecological conservation outcomes, such as increased whale shark numbers and/or increased range. In fact, whale shark declines have been observed at several tourism sites (Mozambique, Belize, Seychelles, Thailand). A recent study of local fishers involved in whale shark tourism activities at various sites in the Philippines did find that participation in whale shark tourism resulted in improved environmental awareness and attitudes as well as pro-conservation behaviors (Ziegler et al. 2020). For example, many of the locals were afraid of the sharks prior to tourism and would stop fishing activities altogether when the sharks appeared, and return to the village, because they were afraid the sharks would tip over their boats. Others would get angry because the sharks would get caught in their nets and destroy them. These fishers would sometimes hit the sharks with their paddles or use stones to scare them away from their nets. Others still simply saw no value in the sharks and would throw rocks or live dynamite at them, as well as use harpoons to ride the sharks (Ziegler et al. 2020). Since tourism activities started, however, none of these behaviors occur. Some fishers now see the intrinsic value (i.e., that sharks have value just by existing) of whale sharks and appreciate their presence in their waters, while others treat the sharks well because of their economic value to the local community (Ziegler et al. 2020). These improved pro-conservation attitudes, awareness, and behaviors towards the sharks have also translated into pro-conservation behaviors towards the ocean and greater environment, such as cleaning up the beaches when they see trash and not throwing their trash on the beach or in the ocean and using more environmentally friendly fishing techniques (e.g., no more cyanide or dynamite fishing, muro ami – smashing coral heads with heavy rocks to flush out fish, buzo – hookah fishing with oxygen, or use of spearguns; Ziegler et al. 2020). This change in attitudes and behaviors to protect whale sharks and the ocean more broadly is largely due to participating in whale shark tourism activities, but also potentially due to workshops run by local NGOs and the government explaining whale shark biology, threats, and conservation needs. Further research is needed to better understand the relationship between working in whale shark tourism and positive conservation outcomes, as well as if these positive changes translate to the wider community and not just those locals working in tourism.

Whale sharks have also taken on an important cultural significance at many locations, as evidenced by the many murals, festivals, and statues celebrating these sharks at the various locations

(Figure 10.5). Whale shark festivals are held at many locations, including in the Mexican Caribbean (Isla Mujeres since 2007, Holbox since 2015), Donsol, Ningaloo, the Maldives (since 2013), and India (since 2001). The importance of these events on conservation outcomes is, as yet, unknown, although the socio-psychological benefits of marine protected areas for local communities may in

Figure 10.5 Cultural importance of whale sharks. (a) Whale shark mural at a restaurant in Koh Phangan, Thailand (P. Dearden); (b, c) participants in the 2017 Donsol Butanding Festival, Philippines (Credit: Jenny Hardy / LAMAVE); (d) whale shark mural in Utila, Honduras (D. Hughes); (e) Billboard from Mafia Island, Tanzania, which states, in Swahili, "Let us protect the whale shark. Their existence is a benefit to us" (M. de Haan); and (f) children dressed in whale shark costumes dancing during the 2017 Donsol Butanding Festival, Philippines (credit: Natalie Hancock / LAMAVE).

fact exceed the more extrinsic economic and fishery benefits that are often emphasized (P. Dearden unpubl. data).

An important area of research that has not received much attention is that of the beliefs that people in different places associate with whale sharks and marine wildlife more generally (e.g., spiritual/religious, materialism, mutualism, concern for safety, attraction/interest, respect; Dayer et al. 2007) and whether these beliefs can shift towards more protectionist views over time due to the presence of tourism. For example, Bajo people in Indonesia are not allowed to catch whale sharks and must release them from their nets if they are caught accidentally, because they believe the sharks are protected by spirits which can come to the aid of fishers during a time of need (Stacey et al. 2012). At Mafia Island in Tanzania, fishers view whale sharks as "friends," despite the damage the sharks can cause to their nets (Box 10.1). In Oslob (Philippines), locals believe a spirit linked to black magic guards the whale sharks in their waters, and if they do not take good care of the sharks, this spirit will transfer the sharks out of their waters (Ziegler 2019). Ziegler et al. (2021) compared the wildlife value orientations (i.e., the pattern of direction and intensity of beliefs that an individual holds towards wildlife) of locals working at four community-based whale shark tourism sites in the Philippines and found that those sites with active tourism had more protectionist views towards marine wildlife than the site with a failed whale shark tourism venture, suggesting that the presence of tourism had resulted in a positive shift in their basic beliefs towards marine wildlife. Those respondents with more protectionist views also reported more positive conservation outcomes, including changes in attitudes and behaviors to protect whale sharks (Ziegler et al. 2021). More social science research is needed to understand how tourism can affect how local people perceive and value marine wildlife.

10.4.3 Building Environmental Awareness

Environmental interpretation and outreach can positively influence tourists' perceptions and attitudes towards conservation, as well as their pro-conservation behavioral intentions and behaviors, which may lead to long-term changes in their attitudes and behaviors (e.g., Ballantyne et al. 2011; Ardoin et al. 2015; Filby et al. 2015; Apps et al. 2018; Sutcliffe and Barnes 2018; Hoberg et al. 2020). These social conservation outcomes may then translate into improved ecological conservation outcomes (Orams 1996).

Studies have found mixed success in the implementation of interpretation (de Vasconcellos Pegas et al. 2013). Mayes et al. (2004) compared the social conservation outcomes of two wild dolphin feeding programs in Australia and found that the one with better environmental interpretation resulted in improved pro-environmental beliefs, attitudes, and behavioral intentions among its participants. Other studies have found that environmental interpretation leads to increased knowledge, but no changes in attitudes or behaviors (see Ardoin et al. 2015 for review). Effective interpretation design is critical for achieving the desired social and ecological conservation outcomes. Increased information, regardless of its quality, does not always result in behavioral change (Fujitani et al. 2016); neither does the accumulation of knowledge (Orams 1994). Research into what makes for a memorable and effective interpretation experience highlights the importance of human connection in ensuring increased pro-environmental behaviors or intentions (Ballantyne et al. 2011; Jacobs and Harms 2014). Topics related to connecting with humans within wildlife tourism include reproduction, birth, death, and social relationships (Lück 2015; Orams 1996). For example, Jacobs and Harms (2014) found a causal link between the use of emotions during interpretation on a whale watching tour and an increased likelihood of participating in marine conservation actions (e.g., donating money, encouraging others to protect wildlife). The authors also found that experiencing the whales without interpretation had no effect on pro-environmental behavioral intentions, highlighting the importance of interpretation in ensuring the larger social and ecological conservation outcomes are achieved. However, this study did not assess if short-term changes in behavioral

BOX 10.1   BEFRIENDED FOE: WHALE SHARKS
AND THEIR VALUE TO MAFIA'S FISHERS

By: Mariam de Haan

In southern Tanzania lies the Mafia archipelago, a collection of eight islands that is heralded as a critical site for biodiversity. It is also the only place in Tanzania where one can view and swim with whale sharks during the months of October to February. These same waters along the coast of Kilindoni are also important fishing grounds for the people of Mafia. A local fisherman best summed up the relationship between fishers and whale sharks when he said

> [Whale sharks] attract a lot of tourism to Mafia. The whale shark harms us. This net has been torn by them three times. They say that it is us who tease the whale shark to the point that they do this. To us, the whale shark is a friend. Wherever the whale shark is, there is fish. We cannot harm the whale shark because the whale shark is the friend of the fishers. However, for the government to understand that is very difficult. It has even reached the point where the parliament said that during the whale shark season, fishers should not fish until the season ends.

The value that the fishers place upon whale sharks is influenced by those who regulate the seas, increasing tourism, and the interactions they have with the animal. Naturally, it is easier to regulate the activities of fishers than the presence and movements of whale sharks. As such, much of the regulation is directed towards modifying and classifying fishing. For example, a proposed whale shark management strategy put forth by the government and WWF includes the 6-month closure of fishing activities during the whale shark tourism season in order to address conflict between fishers and tourism activities. Complications arise with the representation of fishers at formal and informal meetings because those who speak on behalf of fishers currently hold questionable legitimacy within the community. In addition, the perception that marine park and government employees hold of fishers as people who aim to harm and hurt whale sharks is questionable. The damage that fishers attribute to whale sharks is undeniable. When asking them about whale sharks, fishers will complain about the large gashes in their net that the whale sharks have caused as well as the sheer strength the animals possess. However, amidst these complaints, fishers will point out that they are friends with the animal. Furthermore, through the terms they use, they include the whale sharks within their community as well as using narratives to explain a relationship of synergy between man and shark. Yet, the future of this relationship is uncertain. Currently, whale shark tourism is tolerated by fishers because they share the seas with the tour boats and do not mind telling the guides where they spotted whale sharks. Nonetheless, fishers are concerned that whale shark tourism is being given priority above fishing. At the same time, each industry holds its important place within Mafia. Tourism, be it from whale sharks or elsewhere, is indeed more lucrative, but fishing is a much more far reaching industry which holds both societal and cultural importance in Mafia. Both tourism, or rather the competition that tourism presents, and the regulation of fishing activities in local waters influence the relationship that fishers have with whale sharks, which in turn affects the value they place upon whale sharks. Due to the close and daily contact fishers have with whale sharks, the value they place upon them is of particular significance if Mafia is to remain "the home of the whale sharks."

intentions translated into long-term changes in behaviors once participants returned home. Thus, it is unknown if the increase in positive behavioral intentions resulted in improved ecological conservation outcomes for whales or the marine environment.

Some studies have demonstrated that effective interpretation can reduce negative behaviors during the tour itself (Medio et al. 1997). For example, Orams and Hill (1998) assessed the effectiveness of an interpretation program at a wild dolphin feeding site at Tangalooma, Australia, in reducing negative impacts on the dolphins (e.g., touching dolphins) and found a significant reduction in all negative behaviors after the implementation of the interpretation program. Camp and Fraser (2012) found that in-depth environmental interpretation resulted in the significant reduction of diver contacts with coral reefs. Reduced tourism impacts may translate into positive ecological conservation outcomes. However, this component was not included in the studies. Furthermore, these behavioral changes, although beneficial for the focal species during the encounter, may not necessarily translate into longer-term changes in pro-environmental behaviors once the participants return home, such as volunteering for conservation organizations, participating in beach clean-ups, or donating money for conservation. Thus, larger-scale ecological conservation outcomes may not be realized.

Research in the field of marine wildlife tourism highlights the importance of providing examples of pro-environmental actions that could be taken once participants return home (e.g., participate in beach clean-ups, donate money, volunteer) and/or opportunities to act during the tour itself (e.g., participate in research, clean-up, join a conservation organization, petitions) to improve the effectiveness of interpretation in changing pro-environmental behaviors (e.g., Ballantyne et al. 2011; Filby et al. 2015). Lück (2015) found that tourists explicitly requested this type of information be included during tours, with several respondents noting that they were provided information regarding human impacts on the environment and focal species, but nothing on how to help. Hughes et al. (2011) found that those participants provided with post-trip resources were significantly more likely to take pro-conservation actions. Other studies have noted the importance of explaining why behavioral changes are necessary to help improve compliance and therefore conservation outcomes (Camp and Fraser 2012; Medio et al. 1997; Orams 1996; Orams and Hill 1998). These findings suggest that if participants are unaware of the harmful effects of their behavior (e.g., Dearden et al. 2007) or ways they can help address conservation issues, ecological and social conservation outcomes are unlikely to be realized.

The presence and quality of interpretation at whale shark tourism sites varies greatly; some sites do not offer any interpretation, while the quality varies greatly at those sites that do offer some form of interpretation. At many sites, including Madagascar, Southern Leyte, and Saudi Arabia, researchers interact directly with tour participants, as they have a working partnership with local operators, and therefore are able to answer any questions the tourists have and provide in-depth information on research and conservation needs. Those operators who do not have researchers on their vessels at these sites are often unable to provide high-quality interpretation. On the global survey undertaken for this chapter, only four sites reported excellent interpretation – Australia (see Box 10.2), the Galapagos, Saudi Arabia, and Madagascar. In Saudi Arabia and Madagascar, however, researchers are directly involved in providing interpretation during tours, suggesting that tourists get very different experiences depending on whether researchers are present or not. There is clearly much room for improvement in terms of the quality of interpretation being provided at whale shark tourism sites.

10.4.4 Tourist Satisfaction

Unregulated tourism growth is an important concern at many sites and can negatively impact the tourist experience and financial viability of the site. Boat and swimmer crowding are therefore a pressing concern at many sites (Table 10.1 and Figure 10.6). However, only a few sites have assessed issues with crowding: Oslob, Donsol, Holbox, Ningaloo, and Gladden Spit. Ziegler et al. (2016a)

Figure 10.6   Swimmers crowding around a whale shark in the Maldives. (Photo: Simon Pierce.)

assessed perceived crowding of whale shark tour participants in Holbox (Mexico) and found that nearly half of respondents felt crowded. Furthermore, the authors found that the number of boats in the whale shark viewing area may have a greater influence on perceived crowding than the number of swimmers in the water. Ziegler et al. (2019a) found similar outcomes in Oslob (Philippines), where individual respondents reported feeling more crowded due to the number of boats than the number of swimmers during the same experience. Perceived crowding negatively affected not only satisfaction with the tour, and likelihood to recommend the tour to others, but also tourists' perceptions of the impacts that tourism activities have on the local community, tourists, sharks, and wider marine environment, and their support for management interventions that would limit the number of boats per shark and in the area, as well as the regulation or banning of whale shark feeding at the site.

BOX 10.2   INTERPRETATION AT NINGALOO REEF, AUSTRALIA

Ningaloo, Australia, provides a key example of the type of interpretation program that could be mandatory at all whale shark tourism sites. Furthermore, the development and implementation of this program exemplifies how adaptive management can be used to improve the tourism experience. A 2005 tourist survey found that 25% of tour participants at Ningaloo reported a desire for more general information on the biology and ecology of whale sharks during the tour (E. Wilson, pers. comm., 2018). In response, the Department of Biodiversity, Conservation, and Attractions (DCBA) developed a whale shark education and interpretation program to train whale shark operators, as part of their operator license conditions, as well as to educate tourists participating in these tours. Tour operator training includes an eLearning component, an online assessment, and an in-person workshop in Exmouth or Coral Bay – the entry points for whale shark tours at Ningaloo (E. Wilson, pers. comm., 2018). Once completed, the training results in certification good for 5 years and is mandatory for all boat-based employees. The materials provided during training include videos, audio, fact sheets, scientific papers, and quizzes. DBCA also provides operators with brochures, posters, flipcharts, and fact sheets for distribution in their shops and on-board the vessels. Topics covered include whale shark biology, distribution, feeding, habits, migration, management, threats, and tourist conservation actions. Studies suggest this type of interpretation is more likely to result in improved social conservation outcomes, including increased awareness of the whale shark's conservation needs, as well as potential long-term changes in tourists' knowledge, attitudes, and behavior towards whale sharks and the larger marine environment.

A similar crowding issue was also identified at Ningaloo. Davis et al. (1997) found that swimmer crowding was reduced by increasing the minimum viewing distance between swimmers and the sharks, along with a reduction in physical contact rates with the sharks. Catlin and Jones (2010) also identified the number of boats as a growing concern for whale shark tourism at Ningaloo, Australia, despite a cap on the number of licenses at 15, due to the growing trend of sharing sharks among multiple boats. Sharing sharks is a growing practice at many sites when shark sightings are low, regardless of regulations in place limiting the number of boats and or swimmers allowed to interact with them at one time (e.g., Donsol, Holbox, Honduras, Isla Mujeres). For example, Ziegler et al. (2012) observed over 30 boats around a single shark on a particularly bad day in Holbox, with up to 20 people in the water at one time, despite restrictions of one boat per shark and only two swimmers at one time. Operators appeared to be more interested in guaranteeing a whale shark sighting than following the rules.

Many sites do not have permit requirements for offering whale shark tours (e.g., Honduras, Madagascar, Leyte, Thailand, Maldives, Mozambique, Oman, Saudi Arabia, Tanzania, Djibouti, Indonesia) or, if they do, do not limit the number of permits allowed (e.g., Mexico) or do not enforce the regulations in place (e.g., Donsol). Some locations limit the number of boats allowed in the viewing area at one time in order to prevent crowding issues. For example, in Donsol, only 30 boats are allowed out at one time. However, on holidays and weekends, when demand is highest, this rule is rarely respected (J. Ziegler, pers. obs.; Quiros 2005). On such occasions, the number of boats is only restricted by how many boats, licensed or not, that can be found to offer tours. Such practices not only detract from the tourism experience but may also put undue pressure on the sharks.

Numbers of whale sharks also fluctuate wildly over time and have strongly influenced visitation at sites in Belize, Mexico, Philippines, Thailand, Seychelles, and Mozambique (Carne 2005; Theberge and Dearden 2006; Cardenas-Torres et al. 2007; Rohner et al. 2013; Araujo et al. 2017). While the Seychelles have experienced a complete loss of the whale sharks and their associated tourism industry (D. Rowat, pers. comm., 2018), other sites have been affected by unpredictable seasons and/or shark numbers. For example, in 2004, the likelihood of sighting a whale shark at Gladden Spit, Belize, declined to less than 20% from over 80% in 1999 (Carne 2005). Similarly, Rohner et al. (2013) identified a 79% decline in whale shark sightings at Tofo Beach, Mozambique, from 2005 to 2011. Theberge and Dearden (2006) identified a 96% drop in whale shark numbers in Phuket, Thailand, from 1998 to 2001, although recent reports suggest whale shark sightings have increased again in the last few seasons (P. Manopawitr, pers. comm., 2019). Donsol, Philippines, meanwhile, experienced a decline in whale shark sightings during the 2013 and 2014 seasons. Tourist visitation fell from a high of 27,159 in 2012 to a low of 12,911 in 2014. The whale sharks returned to previous numbers during the 2015–2017 seasons, but tourism numbers struggled as word spread of the missing sharks, with only 14,191 visitors in 2017.

10.4.5 Is Whale Shark Tourism Ecotourism?

Ecotourism is predicated upon the tourist activities being essentially non-consumptive in nature. Duffus and Dearden (1990) define non-consumptive wildlife tourism as "a human recreational engagement with wildlife wherein the focal organism is not purposefully removed or permanently affected by the engagement." Non-consumptive wildlife tourism, however, will, by its very nature (i.e., people in close proximity to wild animals), have impacts on the focal organism or environment. The extent of these impacts, and whether they affect the long-term survival of the species, is dependent on management at the site (Duffus and Dearden 1990). Improperly managed marine wildlife tourism can have significant negative impacts on both the conservation status and welfare of targeted species, including the loss of individuals through injury, death, and disease, short- and long-term behavioral changes, stress or negative physiological responses, altered feeding and/or reproductive behavior, and habitat alteration and or loss (e.g., Semeniuk et al. 2009; Parsons 2012; Corcoran et al. 2013; Araujo et al. 2014, 2017). If these impacts translate into long-term changes in the animals' health and fitness, marine wildlife tourism may result in population decline (Parsons 2012).

In light of the literature outlining the negative impacts of marine wildlife tourism on the focal species, some authors argue that marine wildlife tourism is actually a consumptive use of wildlife (e.g., Neves 2010). For example, Higham et al. (2016) argue that declines in a dolphin population in Fiordland, New Zealand, attributed to reduced calf survival because of exposure to repeated tour boat interactions, can be considered a consumptive use of that species. The authors conclude that whale watching, and wildlife tourism in general, should be considered "a form of non-lethal exploitation, which may impact animal morbidity (e.g., sub-lethal anthropogenic stress) and mortality (e.g., vessel strikes)" (Higham et al. 2016).

Whether whale shark tourism can be considered ecotourism is site dependent; some sites meet the definition, but most clearly do not. As Neves (2010) noted with the whale watching industry, whale shark tourism business models vary greatly across the globe. Those business models that prioritize maximizing revenue over the conservation of the species may be considered consumptive, as they are not attempting to minimize the real and potential negative impacts of tourism activities on the sharks. Two examples serve to illustrate these differing approaches.

Provisioned whale shark tourism activities in Oslob (Philippines) can be considered a form of mass tourism. Tourism operations at this location are designed to get as many people as possible in the water with the sharks, with little regard for the well-being of the sharks or the quality of the tourist experience. On an average day, there are over 40 outrigger boats tied to each other in an area no bigger than 480 m × 170 m, each carrying six to eight people, or more than 280 snorkelers, not including scuba divers below, and feeder boats moving in between the lines of tourist boats with sharks in tow. Viewing is possible from 6 am to 12 pm, and interactions are limited to 30 minutes per boat. Most "swimmers" do not, in fact, know how to swim. Local authorities state that there were a minimum of 1,000 people coming through this site daily in 2018 (Ziegler et al. 2019a), with a peak of 2,787 tourists coming on a single day in February 2019 (C. Legaspi, pers. comm., 2019). Legaspi et al. (2020) used in-water observations to estimate the number of swimmers per shark and found a mean of 17.3 swimmers within 10 m of a single shark at the site, with a range of 2–55 swimmers, while the local ordinance stipulates only six swimmers should be allowed per shark. This management approach is having a negative impact on the tourist experience at the site, a 2016 survey identifying that 79% of respondents felt crowded due to the number of swimmers in the water and 96% felt crowded due to the number of boats in the viewing area (Ziegler et al. 2019a). Based on the carrying capacity categories in Vaske and Shelby (2008, p.120), Oslob is considered "overcapacity" in terms of number of swimmers and "greatly overcapacity" in terms of number of boats. Virtually no environmental interpretation is provided for tourists. For example, in a 2016 tourist survey, 96% of respondents reported receiving a briefing prior to their whale shark experience, but only 50% said they were aware of any interaction guidelines in Oslob (J. Ziegler, unpubl. data). There is no real enforcement of regulations, with 93% of swimmers within 2 m of sharks (Legaspi et al. 2020), suggesting there is a very high likelihood of physical contact between swimmers and sharks (Schleimer et al. 2015). Over half of respondents in the 2016 survey felt that snorkeler behaviors were a problem at the site, including swimmers not respecting the minimum viewing distance, and chasing and/or harassing the sharks (Ziegler et al. 2019b). Thematic analysis of TripAdvisor reviews found that getting close to – and touching – whale sharks is a highlight for many tourists, with 40% of respondents admitting to touching a shark in a 2014 survey (Ziegler et al. 2016b). Furthermore, studies indicate that provisioning whale sharks at this site is having a negative impact on both the sharks and surrounding coral reefs (Araujo et al. 2014, 2020; Schleimer et al. 2015; Thomson et al. 2017; Wong et al. 2019; Legaspi et al. 2020). This whale shark tourism business model may be considered consumptive in nature, as money appears to have taken precedence over minimizing tourism impacts, enhancing whale shark conservation, or the tourist experience.

Ningaloo, on the other hand, has prioritized managing its tourism industry within sustainable limits – the number of licenses is limited to 15, tour operators are required to complete training prior to receiving a license, and licenses can be revoked if rules are broken or if a better operator applies during the license renewal process. There is legal enforcement of the industry, appropriate monitoring (e.g., licensed tour boats are required to report their activities to the government via electronic monitoring system on-board), and the social and biological impacts are well studied (Davis et al. 1997; Davis and Tisdell 1999; Catlin and Jones 2010; Catlin et al. 2012; Sanzogni et al. 2015; Raudino et al. 2016; Lester et al. 2019, 2020). This industry is being managed to ensure the activities have minimal impact on the sharks. Effective management is key to transforming whale shark watching to a genuine ecotourism activity.

10.5 MANAGEMENT CHALLENGES

Given the discussion above, there are obviously considerable challenges in managing whale shark tourism so that it can be rightfully called an ecotourism activity. The survey of 22 wild sites and five provisioned sites identified the following management issues as being of serious concern: (a) social and environmental impacts (e.g., perceived crowding, tourists touching sharks, sharks showing avoidance behaviors, negative impacts on the sharks and environment); (b) the rules not being followed; and (c) a noticeable decline in whale shark sightings at a number of sites. Only six experts felt that at least part of their site was being managed effectively (Ningaloo, Galapagos, Tubbataha, Koh Tao, Madagascar, and Teluk Saleh).

Managing whale shark tourism activities at most sites is difficult due to the remoteness of the location, the large physical area over which the activity occurs, the lack of resources (staff, money, boat) to monitor and enforce the rules, and the highly migratory nature of the focal species. Many sites only have voluntary codes of conduct in place, while some sites with legally enforceable regulations lack the resources to enforce those regulations (Table 10.1). Noncompliance is therefore an important issue for whale shark tourism. In Utila, Honduras, for example, tourists on dive boats must pay the boat captain a mandatory $20 fee if they wish to swim with whale sharks sighted during surface intervals (D. Hughes, pers. comm., 2018). The incentive for the captains is therefore to get as many people as possible in the water to see the sharks without any regard for correct vessel approach limits or swimmer numbers as they can make more from a single encounter with a whale shark than their entire weekly wages (D. Hughes, pers. comm., 2018). Nearly 90% of observed encounters involved violations of the code of conduct with up to seven boats within 200 m of a shark and up to 33 people observed in the water at once on April 19, 2017, despite guidelines limiting swimmer numbers to eight and boats within 200 m to one (D. Hughes, pers. comm., 2018). A recent decline in whale sharks sightings may be due to tourism activities, although other factors cannot be ruled out. Despite these issues, there is little interest on the part of operators or government to make the guidelines legally enforceable at the site (D. Hughes, pers. comm., 2018).

At Ningaloo, considered the gold standard for whale shark tourism management, operator activities are monitored using electronic monitoring systems on-board each vessel that record trip information such as the encounter length, the GPS coordinates of the whale shark encounter, and information regarding the sharks (e.g., sex, length, new or shared shark; Anderson et al. 2014). The Western Australia Department of Parks and Wildlife also carries out regular flights over the interaction area to monitor compliance to the code of conduct during the whale shark season (Sanzogni et al. 2015). This level of enforcement is difficult at most whale shark tourism sites, as most do not have the resources to afford such monitoring practices.

**BOX 10.3 INNOVATIONS IN MANAGEMENT OF WHALE
SHARK TOURISM IN BAHIA LA PAZ, MEXICO**

Bahia La Paz on the Pacific coast of Mexico is an important feeding ground for juvenile whale sharks and the site of a growing whale shark tourism industry. First established in the early 1990s, whale shark tourism in Bahia La Paz was not regulated until 2006. Prior to the 2017 season, the main management concerns at this site included high injury rates from boat strikes, illegal vessels providing whale shark tours in the viewing area, and high crowding rates from too many boats in the viewing area at one time (D. Ramírez-Macías, pers. comm., 2018). In order to address these issues, the Mexican government made several key changes to the management plan during the 2017–2018 season. Firstly, World Wildlife Fund for Nature Mexico and TELMEX Telcel Foundation donated 102 GPS devices that use cellular technology to allow real-time tracking of vessel speed, as well as entry, permanence, and distribution, within the whale shark viewing area in Bahia La Paz (WWF 2019). These GPS devices allow the government to enforce the code of conduct at the site, including vessel speed (must not exceed seven knots) and number of vessels in the area (WWF 2019). In addition to the GPS devices, the government also added VHF radio monitoring to control access to the area, as well as a patrolling strategy in partnership with local observers and non-governmental organizations to enforce boat numbers in the area, and a new collaborative training course for captains and guides emphasizing the code of conduct and visitor experience (D. Ramírez-Macías, pers. comm., 2019). At the moment, boats are limited to 14 at one time in the viewing area, with five people allowed per shark, although researchers are testing an adaptive carrying capacity approach to determine how many boats should be allowed to go out per day based on the number of whale sharks sighted (Ramírez-Macías 2019). Operators that violate the code of conduct may now face fines of up to 150,000 pesos (~US$7,500) and loss of their license (WWF 2019). In 2018, the whale shark viewing area was also declared an "área de refugio" – a type of marine reserve – which requires a management plan for the species and its habitat, as well as mandatory training for captains and guides offering whale shark tours (WWF 2019). The primary goal of these changes in management was to reduce boat strike incidents with the sharks, as well as control the number of boats in the viewing area (Ramírez-Macías, pers. comm., 2019). For example, 62% of whale sharks at the site showed signs of injuries from boat strikes during the 2016–2017 season. After the program was implemented, however, injury rates dropped to 52% during the 2017–2018 season and 58% in the 2018–2019 season (WWF 2019). Over half of the observed shark injuries during the 2018–2019 season were due to abrasions – considered to be a result of tourist–shark interactions. This injury rate highlights the need for better enforcement of the boat–shark interactions, including enforcing minimum distances between boats and sharks, as well as the number of boats per shark (D Ramírez-Macías, pers. comm., 2019). There are therefore plans for the patrols to focus on such breaches of the code of conduct during the upcoming 2019–2020 whale shark season (D. Ramírez-Macías, pers. comm., 2019). Innovations in the management of the whale shark tourism industry in Bahia La Paz highlight the importance of adaptive management approaches and government partnerships with research organizations, businesses, and non-governmental organizations, to reduce negative impacts on the focal species.

10.6 MANAGEMENT BEST PRACTICES

Effective management is needed to encourage the sustainability of marine wildlife tourism and to ensure that they are non-consumptive in nature. There is large variation in management practices in whale shark tourism around the world. This section will bring together some aspects of evolving best practices.

10.6.1 Managing Impacts

Duffus and Dearden (1990) suggest using limits of acceptable change (LAC) to set sustainable thresholds for wildlife tourism. Such an approach assesses the impacts of an activity on a target species or environment (e.g., perceived crowding or environmental impacts), uses social and ecological evaluative standards to determine acceptable levels of impact, and then implements monitoring and management interventions to ensure that these limits are not exceeded. Interventions range from capping the number of visitors or boats through to changing visitor and or boat behavior. This approach differs from the traditional carrying capacity approach as the activity is managed to meet desired conditions instead of a specific number of participants (Manning 1999). The focus of LAC is on managing the extent to which the resource and experiential quality can be compromised, rather than the extent to which recreational use should be restricted. This is the ideal scientific way to manage the activity and involves setting both ecological and social limits (see Roman et al. 2007; Bentz et al. 2016).

There are, however, several challenges related to applying a full LAC approach to whale shark management. The approach relies upon having baseline data and measuring deviation from that baseline. For example, if 100 whale sharks used to feed in a given area before tourism activities and only 50 visited after the start of tourism, then this would strongly suggest that tourism was having an unacceptable impact on whale shark behavior and distributions. However, difficulties arise due to the lack of available baseline data against which to assess change and the lack of monitoring to establish changes. Furthermore, observed changes may be the result of other factors, such as environmental change (e.g., Sanzogni et al. 2015).

The approach is iterative, and, if negative impacts exceed limits, then additional management interventions will be required. However, in reality, it is often difficult to downsize the industry due to socio-political constraints. For example, management authorities at Isla Mujeres, Mexico, tried to reduce the number of boats in the whale shark viewing area while still allowing for the continued growth of the industry. They doubled the number of permits, but mandated operators work one day on/one day off. The strategy led to significant backlash from the operators, who in response threatened to shut down the main shipping port (Ziegler et al. 2016a). In addition, the future growth of whale shark tourism is likely to primarily be in developing nations where regulatory processes and enforcement ability to manage these activities are limited.

In lieu of such site-specific information, a more generic approach is to establish guidelines that may mitigate negative impacts. These may be legally codified, but most are voluntary codes of conduct. In effect, there may be little difference between the two in terms of effectiveness; strong peer monitoring on the water may be a more powerful tool for deterrence in some instances than distant legal rules. However, voluntary codes lack the penalties associated with legal prescriptions.

10.6.2 Best Practices for Whale Shark Tourism

Most whale shark tourism sites have at least a voluntary code of conduct for tourist/operator behavior based on the regulations in place at Ningaloo[2], which is considered the gold standard for

---

[2] https://www.dpaw.wa.gov.au/plants-and-animals/animals/whale-sharks?showall=&start=2

Table 10.6   Best Practices for Whale Shark Tourism Activities

Swimmer Behaviors	Tour Operator Behaviors	General Management
Do not touch or ride sharks	Maintain a speed of 8 mph or less in areas known to have whale sharks	Monitoring of ecological, social and economic impacts of tourism activities
Do not restrict its normal movement or behavior, including impeding its path	Inform tourists of code of conduct	Design effective management interventions
Do not free dive, especially near the head of the shark	Provide in-depth interpretation including whale shark threats and conservation needs	Limit the number of permits/ licenses issued to tour operators
No flash photography	Motorized vessels should remain at least 20 m from sharks	Provide training for guides and captains, including safety, tourist needs, shark ecology/biology, language
No use of jet ski or underwater propulsion device	Only one boat should be allowed to interact with a shark at one time; other boats should wait outside a 250 m exclusion radius	Enforce rules and regulations to ensure codes of conduct are being followed
No splash entry	Limit the number of swimmers allowed in the water at one time with a shark; varies by site	Facilitate equitable distribution of benefits
Maintain a distance of at least 3 m from head and 4 m from tail (with some site variability)	Do not block the path of the shark with the boat	Use precautionary approach
	Limit amount of time boats interact with sharks	

whale shark tourism. A universal code is therefore hypothetically or actually in place at most sites for tourist/operator behavior[3]; but in many sites, there is no enforcement. Table 10.6 outlines the key regulations that we think should be included at every site.

However, there is still a need to take variability in conditions into account. For example, if tourists at sites with poor underwater visibility (e.g., Donsol, Holbox) had to abide by the minimum 3 m viewing distance in effect in Australia, they would not be able to see the sharks. Guides in Donsol therefore encourage swimmers to approach within 1 m of the sharks (Quiros 2005). In Holbox, the guidelines allow swimmers to approach within 2 m of the sharks to address issues of poor water quality, although this rule is frequently broken (J. Ziegler, pers. obs.).

In addition to on-the-water management, there are additional best practices that management should aim towards. These are summarized in Table 10.6 "General Management," and several have been touched upon already. The main role of management is to ensure that the site is managed effectively to ensure a sustainable industry. To accomplish this, it is necessary to design and implement effective management interventions that can be enforced. Ongoing monitoring of the sharks and the activity is essential to achieve this. Tourism monitoring can also contribute to identification of whale shark sighting trends and distribution (e.g., Theberge and Dearden 2006) which in turn informs adaptive management. If site managers are unable to undertake this task, then they might look at partnerships with research organizations and NGOs for help. Other important issues include working with local communities to facilitate the equitable distribution of benefits and providing interpretation training, although management authorities may not be the only or the best source for this training.

---

[3] Nonetheless establishment of a global code of conduct was one of the actions called for at the 12th CoP for the Convention on Migratory Species in Manila, 2017

A potential solution would be to create a similar program to the United Nations Environment Programme and Reef-World's Green Fins initiative, which provides an internationally recognized code of conduct to promote sustainable coral reef dive and snorkel tourism (Hunt et al. 2013). This program includes auditing all participating dive centers on a yearly basis, as well as providing education and training to ensure that these centers are minimizing their impacts on the environment and contributing to coral reef conservation. A similar program could be set up for whale shark tourism using the best practices highlighted in Table 10.6. If non-governmental organizations, research organizations, governments, and other stakeholders work together to pool resources, it would be feasible to create such a program for whale sharks. Creating global standards for whale shark tourism is important, especially considering the increased popularity of this activity across the globe, and the fact that most sites are struggling to meet ecotourism standards. Furthermore, if such a program can be done through the Convention on Migratory Species, it may create the political pressure necessary for the signatory countries to start implementing and enforcing regulations to help protect these sharks.

10.7 CONCLUSION

Whale shark tourism has grown rapidly in the last 10 years and is now worth a conservative $1.9 billion worldwide, with the non-captive industry worth over $139 million. Tourism can be an important IBC approach to protect whale sharks. However, there is a lot of variability in the quality of management at sites across the globe, with few achieving the rigorous requirements to attain "ecotourism" status. There needs to be a move towards implementing and enforcing best practices at all sites, so this activity is sustainable and helps to support broader conservation goals for this endangered species. Further research is also needed to better understand the link between whale shark tourism and improved social or ecological conservation outcomes. Little social science research has been completed with respect to whale shark tourism, but it can provide critical insights into the conservation value of whale shark tourism (Duffus and Dearden 1993; Bennett 2016; Ziegler et al. 2020, 2021). It is important, however, that the methods used are rigorous (e.g., mixed methods, sufficient sample sizes; Teel et al. 2018) and appropriate for the research questions being addressed.

REFERENCES

Alam, M. T., R. A. Petit III, T. D. Read, and A. D. M. Dove. 2014. The complete mitochondrial genome sequence of the world's largest fish, the whale shark (*Rhincodon typus*), and its comparison with those of related shark species. *Gene* 539: 44–49.

Anderson, D. J., H. T. Kobryn, B. M. Norman, L. Bejder, J. A. Tyne, and N. R. Loneragan. 2014. Spatial and temporal patterns of nature-based tourism interactions with whale sharks (*Rhincodon typus*) at Ningaloo Reef, Western Australia. *Estuarine, Coastal and Shelf Science* 148:109–19.

Anna, Z., and D. S. Saputra. 2017. Economic valuation of whale shark tourism in Cenderawasih Bay National Park, Papua, Indonesia. *Biodiversitas* 18:1026–34.

Apps, K., K. Dimmock, and C. Huveneers. 2018. Turning wildlife experiences into conservation action: Can white shark cage-dive tourism influence conservation behaviour? *Marine Policy* 88:108–15.

Araujo, G., Labaja, J., Snow, S., et al. 2020. Changes in diving behaviour and habitat use of provisioned whale sharks: Implications for management. *Scientific Reports* 10:16951.

Araujo, G., A. Lucey, J. Labaja, et al. 2014. Population structure and residency patterns of whale sharks, *Rhincodon typus*, at a provisioning site in Cebu, Philippines. *PeerJ* 2:e543.

Araujo, G., F. Vivier, J. J. Labaja, et al. 2017. Assessing the impacts of tourism on the world's largest fish *Rhincodon typus* at Panaon Island, Southern Leyte, Philippines. *Aquatic Conservation: Marine and Freshwater Ecosystems* 27:986–94.

Archabald, K., and L. Naughton-Treves. 2001. Tourism revenue-sharing around national parks in Western Uganda: Early efforts to identify and reward local communities. *Environmental Conservation* 28:135–49.

Ardoin, N. M., M. Wheaton, A. W. Bowers, et al. 2015. Nature-based tourism's impact on environmental knowledge, attitudes, and behavior: A review and analysis of the literature and potential future research. *Journal of Sustainable Tourism* 23:838–58.

Ballantyne, R., K. Hughes, J. Lee, et al. 2018. Visitors' values and environmental learning outcomes at wildlife attractions: Implications for interpretive practice. *Tourism Management* 64:190–201.

Ballantyne, R., J. Packer, and J. Falk. 2011. Visitors' learning for environmental sustainability: Testing short-and long-term impacts of wildlife tourism experiences using structural equation modelling. *Tourism Management* 32:1243–52.

Bateman, P. W., and P. A. Fleming. 2017. Are negative effects of tourist activities on wildlife over-reported? A review of assessment methods and empirical results. *Biological Conservation* 211:10–19.

Bejder, L., A. Samuels, H. Whitehead, H. Finn, and S. Allen. 2009. Impact assessment research: Use and misuse of habituation, sensitisation and tolerance in describing wildlife responses to anthropogenic stimuli. *Marine Ecology Progress Series* 395:177–85.

Bennett, N. J. 2016. Using perceptions as evidence to improve conservation and environmental management. *Conservation Biology* 30:582–92.

Bentz, J., P. Dearden, E. Ritter, and H. Calado. 2014. Shark diving in the Azores: Challenge and opportunity. *Tourism in Marine Environments* 10:71–83.

Bentz, J, F. Lopes, H. Calado, and P. Dearden. 2016. Sustaining marine wildlife tourism through linking limits of acceptable change and zoning in the wildlife tourism model. *Marine Policy* 68:100–07.

Brockelman, W. Y., and P. Dearden. 1990. The role of nature trekking in conservation: A case-study in Thailand. *Environmental Conservation* 17:141–48.

Brooks, J., K. A. Waylen, and M. B. Mulder. 2013. Assessing community-based conservation projects: A systematic review and multilevel analysis of attitudinal, behavioral, ecological, and economic outcomes. *Environmental Evidence* 2:2.

Cagua, E. F., N. M. Collins, J. Hancock, and R. Rees. 2014. Visitation and economic impact of whale shark tourism in a Maldivian marine protected area. *PeerJ PrePrints*. doi:10.7287/PEERJ.PREPRINTS.360V1.

Camp, E., and D. Fraser. 2012. Influence of conservation education dive briefings as a management tool on the timing and nature of recreational SCUBA diving impacts on coral reefs. *Ocean & Coastal Management* 61:30–37.

Cardenas-Torres, N., R. Enríquez-Andrade, and N. Rodríguez-Dowdell. 2007. Community-based management through ecotourism in Bahia de Los Angeles, Mexico. *Fisheries Research* 84:114–18.

Carne, L. 2005. Whale shark tourism at Gladden Spit Marine Reserve, Belize, Central America. In *The First International Whale Shark Conference: Promoting International Collaboration in Whale Shark Conservation, Science, and Management*, eds. T. R. Irvine, and J. K. Keesing, 70.

Catlin, J., M. Hughes, T. Jones, et al. 2013. Valuing individual animals through tourism: Science or speculation? *Biological Conservation* 157:93–98.

Catlin, J., and R. Jones. 2010. Whale shark tourism at Ningaloo Marine Park: A longitudinal study of wildlife tourism. *Tourism Management* 31:386–94.

Catlin, J., T. Jones, and R. Jones. 2012. Balancing commercial and environmental needs: Licensing as a means of managing whale shark tourism on Ningaloo Reef. *Journal of Sustainable Tourism* 20:163–78.

Chaigneau, T., and T. M. Daw. 2015. Individual and village-level effects on community support for Marine Protected Areas (MPAs) in the Philippines. *Marine Policy* 51:499–506.

Chen, V. Y., and M. J. Phipps. 2002. *Management and Trade of Whale Sharks in Taiwan*. Taipei: TRAFFIC-East Asia.

Chiu, A., and C. Lo. 2008. Offer of HK$20,000 convinces fisherman to free whale shark. *South China Morning Post*, June 7, 2008. http://www.scmp.com/article/640647/offer-hk20000-convinces-fisherman-free-whale-shark.

Clarke, S. 2004. *Shark Product Trade in Hong Kong and Mainland China and Implementation of the CITES Shark Listings*. Hong Kong: TRAFFIC East Asia.

Corcoran, M. J., B. M. Wetherbee, M. S. Shivji, et al. 2013. Supplemental feeding for ecotourism reverses diel activity and alters movement patterns and spatial distribution of the southern stingray, *Dasyatis americana*. *PloS One* 8:e59235.

Davis, D., S. Banks, A. Birtles, et al. 1997. Whale sharks in Ningaloo Marine Park: Managing tourism in an Australian marine protected area. *Tourism Management* 18:259–71.

Davis, D., and C. A. Tisdell. 1999. Tourist levies and willingness to pay for a whale shark experience. *Tourism Economics* 5:161–74.

Dayer, A. A., H. M. Stinchfield, and M. J. Manfredo. 2007. Stories about wildlife: Developing an instrument for identifying wildlife value orientations cross-culturally. *Human Dimensions of Wildlife* 12:307–15.

de la Parra Venegas, R., R. Hueter, J. González Cano, et al. 2011. An unprecedented aggregation of whale sharks, *Rhincodon typus*, in Mexican coastal waters of the Caribbean Sea. *PLoS One* 6:e18994.

de Vasconcellos Pegas, F., A. Coghlan, A. Stronza, and V. Rocha. 2013. For love or for money? Investigating the impact of an ecotourism programme on local residents' assigned values towards sea turtles. *Journal of Ecotourism* 12:90–106.

Dearden, P., M. Bennett, and R. Rollins. 2006. Dive specialization in Phuket: Implications for reef conservation. *Environmental Conservation* 33:353–63.

Dearden, P., M. Bennett, and R. Rollins. 2007. Perceptions of diving impacts and implications for reef conservation. *Coastal Management* 35:305–17

Dearden, P., K. Topelko, and J. Ziegler. 2008. Tourist interactions with sharks. In *Marine Wildlife and Tourism Management*, ed. J. Higham and M. Lück, 66–90. Cambridge: CABI.

Duffus, D. A., and P. Dearden. 1990. Non-consumptive wildlife-oriented recreation: A conceptual framework. *Biological Conservation* 53:213–31.

Duffus, D., and P. Dearden. 1993. Recreational use, valuation and management of killer whales (*Orcinus orca*) on Canada's Pacific coast. *Environmental Conservation* 20:149–56.

Eshoo, P. F., A. Johnson, S. Duangdala, and T. Hansel. 2018. Design, monitoring and evaluations of a direct payments approach for an ecotourism strategy to reduce illegal hunting and trade of wildlife in Lao PDR. *PLoS One* 13:e0186133.

Filby, N. E., K. A. Stockin, and C. Scarpaci. 2015. Social science as a vehicle to improve dolphin-swim tour operation compliance. *Marine Policy* 51:40–47.

Fujitani, M. L., A. McFall, C. Randler, and R. Arlinghaus. 2016. Efficacy of lecture-based environmental education for biodiversity conservation: A robust controlled field experiment with recreational anglers engaged in self-organized fish stocking. *Journal of Applied Ecology* 53:25–33.

Gallagher, A. J., and N. Hammerschlag. 2011. Global shark currency: The distribution, frequency, and economic value of shark ecotourism. *Current Issues in Tourism* 14:797–812.

Godinez, A. M., and E. J. Fernandez. 2019. What is the zoo experience? How zoos impact a visitor's behaviors, perceptions, and conservation efforts. *Frontiers in Psychology* 10:1746.

Graham, R. T. 2003. Behaviour and conservation of whale sharks on the Belize Barrier Reef. PhD diss., Univ. of York.

Graham, R. T. 2004. Global whale shark tourism: A "golden goose" of sustainable and lucrative income. *Shark News* 16:8–9.

Graham, R. T., and C. M. Roberts. 2007. Assessing the size, growth rate and structure of a seasonal population of whale sharks (*Rhincodon typus* Smith 1828) using conventional tagging and photo identification. *Fisheries Research* 84:71–80.

Griffiths, J. 2016. Fishers arrested after whale shark's death sparks online outrage. *CNN*, May 12, 2016. https://www.cnn.com/2016/05/12/asia/china-beihai-whale-shark/index.html (accessed January 15, 2021).

Hargreaves-Allen, V. 2009. *Is Marine Conservation a Good Deal? The Value of a Protected Reef in Belize*. Sebastopol: Conservation Strategy Fund.

Haskell, P. J., A. McGowan, A. Westling, et al. 2015. Monitoring the effects of tourism on whale shark *Rhincodon typus* behaviour in Mozambique. *Oryx* 49:492–99.

Higham, J. E., L. Bejder, S. J. Allen, et al. 2016. Managing whale-watching as a non-lethal consumptive activity. *Journal of Sustainable Tourism* 24:73–90.

Higham, J. E. S., and E. J. Shelton. 2011. Tourism and wildlife habituation: Reduced population fitness or cessation of impact? *Tourism Management* 32:1290–98.

Hilton, P. 2016. Indonesia busts illegal wildlife trafficking ring, releases captive whale sharks. *The Coral Triangle*, June 6, 2016. http://thecoraltriangle.com/content/2016/06/indonesia-busts-illegal-wildlife-trafficking-ring-releases-captive-whale-sharks (accessed January 15, 2021).

Hilton, P., and A. Hofford. 2014. Planet's biggest slaughter of whale sharks exposed in PuQi, Zhejiang Province, China. WildLifeRisk. https://web.archive.org/web/20160503104157/http://wildliferisk.org/press-rel ease/ChinaWhaleSharks-WLR-Report-ENG.pdf.

Hoberg, R., L. Kannis-Dymand, K. Mulgrew, V. Schaffer, and E. Clark. 2020. Humpback whale encounters: Encouraging pro-environmental behaviours. *Current Issues in Tourism.* doi:10.108 0/13683500.2020.1808597.

Holmes, C. M. 2003. The influence of protected area outreach on conservation attitudes and resource use patterns: A case study from western Tanzania. *Oryx* 37:305–15.

Hughes, K., J. Packer, and R. Ballantyne. 2011. Using post-visit action resources to support family conservation learning following a wildlife tourism experience. *Environmental Education Research* 17:307–28.

Hunt, C. V., J. J. Harvey, A. Miller, et al. 2013. The Green Fins approach for monitoring and promoting environmentally sustainable scuba diving operations in South East Asia. *Ocean and Coastal Management* 78:35–44.

Huveneers, C., M. G. Meekan, K. Apps, et al. 2017. The economic value of shark-diving tourism in Australia. *Reviews in Fish Biology and Fisheries* 27:665–80.

Ilyas, F. 2014. Another whale shark caught, sold, and chopped. *DAWN*, March 27, 2014. https://www.dawn. com/news/1095778 (accessed January 15, 2021).

Jacobs, M. H., and M. Harms. 2014. Influence of interpretation on conservation intentions of whale tourists. *Tourism Management* 42:123–31.

Jensen, E. A., A. Moss, and M. Gusset. 2017. Quantifying long-term impact of zoo and aquarium visits on biodiversity-related learning outcomes. *Zoo Biology* 36:294–97.

Lee, J. L. 2014. Slaughterhouse said to process "horrifying" number of whale sharks annually. *National Geographic*, January 30, 2014. https://news.nationalgeographic.com/news/2014/01/140129-whale-shark-endangered-cites-ocean-animals-conservation/ (accessed January 15, 2021).

Lee, T. M., N. S. Sodhi, and D. M. Prawiradilaga. 2009. Determinants of local people's attitude toward conservation and the consequential effects on illegal resource harvesting in the protected areas of Sulawesi (Indonesia). *Environmental Conservation* 36:157–70.

Legaspi, C., J. Miranda, J. Labaja, et al. 2020. In-water observations highlight the effects of provisioning on whale shark behaviour at the world's largest whale shark tourism destination. *Royal Society Open Science* 7: 200392.

Lester, E., M. G. Meekan, P. Barnes, et al. 2020. Multi-year patterns in scarring, survival and residency of whale sharks in Ningaloo Marine Park, Western Australia. *Marine Ecology Progress Series* 634: 115–25.

Lester, E., C. W. Speed, D. Rob, et al. 2019. Using an electronic monitoring system and photo identification to understand effects of tourism encounters on whale sharks in Ningaloo Marine Park. *Tourism in Marine Environments* 14:121–31.

Li, W., Y. Wang, and B. Norman. 2012. A preliminary survey of whale shark *Rhincodon typus* catch and trade in China: An emerging crisis. *Journal of Fish Biology* 80:1608–18.

Lück, M. 2015. Education on marine mammal tours–But what do tourists want to learn? *Ocean and Coastal Management* 103:25–33.

Lupton, J. 2008. The value of whale sharks and manta rays to divers at Tofo beach, Mozambique and their willingness to pay tourist fees. MSc Newcastle Univ.

Manning, R. 1999. Crowding and carrying capacity in outdoor recreation: From normative standards to standards of quality. In *Leisure Studies: Prospects for the Twenty-First Century*, eds. T. L. Burton, and E. L. Jackson, 323–34. Virginia: Venture Publishing.

Master, F. 2018. Tidal wave of Chinese marine parks fuels murky cetacean trade. *Reuters*, September 19, 2018. https://www.reuters.com/article/us-china-marineparks-insight/tidal-wave-of-chinese-marine-par ks-fuels-murky-cetacean-trade-idUSKCN1M00OC (accessed January 15, 2021).

Mayes, G., P. Dyer, and H. Richins. 2004. Dolphin-human interaction: Pro-environmental attitudes, beliefs and intended behaviours and actions of participants in interpretation programs: A pilot study. *Annals of Leisure Research* 7:34–53.

Mbaiwa, J. 2011. The effects of tourism development on the sustainable utilization of natural resources in the Okavango Delta, Botswana. *Current Issues in Tourism* 14: 251–73.

McConnell, T. 2017. 'They're like the mafia': The super gangs behind Africa's poaching crisis. *The Guardian*, August 19, 2017. https://www.theguardian.com/environment/2017/aug/19/super-gangs-africa-poaching-crisis (accessed January 15, 2021).

McCoy, E., R. Burce, D. David, et al. 2018. Long-term photo-identification reveals the population dynamics and strong site fidelity of adult whale sharks to the coastal waters of Donsol, Philippines. *Frontiers in Marine Science* 5:271.

Medio, D., R. F. G. Ormond, and M. Pearson. 1997. Effect of briefings on rates of damage to corals by scuba divers. *Biological Conservation* 79:91–95.

Miranda, J. A., N. Yates, A. Agustines, et al. 2020. Donsol: An important reproductive habitat for the world's largest fish *Rhincodon typus*? *Journal of Fish Biology*. doi:10.1111/jfb.14610.

Montero-Quintana, A. N., Vázquez-Haikin, J.A., Merkling, T., et al. 2020. Ecotourism impacts on the behaviour of whale sharks: An experimental approach. *Oryx* 54:270–75.

Neves, K. 2010. Cashing in on cetourism: A critical ecological engagement with dominant e-NGO discourses on whaling, cetacean conservation, and whale watching. *Antipode* 42:719–41.

Newport, K. 2014. Chinese fishers unintentionally catch a massive 2-ton whale shark. *Bleacher Report*, August 5, 2014. http://bleacherreport.com/articles/2152747-chinese-fishers-unintentionally-catch-massive-2-ton-whale-shark (accessed January 15, 2021).

Norman, B. M. 1999. Aspects of the biology and ecotourism industry of the whale shark *Rhincodon typus* in North-Western Australia. MSc thesis, Murdoch Univ.

Norman, B., and J. Catlin. 2007. *Economic Importance of Conserving Whale Sharks*. Australia: International Fund for Animal Welfare.

Norman, B., J. A. Holmberg, Z. Arzoumanian, et al. 2017. Undersea constellations: The global biology of an endangered marine megavertebrate further informed through citizen science. *BioScience* 67: 1029–43.

Obura, D., D. Rowat, A. Nicholas, and S. Pierce. 2013. Whale sharks in the Indian Ocean. *Swara* 2013:26–31.

Orams, M. 1994. Creating effective interpretation for managing interaction between tourists and wildlife. *Australian Journal of Environmental Education* 10:21–34.

Orams, M. B. 1996. Using interpretation to manage nature-based tourism. *Journal of Sustainable Tourism* 4:81–94.

Orams, M. B., and G. J. Hill. 1998. Controlling the ecotourist in a wild dolphin feeding program: Is education the answer? *The Journal of Environmental Education* 29:33–38.

Paddenburg, T. 2010. Giant whale sharks butchered for Asian fish fin market. *Perth Now*, June 5, 2010. https://www.perthnow.com.au/news/giant-whale-sharks-butchered-for-asian-fish-fin-market-ng-7a58ace420deccb11b71190cdd074657 (accessed January 15, 2021).

Parsons, E. C. M. 2012. The negative impacts of whale watching. *Journal of Marine Biology* 2012:1–9.

Patroni J., G. Simpson, and D. Newsome. 2018. Feeding wild fish for tourism – a systematic quantitative literature review of impacts and management. *International Journal of Tourism Research* 20:1–13.

PDO. 2016. Whale shark killed and sold in Guangxi. *People's Daily Online*, May 11, 2016. http://en.people.cn/n3/2016/0511/c98649-9056147.html (accessed January 15, 2021).

Pearl, H. 2016. Diver begs Indonesian fisherman not to slaughter a baby whale shark – as locals climbing on its back eventually accept $75 to let harmless creature go. *Daily Mail*, August 24, 2016. http://www.dailymail.co.uk/news/article-3757382/Diver-begs-life-whale-shark-caught-Indonesian-fisherman.html (accessed January 15, 2021).

Penketh, L., A. Schleimer, J. Labaja, et al. 2020. Scarring patterns of whale sharks, *Rhincodon typus*, at a provisioning site in the Philippines. *Aquatic Conservation: Marine and Freshwater Ecosystems*: https://doi.org/10.1002/aqc.3437.

Pierce, S. J., A. Méndez-Jiménez, K. Collins, et al. 2010. Developing a code of conduct for whale shark interactions in Mozambique. *Aquatic Conservation: Marine and Freshwater Ecosystems* 20:782–88.

Quiros, A. 2005. Whale shark 'ecotourism' in the Philippines and Belize: Evaluating conservation and community benefits. *Tropical Resources* 24:42–48.

Quiros, A. L. 2007. Tourist compliance to a Code of Conduct and the resulting effects on whale shark (*Rhincodon typus*) behavior in Donsol, Philippines. *Fisheries Research* 84:102–08.

Ramírez-Macías, D. 2019. Key elements for managing whale shark tourism in the Gulf of California. Poster presented at the *5th International Whale Shark Conference* Ningaloo, Australia, May 28–31.

Raudino, H., D. Rob, P. Barnes, et al. 2016. Whale shark behavioural responses to tourism interactions in Ningaloo Marine Park and implications for future management. *Conservation Science Western Australia* 10:2.

Read, T. D., R. A. Petit III, S. J. Joseph et al. 2017. Draft sequencing and assembly of the genome of the world's largest fish, the whale shark: *Rhincodon typus* Smith 1828. *BMC Genomics* 18:532.

Reinl, J. 2013. Kenya whale shark safari swims in controversy. *Al Jazeera*, February 6, 2013. https://www.aljazeera.com/indepth/features/2013/02/20132149933476592.html (accessed January 15, 2021).

Rohner, C. A., S. J. Pierce, A. D. Marshall, S. J. Weeks, M. B. Bennett, and A. J. Richardson. 2013. Trends in sightings and environmental influences on a coastal aggregation of manta rays and whale sharks. *Marine Ecology Progress Series* 482:153–68.

Roman, G. S. J., P. Dearden, and R. Rollins. 2007. Application of zoning and "limits of acceptable change" to manage snorkelling tourism. *Environmental Management* 39:819–30.

Rowat, D., and K. S. Brooks. 2012. A review of the biology, fisheries and conservation of the whale shark *Rhincodon typus*. *Journal of Fish Biology* 80:1019–56.

Rowat, D., and U. Engelhardt. 2007. Seychelles: A case study of community involvement in the development of whale shark ecotourism and its socio-economic impact. *Fisheries Research* 84:109–13.

Rowat, D., M. G. Meekan, U. Engelhardt, B. Pardigon, and M. Vely. 2007. Aggregations of juvenile whale sharks (*Rhincodon typus*) in the Gulf of Tadjoura, Djibouti. *Environmental Biology of Fishes* 80:465–72.

Salafsky, N., H. Cauley, G. Balachander, et al. 2001. A systematic test of an enterprise strategy for community-based biodiversity conservation. *Conservation Biology* 15:1585–95.

Sanzogni, R. L., M. G. Meekan, and J. J. Meeuwig. 2015. Multi-year impacts of ecotourism on whale shark (*Rhincodon typus*) visitation at Ningaloo Reef, Western Australia. *PloS One* 10:e0127345.

Schleimer, A., G. Araujo, L. Penketh, et al. 2015. Learning from a provisioning site: Code of conduct compliance and behaviour of whale sharks in Oslob, Cebu, Philippines. *PeerJ* 3:e1452.

Semeniuk, C. A., S. Bourgeon, S. L. Smith, and K. D. Rothley. 2009. Hematological differences between stingrays at tourist and non-visited sites suggest physiological costs of wildlife tourism. *Biological Conservation* 142:1818–29.

Sharpley, R. 2006. Ecotourism: A consumption perspective. *Journal of Ecotourism* 5:7–22.

Skibins, J. C., R. B. Powell, and J. C. Hallo. 2013. Charisma and conservation: Charismatic megafauna's influence on safari and zoo tourists' pro-conservation behaviours. *Biodiversity and Conservation* 22:959–82.

Sorice, M. G., C. S. Shafer, and D. Scott. 2003. Managing endangered species within the use/preservation paradox: Understanding and defining harassment of the West Indian manatee (*Trichechus manatus*). *Coastal Management* 31:319–38.

Spiteri, A., and S. K. Nepal, 2006. Incentive-based conservation programs in developing countries: A review of some key issues and suggestions for improvements. *Environmental Management* 37:1–14.

Stacey, N. E., J. Karam, M. G. Meekan, et al. 2012. Prospects for whale shark conservation in eastern Indonesia through Bajo traditional ecological knowledge and community-based monitoring. *Conservation and Society* 10:63–75.

Sutcliffe, S. R., and M. L. Barnes. 2018. The role of shark ecotourism in conservation behaviour: Evidence from Hawaii. *Marine Policy* 97:27–33.

Teel, T. L., C. B. Anderson, M. A. Burgman, et al. 2018. Publishing social science research in *Conservation Biology* to move beyond biology. *Conservation Biology* 32:6–8.

The Strait Times. 2018. Marine parks get popular in China. *The Strait Times*, September 24, 2018. https://www.straitstimes.com/lifestyle/entertainment/marine-parks-get-popular-in-china (accessed January 15, 2021).

Theberge, M. M., and P. Dearden. 2006. Detecting a decline in whale shark *Rhincodon typus* sightings in the Andaman Sea, Thailand, using ecotourist operator-collected data. *Oryx* 40:337–42.

Thomson, J. A., G. Araujo, J. Labaja, E. McCoy, R. Murray, and A. Ponzo. 2017. Feeding the world's largest fish: Highly variable whale shark residency patterns at a provisioning site in the Philippines. *Royal Society Open Science* 4:170394.

Topelko, K., and P. Dearden. 2005. The shark watching industry and its potential contribution to shark conservation. *Journal of Ecotourism* 4:108–28.

Vafeiadis, M. 2012. Huge whale shark sold for $2,200 in Pakistan. *The Christian Science Monitor*, February 8, 2012. https://www.csmonitor.com/Science/2012/0208/Huge-whale-shark-sold-for-2-200-in-Pakistan (accessed January 15, 2021).

Vaske, J. J., and L. B. Shelby. 2008. Crowding as a descriptive indicator and an evaluative standard: Results from 30 years of research. *Leisure Sciences: An Interdisciplinary Journal* 30:111–26.

Winter, C. 2018. Chinese gangs fuel illegal South Africa abalone trade: Report. *Deutsche Welle*, September 19, 2018. https://www.dw.com/en/chinese-gangs-fuel-illegal-south-africa-abalone-trade-report/a-45567801 (accessed January 15, 2021).

Withnall, A. 2014. Endangered whale shark caught by Chinese fishers. *Independent*, August 3, 2014. https://www.independent.co.uk/news/world/asia/endangered-whale-shark-caught-by-chinese-fishers-9644885.html (accessed January 15, 2021).

Wong, C. W. M., I. Conti-Jerpe, L. J. Raymundo, et al. 2019. Whale shark tourism: Impacts on coral reefs in the Philippines. *Environmental Management* 63:282–91.

World Wildlife Fund (WWF). 2019. Gracias a tecnología GPS reducen 16% lesiones de tiburones ballena en la Bahía de La Paz. Accessed October 19, 2019 from: https://www.wwf.org.mx/noticias/noticias_golfo_de_california.cfm?uNewsID=352231.

Yaptinchay, A. A. 1999. Marine wildlife conservation and community-based ecotourism. In *Proceedings of Conference – Workshop on Ecotourism, Conservation and Community Development*, 90–99. VSO Publication.

Zenteno, I. Y. 2007. Análisis de alternativas económicas sostenibles en la isla de Holbox como sitio de influencia en el Arrecife Mesoamericano. Proyecto ICRAN-MAR Modelo de negocios para determiner la viabilidad financiera de la actividad económica del Tiburon ballena en Holbox. Prepared for WWF Central America.

Ziegler, J. 2010. Assessing the sustainability of whale shark tourism: A case study of Isla Holbox, Mexico. MSc thesis, Univ. of Victoria.

Ziegler, J. 2019. Conservation outcomes and sustainability of whale shark tourism in the Philippines. PhD diss., Univ. of Victoria.

Ziegler, J. A., G. Araujo, J. Labaja, et al. 2019a. Measuring perceived crowding in the marine environment: Perspectives from a mass tourism "swim-with" whale shark site in the Philippines. *Tourism in Marine Environments* 14:211–30.

Ziegler, J. A., G. Araujo, J. Labaja, et al. 2020. Can ecotourism change community attitudes towards conservation? *Oryx*. doi:10.1017/S0030605319000607.

Ziegler, J. A., G. Araujo, J. Labaja, et al. 2021. Exploring the wildlife value orientations of locals working in community-based marine wildlife tourism in the Philippines. *Tourism in Marine Environments*. doi:10.3727/154427321X16101028725332.

Ziegler, J., P. Dearden, and R. Rollins. 2012. But are tourists satisfied? Importance-performance analysis of the whale shark tourism industry on Isla Holbox, Mexico. *Tourism Management* 33:692–701.

Ziegler, J. A., P. Dearden, and R. Rollins. 2016a. Participant crowding and physical contact rates of whale shark tours on Isla Holbox, Mexico. *Journal of Sustainable Tourism* 24:616–36.

Ziegler, J. A., J. N. Silberg, G. Araujo, et al. 2018. A guilty pleasure: Tourist perspectives on the ethics of feeding whale sharks in Oslob, Philippines. *Tourism Management* 68:264–74.

Ziegler, J. A., J. N. Silberg, G. Araujo et al. 2019b. Applying the precautionary principle when feeding an endangered species for marine tourism. *Tourism Management* 72:155–58.

Ziegler, J. A., J. Silberg, P. Dearden, and A. Ponzo. 2016b. Human dimensions of whale shark provisioning in Oslob, Philippines. *Paper Presented at the Meeting of the Fourth International Whale Shark Conference*, Doha, Qatar.

CHAPTER 11

Global Threats to Whale Sharks

David Rowat
Marine Conservation Society Seychelles (ret.)

Freya Womersley
Marine Biological Association

Bradley M. Norman
ECOCEAN, Inc.

Simon J. Pierce
Marine Megafauna Foundation

CONTENTS

11.1 INTRODUCTION

Following an estimated decline of more than 50% across their global distribution, whale sharks were downgraded to globally Endangered on the IUCN Red List of Threatened Species in 2016 (Pierce and Norman 2016). Prior to the 1980s, whale sharks were rarely even seen (Silas 1986; Wolfson 1986; Taylor 2019), so what happened over the past few decades? Answering that question, and suggesting some practical measures to improve the current situation, are the primary objectives of this chapter. We also point out some future concerns for whale sharks in a world where our own population is still increasing, and our activities are changing the very climate of this planet we call home.

Despite years of study, whale sharks are still an enigmatic species (Rowat and Brooks 2012). Their long potential lifespan (Perry et al. 2018; Meekan et al. 2020; Ong et al. 2020) and high mobility mean that individual sharks routinely swim through multiple states and jurisdictions (Eckert et al. 2002; Rowat and Gore 2007; Hueter et al. 2013; Guzman et al. 2018), so individual sharks will likely be exposed to different suites of potential threats. This makes it difficult to quantify the population-level impacts of human activities. Furthermore, dead whale sharks generally sink without trace, leaving no direct evidence for us to determine the cause of death (Rowat 2010). While we can extrapolate from published records in some cases, such as for commercial fisheries catches, most whale sharks killed by fast-moving ships or ingested plastic (see Chapter 13) probably go unrecorded. Instead, we have to make estimate mortality from other clues, such as the scarring present on live whale sharks, or the projected overlap between known whale shark foraging habitat and oil spills, to investigate the impacts of these insidious threats.

These complications are not just a challenge for whale shark scientists; organizing and prioritizing a diverse array of threats for management attention is a common goal across researchers working on wide-ranging species. To assist this effort, the IUCN has developed a cross-species "Threats Classification Scheme" (IUCN 2020) that provides a standard terminology and categorization for the various threats that can affect whale sharks (and other animals). We apply that framework later in this chapter to help inform conservation planning efforts going forward.

11.2 WHY ARE WHALE SHARKS IN DANGER?

Whale sharks are huge, growing to around 18–20 m in length (Rowat and Brooks 2012), and long-lived, with the largest sharks likely exceeding 100 years in age (Perry et al. 2018). Individual sharks grow slowly and take decades just to reach reproductive maturity (Norman and Morgan 2016; Matsumoto et al. 2019; Meekan et al. 2020; Ong et al. 2020). However, they are relatively small and defenseless when born, measuring only 40–60 cm in length. Without the thick skin and size advantage of adults, young whale sharks face many potential predators, such as large sharks, marine mammals, and billfish, and are likely to have a high natural mortality rate over the first few months or years of life (Rowat and Brooks 2012). If young sharks survive these early years and attain a length of a few meters, such as those measuring 3–4 m that seem to start joining feeding constellations (Chapter 7; Norman et al. 2017), they probably have a good chance of surviving to adulthood, having simply outgrown the mouths of many potential predators. As a species, the whale shark has persisted in more-or-less its modern form for around 23 million years (Cicimurri and Knight 2009; Carrillo-Briceño et al. 2020), out-surviving some of their probable – and truly awesome – natural predators, such as "megalodon" *Otodus megalodon*, a giant lamniform shark, and the ancestral sperm whale *Livyatan melvillei*, whose teeth were over 30 cm in height. Clearly, this "conservative" life-history strategy has proven to be a successful way for whale sharks to live in the tropical open ocean.

Unfortunately, the threats created by humans are far more problematic for whale sharks than those of natural origins. Over an evolutionary timescale, whale sharks can clearly adapt to substantial variation in oceanographic conditions, prey distribution, and the diversity of predatory species they must evade. Human influences, however, have entered their world over a much shorter timescale. Global trade by sea has increased by over 400% since just 1970, while the size and speed of ships has also been increasing (Erbe et al. 2020). Plastics only entered widespread use in the 1950s, and production had increased to 322 million metric tons per year by 2015 (Worm et al. 2017). The prevalence of waste plastic in the ocean has followed the same trajectory. These huge changes to the marine environment are unprecedented, and have occurred within the lifespan of many adult whale sharks alive today. As a species, they have not had enough time to change habits that have served them well for millennia or to evolve biological adaptations to help them survive these unnatural threats. Instead, as the creators of these problems, the onus is on us to come up with solutions.

11.3 FISHERIES

Much of the global decline in whale shark numbers in the past few decades can be attributed to overfishing. Despite their huge size, whale sharks are uniquely vulnerable to capture. They seasonally aggregate in certain coastal areas to feed, where they can be predictably found (Chapter 7). They often move slowly at the surface, where they are easy to locate and intercept. They are placid animals, unafraid of boats, and unaggressive towards people. That means whale sharks can be, have been, and in some cases still are caught with centuries-old methods, such as the use of large harpoons or gaff-hooks from row or sailing boats, and techniques that were developed for hunting whales in places like Indonesia (Egami and Kojima 2013) and the Philippines (Alava and Dolumbalo 2002; Acebes 2013). More recently, commercial demand for whale shark meat, fins, and other by-products has led fishers to target the species using more recent technology and methods. Sharks, including whale sharks, are also a common bycatch (that is, accidental catch) of fisheries using huge nets to target other species. It makes some sense, then, to group the discussion of human fishers and their interactions with whale sharks into three categories: historical catches, commercial fisheries, and bycatch.

11.3.1 Historical Catches

Whale sharks have probably been caught, occasionally at least, since people have had the capacity to do so. The original 1828 description of the species was, after all, based on a juvenile shark that was harpooned off South Africa (Smith 1828). Whale sharks have been caught commercially in fisheries since at least the 1850s, including for the international export of their fins (e.g. from India to China, see Bean 1907). Before the 1980s, though, they were rarely targeted (Silas 1986; Colman 1997a; Pravin 2000; Hsu et al. 2012; Acebes 2013), but rather harpooned opportunistically or landed as bycatch. Whale shark muscle tissue has a high water content and has generally been considered unpalatable, or at least of low value across much of their distribution (Silas 1986; Colman 1997a; Acebes 2013). Some whale shark carcasses were even towed to sea and discarded in India (Silas 1986). The liver has long been viewed as useful, however, and was the primary reason for catches being landed in many areas as the liver oil could be used as a preservative for the timber used in boat hulls (Silas 1986; Anderson and Ahmed 1993; Hanfee 2001; Akhilesh et al. 2013). Some whale sharks were taken specifically for their liver in India as early as 1955–1960 (Hanfee 2001).

These small-scale fisheries were not necessarily insignificant in terms of localized impact. Fishers in the Maldives used to take 20–30 whale sharks per year for their oil, but reported declining catches during the 1980s to early 1990s (Anderson and Ahmed 1993), and the fishery was banned in 1995 due to concerns for the species (although occasional hunting may have continued following protection; Riley et al. 2009). Fishermen off the state of Gujarat in India harpooned about 40 sharks for liver oil extraction over just a few days in 1982, leading to a review of the relatively small-scale catches in the country being subtitled, "Is the species endangered or vulnerable?" (Silas 1986). In hindsight, this question certainly foreshadowed the events that were to come.

11.3.2 Commercial Fishing and International Trade

Prior to 1985, there was little demand for whale shark meat in Taiwan. Specimens of several tons in weight sold at US$200–300 (Chen and Phipps 2002). No dedicated fishery was present, although whale sharks were caught regularly as a bycatch in set-net fisheries (Chen et al. 1997). A meat fishery developed during the 1990s, however, with annual catches accelerating to 272 individuals in 1996, derived from both set-net and harpoon fisheries. Whale shark became the most expensive shark meat available in Taiwan by 1997, reaching prices of USD$13.93/kg. A small 2 t whale shark could by then fetch USD$14,000, with a larger 10 t shark selling for around USD$70,000 in 1997 (Chen et al. 1997). Catches declined from this peak to 80–100 sharks throughout the country each year after 1997 (Hsu et al. 2012), perhaps due to local stock depletion. The annual volume of whale shark meat traded more than doubled between 1998 and 2000, however, to 60 t in 2000 (Chen and Phipps 2002). Market surveys in 2001 indicated that catches were being under-reported in official statistics and that significant quantities of whale shark meat were also being imported through unofficial channels (Chen and Phipps 2002). Following the introduction of specific export codes for whale shark meat in 2001, no imports and 2 t of exports were recorded over the following year – mostly to Spain, and valued at USD$1.15/kg (Chen and Phipps 2002). Concern for the status of the species led to total allowable catch quotas being steadily reduced down to zero sharks over the 2001–2007 period (Hsu et al. 2012). A total of 693 sharks was caught in Taiwan between 2001 and 2008 (Hsu et al. 2012).

The dramatic increase in value for whale shark meat led to buyers seeking international imports to supply this demand (Hanfee 2001; Chen and Phipps 2002; Acebes 2013). In the Bohol Sea of the Philippines, this led to former marine mammal hunters switching to whale sharks, and the improved profit enabled boat owners to invest in larger, faster vessels (Acebes 2013). From 1990 to 1997, 624–627 whale sharks were caught from four of the primary fishing sites in the Philippines (Alava and Dolumbalo 2002). When whale sharks were protected in the Maldives and the Philippines in 1998, Indian exporters realized that whale shark exports could prove lucrative and encouraged more fishers there to start targeting the species (Hanfee 2001). Between 1889 and 1998, 1,974 whale sharks were recorded as landed in India (Pravin 2000). Surveys in Gujarat from 1999 to 2000 recorded about 600 whale sharks landed, with some additional reports received following the cessation of on-site catch monitoring (Hanfee 2001). Following a conservation campaign, whale sharks were legally protected in Indian territorial waters in 2001 (Akhilesh et al. 2013).

Whale shark meat from mainland China was probably also exported to the Taiwanese market (Chen and Phipps 2002). A 2012 paper on the "emerging crisis" represented by the Chinese whale shark fishery was compiled through literature and media reports along with around 1,000 fisher surveys from 2008 to 2010 (Li et al. 2012). This was followed by an NGO exposé on the "planet's biggest slaughter of whale sharks" at Puqi markets in Zhejiang province, which processed at least 600 whale sharks per year based on investigations from 2010 to 2013 (WildLifeRisk 2014). Li et al. (2012) found that while whale sharks were not necessarily targeted off mainland China, there was a large bycatch estimated at more than 1,000 individuals per annum due to the intense fishing pressure of the Chinese fleet. Whale sharks were considered to be a high-value catch in the fishery, raising the prospect of their being actively targeted in the future (Li et al. 2012). After the meat, the most profitable product was the liver oil, which was extracted for use in cosmetics and health supplements (WildLifeRisk 2014). Whale sharks are technically protected in China by virtue of their 2002 listing on Appendix II of the Convention on International Trade in Endangered Species (CITES), which triggered domestic legislation that requires a special permit to allow catches. Enforcement of this rule has been minimal, however (Li et al. 2012), and shark by-products were still being exported internationally in contravention of CITES (WildLifeRisk 2014). Whale shark filter pads also have commercial value in Chinese trade, likely as a substitution for the gill-rakers of mobulid rays (O'Malley et al. 2017; Steinke et al. 2017). At the time of writing, the current status of the Chinese whale shark fishery is unclear. Reports from Puqi, which remains the most important processing hub for sharks in China, suggest that the scale of the shark fishing industry overall has substantially contracted in recent years (Wu 2016). This may indicate a decline in the number of whale sharks being caught, although small quantities of whale shark products have been reported from Chinese trade surveys fairly recently (O'Malley et al. 2017; Steinke et al. 2017).

Whale shark fins can also have high trade value (McClenachan et al. 2016). While international trade in whale shark fins has taken place for well over a century (Bean 1907), whale shark fins were regarded as low value in Taiwan in the 1990s due to poor quality and the difficulty of their preparation (Chen and Phipps 2002). Demand for whale shark fins within the broader shark fin trade was minimal, although they were sometimes sold as display fins for shark fin soup restaurants (Chen and Phipps 2002). The survey of the Indian fishery by Hanfee (2001), however, found that the fins were a high-value export product to Hong Kong, Singapore, Taiwan, and the USA; fishers were paid up to US$1,100 per set of fins in 1999. Chinese surveys in the past decade found that whale shark fins continued to demand high prices (Li et al. 2012; WildLifeRisk 2014). Little information is available on the international trade in whale shark fins after the species was listed on CITES from 2002, which required fishing states to demonstrate that exports originated from a sustainably managed population. Although there is no record of whale shark trade in the CITES Wildlife Trade Database (http://trade.cites.org), and no efforts to certify any country's whale shark fishery as sustainable, occasional reports of whale shark fins being taken in various countries, including the Maldives (Riley et al. 2009), Indonesia (White and Cavanagh 2007), and Venezuela (Sánchez et al. 2020), and their occasional presence in trade surveys (Pierce and Norman 2016; Steinke et al. 2017), indicate that at least some illegal international trade persisted (and may persist) after these regulations were enacted.

11.3.3 Bycatch in Tuna Fisheries

Whale sharks have been used as indicators of tuna presence in offshore fisheries for many years (Chapter 4; Iwasaki 1970; Rao 1986). Whale sharks are seldom, if ever, caught as a bycatch of "pole and line" fisheries (Matsunaga et al. 2003), but are caught in pelagic gillnet (driftnet) and purse-seine fisheries. Limited information is available on whale shark catches in tuna gillnet fisheries, which are primarily located in the northern Indian and the western Pacific oceans. This is a major international fishery, accounting for approximately 34% of Indian Ocean tuna catches (Anderson et al. 2020), and the Indonesian fleet (which operates mostly in the western Pacific) was estimated to exceed 280,000 vessels in 2011 (Anderson et al. 2020). The gillnets, which are designed for entangling tuna and other pelagic fish, can be over 30 km in length (Anderson et al. 2020). Observers have been placed on a few Pakistan-based vessels or recruited from within working crews to monitor bycatch (Razzaque et al. 2020). Whale sharks are certainly caught in this fishery and, although live release is possible (Razzaque et al. 2020), few other data are available. A concerning bycatch of cetaceans (whales and dolphins) from the Indian Ocean tuna gillnet fishery, estimated at 4.1 million individuals from 1950 to 2018 (Anderson et al. 2020), indicates that this fishery should also be treated as a major threat to whale sharks.

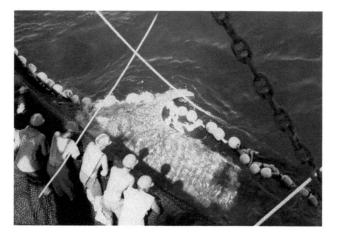

Figure 11.1 Whale shark in purse-seine net prior to release. (Photo reproduced with permission of Indian Ocean Tuna Commission, Victoria, Seychelles.)

Figure 11.2 Whale shark with rope attached (subsequently freed) at Mafia Island, Tanzania. (Photo: Simon Pierce.)

Whale sharks are often encircled in tuna purse-seine fisheries. The sharks can become entangled in these nets, although this has only been recorded in a small percentage of cases in the Indian and Atlantic Oceans (Amandè et al. 2010; Chassot et al. 2009) and care is generally taken to release safely sharks from the nets (Figure 11.1). Nonetheless, more than 60 whale sharks were reported to have been killed between 2007 and 2009 in the western and central Pacific purse-seine fishery (SPC-OFP 2010). Intentional sets of purse-seine nets around whale sharks have been banned within the Pacific Ocean and Indian Ocean, although not yet within the Atlantic (Capietto et al. 2014; Fowler 2016; Escalle et al. 2018). However, a large proportion of encircled whale sharks, as much as 73% in the western and central Pacific (SPC-OFP 2010), are not seen prior to net deployment and are therefore not considered intentional. Safe release of whale sharks using best-practice techniques (Capietto et al. 2014) is now mandatory when whale sharks are accidentally encircled in the Indian and Pacific oceans, and these guidelines are also generally followed by European vessels in the Atlantic (Escalle et al. 2016). Where poor release practices are still used, such as lifting the sharks from the water by their tails, or leaving ropes attached to the sharks following release (Figure 11.2), delayed mortality may still be an issue. An expert survey estimated a 10% post-release mortality rate in the western and central Pacific, although uncertainty around this figure was large (Neubauer et al. 2018). Further work may be needed to get a robust estimate, particularly tagging whale sharks at release from nets for longer-term tracking of mortality, as well as ensuring that best-practice release techniques are being adhered to so as to minimize this threat.

11.3.4 Bycatch in Coastal Net Fisheries

The accidental entanglement of whale sharks in fishing gear can be a major problem in some areas, especially in nets (Nair et al. 1986; Rajapackiam et al. 1994; Rajapackiam and Mohan 2006; White and Cavanagh 2007; Das et al. 2010). After the ban on targeted fishing of whale sharks in India, 79 individuals were accidentally captured and landed between 2001 and 2011 (Akhilesh et al. 2013). Bycatch in coastal gillnet and ring-net fisheries is likely to occur through much of the whale shark's distribution (Figure 11.3; Pajuelo et al. 2018; Rohner et al. 2018; Sánchez et al. 2020), but most catches appear to go unreported unless specific surveys are conducted (such as Temple et al. 2019).

11.3.5 Can Whale Sharks be Fished Sustainably?

The listing of whale sharks on Appendix II of CITES does not ban fishing as such, but does require that signatory states demonstrate that any exports were derived from a sustainably managed population. This requires the release of a formal "no detriment" finding to confirm that such trade is monitored and does not pose a risk to the conservation status of the species. Given that whale sharks are not protected by legislation in several important range states (see Chapter 12), it is worth considering whether sustainable fishing is actually possible for the species.

A global review by Simpfendorfer and Dulvy (2017) found that "with strong science-based management, most shark species have the potential to support sustainable fisheries," but this statement came with extensive caveats. They identified only a few species of sharks and their relatives that represented sustainably managed fisheries, and these were exclusively fisheries based in developed countries, specifically Australia, Canada, New Zealand, and the USA. At that time, only about 4% of the global take was managed for sustainability (Simpfendorfer and Dulvy 2017). The International Agreement relating to the Conservation and Management of Straddling Fish Stocks and Highly Migratory Fish Stocks (see Chapter 12) is a component of the United Nations Convention

Figure 11.3 Whale sharks inside a ring-net at Mafia Island, Tanzania. (Photo Simon Pierce.)

on the Law of the Sea. The whale shark is listed as a "Highly Migratory Species" on Annex I of the Convention. The Agreement mandates that management must take a precautionary approach and be based on the best available scientific information. The Food and Agriculture Organization of the United Nations (FAO) assessed the Agreement's effectiveness for a Review Conference in 2016 (FAO 2016) and reported that "given its life-history characteristics, the whale shark is expected to have low resilience to exploitation, but the state of stocks remains uncertain in most areas. Unless demonstrated otherwise, it is prudent to consider the species as being fully exploited globally."

There are several arguments against targeted fishing for whale sharks. First, while they likely have a high juvenile mortality, which hypothetically enables the take of small sharks that "would likely die anyway," in reality whale sharks of <3 m length are rarely seen in either fishery catches or known constellation sites (Chapter 7; Rowat and Brooks 2012; Norman et al. 2017). Second, biological data on the species indicate slow growth and a late age of maturity, with an assumed low natural mortality rate once the sharks reach a few meters in length (see Chapter 2). This makes the species highly vulnerable to decline if mortality from fisheries is present, reflected by its globally endangered status (Pierce and Norman 2016). Third, most past whale shark fisheries, and the location of many contemporary constellations, are under the jurisdiction of countries that are not known for sustainable, data-driven fisheries management. Many of these countries have now protected the species, nonetheless. Fourth, while whale sharks have had a high value in some domestic fisheries and in international trade, they have generally been taken primarily as an opportunistic bycatch rather than specifically targeted. That would mean that, even within a highly depleted whale shark population that would not be economically viable as a targeted fishery, animals could still be taken opportunistically, during the course of normal fishing activities, without being subject to normal cost/benefit business considerations. This scenario has been demonstrated in other shark species, such as the now critically endangered angel sharks *Squatina* spp. (Lawson et al. 2020) and could further heighten the extinction risk for whale sharks. All in all, we think that both a biological and practical assessment of whale shark catches makes the prospect of sustainable fishing infeasible for the species.

11.4 BOAT STRIKES

Boat strikes or collisions are defined as any impact between any part of a watercraft (most commonly the propeller, bow, or hull) and a live marine animal (Peel et al. 2018). They became apparent as a potential issue for whale sharks in the early 1900s, when a number of anecdotes described interactions between whale sharks and the ocean-going vessels of the day (e.g. Gudger 1927, 1937a, b, c, 1938). In fact, most published observations of whale sharks in the early years involved either a collision event or reference to close encounters between whale sharks and vessels. Early collision accounts described many boat strike instances occurring across the world, from Socotra Island in the Indian Ocean, to the Gulf of Mexico. Descriptions of these instances provided detailed, albeit informal, snapshots of how whale sharks responded to boats and of the disastrous consequences if a collision should occur. In many of these cases, whale sharks did not survive the impact or swam away with severe injuries. The boats were often notably damaged too (Gudger 1938). Through Gudger's accounts, whale sharks gained a reputation as "sluggish leviathans" that "fear not shark, man nor ship" (Gudger 1940, 1941) as it became clear that boat strikes were a significant threat to the species. Although collision events certainly still occur today (Figure 11.4), there is no coordinated recording network, and vessels are so much larger and more powerful now that impacts are unlikely even to be noticed (Rowat 2010, Erbe et al. 2020). We still know little about an ongoing threat that has been well recognized for almost a century.

Whale sharks are negatively buoyant. If a shark is struck by a vessel and dies, its body is likely to sink and is unlikely to be found or recorded (Speed et al. 2008). This makes quantifying the scale

Figure 11.4   A whale shark with a severe propeller injury off Qatar. (Photo: David P. Robinson.)

Figure 11.5   Shipping vessel traveling close to a whale shark constellation off Isla Mujeres, Mexico. (Photo: Simon Pierce.)

of boat strike mortality challenging, and also helps explain why documented collision events are so rare. Considering their habit of swimming close to the surface (Rowat and Brooks 2012), their use of coastal aggregation sites where vessels also occur, and the increasing size, speed, and quantity of oceangoing vessels globally (Figure 11.5; Erbe et al. 2020), boat strike may well be the largest source of mortality today (see Section 11.8). In areas where there are large predictable constellations that are also heavily trafficked by vessels, such as the Gulf of Mexico (de la Parra et al. 2011), Mexican Caribbean (Hueter et al. 2013), and the Arabian Gulf (Robinson et al. 2013), whale sharks may be at significant risk of collision when traversing shipping routes or port entrances (Pirotta et al. 2019).

Without formal collision accounts, an alternative way to quantify the prevalence of boat strikes within a population is to use whale shark injuries and scars as a proxy measure (Figure 11.6). Injuries resulting from boat strikes are commonly reported by regional monitoring programs based at known constellation sites (Taylor 1989, 2019; Arzoumanian et al. 2005; Meekan et al. 2006; Rowat et al. 2007; Speed et al. 2008), and in some areas, these injuries have been increasing in recent years (Lester et al. 2020). We can estimate what type of event led to an injury by examining wound characteristics (Table 11.1). For example, clean-edged deep cuts in the skin with a parallel,

Figure 11.6   Whale shark with propeller scarring in the Maldives. (Photo: Naushad Mohamed of Island Divers, courtesy of the Maldives Whale Shark Research Programme.)

repeating pattern are indicative of a strike from a vessel's propeller, whereas strikes from a vessel hull can be characterized by large wounds including blunt traumas or large areas of removed skin (abrasions) that are unlikely to result from any other natural or human cause (Rommel et al. 2007). Although whale sharks have remarkable healing capabilities (Womersley et al. 2021), even short-term injuries can have a potential impact on individual fitness, or induce changes in physiology and/or behavior, and should be mitigated where possible (see Speed et al. 2008 for a review).

A major caveat of this approach is that it is generally applied to known constellation areas, where tourism and other small boat traffic is often common. A lot of the injuries and scars recorded at these locations are then, unsurprisingly, linked to impacts from small vessels and propellers, as shown by the small pitch and the close parallel pattern of the cuts (Araujo et al. 2014). While some studies have suggested that all types of scarring, including those linked to collisions, have no influence on mortality (Speed et al. 2008, Lester et al. 2020), this approach is unlikely to capture the threat posed by all vessel activity, particularly that of larger commercial cargo and tanker ships, which are presumably more likely to cause short-term mortality. Nonetheless, these population-level efforts remain a useful complement to other potential modeling initiatives, as put forward below.

Global shipping maps (Halpern et al. 2015, Pirotta et al. 2019; Sequeira et al. 2019) show that satellite-tracked vessel traffic is extensive across both coastal and oceanic waters, which suggests that whale sharks may be at risk of collision across broad portions of their range. There is no metric for the number of whale sharks currently killed by vessel collisions, nor is there a coordinated reporting system to capture such information. By comparison, there has been a significant body of work on the issue of vessel strikes for marine mammals (Laist et al. 2001, Schoeman et al. 2020). Most marine mammals float when killed, and, as a result, boat strike mortalities in these species are noted more frequently. Nonetheless, the International Whaling Commission (IWC), which compiles reports on marine mammal vessel strikes, states that for every fatality recorded, there are also many that go unnoticed (https://iwc.int/ship-strikes). As few as 17% of marine mammals fatally struck by vessels wash ashore or are reported (Rockwood et al. 2017). For some species, it is possible to model the likelihood of vessel collisions and associated mortality based on known habits and migratory paths and the large vessel usage of the waters concerned (Hazel and Gyuris 2006; Vanderlaan and Taggart 2007; Rockwood et al. 2017; Keen et al. 2019). With this information, management and mitigation measures can be proposed, such as moving shipping lanes (Rockwood et al. 2017) or imposing speed limits (Vanderlaan and Taggart 2007). This type of modeling could be similarly applied to whale sharks to plan for spatial mitigation measures.

Traffic Separation Schemes (TSS), Areas to Be Avoided (ATBA), Dynamic Management Areas (DMA), and Particularly Sensitive Sea Areas (PSSA) are all vessel-related mitigation measures that have had a positive influence on marine mammal populations, particularly in US waters (Schoeman et al. 2020). For example, NOAA Fisheries coordinates a program to reduce the risk of vessel collisions with critically endangered North Atlantic right whales *Eubalaena glacialis*, for which vessel collision has been highlighted as a leading cause of population decline (Kraus et al. 2005). Besides minimum safe distances and maximum speed limits, vessels must report when in a "Mandatory Ship Reporting" area by giving their location, destination, and speed. Vessel operators receive a return message with locations of current North Atlantic right whale aggregations and other pertinent information (Silber et al. 2015). Such a system could well be a valuable addition to whale shark conservation in areas where individuals aggregate in large numbers, but will require the development of methods to accurately capture the aggregation's location, as well as a call-center facility. One potential approach for whale sharks could be the use of near-real-time tracking using satellite transmitters that can relay the location of sharks to local resource managers. Adaptive conservation methods using near-real-time tracking or dynamic species distribution models with high spatio-temporal resolution, such as initiatives like TurtleWatch (NOAA 2020a) and WhaleWatch (NOAA 2020b), could be an excellent way of mitigating boat strike on local scales (Blondin et al. 2020). However, it is equally important to distill high-resolution data to refine patterns of vessel activity in order for these approaches to work, which raises questions like; should we remotely monitor all vessels, including recreational and tourist boats, operating in protected areas and around whale shark aggregations? There are still many questions related to the best approach for mitigating collisions between whale sharks and vessels. With increased data sharing, vessel tracking, and conservation awareness, a clear goal for whale shark recovery should be to formally develop and enforce boat strike mitigation measures in appropriate areas.

11.5 TOURISM INTERACTIONS

11.5.1 Behavioral Change and Physical Impacts

Whale shark tourism has grown dramatically since the first predictable whale shark constellation was discovered off Ningaloo Reef in Western Australia in the 1990s (Taylor 1989, 2019; and see Chapter 10). Tourism can serve as an excellent tool to connect interested members of the public to the life of whale sharks, which can promote further understanding, increase empathy, and ultimately improve protection of this endangered species (Ziegler et al. 2012). It can also provide an important source of income to local communities, some of which may depend on the revenue that whale shark tours generate (Cagua et al. 2014). Global whale shark tourism has allowed for a huge network of citizen scientists to contribute to monitoring the species, wherein whale shark images from around the world can be collected and used for research and conservation outputs (Norman et al. 2017). There is, however, the potential for negative impacts from tourism activities, as is the case for other elasmobranch species (Fitzpatrick et al. 2011; Corcoran et al. 2013; Huveneers et al. 2013). As the industry grows, we are learning more about these threats, which mostly relate to the physical and behavioral impacts that large numbers of boats and swimmers can have on individual whale sharks, as well as how they can be avoided or minimized (Gallagher and Huveneers 2018).

As noted earlier, the threat of boat collisions can be increased by wildlife tourism activities leading to greater numbers of vessels in high-use whale shark habitats (Cagua et al. 2014). As tourism has increased in popularity in recent years, the frequency of whale sharks with boat-related injuries at specific constellation sites has become a key metric in monitoring studies (Lester et al. 2020), and mitigation measures have been developed to minimize the potential for negative physical impacts. In some areas, legislated or voluntary Codes of Conduct for operators have been developed to ensure minimum distances are maintained between boats, swimmers, and the sharks (Chapter 10, Pierce et al. 2010). Compliance to these rules or guidelines can help to reduce the chance of collisions or other negative impacts (Quiros 2007). Engagement activities aimed at local communities, and tour operators can usefully educate stakeholders on the importance and benefits of following such procedures. In areas such as Ningaloo, where strict licensing rules for tourism operators and encounter regulations have been in place for many years (Colman 1997b; Davis et al. 1997; Mau and Wilson 2008), the frequency of vessel injuries is relatively low compared to some other locations (Lester et al. 2020), which provides support for developing regulations and enforcement strategies to limit injuries at other aggregation sites.

In addition to obvious injuries, stress may also be an important tourism-related factor to consider for whale sharks. In Donsol in the Philippines, where tourism numbers rose from 900 in 1998 to 27,000 in 2012 (Quiros 2007; McCoy et al. 2018), tourists have been involved in touching individuals, obstructing swimming routes, using flash photography, or free-diving down to the shark. These activities can induce evasive behaviors, such as abrupt changes of direction, banking, diving or violent shuddering (Quiros 2007), distracting the shark from its natural behaviors, such as feeding. A similar pattern of impact has been seen in the constellation based off Tofo, in Mozambique, where the duration of encounters dropped significantly when an avoidance behavior was noted; this suggests that tourism interactions can have a direct impact on short-term behaviors (Haskell et al. 2015). This was also seen off Panaon Island, Southern Leyte, in the Philippines, where whale shark proximity to motor vessels and tourism interactions in deep water were found to significantly shorten interaction times (Araujo et al. 2017). This is a key concern because avoidance responses may have energetic costs for whale sharks. When the cost of whale watching boat avoidance was quantified for minke whales *Balaenoptera acutorostrata* in Iceland, respiration rates were higher during interactions with tourist boats (at any given speed), which suggested that boat presence elicited a stress response in the animals (Christiansen et al. 2014). This resulted in a 23.2% increase in estimated energy expenditure and an increased swimming speed (which resulted in a further 4.4% increase in estimated energy expenditure) during whale watching interactions (Christiansen et al. 2014). Recent work on whale sharks suggests that foraging bouts generally last several hours (Cade et al. 2020) and that interruptions to foraging during critical feeding periods may indeed represent substantial energetic costs to these endangered species.

Tourism has been linked to fluctuating hormonal stress levels in other species, including yellow-eyed penguins *Megadyptes antipodes* in New Zealand (Ellenberg et al. 2007) and a range of marine mammals (New et al. 2015). While short-term stress responses can often be beneficial, allowing individuals to better respond to perceived threats or dangers (Reeder and Kramer 2005), chronic stress has been shown to be maladaptive (Romero and Butler 2007). Prolonged disturbance related to swimmers and boats may conceivably cause chronic stress in whale sharks; the resulting hormonal response could hypothetically limit reproduction, suppress growth, and result in compromised immune system function (Romero and Butler 2007). On the other hand, the lack of elevated stress hormone (cortisol) recorded in humpback whales *Megaptera novaeangliae* occupying highly trafficked areas was potentially indicative of these whales being habituated to vessel traffic (Teerlink et al. 2018), which highlights the complexity of this issue. Either way, habituation or stress could have serious negative implications for both individual whale sharks and populations. More research is needed to ascertain if stress is linked to boat and swimmer disturbance around this species.

One of the challenges faced in behavioral research of whale sharks is that, in most cases, the observers have themselves been within the sensory range of the sharks. This could in itself influence the shark's responses. To avoid this, observations from an aircraft have been used to separate the observer from the encounter. Eighty whale shark encounters were observed using this method during a study based at Ningaloo between 2007 and 2009. While whale sharks continued to exhibit neutral behaviors such as surface swimming, swimming at depth, and resurfacing in the presence of a vessel, they changed direction twice as frequently when a vessel was present (Raudino et al. 2016). In the Seychelles, when observed from an aircraft, whale sharks spent significantly shorter times swimming at the surface when either a boat or a swimmer was present (Bluemel et al. 2013). Although Codes of Conduct can alleviate some of the potential behavioral impacts of tourism described above, more research is still needed in order to build upon these baseline data and hypothetical concepts to inform tourism management.

11.5.2 Artificial Provisioning and Artisanal Captivity

A relatively new issue in whale shark tourism is that of provisioning, the controlled feeding of whale sharks to increase their residency to an area. So far, these tourism sites are primarily in Asian waters and operate on relatively small scales (Himawan et al. 2015; Thomson et al. 2017; Araujo et al. 2020; Penketh et al. 2020). One of the first examples occurred in 2010 when it was discovered that whale sharks were aggregating around local fishing platforms called "bagans," in the southern part of Cenderawasih Bay Marine Park, West Papua, Indonesia. These platforms lower huge nets at night and suspend bright lights above the water to attract thousands of tiny baitfish or "puri." Nets are raised through the night, and the fish are harvested for food or use as bait. This process often attracts whale sharks. Later, the fishers began to throw the sharks a handful of fish during the day to keep them out of the way of handlining efforts. This slowly developed into a tourism activity when passing SCUBA liveaboards would come to swim with the sharks in exchange for cash, creating a "provisioning" tourism industry (Himawan et al. 2015). A much larger whale shark provisioning site operates at Oslob on Cebu Island in the Philippines, where whale sharks are hand fed krill by fishers and tourism operators on a daily basis (Araujo et al. 2014), attracting hundreds of thousands of tourists annually and comprising a significant boost to the local economy (Chapter 10).

The behavioral impacts of provisioning whale sharks at Oslob have been investigated by studies on 208 individual sharks over 3 years of monitoring (Araujo et al. 2014; Thomson et al. 2017). These sharks showed a diversity of residence patterns, with 21 seasonally resident and 9 year-round residents, demonstrating that provisioning can affect whale shark residency and movement (Thomson et al. 2017). Scarring incidence was found to be significantly higher in sharks that regularly visited the provisioning site in Oslob compared to other non-provisioned sites globally, and analysis of scarring over time showed that highly resident whale sharks accumulated scars during periods when they were consistently re-sighted (Penketh et al. 2020). Frequenting provisioning sites may alter energy expenditure in whale sharks, as is the case for other elasmobranch species (Barnett et al. 2016). Recent research found that the amount of time whale sharks spent at the surface shifted depending on whether they were near to the provisioning site and that dive profiles were linked to the timing of provisioning activities (Araujo et al. 2020). It appears that sharks dived deeper to cool down following prolonged periods feeding in warmer surface waters (Araujo et al. 2020). The authors drew attention to the potential fitness implications for individual whale sharks and estimated a 7.2% higher metabolic rate for sharks that frequent the provisioning site (Araujo et al. 2020). Providing food to wild animals is generally thought to alter natural behaviors, promote habituation, and potentially compromise individuals' overall health (Orams 2002), and these factors should be considered by local decision-makers.

Whale sharks have been kept in sea-pens off some Asian countries for many years. It is difficult to establish the scale of sea-pen-based tourism today. The only formal study was a theoretical investigation on whether sea-pens were a viable way to create a whale shark tourism attraction in Taiwan (Cruz et al. 2013). There are clear ethical considerations associated with the maintenance of whale sharks in a captive environment (discussed in Chapters 9 and 10). While whale sharks demonstrably can be habituated to a life in large aquaria, considerable expertise and resources are required for success (Chapter 9). The sea-pens used for tourism purposes have typically been situated in shallow coastal areas, and the conditions in which whale sharks are kept have generally been unsuitable for anything other than short-term care, with little consideration of the sharks' dietary or physiological requirements.

11.6 MARINE POLLUTION

Plastic pollution has become ubiquitous in the world's oceans. While the impact of plastic debris on various marine life is well documented, the effects on whale sharks are less well known (Germanov et al. 2018). Modeled maps of plastic density have revealed that ocean plastic is globally distributed, with millions of tons and trillions of individual plastic pieces floating in the marine environment (Eriksen et al 2014; Sequeira et al. 2019). As it degrades, plastic forms smaller and smaller particles, which ultimately end up as "microplastics" that can be inadvertently ingested by sea-life of all forms, especially when covered in a layer of biofouling (Arthur et al. 2009). A recent study showed that biofouling can make plastic smell like food, which increases the chances of consumption by marine life (Pfaller et al. 2020). The large bodies of whale sharks suggest they are vulnerable to ingesting a range of plastic sizes (Jâms et al. 2020), from the smallest microplastics to large pieces, over 10 cm long. At this stage, however, documented cases of plastic-induced fatalities or illness are rare. One case involved a necropsy of a dead whale shark that was found to have ingested a plastic straw that had hardened under the influence of gastric fluids and caused a punctured gut, multiple ulcers, and bleeding, with fatal consequences (Haetrakul et al. 2009). A whale shark, found alive but beached in the Philippines, was found during post-mortem examination to have litter lodged in its gills and plastic in its stomach (Abreo et al. 2019). Necropsy of a whale shark from Japan found that a plastic plate had become lodged in the shark's pylorus (Matsumoto et al. 2017, and see Chapter 13). Necropsy of a whale shark found stranded south of Kota Kinabalu in Sabah, Malaysian Borneo, identified a plastic bag obstructing the lower gastrointestinal tract as the probable cause of death (Lee 2019, Sen Nathan pers. comm.). Ingested plastic debris were also found within a stranded whale shark in Brazil (Sampaio et al. 2018). Other studies have found that whale sharks are vulnerable to accidentally ingesting large quantities of microplastics while filter feeding in polluted areas, with up to approximately 137 pieces per hour reported from Java in Indonesia (Germanov et al. 2019). Another route of microplastic ingestion is likely to be in the zooplankton that the sharks are feeding on, as plankton have also been shown to ingest microplastic particles (Desforges et al. 2015). Some of these microplastics can travel through the gut of the animal and be excreted without further change. However, some can adhere to the plastic surface, which may be released in the gut and taken up by the animal during digestion (Takada 2009). Mediterranean fin whales *Balaenoptera physalus* and basking sharks *Cetorhinus maximus* were both found to have the plasticizer DEHP, commonly found in PVC products, present in their tissues (Fossi et al. 2014). Higher concentrations of DEHP were found in the muscles of the basking shark when compared to the blubber of the fin whale (Fossi et al. 2014). There is a large body of research going back into the 1980s on contaminants in marine mammals that may indicate what to expect with future investigations of ecologically similar whale sharks – though as endotherms these species consume far greater quantities of prey than do whale sharks, and may therefore accumulate more contaminants. Early work indicated that marine mammals in the Northern Hemisphere had higher levels of pollutants than did those in the Southern Hemisphere (Tanabe et al. 1983). Based on prey type alone, whale sharks should be similar to the baleen whales, but body size also seems to affect pollutant uptake; minke whales *B. acutorostrata* weighing 10 t were found to have 50% more DDT (dichlorodip henyltrichloroethane) in their blubber than did fin whales weighing 80 t, although both species eat krill (Henry and Best 1983). Investigation of anthropogenic pollutants, such as heavy metals, PCBs and DDT, in whale sharks has begun only recently, with Fossi et al. (2017) documenting the presence of PCBs and DDT in shark skin tissue in the Gulf of California, as well as traces of polybrominated diphenyl ethers (PBDEs), which may have had a plastic source.

Other forms of marine pollution may also affect whale sharks. The "Deepwater Horizon" oil spill in 2010 affected a known whale shark habitat within the northern Gulf of Mexico (Campagna et al. 2011; Frias-Torres and Bostater 2011). This may have caused mortalities or changes in movement behavior in the regional whale shark population (Hueter et al. 2013). Similarly, "Red tides," caused by toxic blooms of *Karenia* spp. dinoflagellates that are associated with nutrient run-off, are increasing in frequency along the southern US coast (Brand and Compton 2007). These often result in mortality of coastal marine species, including sharks (Flewelling et al. 2010), and a probable whale shark mortality from this cause has been reported from Florida (Furby 2018). Whale sharks are also vulnerable to entanglement in fishing gear, but as yet most accounts of these instances are linked to informal social media reports (Parton et al. 2019). A recent review of this issue based on sources compiled from Twitter found numerous tweets regarding whale shark entanglement, while few have been reported in the published literature (Parton et al. 2019). This area of study would benefit from a formalized recording network to help establish the full scope and scale of the threat of entanglement in whale sharks.

11.7 CLIMATE CHANGE

While empirical data are somewhat lacking, we can make some informed predictions on the potential consequences of climate change on whale sharks based on what we know of the species' physiological tolerances, behavior, and spatial distributions. Whale sharks are ectothermic and therefore need to thermoregulate their body temperature depending on their surrounding environment. They return to surface waters to warm up after deep dives into colder water (Thums et al. 2013) or move into deeper, cooler waters after feeding for prolonged periods in the warm surface layer (Robinson et al. 2017, Araujo et al. 2020). It is likely, therefore, that future temperature increases will influence the vertical movements of this species. The potential impacts of ocean warming may also result in a broadening of the horizontal range of whale sharks into waters that were previously too cold for regular use; already there have been increasing sightings of whale sharks in "new" locations such as the Azores, and even mainland Europe, significantly further north in the Atlantic than they were known to occur previously, which suggests a range expansion (Rodrigues et al. 2012; Afonso et al. 2014). Species distribution modeling for different climate change scenarios has suggested that there may be a slight shift of suitable whale shark habitat towards the poles in response to changes in sea surface temperature, although this could be accompanied by an overall range contraction if whale sharks previously suitable tropical habitats are no longer habitable (Sequeira et al. 2014).

Changes in sea temperature may also affect the distribution of prey resources, indirectly affecting whale shark populations. Plankton taxa respond differently to thermal shifts; some taxa are likely to expand their ranges, while others may contract. This could prompt a major restructure of plankton communities, potentially modifying entire food webs (Chivers et al. 2017). Several whale shark constellations have been linked to teleost spawning events, such as in the mackerel tuna *Euthynnus affinis* (Robinson et al. 2013), and these species may also be affected by increased sea temperatures and subject to range shifts (Wabnitz et al. 2018). Future changes in prey populations will lead to changes in whale shark movements and distribution. In marine mammals, the primary influence of climate change is thought to be changes in prey distribution; more mobile species can then adapt and respond to ocean change more readily than can those with narrower ecological niches, such as polar specialists, which are especially vulnerable (Simmonds and Isaac 2007). Changes in prey availability due to the effects of El Niño have also been linked to high juvenile mortality in seabirds and seals (Würsig et al. 2002) and reduced condition in the Brazilian sharpnose shark *Rhizoprionodon lalandii* (Corsso et al. 2018). These are potential impacts to consider for whale sharks, too. Tracking whale shark and prey responses to broad-scale climatic events like this may be a useful way to examine such hypotheses.

Oceanographic factors other than temperature will also likely impact whale sharks in future climate scenarios. A comprehensive review of the impacts of climate change in coastal marine systems showed that the effects of sea-level rise, fluctuations in ocean circulation and CO_2 concentration, and pH and UV alterations can cause distributional shifts in zonation and biogeographical ranges, changes in species composition and community structure, changes in primary and secondary production, and, ultimately, shifts in population dynamics (Harley et al. 2006). Change in oxygen availability is also gaining traction as a key concern for pelagic sharks in the future (Laffoley & Baxter 2019; Lawson et al. 2019). Many of these factors could have an impact on whale sharks and/or their food resources. As yet there is little dedicated research into this topic, and the continued development of regional ocean climate models will be necessary to inform future whale shark conservation efforts.

11.8 THREAT CLASSIFICATION

The IUCN's Threat Classification Scheme (IUCN 2020) has been developed to help summarize the direct and indirect threats affecting different species to inform assessment of species-level conservation status and to prioritize management actions. The threat impact rating developed by IUCN (Table 11.1) includes "timing," which considers whether the threat has occurred in the past, is ongoing, or is projected to become an issue in the future. The "scope" estimates the proportion of the total population that is affected by the threat. "Severity" projects the overall population-level effects of each threat. While we focus here on assessing the threats to whale sharks at a global level, the suite of threats and their relative importance will vary at a regional scale. Two subpopulations of whale sharks (Atlantic Ocean and Indo-Pacific, respectively) were assessed separately for the 2016 global IUCN Red List assessment (Pierce and Norman 2016), and one regional assessment has been formally published separately for the Arabian Sea region (Dulvy et al. 2017). Other regional assessments may be possible in the future, which is discussed in detail within Chapter 12, and we hope this chapter can help to inform such work.

11.8.1 Energy Production and Mining

This category evaluates the threat posed to whale sharks from extraction or production of non-biological resources. The particular subcategory considered here is *oil and gas drilling*, which includes the threats created by people exploring for, developing, and producing petroleum and other liquid hydrocarbons. A specific and relevant incident that likely affected whale sharks was the *Deepwater Horizon* oil spill that occurred in the Gulf of Mexico in 2010. Although there is

Table 11.1 Threat Impact Scoring System for Ongoing Threats to Whale Sharks over Three Generations (~75 years) in the Future, Based on the IUCN Threat Classification System (2020)

	Severity of Impact				
Scope of Impact	Very Rapid Decline (>30%)	Rapid Decline (20%–30%)	Slow Decline (<20%)	Causes Fluctuation	Negligible Decline
Whole population (>90%)	High	High	Medium	Medium	Medium
Majority of population (50%–90%)	High	Medium	Medium	Medium	Low
Minority of population (<50%)	Medium	Medium	Low	Low	Low

Table 11.2 Summary of Impact Assessments from Ongoing Threats to Whale Sharks over Three Generations (~75 years) in the Future, Based on the IUCN Threat Classification System (2020)

Threat	Severity	Scope	Score
Oil & gas drilling	Negligible	Minority	Low impact
Shipping lanes	Very Rapid	Whole	High impact
Fishing	Rapid	Majority	Medium impact
Recreational activities	Negligible	Minority	Low impact
Agricultural & forestry effluents	Negligible	Minority	Low impact
Garbage & solid waste	Negligible	Minority	Low impact
Habitat shifting & alteration	Negligible	Minority	Low impact

extensive oil and gas production close to significant whale shark aggregation areas, such as the Gulf of Mexico (Hoffmayer et al. 2007; Hueter et al. 2013; McKinney et al. 2017) and the Arabian Gulf (Robinson et al. 2013, 2016, 2017), the regionally restricted nature of the industry means that localized spills are likely to affect a small minority of the global whale shark population, resulting in a negligible decline overall. With that in mind, oil and gas drilling is therefore assessed as *Low Impact* (Table 11.2).

11.8.2 Transportation and Service Corridors

This category focuses on the threat posed by the presence of transport corridors and, in this case, the ships and other vessels that use them. The subcategory of *shipping lanes* is relevant to whale sharks. Shipping is a continuous and ubiquitous activity, and the almost complete coverage of these "marine roads" across the whale shark's tropical and temperate distribution (Pirotta et al. 2019) indicates that shipping is a potential threat to the global population of the species. Severity is unclear due to the lack of mortality data, in turn due to the fact that whale shark carcasses sink, unlike those of true whales. Based on the assessment period used in evaluating these criteria (either 10 years or three generations, whichever is longer), and applying the 25-year generation time estimated for whale sharks in the 2016 Red List assessment, and given the lack of management and mitigation that is currently in place, it seems reasonable to infer that shipping could cause "very rapid declines" i.e. >30% of the global population of adult whale sharks. Global shipping is therefore assessed as *High Impact* (Table 11.2).

11.8.3 Biological Resource Use

This category focuses on the consumptive use of "wild" biological resources, including the effects of both deliberate and unintentional harvesting. The specific subcategory relevant to whale sharks is *fishing and harvesting aquatic resources*. This includes intentional use, wherein whale sharks are the target of small-scale or larger-scale fisheries, as well as unintentional bycatch. While targeted fisheries, such as those previously present in India, the Maldives, the Philippines, and Taiwan, have largely closed as a result of conservation and management efforts, the situation in mainland China is unclear (Chapter 12). Whale sharks are still a regular bycatch in tuna fisheries and coastal net fisheries. Fishing remains a potential threat to the majority (50–90%) of the global whale shark population, particularly in the Indo-Pacific, and we estimate that there could be a rapid decline (20–30%) in the global whale shark population over the next 75 years as a result. Fishing is therefore assessed as *Medium Impact* based on present knowledge (Table 11.2).

11.8.4 Human Intrusions and Disturbance

This category includes other threats from human activities that alter, destroy, or disturb habitats and species. Here, we consider the subcategory of *recreational activities*, which includes people spending time in nature or traveling in boats for recreational reasons, which includes marine tourism. Tourism and leisure boating, whether specifically focused on viewing whale sharks or driving through the habitats they frequent, can certainly result in disturbance and injury to sharks, a sublethal endpoint. As these activities tend to occur in very specific constellation areas close to the necessary infrastructure, at a population level this likely affects a minority of the global whale shark population, so recreation is unlikely to result in significant declines overall. Recreational activities are therefore assessed as *Low Impact* (Table 11.2).

11.8.5 Pollution

This category focuses on the threat posed by the introduction of exotic and/or excess materials or energy to the environment. Here, we are concerned with two subcategories, *agricultural and forestry effluents*, specifically the high nutrient loads caused by run-off that can create toxic algal blooms (red tides), and *garbage and solid waste*, such as plastic pollution. The former issue appears to be localized to the southern USA, based on current information, although the risk factors may well occur more broadly. Marine plastic pollution is a clear emerging threat to whale sharks that we are just beginning to understand. Germanov et al. (2018) compared known hotspots for buoyant microplastic with the whale shark's global distribution and identified sufficient overlap to fall within the "majority" scope used here. At this stage, there are only a few studies showing a confirmed or potential threat to whale sharks, particularly within Southeast Asia (Haetrakul et al. 2009; Abreo et al. 2019; Germanov et al. 2019), leading to an assessed *Low Impact* overall (Table 11.3), but really the threat is data deficient and further studies are urgently needed to quantify the true risk. We have singled out this topic as a research priority in Chapter 13.

11.8.6 Climate Change and Severe Weather

Long-term climate change will almost certainly affect whale sharks via the *habitat shifting and alteration* subcategory used by the IUCN. There may well be a slight expansion of whale sharks' seasonal penetration into higher latitudes, as reported from European waters (Rodrigues et al. 2012; Afonso et al. 2014), although model predictions suggest that whale shark range may slightly contract overall due to reduced suitability of some tropical areas (Sequeira et al. 2014). More study is needed of the potential impacts of climate change on whale sharks and their preferred prey species. Some marine mammal species are projected to be highly vulnerable to declines created by global warming effects, but the most susceptible tend to be specialized for specific habitats (Albouy et al. 2020). At this stage, while climate change may affect a majority of whale sharks, it appears that their broad thermal tolerance and wide distribution may buffer them from direct mortality, resulting in a *Low Impact* assessment here (Table 11.3). The potential for ecosystem-level effects remains a concern, however.

11.8.7 Prioritization of Ongoing Threats

This impact evaluation exercise provisionally identifies shipping and fishing as the two ongoing threats to the global whale shark population that should be the highest priority for conservation intervention at a global level (Table 11.3). If conservation initiatives do not target the reduction of harm created by these activities, then there is a high probability of further whale shark population declines on a global scale, with some regions facing cumulative threats.

11.9 CONCLUSIONS

Targeted hunting was almost certainly the major cause of whale shark mortality over the past few decades. Directed commercial fisheries are subject to economic considerations fishery management, so the potential exists for management regulations to be introduced and enforced. Following the closure of the Taiwanese fishery in 2007 (Hsu et al. 2012), the principal causes of contemporary mortality have become significantly harder to determine. In terms of fisheries, bycatch in gillnets remains a major concern. A lack of management attention and observer coverage, even in large commercial fisheries, makes it difficult to quantify the current level of catches (see the section on data gleaned from RFMOs in Chapter 12), and how "survivable" such interactions might be for whale sharks. Unwanted bycatch species can be inadvertently driven to near extinction because their decline has little or no impact on the broader profitability of the target fishery, which means that there is no reduction of fishing effort as the numbers of the bycatch animal continue to decline. This situation has led to regional extinctions in elasmobranch species that are particularly susceptible to bycatch, such as sawfish (*Pristis* spp.; Dulvy et al. 2016).

While it is possible that commercial fishing of whale sharks still occurs, particularly in China (WildLifeRisk 2014), the number of sharks killed "accidentally" is likely to have overtaken direct exploitation as the leading cause of mortality and population declines. Two of the most concerning indirect causes of whale shark death are those of ship strike and pollution, including microplastics. Both of these threats are difficult to quantify, in either rates of mortality or of sublethal morbidity, but both have a high degree of spatial overlap with whale shark populations (Germanov et al. 2018; Pirotta et al. 2019). Successful mitigation of shipping impacts through management actions has at least been implemented for some whales, such as the North Atlantic right whale in Massachusetts USA, providing potential case studies for whale shark conservation. Marine pollution is less amenable to directed marine conservation and management activities, with the exception of end-point measures such as clean-ups. To achieve meaningful reductions on a global level, mitigation will largely rely on "shutting off the tap": the reduction of plastic use within the supply chain and better waste management infrastructure in developing countries. This is achievable, but in the short-term, it might be most effective to identify and address the regionally specific sources of pollution that affect coastal feeding areas for whale sharks, where they are likely to be most at risk from ingestion and other acute and chronic impacts.

The threats that increasing whale shark tourism poses in its various forms are also a cause for concern, largely due to a lack of understanding of the longer-term impacts of these activities on the species, whether this be through injuries caused by boats (e.g. Meekan et al. 2006, Rowat et al. 2007, Speed et al. 2008), behavioral changes caused by disturbance during feeding (Norman 2002), changes in migratory or feeding habits due to provisioning (Thomson et al. 2017), or captivity (Cruz et al. 2013). Careful regulation of tourism activities can reduce the real and potential threats associated with this sector, and Chapter 10 provides more information on current best-practice within this industry.

This chapter provides an outline of the various threats to whale sharks, but in closing, we need to emphasize that these issues do not each act in isolation (Brook et al. 2008). The current, depleted whale shark population is more affected by, and less able to rebound from, human impacts. A low level of bycatch, for instance, or a few sharks being killed by ships each year, may be enough to ensure continued declines within already-reduced whale shark populations (Bradshaw et al. 2007). Pressure from tourism, and a resulting increased energy expenditure, might be enough to affect the fitness of individual whale sharks (Araujo et al. 2020). Population viability analysis of global whale shark populations would be a useful exercise to inform conservation interventions. While there is substantial cause for optimism in whale shark conservation – the reduction in fisheries impacts has been hard-won, and significantly improves the outlook for the species going forward – there is clearly more work to do.

REFERENCES

Abreo, N. A. S., D. Blatchley, and M. D. Superio. 2019. Stranded whale shark (*Rhincodon typus*) reveals vulnerability of filter-feeding elasmobranchs to marine litter in the Philippines. *Marine Pollution Bulletin* 141:79–83.

Acebes, J. M. V. 2013. Hunting 'big fish': A marine environmental history of a contested fishery in the Bohol Sea. PhD diss., Murdoch University.

Afonso, P., McGinty, N., and M. Machete. 2014. Dynamics of whale shark occurrence at their fringe oceanic habitat. *PloS One* 9:e102060.

Akhilesh, K. V., Shanis, C. P. R., White, W. T., et al. 2013. Landings of whale sharks *Rhincodon typus* Smith, 1828 in Indian waters since protection in 2001 through the Indian Wildlife (Protection) Act, 1972. *Environmental Biology of Fishes* 96:713–22.

Alava, M. N. R., and E. R. Z. Dolumbalo. 2002. Fishery and trade of whale sharks and manta rays in the Bohol Sea, Philippines. In *Elasmobranch Biodiversity, Conservation and Management: Proceedings of the International Seminar and Workshop, Sabah, Malaysia, July 1997*, ed. S. L. Fowler, T. M. Reed, and F. A. Dipper, 132–48. Occasional Paper of the IUCN Species Survival Commission.

Albouy, C., Delattre, V., Donati, G., et al. 2020. Global vulnerability of marine mammals to global warming. *Scientific Reports* 10:548.

Amandè, M. J., Ariz, J., Chassot, E., et al. 2010. Bycatch of the European purse seine tuna fishery in the Atlantic Ocean for the 2003–2007 period. *Aquatic Living Resources* 23:353–62.

Anderson, R. C., Herrera, M., Ilangakoon, A. D., et al. 2020. Cetacean bycatch in Indian Ocean tuna gillnet fisheries. *Endangered Species Research* 41:39–53.

Anderson, R. C., and H. Ahmed. 1993. *The Shark Fisheries of the Maldives*. FAO, Rome, and Ministry of Fisheries, Male, Maldives.

Araujo, G., Labaja, J., Snow, S., et al. 2020. Changes in diving behaviour and habitat use of provisioned whale sharks: Implications for management. *Scientific Reports* 10:16951.

Araujo, G., Lucey, A., Labaja, J., et al. 2014. Population structure and residency patterns of whale sharks, *Rhincodon typus*, at a provisioning site in Cebu, Philippines. *PeerJ* 2:e543.

Araujo, G., Vivier, F., Labaja, J. J., et al. 2017. Assessing the impacts of tourism on the world's largest fish *Rhincodon typus* at Panaon Island, Southern Leyte, Philippines. *Aquatic Conservation: Marine and Freshwater Ecosystems* 27:986–994.

Arthur, C., Baker, J., and H. Bamford. 2009. *Proceedings of the International Research Workshop on the Occurrence, Effects, and Fate of Microplastic Marine Debris*, September 2008. Silver Spring: NOAA Marine Debris Division.

Arzoumanian, Z., Holmberg, J., and B. Norman. 2005. An astronomical pattern-matching algorithm for computer-aided identification of whale sharks *Rhincodon typus*. *Journal of Applied Ecology* 42:999–1011.

Barnett, A., N. L. Payne, J. M. Semmens, and R. Fitzpatrick. 2016. Ecotourism increases the field metabolic rate of whitetip reef sharks. *Biological Conservation* 199:132–36.

Bean, B. A. 1907. The history of the whale shark (*Rhinodon typicus* Smith). *Smithsonian Miscellaneous Collections* 48:139–48.

Blondin, H., B. Abrahms, L. B. Crowder, and E. L. Hazen. 2020. Combining high temporal resolution whale distribution and vessel tracking data improves estimates of ship strike risk. *Biological Conservation* 250:108757.

Bluemel, J. K., French, G. C. A., and D. Rowat. 2013. An aerial view: Insights into the effects of ecotourism on the behavior of whale sharks (*Rhincodon typus*) in Seychelles. *PeerJ PrePrints* 1:e103v1.

Bradshaw, C. J. A., Mollet, H. F., and M. G. Meekan. 2007. Inferring population trends for the world's largest fish from mark-recapture estimates of survival. *The Journal of Animal Ecology* 76:480–89.

Brand, L. E., and A. Compton. 2007. Long-term increase in *Karenia brevis* abundance along the southwest Florida coast. *Harmful Algae* 6:232–52.

Brook, B. W., Sodhi, N. S., and C. J. A. Bradshaw. 2008. Synergies among extinction drivers under global change. *Trends in Ecology & Evolution* 23:453–60.

Cade, D. E., J. J. Levenson, R. Cooper, et al. 2020. Whale sharks increase swimming effort while filter feeding, but appear to maintain high foraging efficiencies. *Journal of Experimental Biology* 223. doi:10.1242/jeb.224402.

Cagua, E. F., N. Collins, J. Hancock, and R. Rees. 2014. Whale shark economics: A valuation of wildlife tourism in South Ari Atoll, Maldives. *Peer J* 2:e515.

Campagna, C., Short, F. T., Polidoro, B. A., et al. 2011. Gulf of Mexico oil blowout increases risks to globally threatened species. *Bioscience* 61:393–97.

Capietto, A., Escalle, L., Chavance, P., et al. 2014. Mortality of marine megafauna induced by fisheries: Insights from the whale shark, the world's largest fish. *Biological Conservation* 174:147–51.

Carrillo-Briceño, J. D., Villafaña, J. A., De Gracia, C., et al. 2020. Diversity and paleoenvironmental implications of an elasmobranch assemblage from the Oligocene–Miocene boundary of Ecuador. *PeerJ* 8:e9051.

Chassot, E., Amandè, M. J., Chavance, P., et al. 2009. Some preliminary results on tuna discards and bycatch in the French purse seine fishery of the eastern Atlantic Ocean. *Collective Volume of Scientific Papers, ICCAT* 64:1054–67.

Chen, C. T., Liu, K.-M., and S.-J. Joung 1997. Preliminary report on Taiwan's whale shark fishery. *TRAFFIC Bulletin* 17:53–57.

Chen, V. Y., and M. J. Phipps. 2002. *Management and Trade of Whale Sharks in Taiwan.* Taipei: TRAFFIC East Asia-Taipei.

Chivers, W. J., Walne, A. W., and G. C. Hays. 2017. Mismatch between marine plankton range movements and the velocity of climate change. *Nature Communications* 8:14434.

Christiansen, F., M. H. Rasmussen, and D. Lusseau. 2014. Inferring energy expenditure from respiration rates in minke whales to measure the effects of whale watching boat interactions. *Journal of Experimental Marine Biology and Ecology* 459:96–104.

Cicimurri, D. J., and J. L. Knight. 2009. Late Oligocene sharks and rays from the Chandler Bridge Formation, Dorchester County, South Carolina, USA. *Acta Palaeontologica Polonica* 54:627–47.

Colman, J. G. 1997a. A review of the biology and ecology of the whale shark. *Journal of Fish Biology* 51:1219–34.

Colman, J. G. 1997b. *Whale Shark Interaction Management, with Particular Reference to Ningaloo Marine Park 1997–2007.* Western Australia: Department of Conservation and Land Management.

Corcoran, M. J., B. M. Wetherbee, M. S. Shivji, et al. 2013. Supplemental feeding for ecotourism reverses diel activity and alters movement patterns and spatial distribution of the southern stingray, *Dasyatis americana. PLoS One* 8:e59235.

Corsso, J. T., Gadig, O. B. F., Barreto, R. R. P., and F. S. Motta. 2018. Condition analysis of the Brazilian sharpnose shark *Rhizoprionodon lalandii:* Evidence of maternal investment for initial post-natal life. *Journal of Fish Biology* 93:1038–45.

Cruz, F. A., Joung, S., Liu, K. M., et al. 2013. A preliminary study on the feasibility of whale shark (*Rhincodon typus*) ecotourism in Taiwan. *Ocean & Coastal Management* 80:100–106.

Davis, D., S. Banks, A. Birtles, et al. 1997. Whale sharks in Ningaloo Marine Park: Managing tourism in an Australian marine protected area. *Tourism Management* 18:259–71.

Das, T., Sundaram, S., Katkar, B. N., and B. B. Chavan. 2010. Accidental capture and landing of whale shark, *Rhincodon typus* (Smith, 1828) and tiger shark, *Galeocerdo cuvier* (Peron and Le Sueur, 1822) by trawlers at New Ferry Wharf, Mumbai. *Marine Fisheries Information Service* 205:17–19.

De la Parra Venegas, R., R. Hueter, J. González Cano, et al. 2011. An unprecedented aggregation of whale sharks, *Rhincodon typus*, in Mexican coastal waters of the Caribbean Sea. *PLoS One* 6:e18994.

Desforges, J. P. W., Galbraith, M., and P. S. Ross. 2015. Ingestion of microplastics by zooplankton in the northeast Pacific Ocean. *Archives of Environmental Contamination and Toxicology* 69:320.

Dulvy, N. K., Robinson, D. P., Pierce, S. J., et al. 2017. Whale shark *Rhincodon typus* Smith, 1828. In *The Conservation Status of Sharks, Rays, and Chimaeras in the Arabian Sea and Adjacent Waters*, edited by R. W. Jabado, P. M. Kyne, R. A. Pollom, et al., 2.1:236. Vancouver: Environment Agency – Abu Dhabi, UAE and IUCN Species Survival Commission Shark Specialist Group.

Dulvy, N. K., Davidson, L. N. K., Kyne, P. M., et al. 2016. Ghosts of the coast: Global extinction risk and conservation of sawfishes. *Aquatic Conservation: Marine and Freshwater Ecosystems* 26:134–53.

Eckert, S. A., Dolar, L. L., Kooyman, G. L., et al. 2002. Movements of whale sharks (*Rhincodon typus*) in South-East Asian waters as determined by satellite telemetry. *Journal of Zoology* 257:111–15.

Egami, T., and K. Kojima. 2013. Traditional whaling culture and social change in Lamalera, Indonesia: An analysis of the catch record of whaling. *Senri Ethnological Studies* 84:155–76.

Ellenberg, U., A. N. Setiawan, A. Cree, et al. 2007. Elevated hormonal stress response and reduced reproductive output in yellow-eyed penguins exposed to unregulated tourism. *General and Comparative Endocrinology* 152:54–63.

Erbe, C., Smith, J. N., Redfern, J. V., and D. Peel. 2020. Impacts of shipping on marine fauna. *Frontiers in Marine Science* 7:637.

Eriksen, M., L. C. M. Lebreton, H. S. Carson, et al. 2014. Plastic pollution in the world's oceans: More than 5 trillion plastic pieces weighing over 250,000 tons afloat at sea. *PloS One* 9:e111913.

Escalle, L., Amandé, J. M., Filmalter, J. D., et al. 2018. Update on post-release survival of tagged whale shark encircled by tuna purse-seiner. *International Commission for the Conservation of Atlantic Tunas* 74:3671–78.

Escalle, L., Murua, H., Amandé, J. M., et al. 2016. Post-capture survival of whale sharks encircled in tuna purse-seine nets: Tagging and safe release methods. *Aquatic Conservation: Marine and Freshwater Ecosystems* 26:782–89.

FAO. 2016. FAO's input to the UN Secretary-General's comprehensive report for the 2016 resumed review conference on the UN Fish Stocks Agreement. http://www.un.org/Depts/los/2016_FAO_Overview.pdf (accessed November 17, 2020).

Fitzpatrick, R., K. G. Abrantes, J. Seymour, and A. Barnett. 2011. Variation in depth of whitetip reef sharks: Does provisioning ecotourism change their behaviour? *Coral Reefs* 30:569–77.

Flewelling, L. J., Adams, D. H., Naar, J. P., et al. 2010. Brevetoxins in sharks and rays (Chondrichthyes, Elasmobranchii) from Florida coastal waters. *Marine Biology* 157:1937–53.

Fossi, M. C., Baini, M., Panti, C., et al. 2017. Are whale sharks exposed to persistent organic pollutants and plastic pollution in the Gulf of California (Mexico)? First ecotoxicological investigation using skin biopsies. *Comparative Biochemistry and Physiology Part C: Toxicology & Pharmacology* 199:48–58.

Fossi, M. C., Coppola, D., Baini, M., et al. 2014. Large filter feeding marine organisms as indicators of microplastic in the pelagic environment: The case studies of the Mediterranean basking shark (*Cetorhinus maximus*) and fin whale (*Balaenoptera physalus*). *Marine Environmental Research* 100:17–24.

Fowler, S. 2016. Gap analysis of activities for the conservation of species listed in Annex 1 under relevant fisheries related bodies. Memorandum of Understanding on the Conservation of Migratory Sharks. http://www.cms.int/sharks/sites/default/files/document/CMS_Sharks_CWG1_Doc_2_1.pdf (accessed November 17, 2020).

Frias-Torres, S., and C. R. Bostater Jr. 2011. Potential impacts of the Deepwater Horizon oil spill on large pelagic fishes. In *Remote Sensing of the Ocean, Sea Ice, Coastal Waters, and Large Water Regions 2011*, 8175:81750F. International Society for Optics and Photonics.

Furby, K. 2018. A red tide ravaging Florida may have killed a whale shark for the first known time. *The Washington Post*, August 3, 2018. https://www.washingtonpost.com/news/speaking-of-science/wp/2018/08/03/a-red-tide-ravaging-florida-may-have-killed-a-whale-shark-for-the-first-known-time/ (accessed November 17, 2020).

Gallagher, A. J., and C. P. M. Huveneers. 2018. Emerging challenges to shark-diving tourism. *Marine Policy* 96:9–12.

Germanov, E. S., Marshall, A. D., Bejder, L., et al. 2018. Microplastics: No small problem for filter-feeding megafauna. *Trends in Ecology & Evolution* 33:227–32.

Germanov, E. S., Marshall, A. D., Hendrawan, I. G., et al. 2019. Microplastics on the menu: Plastics pollute Indonesian manta ray and whale shark feeding grounds. *Frontiers in Marine Science* 6:679.

Gudger, E. W. 1927. A second whale shark, *Rhineodon typus*, impaled on the bow of a steamship. *Zoological Society Bulletin* 30:76–77.

Gudger, E. W. 1937a. A whale shark impaled on the bow of a steamer near the Tuamotus, South Seas. *Science* 8:314.

Gudger, E. W. 1937b. A whale shark rammed by a steamer off Colombo, Ceylon. *Nature* 139:549.

Gudger, E. W. 1937c. A whale shark speared on the bow of a steamer in the Caribbean Sea. *Copeia* 1937:60.

Gudger, E. W. 1938. Four whale sharks rammed by steamers in the Red Sea region. *Copeia* 1938:170–73.

Gudger, E. W. 1940. Whale sharks rammed by ocean vessels: How these sluggish leviathans aid in their own destruction. *New England Naturalist* 7:1–10.

Gudger, E. W. 1941. The whale shark unafraid: The greatest of the sharks, *Rhineodon typus*, fears not shark, man nor ship. *The American Naturalist* 75:550–68.

Guzman, H. M., Gomez, C. G., Hearn, A., and S. A. Eckert. 2018. Longest recorded trans-Pacific migration of a whale shark (*Rhincodon typus*). *Marine Biodiversity Records* 11:8.

Haetrakul, T., Munanansup, S., Assawawongkasem, N., and N. Chansue. 2009. A case report: Stomach foreign object in whaleshark (*Rhincodon typus*) stranded in Thailand. *Proceedings of the 4th International Symposium on SEASTAR 2000 and Asian Bio-Logging Science (The 8th SEASTAR 2000 Workshop)*:83–85.

Halpern, B. S., M. Frazier, J. Potapenko, et al. 2015. Spatial and temporal changes in cumulative human impacts on the world's ocean. *Nature Communications* 6:7615.

Hanfee, F. 2001. Gentle giants of the sea: India's whale shark fishery. TRAFFIC-India, WWF-India.

Harley, D. G., Hughes, A. R., Hultgren, K. M., et al. 2006. The impacts of climate change in coastal marine systems. *Ecology Letters* 9:228–41.

Haskell, P. J., McGowan, A., Westling, A., et al. 2015. Monitoring the effects of tourism on whale shark *Rhincodon typus* behaviour in Mozambique. *Oryx* 49:492–99.

Hazel, J., and E. Gyuris. 2006. Vessel-related mortality of sea turtles in Queensland, Australia. *Wildlife Research* 33:149–54.

Henry, J., and P. B. Best. 1983. Organochlorine residues in whales landed at Durban, South Africa. *Marine Pollution Bulletin* 4:223–27.

Himawan, M. R., Tania, C., Noor, B. A., et al. 2015. Sex and size range composition of whale shark (*Rhincodon typus*) and their sighting behaviour in relation with fishermen lift-net within Cenderawasih Bay National Park, Indonesia. *AACL Bioflux* 8:123–33.

Hoffmayer, E. R., Franks, J. S., Driggers III, W. B., et al. 2007. Observations of a feeding aggregation of whale sharks, *Rhincodon typus*, in the north central Gulf of Mexico. *Gulf and Caribbean Research* 19:69–73.

Hsu, H.-H., Joung, S. J., and K.-M. Liu. 2012. Fisheries, management and conservation of the whale shark *Rhincodon typus* in Taiwan. *Journal of Fish Biology* 80:1595–607.

Hueter, R. E., Tyminski, J. P., and R. de la Parra. 2013. Horizontal movements, migration patterns, and population structure of whale sharks in the Gulf of Mexico and northwestern Caribbean Sea. *PLoS One* 8:e71883.

Huveneers, C., P. J. Rogers, C. Beckmann, et al. 2013. The effects of cage-diving activities on the fine-scale swimming behaviour and space use of white sharks. *Marine Biology* 160: 2863–75.

IUCN. 2020. Threats Classification Scheme: The IUCN Red List of Threatened Species. https://www.iucnredlist.org/resources/threat-classification-scheme (accessed November 30, 2020).

Iwasaki, Y. 1970. On the distribution and environment of the whale shark, *Rhincodon typus*, in skipjack fishing grounds in the western Pacific Ocean. *Journal of the College of Marine Science Technology, Tokai University* 4 37–51.

Jâms, I. B., F. M. Windsor, T. Poudevigne-Durance, et al. 2020. Estimating the size distribution of plastics ingested by animals. *Nature Communications* 11:1594.

Keen, E. M., K. L. Scales, B. K. Rone, et al. 2019. Night and day: Diel differences in ship strike risk for fin whales (*Balaenoptera physalus*) in the California current system. *Frontiers in Marine Science* 6:730.

Kraus, S. D., M. W. Brown, H. Caswell, et al. 2005. North Atlantic right whales in crisis. *Science* 309:561–62.

Laffoley, D., and J. M. Baxter. 2019. *Ocean Deoxygenation: Everyone's Problem*. Gland: IUCN.

Laist, D. W., A. R. Knowlton, J. G. Mead, et al. 2001. Collisions between ships and whales. *Marine Mammal Science* 17:35–75.

Lawson, C. L., L. G. Halsey, G. C. Hays, et al. 2019. Powering ocean giants: The energetics of shark and ray megafauna. *Trends in Ecology and Evolution* 34:1009–21.

Lawson, J. M., Pollom, R. A., Gordon, C. A., et al. 2020. Extinction risk and conservation of critically endangered angel sharks in the Eastern Atlantic and Mediterranean Sea. *ICES Journal of Marine Science* 77: 12–29.

Lee, S. 2019. Plastic bag causes death of whale shark in Sabah. https://www.thestar.com.my/news/nation/2019/02/08/plastic-bag-causes-death-of-whale-shark-in-sabah/ (accessed January 5, 2021).

Lester, E., M. G. Meekan, P. Barnes, et al. 2020. Multi-year patterns in scarring, survival and residency of whale sharks in Ningaloo Marine Park, Western Australia. *Marine Ecology Progress Series* 634:115–25.

Li, W., Wang, Y., and B. Norman. 2012. A preliminary survey of whale shark *Rhincodon typus* catch and trade in China: An emerging crisis. *Journal of Fish Biology* 80:1608–18.

Matsumoto, R., Toda, M., Matsumoto, Y., et al. 2017. Notes on husbandry of whale sharks, >> *Rhincodon typus*>>, in aquaria. In *The Elasmobranch Husbandry Manual II: Recent Advances in the Care of Sharks, Rays and Their Relatives*, eds. Smith, M., D. Warmolts, D. Thoney, R. Hueter, M. Murray, and J. Ezcurra, 15–22. Columbus: Special Publication of the Ohio Biological Survey.

Matsumoto, R., Matsumoto, Y., Ueda, K., et al. 2019. Sexual maturation in a male whale shark (*Rhincodon typus*) based on observations made over 20 years of captivity. *Fishery Bulletin* 117:78–86.

Matsunaga, H., Nakano, H., Okamoto, H. and G. Suzul. 2003. Whale shark migration observed by pelagic tuna fishery near Japan. *National Research Institute of Far Seas Fisheries, Shizuoka* 12:1–7.

Mau, R., and E. Wilson. 2008. Industry trends and whale shark ecology based on tourism operator logbooks at Ningaloo Marine Park. In *The First International Whale Shark Conference: Promoting International Collaboration in Whale Shark Conservation, Science and Management*, eds. T. R. Irvine, and J. K. Keesing, 81. Perth: CSIRO Marine and Atmospheric Research.

McClenachan, L., Cooper, A. B., and N. K. Dulvy. 2016. Rethinking trade-driven extinction risk in marine and terrestrial megafauna. *Current Biology* 26:1640–46.

McCoy, E., Bruce, R., David, D., et al. 2018. Long-term photo-identification reveals the population dynamics and strong site fidelity of adult whale sharks to the coastal waters of Donsol, Philippines. *Frontiers in Marine Science* 5:271.

McKinney, J. A., Hoffmayer, E. R., Holmberg, J., et al. 2017. Long-term assessment of whale shark population demography and connectivity using photo-identification in the western Atlantic Ocean. *PloS One* 12:e0180495.

Meekan, M. G., Bradshaw, C. J. A., Press, M., et al. 2006. Population size and structure of whale sharks (*Rhincodon typus*) at Ningaloo Reef, Western Australia. *Marine Ecology Progress Series* 319:275–85.

Meekan, M. G., Taylor, B. M., Lester, E., et al. 2020. Asymptotic growth of whale sharks suggests sex-specific life-history strategies. *Frontiers in Marine Science* 7:774.

Nair, K. V. S., Jayaprakash, A. A., and V. A. Narayanankutty. 1986. On a juvenile whale shark *Rhincodon typus* Smith landed at Cochin. *Marine Fisheries Information Service* 66:36.

Neubauer, P., Richard, Y., and S. Clarke. 2018. Risk to the Indo-Pacific Ocean whale shark population from interactions with Pacific Ocean purse-seine fisheries. WCPFC-SC14-2018/SA-WP-12 (rev. 2). Western and Central Pacific Fisheries Commission.

New, L. F., A. J. Hall, R. Harcourt, et al. 2015. The modelling and assessment of whale-watching impacts. *Ocean and Coastal Management* 115:10–16.

NOAA. 2020a. TurtleWatch. https://www.fisheries.noaa.gov/resource/map/turtlewatch (accessed November 29, 2020).

NOAA. 2020b. WhaleWatch. https://www.fisheries.noaa.gov/west-coast/marine-mammal-protection/whalewatch (accessed November 29, 2020).

Norman, B. 2002. Review of Current and Historical Research on the Ecology of Whale Sharks (*Rhincodon typus*), and Applications to Conservation Through Management of the Species. Freemantle: Western Australian Department of Conservation and Land Management.

Norman, B. M., Holmberg, J. A., Arzoumanian, Z., et al. 2017. Undersea constellations: The global biology of an endangered marine megavertebrate further informed through citizen science. *Bioscience* 67:1029–43.

Norman, B. M., and D. L. Morgan. 2016. The return of 'Stumpy' the whale shark: Two decades and counting. *Frontiers in Ecology and the Environment* 14:449–50.

O'Malley, M. P., Townsend, K. A., Hilton, P., et al. 2017. Characterization of the trade in manta and devil ray gill plates in China and South-East Asia through trader surveys. *Aquatic Conservation: Marine and Freshwater Ecosystems* 27:394–413.

Ong, J. J. L., Meekan, M. G., Hsu, H.-H., et al. 2020. Annual bands in vertebrae validated by bomb radiocarbon assays provide estimates of age and growth of whale sharks. *Frontiers in Marine Science* 7: 1–7.

Orams, M. B. 2002 Feeding wildlife as a tourism attraction: A review of issues and impacts. *Tourism Management* 23:281–93.

Pajuelo, M., Alfaro-Shigueto, J., Romero, M., et al. 2018. Occurrence and bycatch of juvenile and neonate whale sharks (*Rhincodon typus*) in Peruvian waters. *Pacific Science* 72:463–73.

Parton, K. J., T. S. Galloway, and B. J. Godley. 2019. Global review of shark and ray entanglement in anthropogenic marine debris. *Endangered Species Research* 39:173–90.

Peel, D., J. N. Smith, and S. Childerhouse. 2018. Vessel strike of whales in Australia: The challenges of analysis of historical incident data. *Frontiers in Marine Science* 5:69.

Penketh, L., A. Schleimer, J. Labaja, et al. 2020. Scarring patterns of whale sharks, *Rhincodon typus*, at a provisioning site in the Philippines. *Aquatic Conservation: Marine and Freshwater Ecosystems*. doi:10.1002/aqc.3437.

Perry, C. T., Figueiredo, J., Vaudo, J. J., et al. 2018. Comparing length-measurement methods and estimating growth parameters of free-swimming whale sharks (*Rhincodon typus*) near the South Ari Atoll, Maldives. *Marine and Freshwater Research* 69:1487–95.

Pfaller, J. B., K. M. Goforth, M. A. Gil, et al. 2020. Odors from marine plastic debris elicit foraging behavior in sea turtles. *Current Biology* 30:R213–14.

Pierce, S. J., A. Méndez-Jiménez, K. Collins, et al. 2010. Developing a code of conduct for whale shark interactions in Mozambique. *Aquatic Conservation: Marine and Freshwater Ecosystems* 20:782–88.

Pierce, S. J., and B. Norman. 2016. *Rhincodon typus*. The IUCN Red List of Threatened Species 2016:e. T19488A2365291.

Pirotta, V., Grech, A., Jonsen, I. D., et al. 2019. Consequences of global shipping traffic for marine giants. *Frontiers in Ecology and the Environment* 17:39–47.

Pravin, P. 2000. Whale shark in the Indian coast - Need for conservation. *Current Science* 79:310–15.

Quiros, A. L. 2007. Tourist compliance to a code of conduct and the resulting effects on whale shark (*Rhincodon typus*) behavior in Donsol, Philippines. *Fisheries Research* 84:102–08.

Rajapackiam, S., and S. Mohan. 2006. A giant whale shark (*Rhincodon typus*) caught at Chennai Fisheries Harbour. *Marine Fisheries Information Service* 189:25–26.

Rajapackiam, S., Ameer Hamsa, K. M. S., Balasubramanian, T. S., and H. M. Kasim. 1994. On a juvenile whale shark *Rhincodon typus* caught off Kayalpatnam (Gulf of Mannar). *Marine Fisheries Information Service* 127:14–15.

Rao, G. S. 1986. Note on the occurrence of the whale shark off Veraval Coast. *Marine Fisheries Information Service* 66:30.

Raudino, H., Rob, D., Barnes, P., et al. 2016. Whale shark behavioural responses to tourism interactions in Ningaloo Marine Park and implications for future management. *Conservation Science Western Australia* 10:2.

Razzaque S. A., Khan, M. M., Shahid, U., et al. 2020. Safe handling & release for gillnet fisheries for whale shark, manta & devil rays and sea turtles. IOTC-2020-WPEB16-26_Rev1. Indian Ocean Tuna Commission.

Reeder, D. M., and K. M. Kramer. 2005. Stress in free-ranging mammals: Integrating physiology, ecology, and natural history. *Journal of Mammalogy* 86:225–35.

Riley, M. J., Harman, A., and R. G. Rees. 2009. Evidence of continued hunting of whale sharks *Rhincodon typus* in the Maldives. *Environmental Biology of Fishes* 86:371–74.

Robinson, D. P., Jaidah, M. Y., Jabado, R. W., et al. 2013. Whale sharks, *Rhincodon typus*, aggregate around offshore platforms in Qatari waters of the Arabian Gulf to feed on fish spawn. *PLoS One* 8:e58255.

Robinson, D. P., Jaidah, M. Y., Bach, S., et al. 2016. Population structure, abundance and movement of whale sharks in the Arabian Gulf and the Gulf of Oman. *PLoS One* 11:e0158593.

Robinson, D. P., Jaidah, M. Y., Bach, S., et al. 2017. Some like it hot: Repeat migration and residency of whale sharks within an extreme natural environment. *PLoS One* 12:e0185360.

Rockwood, R. C., Calambokidis, J., and J. Jahncke. 2017. High mortality of blue, humpback and fin whales from modeling of vessel collisions on the U. S. West Coast suggests population impacts and insufficient protection. *PLoS One* 12:e0183052.

Rodrigues, N. V., Correia, J. P. S., Graça, J. T. C., et al. 2012. First record of a whale shark *Rhincodon typus* in continental Europe. *Journal of Fish Biology* 81:1427–29.

Rohner, C. A., Richardson, A. J., Jaine, F. R. A., et al. 2018. Satellite tagging highlights the importance of productive Mozambican coastal waters to the ecology and conservation of whale sharks. *PeerJ* 2018:e4161.

Romero, M. L., and L. K. Butler. 2007. Endocrinology of stress. *International Journal of Comparative Psychology* 20:89–95.

Rommel, S. A., A. M. Costidis, T. D. Pitchford, et al. 2007. Forensic methods for characterizing watercraft from watercraft-induced wounds on the Florida manatee (*Trichechus manatus latirostris*). *Marine Mammal Science* 23:110–32.

Rowat, D. 2010. *Whale Sharks: An Introduction to the World's Largest Fish from One of the World's Smallest Nations, the Seychelles*. Victoria: Marine Conservation Society Seychelles.

Rowat, D., and K. S. Brooks. 2012. A review of the biology, fisheries and conservation of the whale shark *Rhincodon typus*. *Journal of Fish Biology* 80:1019–56.

Rowat, D., and M. Gore. 2007. Regional scale horizontal and local scale vertical movements of whale sharks in the Indian Ocean off Seychelles. *Fisheries Research* 84:32–40.

Rowat, D., Meekan, M. G., Engelhardt, U., et al. 2007. Aggregations of juvenile whale sharks (*Rhincodon typus*) in the Gulf of Tadjoura, Djibouti. *Environmental Biology of Fishes* 80:465–72.

Sampaio, C. L. S., L. Leite, J. A. Reis-Filho, et al. 2018. New insights into whale shark *Rhincodon typus* diet in Brazil: An observation of ram filter-feeding on crab larvae and analysis of stomach contents from the first stranding in Bahia state. *Environmental Biology of Fishes* 101:1285–93.

Sánchez, L., Briceño, Y., Tavares, R., et al. 2020. Decline of whale shark deaths documented by citizen scientist network along the Venezuelan Caribbean coast. *Oryx* 54: 600–601.

Schoeman, R. P., C. Patterson-Abrolat, and S. Plön. 2020. A global review of vessel collisions with marine animals. *Frontiers in Marine Science* 7: 292.

Sequeira, A. M. M., Mellin, C., Fordham, D. A., et al. 2014. Predicting current and future global distributions of whale sharks. *Global Change Biology* 20:778–89.

Sequeira, A. M. M., G. C. Hays, D. W. Sims, et al. 2019. Overhauling ocean spatial planning to improve marine megafauna conservation. *Frontiers in Marine Science* 6:639.

Silas, E. G. 1986. The whale shark (*Rhiniodon typus*) in Indian coastal waters: Is the species endangered or vulnerable? *Marine Fisheries Information Service Technical and Extension Series* 66:1–19.

Silber, G. K., Adams, J. D., Asaro, M. J., et al. 2015. The right whale mandatory ship reporting system: A retrospective. *PeerJ* 3:e866.

Simmonds, M. P., and S. J. Isaac. 2007. The impacts of climate change on marine mammals: Early signs of significant problems. *Oryx* 41:19–26.

Simpfendorfer, C. A., and N. K. Dulvy. 2017. Bright spots of sustainable shark fishing. *Current Biology* 27:97–98.

Smith, A. 1828. Descriptions of new, or imperfectly known objects of the animal kingdom, found in the south of Africa. *South African Commercial Advertiser* 3:2.

SPC-OFP. 2010. Summary information on whale shark and cetacean interactions in the tropical WCPFC purse seine fishery. Western and Central Pacific Fisheries Commission. https://www.wcpfc.int/system/files/WCPFC-2011-IP-01-Rev-1-Whale-shark-cetacean-interactions-(18-Jan-2012).pdf.

Speed, C. W., Meekan, M. G., Rowat, D., et al. 2008b. Scarring patterns and relative mortality rates of Indian Ocean whale sharks. *Journal of Fish Biology* 72:1488–1503.

Steinke, D., Bernard, A. M., Horn, R. L., et al. 2017. DNA analysis of traded shark fins and mobulid gill plates reveals a high proportion of species of conservation concern. *Scientific Reports* 7:9505.

Takada, H. 2009. International pellet watch: Global distribution of persistent organic pollutants (POPs) in marine plastics and their potential threat to marine organisms. In *Proceedings of the International Research Workshop on the Occurrence, Effects, and Fate of Microplastic Marine Debris, September 2008*, eds. C. Arthur, J. Baker, and H. Bamford. Silver Spring: NOAA Marine Debris Division.

Tanabe, S., Mori, T., Tatsukawa, R., and N. Miyazaki. 1983. Global pollution of marine mammals by PCBs, DDTs and HCHs (BHCs). *Chemosphere* 12:1269–75.

Taylor, G. 1989. Whale sharks of Ningaloo Reef, Western Australia: A preliminary study. *Western Australian Naturalist* 18:7–12.

Taylor, G. 2019. *Whale Sharks: The Giants of Ningaloo Reef.* 2nd Revised Edition. Perth: Scott Print.

Teerlink, S., L. Horstmann, and B. Witteveen. 2018. Humpback whale (*Megaptera novaeangliae*) blubber steroid hormone concentration to evaluate chronic stress response from whale-watching vessels. *Aquatic Mammals* 44:411–25.

Temple, A. J., Wambiji, N., Poonian, C. N. S., et al. 2019. Marine megafauna catch in southwestern Indian Ocean small-scale fisheries from landings data. *Biological Conservation* 230:113–21.

Thomson, J. A., Araujo, G., Labaja, J., et al. 2017. Feeding the world's largest fish: Highly variable whale shark residency patterns at a provisioning site in the Philippines. *Royal Society Open Science* 4:170394.

Thums, M., Meekan, M. G., Stevens, J. D., et al. 2013. Evidence for behavioural thermoregulation by the world's largest fish. *Journal of the Royal Society Interface* 10:20120477.

Vanderlaan, A. S. M., and C. T. Taggart. 2007. Vessel collisions with whales: The probability of lethal injury based on vessel speed. *Marine Mammal Science* 23:144–56.

Wabnitz, C. C. C., Lam, V. W. Y., Reygondeau, G., et al. 2018. Climate change impacts on marine biodiversity, fisheries and society in the Arabian Gulf. *PLoS One* 13:e0194537.

White, W. T., and R. D. Cavanagh. 2007. Whale shark landings in Indonesian artisanal shark and ray fisheries. *Fisheries Research* 84:128–31.

WildLifeRisk. 2014. Planet's biggest slaughter of whale sharks exposed in PuQi, Zhejiang Province, China. https://web.archive.org/web/20160503104157/http://wildliferisk.org/press-release/ChinaWhaleSharks-WLR-Report-ENG.pdf (accessed November 18, 2020).

Wolfson, F. H. 1986. Occurrences of the whale shark, *Rhincodon typus* Smith. In *Indo-Pacific Fish Biology: Proceedings of the Second International Conference on Indo-Pacific Fishes*, 208–26. Ichthyological Society of Japan.

Womersley, F., Hancock, J., Perry, C. T., and D. Rowat. 2021. Wound healing capabilities of whale sharks (*Rhincodon typus*) and implications for conservation management. *Conservation Physiology.* doi:10.1093/conphys/coaa120.

Worm, B., Lotze, H. K., Jubinville, I., et al. 2017. Plastic as a persistent marine pollutant. *Annual Review of Environment and Resources* 42:1–26.

Wu, J. 2016. *Shark Fin and Mobulid Ray Gill Plate Trade in Mainland China, Hong Kong and Taiwan.* Cambridge: TRAFFIC.

Würsig, B., Reeves, R. R., and J. G. Ortega-Ortiz. 2002. Global climate change and marine mammals. In *Marine Mammals* eds. P. G. H. Evans, and J. A. Raga, 589–608. New York: Kluwer Academic/Plenum Publishers.

Ziegler, J., P. Dearden, and R. Rollins. 2012. But are tourists satisfied? Importance-performance analysis of the whale shark tourism industry on Isla Holbox, Mexico. *Tourism Management* 33:692–701.

CHAPTER 12

Conservation of Whale Sharks

Simon J. Pierce
Marine Megafauna Foundation

Molly K. Grace
University of Oxford

Gonzalo Araujo
Large Marine Vertebrates (LAMAVE) Research Institute Philippines

CONTENTS

12.1 INTRODUCTION

Whale sharks are huge, placid, and harmless to people. These characteristics have made them a globally iconic species, and swimming with these enormous sharks is a popular focus of marine tourism (Chapter 10). Sadly, these same attributes have also led to whale sharks becoming a target for unsustainable fisheries (Chapter 11). Sighting and abundance trend data indicate that the global whale shark population has been halved over the past 50 years (Pierce and Norman 2016). The primary cause of this decline is likely to be the significant commercial fisheries for the species that began in the 1980s and 1990s, along with bycatch in other net fisheries, ship strikes, and potentially other causes such as pollution.

Fortunately, the whale sharks' increasing risk of extinction has been recognized. National, regional, and international legislative measures have been introduced to protect the species. Here, we summarize the present-day outlook for whale sharks, outline the conservation measures and management frameworks introduced to stem their decline, and provide a provisional roadmap to secure their long-term recovery.

12.1.1 A History of Concern

Historically, commercial and small-scale fisheries have been the dominant threat to whale shark populations. The species has been caught opportunistically in fisheries since at least the 1850s, including for the international export of their fins (e.g., from India to China; Bean 1907). Before the 1980s, though, they were rarely targeted (Silas 1986; Colman 1997; Pravin 2000; Hsu et al. 2012; Acebes 2013). Nevertheless, in the first review of whale shark biology and fisheries in India, Silas (1986) correctly noted, '... I would not consider the whale shark as an endangered species at this point of time, but a highly vulnerable one.' His perspective was, unfortunately, prescient.

Whale shark meat came to be regarded as a marketable delicacy in Taiwan in the mid-1980s (Chen et al. 1997; Hsu et al. 2012). This pushed up consumer prices and, as a result, dramatically increased the value of the species to the fishing industry. Before 1985, a whale shark weighing 'several tons' would fetch between US$200–300 at auction in Taiwan. By 1997, a 10-tonne shark could sell for US$70,000 wholesale (Chen et al. 1997). Fishing effort scaled up accordingly, leading to increasing catches followed by rapid declines (Hsu et al. 2012). Chen et al. (1997) reported that 'although information is too sketchy to conclude with any certainty that whale shark populations off Taiwan are declining, anecdotal evidence, paired with recent information on the species' reproductive patterns, give cause for concern.' The Taiwanese government began introducing monitoring and management measures for whale sharks in 2001, affording them full protection in 2007 (Hsu et al. 2012).

Before the late 1980s, whale sharks had been caught occasionally in India and the Philippines (Silas 1986; Hanfee 2001; Acebes 2013), and sometimes retained for local consumption or to use the sharks' liver oil for waterproofing boats (Hanfee 2001; Alava and Dolumbalo 2002). When the Taiwanese consumer demand was not being met by local catches, Indian and Philippines fisheries changed their focus to supply this new market. The fast development of the Indian whale shark fishery has been well documented. From 1991, an entire mechanized fleet began targeting the sharks at what was clearly an important aggregation area off the Gujarat coast, where 'schools' of up to 300–500 sharks were reported (Hanfee 2001). Both large adults and juvenile sharks were caught in this fishery (Pravin 2000; Hanfee 2001; Borrell et al. 2011). Pravin et al. (1998) summarized the situation for whale sharks as 'the present level of exploitation may lead to its extinction very soon.' Fortunately, the species was formally protected in 2001 following lobbying efforts by Indian conservationists. Since then, most whale sharks incidentally caught in net fisheries have been released alive (Akhilesh et al. 2013).

Meanwhile, in the Philippines, Alava and Dolumbalo (2002) wrote in 1997 that 'the current exploitation rate is increasing so rapidly that a recommendation for a national ban on the fishery may be needed.' Species-level protection was indeed implemented in 1998 following negative publicity around several whale sharks that were killed at Donsol in Sorsogon province (Alava and Dolumbalo 2002; Acebes 2013), which has since become a significant whale shark tourism destination (McCoy et al. 2018; Chapter 10). Few sharks have been caught in the country after protection was enacted (Acebes 2013; Araujo et al. 2019).

The only major commercial fishery for whale sharks to have persisted into the past decade is in China. Li et al. (2012), in a paper titled 'an emerging crisis,' discussed the increasing catches of the species in southern China, to an estimated 1,000 sharks per year. They reported that 'the trend towards increased exploitation of *R. typus* and other shark species is likely to continue and intensify further. Since catch and trade of sharks is unregulated and even encouraged in China, intensive hunting has the potential to become ecologically and economically unsustainable.' One of the larger fish factories in the town of Puqi was estimated to be processing over 600 whale sharks per year around this time (WildLifeRisk 2014). The current status of this whale shark fishery is unclear. More recent reports from Puqi, the most important processing hub for sharks in China, suggest that the scale of the shark fishing industry overall has substantially contracted in recent years (Wu 2016), although small quantities of whale shark products have been reported from trade surveys more recently (O'Malley et al. 2017; Steinke et al. 2017).

12.1.2 Contemporary Threats

To the best of our knowledge – and with the possible exception of the Chinese fishery – targeted fishing for whale sharks is now rare. Whale sharks do, unfortunately, still face a number of direct and ongoing threats, particularly ship strikes, bycatch, and marine pollution. Chapter 11 provides a more detailed assessment on the contemporary human threats to the species. Other challenges, such as climate change, prey depletion, and habitat degradation, are also important issues to consider, and they are discussed in Chapter 11. Here, we focus on the documented sources of whale shark mortality.

Ship strikes have been a documented source of whale shark deaths for many years (Gudger 1938). Rapid increases in both the speed and quantity of marine traffic mean that mortality from ship strikes has probably supplanted fisheries as the main contemporary threat to whale sharks through much of their distribution (Chapter 11; Pirotta et al. 2019). Direct records of mortality are rare, as the sharks sink if killed, but the frequency of injuries from small and large vessels seen in live whale sharks (e.g., Boldrocchi et al. 2020; Lester et al. 2020; Rohner et al. 2020) suggest a high prevalence of ship strikes in some areas. These documented injuries are likely to be the 'tip of the iceberg' in terms of the real mortality risk, as whale sharks are unlikely to survive propeller or impact wounds from large vessels.

Whale sharks are often caught accidentally in large nets set for other species (Pierce and Norman 2016). While some are released alive, others are dead when found or killed for their meat or fins (Akhilesh et al. 2013; Pajuelo et al. 2018; Rohner et al. 2018; Wyatt et al. 2019; Sánchez et al. 2020). This is likely to occur across much of their distribution, with frequent reports coming from gillnet fisheries (i.e., Akhilesh et al. 2013; Barbosa-Filho et al. 2016; Pajuelo et al. 2018; Rohner et al. 2018, 2020), and the major tuna gillnet fishery in the Arabian Sea may still pose a significant threat (Shahid et al. 2015; Moazzam and Nawaz 2017).

Whale sharks are a common bycatch in tuna purse-seine fisheries (Clarke 2015; Escalle et al. 2016b; Román et al. 2018). Whale sharks, which are often associated with tuna in oceanic waters (Chapter 4), are encircled in huge nets along with the tuna species that are the intended catch. While the whale sharks are usually released, a few are accidentally killed (Clarke 2015; Román et al. 2018), although more recent application of safe release practices appears to minimize at least short-term mortality (Capietto et al. 2014; Escalle et al. 2016a, 2018a). Where poor release practices are used, though, such as lifting the sharks from the water by their tails or leaving ropes attached to the sharks following release, longer-term mortality may still be an issue. An expert survey estimated a 10% post-release mortality rate in the Western Central Pacific, although uncertainty was large (Neubauer et al. 2018). Further work, particularly tagging whale sharks at release for longer-term tracking, may be needed to get a more robust estimate.

It has become clear in recent years that marine pollution is also a threat to whale sharks (Chapter 11). The Deepwater Horizon oil spill in 2010 in the northern Gulf of Mexico affected a known whale shark habitat (Campagna et al. 2011; Frias-Torres and Bostater 2011), potentially causing mortalities or changes in movement behavior (Hueter et al. 2013). 'Red tides,' caused by toxic blooms of *Karenia* spp. dinoflagellates, are associated with nutrient run-off and are increasing in frequency along the southern US coast (Brand and Compton 2007). These often result in shark kills (Flewelling et al. 2010), among many other marine species, and the first probable whale shark mortality from this cause was reported in Florida in 2018 (Furby 2018). Plastic pollution is a significant, ubiquitous threat to ocean health, and filter-feeding elasmobranchs are particularly vulnerable (Germanov et al. 2018). Whale sharks can accidentally ingest large quantities of microplastics while feeding in some areas, with up to about 137 pieces per hour reported from Java in Indonesia (Germanov et al. 2019). Whale shark mortalities from plastic ingestion have been reported from Japan (Matsumoto et al. 2017), Malaysia (Lee 2019, Sen Nathan pers. comm.), the Philippines (Abreo et al. 2019), and Thailand (Haetrakul et al. 2009), and a variety of other sublethal effects are possible, such as endocrine disruption or toxicosis (Germanov et al. 2018). Entanglement, particularly in discarded or lost fishing gear, is also a likely source of mortality (Wilcox et al. 2016; Parton et al. 2019).

12.1.3 The IUCN Red List of Threatened Species

Concern for whale sharks, and the increasing public awareness of their overexploitation and ongoing decline, have been reflected in the classification of their global extinction risk. The IUCN

Red List of Threatened Species is widely regarded and cited as a respected and standardized international assessment of conservation status for each species. Whale sharks were first categorized as Indeterminate on the Red List in 1990, and again when re-examined in 1994. At that time, this classification was applied to species that were thought to be Endangered, Vulnerable, or Rare, but where sufficient data for a specific categorization were lacking (Colman 1997). The species was reassessed as Data Deficient in 1996, following updated terminology. Whale sharks were first listed as Vulnerable (a globally 'Threatened' status) in 2000 (Norman 2000), and again when reassessed in 2005 (Norman 2005), based largely on the improved information on targeted fisheries for the species that was available by that time.

The contemporary assessment by Pierce and Norman (2016) used genetic data, supported by photo-ID and tagging studies, to subdivide the global population into Atlantic and Indo-Pacific subpopulations, addressing each separately for the first time. The Indo-Pacific subpopulation was classed as Endangered (based on technical criteria A2bd+4bd) based on an inferred (and ongoing) $\geq 63\%$ reduction in population size over the past three generations, where generation time was estimated to be 25 years. The Atlantic subpopulation was assessed as Vulnerable (A2b+4b), assuming a $\geq 30\%$ population reduction over three generations, based on declines in sighting indices. As the bulk of the global whale shark population (then estimated at 75%) was thought to be in the Indo-Pacific, the overall assessment of extinction risk was weighted towards that region. The whale shark was therefore classed as globally Endangered (A2bd+4bd). The justification noted that, in the absence of conservation action, whale shark population declines were likely to continue.

Following Pierce and Norman (2016), whale sharks were subsequently assessed as Endangered (C1) in the Arabian Sea region (Dulvy et al. 2017) based on the small regional population documented by a connectivity study (Robinson et al. 2016), coupled with declining sightings in the surrounding Western Indian Ocean, specifically in the Mozambique Channel and Seychelles (Pierce and Norman 2016).

The IUCN Red List assessment is an advisory rather than a legislative instrument (i.e., it has no enforcement power). However, this improving knowledge of whale shark threats and declines has been accompanied by, and has in some cases triggered, a number of formal conservation and management measures that either include or are otherwise relevant to the species. For this reason, future changes in the Red List status of whale sharks can also be viewed as a bellwether of international conservation interventions.

12.2 INTERNATIONAL LEGAL AND MANAGEMENT FRAMEWORKS

The United Nations Conference on the Human Environment (1972) formally recognized the need for states to cooperate in the conservation and management of animals that migrate across national boundaries, or between political jurisdictions and the high seas. That led to the establishment of several important intergovernmental wildlife and fisheries treaties that have the potential to influence present-day whale shark conservation, which we summarize below.

12.2.1 United Nations Convention on the Law of the Sea

The United Nations Convention on the Law of the Sea (UNCLOS) was developed in 1982 to provide a comprehensive policy framework to govern all uses of the marine environment and the resources found therein (United Nations 2020a). The whale shark is listed as a 'Highly Migratory Species' on Annex I of the Convention.

The Agreement relating to the Conservation and Management of Straddling Fish Stocks and Highly Migratory Fish Stocks is a component of UNCLOS (United Nations 2020b). 'The Agreement' outlines principles for the conservation and management of fisheries resources that occur both in and outside the exclusive economic zone (EEZ) of individual countries. The Agreement mandates that management must take a precautionary approach and be based on the best available scientific information.

The Food and Agriculture Organization of the United Nations (FAO) reviewed the Agreement's effectiveness in securing the conservation and management of highly migratory fish stocks for a review conference in 2016 (FAO 2016). The FAO reported that no whale shark catches had been reported to the FAO Global Capture Production database but concluded that, 'Unless demonstrated otherwise, it is prudent to consider the species as being fully exploited globally.' A subsequent (mid-2020) check of this database revealed that no catches of whale sharks had been added to this dataset (FAO 2020a). To date, we are not aware of any specific management initiatives enacted through UNCLOS that have included the whale shark.

12.2.2 Convention on the Conservation of Migratory Species of Wild Animals

The Convention on the Conservation of Migratory Species of Wild Animals (CMS), signed in 1979, is managed by the United Nations Environment Program. The CMS specifically focuses on the conservation of migratory species, both marine and terrestrial, along with their habitats and migration routes. As such, CMS provides a legally non-binding framework for coordinating conservation measures across a species' range. There are two Appendices, with distinct obligations relating to each (species can be listed on both Appendices I and II simultaneously). Appendix I comprises migratory species that are at risk of extinction, generally corresponding with the IUCN Red List classification of Endangered or Critically Endangered (CMS 2020a). Countries that have signed the CMS that are within the range of Appendix I species should '… endeavor to strictly protect them by: prohibiting the taking of such species, with very restricted scope for exceptions; conserving and where appropriate restoring their habitats; preventing, removing or mitigating obstacles to their migration and controlling other factors that might endanger them' (CMS 2020a). Appendix II covers migratory species that have an 'unfavorable conservation status' and would benefit from international cooperation to improve their conservation and management (CMS 2020a).

The whale shark was listed in Appendix II of CMS in 1999 and, following its Endangered listing on the IUCN Red List, was added to Appendix I in 2017. The CMS had 130 Parties as of late 2019, with many whale shark range states included (CMS 2020a). A Concerted Actions document was submitted by the lead proponent, the Philippines, to accompany the Appendix I proposal in 2017 (CMS 2017a, b). This proposed several objectives for research and monitoring, noted the need for tourism guidelines, recommended the presence of onboard observers in fisheries that may have whale shark bycatch, and outlined legislative requirements for improved whale shark conservation (CMS 2017b). Efforts are underway to fulfil research recommendations in the Philippines, to develop global tourism standards, and to assist identified priority countries with required legislation (Sea Shepherd Legal 2019a, 2019b; LAMAVE 2019).

Since 2010, the whale shark has also been included in Annex I of the CMS Memorandum of Understanding on the Conservation of Migratory Sharks (Sharks MOU), which has been signed by 48 States (CMS 2020b). Some of these States, including the USA, are signatories to the MOU but not the CMS itself. The Sharks MOU is a legally non-binding instrument that aims to facilitate favorable conservation status for migratory sharks, following the aims set out for Appendix II species.

12.2.3 Convention on International Trade in Endangered Species of Wild Fauna and Flora

As noted earlier and in Chapter 11, targeted fishing to supply international markets has been a major driver of the decline in whale shark numbers, particularly in the Indo-Pacific subpopulation

(Pierce and Norman 2016). The Convention on International Trade in Endangered Species of Wild Fauna and Flora (CITES) was established to protect species of wild fauna and flora from over-exploitation resulting from international trade. Appendix I includes species that are threatened with extinction, and hence commercial trade in these species is prohibited. Appendix II includes species which, although not necessarily threatened with immediate extinction, have the potential to decline further unless trade is regulated. The whale shark was listed on Appendix II of CITES in 2002.

The international trade of a listed species is controlled by the exporting state. This requires the release of a formal 'no detriment' finding to confirm that such trade is monitored and does not pose a risk to the conservation status of the listed species. This generally requires fishing states to demonstrate that exports originate from a sustainably managed population, although exceptions have been identified (Sea Shepherd Legal 2019a). There has not been a 'no detriment' finding issued for whale sharks by any country to date, and we are not aware of any attempts to develop a submission. However, the continued presence of whale shark fins and gill plates in trade – despite no records in the CITES Wildlife Trade Database (http://trade.cites.org/) – suggests that some level of illegal trade is still occurring outside the CITES permit system (Pierce and Norman 2016; Steinke et al. 2017; Haque et al. 2019).

12.2.4 Convention on Biological Diversity

The Convention on Biological Diversity (CBD), adopted in 1992, aims to conserve biodiversity and to promote the sustainable, fair, and equitable use of its benefits. Parties are required to develop or adopt a national strategy, plan, or program for the conservation and sustainable use of biological diversity within their jurisdiction. They are also required to monitor components of biodiversity that are important for conservation and to identify and monitor activities that are likely to have adverse effects on the conservation and sustainable use of biodiversity.

The tenth meeting of the Conference of the Parties for CBD, held in Aichi, Japan, in 2010, adopted a revised and updated Strategic Plan for Biodiversity, including the 'Aichi Biodiversity Targets' for 2011–2020 (CBD 2010). The most directly relevant to whale sharks, Target 6, states that 'by 2020 all fish and invertebrate stocks and aquatic plants are managed and harvested sustainably, legally and applying ecosystem based approaches, so that overfishing is avoided, recovery plans and measures are in place for all depleted species, fisheries have no significant adverse impacts on threatened species and vulnerable ecosystems and the impacts of fisheries on stocks, species and ecosystems are within safe ecological limits.'

Since that meeting, 154 Parties have submitted their National Biodiversity Strategies and Action Plans (International Year of Biodiversity 2010). At least one country, Qatar, has specifically addressed whale shark conservation within their national plan (Qatar Ministry of Environment 2014). While the CBD's 2020 assessment of overall progress reported that Target 6 has not been achieved, there has been progress made towards some objectives that are relevant to whale shark conservation, such as an increasing number of formal marine protected areas (MPAs) (CBD 2020).

12.2.5 FAO International Plan of Action for the Conservation and Management of Sharks

The FAO Conference adopted the International Plan of Action for the Conservation and Management of Sharks (IPOA-Sharks) in 1999. This non-binding plan encouraged countries to assess the state of shark stocks in their EEZ and those fished on the high seas. If directed or non-target shark fisheries were identified, a National Plan of Action (NPOA) for the conservation and management of shark stocks was required. A brief biennial summary and report on activities are requested by FAO.

The IPOA-Sharks also states that, 'Where transboundary, straddling, highly migratory and high seas stocks of sharks are exploited by two or more States, the States concerned should strive to ensure effective conservation and management of the stocks.' States are encouraged to develop regional plans through regional and sub-regional fisheries management organizations or similar arrangements. Several whale shark range states have prepared their own NPOA on shark fisheries management and conservation, and some regional plans have also been submitted (available at FAO 2020b). While these do not deal solely with whale sharks, the species is implicitly incorporated into such plans.

12.2.6 Regional Fisheries Management Organizations

As noted earlier, whale sharks are a regular bycatch in tropical tuna fisheries. Globally, these fisheries are managed by four Regional Fisheries Management Organizations (RFMOs): the Indian Ocean Tuna Commission (IOTC), Western and Central Pacific Fisheries Commission (WCPFC), Inter-American Tropical Tuna Commission (IATTC) in the Eastern Pacific, and the International Commission for the Conservation of Atlantic Tunas (ICCAT). Together, these four RFMOs span the whale sharks' circum-tropical distribution.

Intentional sets of purse-seine nets around whale sharks have been banned by IOTC, WCPFC, and IATTC, although not yet by ICCAT (Capietto et al. 2014; Fowler 2016; Escalle et al. 2018a). Even where catches are unintentional, however, a large proportion of entangled whale sharks (73% in WCPFC waters; SPC-OFP 2010) are not seen before the nets are deployed. The IATTC, IOTC, and WCPFC require that best practices for safe release are followed when whale sharks are accidentally encircled (i.e., Capietto et al. 2014), and these are also generally followed by European vessels in the ICCAT region (Escalle et al. 2016a).

In the IOTC region, in particular, gillnet fisheries for tuna also catch whale sharks (Shahid et al. 2015; Razzaque et al. 2020). These fisheries have had low observer coverage, and total catch reports are not available. A change to subsurface gillnets in the Pakistan-based fleet has reduced bycatch overall, but not for whale sharks (Moazzam and Nawaz 2017). A guide for safely releasing whale sharks caught in the Arabian Sea gillnet fishery has been created (Razzaque et al. 2020).

12.3 NATIONAL LEGAL PROTECTIONS

12.3.1 Species Protection

A number of range states have specific legal protections in place for whale sharks. Some countries have prohibited any commercial fishing of sharks in their EEZ, creating a 'shark sanctuary.' Since this encompasses protection for whale sharks, these countries are also included here.

National- or territory-level protection measures for whale sharks are known to be in place in American Samoa, Australia, Bahamas, Bangladesh, Belize, Brazil, British Virgin Islands, Cabo Verde, Cambodia, Caribbean Netherlands, Cayman Islands, Chagos Archipelago (UK), China, Congo-Brazzaville, Cook Islands, Costa Rica, Curacao, Djibouti, Dominican Republic, Ecuador, Egypt, El Salvador, French Polynesia, Grenada, Guatemala, Guadeloupe, Guyana, Honduras, Hong Kong, Indonesia, India, Israel, Kuwait, Maldives, Malaysia, Marshall Islands, Mayotte (France), Mexico, Micronesia, Mozambique, Myanmar, New Caledonia, New Zealand, Nicaragua, Pakistan, Palau, Panama, Peru, Philippines, Reunion, Saudi Arabia, Seychelles, South Africa, St Helena Island (UK), St Maarten, Taiwan, Tanzania, Thailand, Tokelau, United Arab Emirates, United States, and Wallis and Futuna (France).

Notable whale shark range states in which national protection is *not* known to be in place at the time of writing include Colombia, Cuba, Gabon, Iran, Iraq, Japan, Kenya, Madagascar, Oman, and Qatar.

12.3.2 Marine Protected Areas

No-take MPAs, often termed marine reserves, are defined here as 'protected areas where shark fishing is banned.' Whale sharks are present, seasonally or year-round, in many MPAs through their distribution, including some that have been purposefully set up to protect important whale shark habitats, such as the Whale Shark Biosphere Reserve in Mexico (see Section 12.4.2). Some important MPAs for whale sharks in countries in which the species is otherwise unprotected include Malpelo Island in Colombia, Jardines de la Reina in Cuba, and the Ankivonj and Ankarea MPAs in Madagascar. There is also a significant MPA network along the coast of Gabon, although the protection offered for sharks is unclear. The tendency of whale sharks to return predictably to certain areas to feed, or as 'transit' sites, makes MPAs a useful management tool even for this wide-ranging species (Rigby et al. 2019b). This is discussed more in a later section.

12.4 WHALE SHARKS AS CONSERVATION ICONS

The IUCN Red List status of a species is often considered by governments and management authorities when prioritizing conservation measures. Ultimately, though, the outcome of these discussions is also influenced by political factors. As an example, several large marine species that are at higher extinction risk than whale sharks, including Critically Endangered southern bluefin tuna *Thunnus maccoyii* and scalloped hammerhead sharks *Sphyrna lewini*, are still targeted by international fisheries due to their high commercial value (Collette et al. 2011; Rigby et al. 2019a).

As documented above, many of the multilateral measures for protecting whale sharks were in place prior to the species being formally classed as Endangered by the IUCN in 2016. The whale shark is, therefore, better-protected than might have been predicted based purely on the data-driven assessment of extinction risk through time. To address this paradox, it is worth noting the broader context of conservation decision-making.

12.4.1 Flagship Species

To counter the economic and associated political advantages possessed by commercial industries, conservation organizations rely on their own public awareness and lobbying efforts for both fundraising and to apply pressure on decision-makers. One of the outputs of this practical marketing philosophy has been the employment of 'flagship species' to front conservation initiatives. The World Wide Fund for Nature (WWF), a large international conservation organization that has been involved with whale shark research and conservation initiatives in several countries, defines flagship species as

… a species selected to act as an ambassador, icon or symbol for a defined habitat, issue, campaign or environmental cause. By focusing on, and achieving conservation of that species, the status of many other species which share its habitat – or are vulnerable to the same threats - may also be improved. Flagship species are usually relatively large and considered to be 'charismatic' in western cultures

(WWF 2020a)

WWF has explicitly adopted this flagship approach to global species conservation through its use of 'priority species clusters' (WWF 2020b). The Wildlife Conservation Society (WCS), another large international conservation non-profit, takes a similar approach through its identification of 'global priority' species (WCS 2020). Classic examples of flagship species, used widely by international conservation organizations (including both WWF and WCS), include large mammal species such as tigers and elephants (Smith et al. 2012). Courchamp et al. (2018) used large-scale

Table 12.1   The Ten Most Charismatic Animals, as Provided by Courchamp et al. (2018).

Common Name (Page Name)	WWF 'Cluster'	WCS Global Priority	Pageviews (2019)
Lion	Big cats	Lions	2,047,698
Tiger	Big cats	Tigers	1,894,592
Panda (Giant panda)	Bears	–	1,488,640
Elephant	Elephants	African elephants, Asian elephants	1,454,444
Gray wolf (Wolf)	–	–	1,410,391
Polar bear	Bears	–	1,284,154
Cheetah	Big cats	Cheetahs	1,137,410
Giraffe	–	–	1,094,199
Leopard	Big cats	Snow leopards	1,043,792
Gorilla	Great apes	Gorillas	946,625

The common name is as provided by those authors, with 'page name' representing the corresponding Wikipedia page for that species or group, if different. Pageviews for 2019 were obtained from Wikipedia's 'Pageviews Analysis' tool.

public surveys, zoo exhibits, and animated movie coverage to identify the Western public's 'most charismatic animals.' The top 10, many of which are actually species groups, are presented above in Table 12.1. As expected, there is a significant overlap between these popular and well-known species and those chosen as flagships by WWF and WCS.

Whale sharks are often referred to as a flagship species, but we are not aware of any studies that explicitly examine that view. To enable a broader comparison between species, and to test that assertion with respect to whale sharks, we tabulated the pageviews for each species group on Wikipedia (for the entire year of 2019) from the Courchamp et al. (2018) results. This metric does not, of course, capture all web searches for each species, but as Wikipedia is one of the world's most popular websites and generally ranks high in search engine results, it is a useful independent indicator of the public awareness and interest in wildlife species (Roll et al. 2016; McGowan et al. 2020). Wikipedia views do indeed emphasize the interest in these animals (Table 12.1), with most receiving over a million pageviews through the year.

Among these results, only the polar bear is (at least technically) a marine species. We are not aware of any peer-reviewed exercise ranking the popularity of strictly marine species. McClain et al. (2015) published an analysis of the largest marine species, including both invertebrates and vertebrate species. Whale sharks are on this list, of course, as the largest fish species. As noted above, size is often an important criterion in public interest, so we used this list to assess page views for all the species (and species groups) documented therein. The top 10 from this exercise are listed in Table 12.2.

The use of flagship species leans upon their familiarity and popularity in the target audience, rather than their own conservation status, to garner public and financial support for ecosystem-level conservation programs (Veríssimo et al. 2011; McGowan et al. 2020). We can see here that whale sharks rank highly in public online interest and do therefore fit the core criteria for use as a conservation flagship. This result is supported by the WWF's inclusion of sharks and rays as one of their priority species clusters, while WCS also includes sharks as a global priority (WCS 2020).

From a marketing perspective, whale sharks have a lot of distinct advantages. The species is relatively familiar and appealing (Gallagher and Hammerschlag 2011; Veríssimo et al. 2018; Mazzoldi et al. 2019), they are harmless to people, and whale shark presence can provide considerable economic benefit through tourism (see Chapter 10). There are many contemporary indicators of high public awareness, cultural penetration, and pride. Whale shark festivals are held annually

Table 12.2 Ten Large and Popular Marine Species

Common Name	WWF 'Cluster'	WCS Global Priority	Pageviews (2019)
Blue whale	Cetaceans	Whales	1,653,507
Great white shark	Sharks and rays	Sharks	1,271,688
Greenland shark	Sharks and rays	Sharks	860,591
Sperm whale	Cetaceans	Whales	778,052
Whale shark	Sharks and rays	Sharks	689,228
Giant squid	–	–	675,811
Ocean sunfish	–	–	667,878
Colossal squid	–	–	591,994
Walrus	–	–	541,176
Oarfish	–	–	503,129

Common names are as provided by McClain et al. (2015), which were also used for the corresponding Wikipedia page for that species or group. Pageviews for 2019 were obtained from Wikipedia's 'Pageviews Analysis' tool.

in some countries, such as the Philippines and Mexico, and International Whale Shark Day is held annually on 30 August. The species is featured on banknotes in Djibouti, the Maldives, and the Philippines, and they have featured on vehicle license plates in Quintana Roo, Mexico, and in Georgia, USA. 'Destiny' the whale-speaking whale shark featured as a major character in a 2016 Pixar/Disney movie, Finding Dory. There is even an approximately quadrennial 'international whale shark conference,' demonstrating a high level of research interest. Not least, here we are writing a book on the species! These flagship attributes have undoubtedly benefited whale sharks. International and national media attention on whale sharks has been a significant impetus for conservation and management measures in multiple countries (Graham 2007; Mau 2008; Acebes 2013; Akhilesh et al. 2013). But, in practice, does whale shark conservation also benefit other marine species?

12.4.2 Umbrella Species

A separate, but related concept in conservation marketing is the use of 'umbrella species.' Simply defined, umbrella species are 'those whose conservation will also conserve other species' (Zacharias and Roff 2001). The presence of charismatic species, including whale sharks, is routinely used as a justification for the creation of MPAs. Specific examples include Gladden Spit in Belize, where the presence of whale sharks was used to argue for the protection of a teleost spawning site (Heyman et al. 2001); the Reserva de la Biosfera Tiburón Ballena, or Whale Shark Biosphere Reserve, which protects an upwelling zone near Isla Holbox in Mexico (de la Parra Venegas et al. 2011; Ibarra-García et al. 2017); and the South Ari MPA in the Maldives, which protects an area of outer coral atoll where whale sharks are frequently encountered. The species' penchant for seasonally productive upwelling and spawning areas enhances the potential for broader ecosystem-level benefits (Erisman et al. 2017), as these areas tend to have particularly high levels of biodiversity, although the high interest in viewing the sharks can also lead to significant tourism pressure on these small sites (Heyman et al. 2010; Cagua et al. 2014).

The presence of whale sharks has had a demonstrable positive influence on the establishment, and ongoing management, of globally significant MPAs. A quick examination of the 50 Marine World Heritage Areas listed by UNESCO for their 'Outstanding Universal Value' (UNESCO 2020) identifies that the presence of whale sharks has been explicitly noted as a listing justification for six: the Ningaloo Coast of Western Australia, the Malpelo Fauna and Flora Sanctuary off Colombia, Cocos Island National Park off Costa Rica, Galapagos Islands of Ecuador, Tubbataha Reefs Natural

Park in the Philippines, and iSimangaliso Wetland Park in South Africa. There is considerable support and precedent, then, for viewing (and promoting) the whale shark as a globally relevant umbrella species for ecosystem-level marine conservation initiatives. Germanov et al. (2018) also suggested that filter-feeding megafauna could be used as umbrella species for stimulating action on marine plastic reduction.

12.4.3 Key Biodiversity Areas

The broader ecosystem-level benefits of whale shark conservation could be expanded further by explicitly identifying Key Biodiversity Areas (KBAs) in which the species is present and promoting the sharks as flagships to enhance protection for these sites. Recognized KBAs are 'sites contributing significantly to the global persistence of biodiversity' (CBD 2020). The KBA assessment framework was created by a Joint Task Force on Biodiversity and Protected Areas, formed by the IUCN Species Survival Commission and IUCN World Commission on Protected Areas, to unify conservation efforts for threatened species and key ecosystems (IUCN 2016). However, more than 50% of recognized marine KBAs have no formal protection in place (CBD 2020).

Several important sites for whale sharks have already been identified as KBAs. As yet, only the Galapagos Marine Reserve in Ecuador (which is also a World Heritage Area, as above) appears to have specifically included the whale shark in a published planning process for KBAs (Edgar et al. 2008). Some identified KBAs, such as the Daymaniyat Islands and the Musandam regions in Oman, and the northwestern Madagascar area (Key Biodiversity Areas 2020), could also provide an avenue for the enhanced protection of whale sharks in countries that do not presently have species-level legislation in place.

Whale shark presence can also be included in the assessment criteria for listing future KBAs. Criteria A1 (threatened species) and D1 (demographic aggregations) for KBA identification deal specifically with threatened species, based on the IUCN Red List, and can include important aggregation sites for these species across life stages (IUCN 2016). Many of the whale shark constellations, including the feeding and breeding areas identified in Chapter 7, qualify for inclusion as KBAs.

12.5 THE ROAD TO RECOVERY

While recent decades have been bleak for whale sharks, they are now one of the best known and well protected of all shark species. A lot of work remains to be done, but there is cause for optimism going forward. With that in mind, when can we consider whale sharks to be fully recovered? In other words, when is the species not just safe from extinction, but restored close to its pre-decline numbers and ecological role? And how should we monitor progress towards this goal?

12.5.1 The IUCN Green Status of Species

"Recovery" is a contentious term in conservation circles (Akçakaya et al. 2018). It has been linked to an improvement in status on the IUCN Red List, that is, moving from a "threatened" category – Vulnerable, Endangered, or Critically Endangered – to Least Concern. However, even if their overall number is increasing, the extant population of a 'Least Concern' species can still be hugely reduced compared to their historical abundance. The species may also be found in only a small portion of their prior range, rendering them functionally extinct, in that they are not present to perform their pre-impact ecological role (Akçakaya et al. 2018, 2020).

An effort to formalize a practical definition of species recovery is now underway through the development of the 'IUCN Green Status of Species,' an initiative that has been in progress since

2012 (Akçakaya et al. 2018; IUCN in review). The Green Status aims to complement the current IUCN Red List classification by providing an evaluation of current status (both in terms of extinction risk and progress toward recovery) and measuring the success of conservation efforts. The Green Status approach is, like the Red List, intended to be broadly useful across taxonomic and ecological groups. An initial test of the method on 181 species indicates that the Green Status assessment is applicable to wide-ranging marine taxa (e.g., great white shark *Carcharodon carcharias*, gray whale *Eschrichtius robustus*; Grace et al. in review). Following this framework here, we can present a first pass at determining the current Green Status of whale sharks and thus their long-term recovery potential.

The Green Status process and associated metrics are presently in active development (Stephenson et al. 2020; IUCN in review). As it stands, the Green Status assesses species against three essential aspects of recovery – representation, viability, and functionality (Akçakaya et al. 2018). A species can be considered *fully recovered* if

1. It is present in all parts of its range, including those that are no longer occupied, but where the species lived prior to major human impacts/disruption. This requires knowledge of the species' pre-impact distribution and division of the distribution into parts (spatial units) in which its status can be assessed separately.
2. The species is viable (i.e., not threatened with extinction) in each spatial unit.
3. The species is also performing its ecological functions in each spatial unit.

These factors contribute towards a "Green Score," ranging from 0% to 100%. Calculation of the Green Score is achieved by assessing each spatial unit as one of four options:

- Absent: Does not occur, but did historically.
- Present: Occurs, but is not at a Viable or Functional level.
- Viable: Low risk of regional extinction (i.e., a Red List assessment of Least Concern, or Near Threatened and not declining), but not Functional.
- Functional: Population size, density, and structure allow for full ecological function and/or roles. A species must qualify as Viable in order for the Functional label to be assigned (with rare exceptions; IUCN in review).

The sum of the numerical values assigned to the observed states (Table 12.1) is then divided by the highest possible value (i.e., a Functional population in each spatial unit) and multiplied by 100 to calculate a Green Score (G):

$$G = \frac{\sum W_s}{W_F \times N} \times 100$$

where W_s is the weight of the state in the spatial unit. W_F is the weight of the "Functional" category. N is the number of spatial units.

The maximum score is 100% for a species that is Functional in all spatial units, representing the fully recovered state. The minimum score is 0% for a species that is Absent throughout its natural range (i.e., extinct in the wild). As outlined in Table 12.3, a spatial unit can only be assessed as Viable or Functional if the population in question is considered to be Least Concern or Near Threatened (but no longer in decline) based on Regional Red List criteria.

The Green Score that reflects the status of a species at the time of assessment is called the "Species Recovery Score" (SRS), which can be tracked over time to demonstrate progress towards the fully recovered state (IUCN in review).

Table 12.3   Description of Possible Species States in a Spatial Unit, Based on Akçakaya et al. (2018) and IUCN (in review)

State (Default)	Default Weight	Fine-Resolution State (Optional, e.g., If One or Few Spatial Units)	Fine-Resolution Weight (Optional)
Absent	0	Absent	0
Present	3	Present-CR	1.5
		Present-EN	2.5
		Present-VU	3.5
		Present-NT with continuing decline	4.5
Viable	6	Viable-NT without continuing decline	5.5
		Viable-LC	6.5
Functional	9[a]	Functional in <40% of SU	8
		Functional in 40%–70% of SU	9
		Functional in >70% of SU	10[a]

[a] when using this set of weights for the assessment, this is the highest value that should be multiplied by the number of spatial units in the denominator.

Assessors can choose to use either the default or fine-resolution weights in their assessment; fine-resolution weights incorporate the Red List status of the species in the spatial unit. The full definition of terms used to describe states is available in IUCN (in review).

12.5.2 Building a Green Status Assessment

12.5.2.1 Whale Shark Distribution and Spatial Units

The Green Status of Species attempts to avoid a 'shifting baseline' by anchoring the definition of the fully recovered state to the species' pre-impact distribution (Akçakaya et al. 2018; Grace et al. 2019). The whale shark is one of few circum-global marine species. Aside from opportunistic sighting reports, minimal data are available on whale sharks from before 75 years ago (1945). While activities that have contributed to the species' population decline, such as fisheries and shipping, existed prior to 1945 (Bean 1907; Gudger 1941), the whale shark has not suffered any clear range contraction due to human pressures. Therefore, the range of the whale shark considered for Green Status assessment is the same as its current range. The seasonal extent of whale shark distribution appears to be limited more by physiological constraints tied more to sea surface temperature (Afonso et al. 2014) than a lack of suitable habitat or prey. However, model projections suggest that the overall range of the whale shark may slightly decrease by 2070 with predicted warming trends, due to a decrease in habitat suitability in some areas of their existing distribution (Sequeira et al. 2014), although whale sharks can tolerate warm surface temperatures if a depth refuge is available (Robinson et al. 2017).

The next step is to identify spatial units within the enormous distribution of this species. The Atlantic and Indo-Pacific subpopulation split used for the current Red List assessment (Pierce and Norman 2016) has been broadly supported by genetic studies (Chapter 5). The possibility that finer-scale population division exists, particularly in the Indo-Pacific (Norman et al. 2017; Prebble et al. 2018; Yagishita et al. 2020), is a current subject of research. For the purposes of this Green Status assessment, we acknowledge then that our chosen spatial units do not necessarily represent genetically distinct populations of whale sharks. Rather, we are subdividing whale sharks into more tractable geographical units for research and conservation initiatives based on known regional connectivity, differing threat profiles, and the existence of relevant management frameworks (i.e., RFMOs), since these have been created to facilitate a unified work plan on fisheries in international waters. Based on our current knowledge of whale shark ecology and their threats, we consider the following spatial units for Green Status assessment:

1. Atlantic Ocean (ICCAT region).
2. Western Indian Ocean (IOTC region).
3. Arabian Sea (IOTC region).
4. Southeast Asia (incorporating Western Australia and the Western Central Pacific; IOTC/WCPFC regions).
5. Eastern Pacific Ocean (IATTC).

Again, we do not here imply that clear boundaries are present, or necessarily exist, between these regions. Some of these spatial units could well be combined, or new units identified, as knowledge improves. We also note that the IUCN is formalizing a process for reviewing and publishing Green Status of Species assessments at the time of writing (IUCN in review). The spatial units used for the first formal whale shark assessment may, therefore, end up being different from those described here.

12.5.2.2 *Contemporary Regional Red List Assessments*

As noted earlier, the whale shark is currently listed as Endangered globally (criteria A2bd+4bd) and in the Indo-Pacific (A2bd+4bd), but Vulnerable (A2b+4b) in the Atlantic Ocean (Pierce and Norman 2016). While additional relevant information has been published since these assessments were completed, such as gillnet bycatch information from Pakistan (Razzaque et al. 2020), Peru (Pajuelo et al. 2018), and Venezuela (Sánchez et al. 2020), these new data do not appear to significantly change the current classifications. A separate regional Red List assessment was completed for whale sharks in the Arabian Sea, classifying them as Endangered (C1) due to high levels of fishing pressure in the region (Dulvy et al. 2017). The three 'new' spatial units have not been formally assessed using IUCN regional Red List criteria, so we can provide provisional classifications here.

12.5.2.3 *Western Indian Ocean*

We consider the Western Indian Ocean (WIO) spatial unit to extend south from East Africa to South Africa, and east to the island nations of the Seychelles and Madagascar. Limited direct connectivity has been documented between the known whale shark feeding areas in this region (Brooks et al. 2010; Andrzejaczek et al. 2016; Norman et al. 2017; Diamant et al. 2018; Prebble et al. 2018). Sharks in these constellations were mostly juveniles, however. The relatively high densities of whale sharks present in the open ocean of the Mozambique Channel (Sequeira et al. 2012; Escalle et al. 2016b), and recent data on genetic homogeneity between Red Sea and Tanzanian sharks (R. Hardenstine pers. comm.), suggest that some mixing is likely to occur, potentially of adults during the 'offshore' phase of their life (Chapter 7).

No large-scale commercial fisheries targeting whale sharks are known from the WIO. The connectivity of sharks in this spatial unit to those caught in current and historical whale shark fisheries in the Arabian Sea, India, and the Maldives (as described in Chapter 11) is unknown. There is a low contemporary bycatch of the species from small-scale fisheries in WIO countries (Rohner et al. 2018; Temple et al. 2019). Whale sharks are still caught, or at least encircled, within WIO tuna purse-seine fisheries, but few sharks die during capture or shortly afterwards when 'safe release' practices are used (Capietto et al. 2014). That said, one of us (SJP) has seen several sharks in this area with attached tail-ropes, likely as a result of poor purse-seine release practices, which could lead to mortality in those individuals. Using 'tow ropes' in such a way is explicitly prohibited in the Pacific region by IATTC (Sea Shepherd Legal 2019a), but apparently not within the WIO. Shipping may also pose a risk to WIO whale sharks (Speed et al. 2008; Pirotta et al. 2019), as may plastic ingestion (Germanov et al. 2018).

There have been significant declines in sightings noted from southern Mozambique, the Mozambique Channel, and also the Seychelles ((Pierce and Norman 2016; D. Rowat pers. comm.).

A decline has also been noted in Kenyan sightings (Oddenyo et al. 2018). Recent analyses show a stable abundance trend in Tanzania and Madagascar over the past few years (SJP unpubl. data). Since the declining trends in the WIO contributed to the overall Endangered listing, and recent data are insufficient to warrant any change, we estimate the whale shark to be Endangered (A2b) in the WIO.

12.5.2.4 Southeast Asia

Southeast Asia, incorporating Western Australia and the Western & Central Pacific, is treated here as a single spatial unit based on ongoing tracking and photo-identification studies that demonstrate routine movement of whale sharks through much of this area (Chapter 6, and see also Araujo et al. 2020; Meyers et al. 2020). The targeted whale shark fisheries in Taiwan, southern China, and the Philippines have depleted the whale shark population in this region (Pierce and Norman 2016). There is substantial evidence for a reduction of large adult sharks, decreases in catches per unit effort from fisheries, and declining sightings per unit effort from fisheries- and tourism-based studies (summarized in Pierce and Norman 2016).

Other threats remain present, such as plastic pollution (Germanov et al. 2018, 2019; Abreo, et al. 2019; Lee 2019), ship strikes (Pirotta et al. 2019; Lester et al. 2020), along with bycatch in tuna purse-seine fisheries (Clarke 2015; Neubauer et al. 2018). This spatial unit has a relatively high tuna purse-seine effort (Neubauer et al. 2018). A question mark remains around the current catch levels in mainland China. A major processing hub for this fishery, the town of Puqi (Li et al. 2012; WildLifeRisk 2014), has significantly reduced in capacity in recent years (Wu 2016). This may indicate a decline in the number of whale sharks being caught. Whale shark bycatch in gillnets is likely to be a concern through the region (Wyatt et al. 2019; GA unpubl. data) and perhaps especially within the pelagic gillnet fishery (Anderson et al. 2020).

At present, we estimate the whale shark to be Endangered (A2bd+4bd) in this region. While there are some encouraging signs of increasing whale shark sightings in certain areas, such as off Western Australia (Holmberg et al. 2009; B. Norman pers. comm.), and around Donsol and Palawan in the Philippines (GA unpubl. data), a decline in whale shark sightings has been noted across much of the Pacific Ocean based on records from tuna fisheries (Neubauer et al. 2018; Román et al. 2018).

12.5.2.5 Eastern Pacific Ocean

There are few trend data available from the Eastern Pacific (EP). Interviews with long-term dive guides provide some evidence for stable sightings of whale sharks around Ecuador's Galapagos Islands (Peñaherrera-Palma et al. 2018). Whale sharks are frequent seasonal visitors to the northern-most islands and occasionally seen around the main island group (Acuña-Marrero et al. 2014; Hearn et al. 2016). Similarly, data from scuba diver logbooks from Cocos Island, off Costa Rica, showed an increase in the occurrence of whale sharks of approximately 4.5% per year from 1993-2013, although the species was only recorded on 2% of dives overall across that period (White et al. 2015). However, broad-scale trend analysis from EP tuna fisheries data indicates a general decline from 2003 to 2016 (Román et al. 2018).

No significant targeted fisheries for whale sharks are known to have operated in the EP itself. A small amount of bycatch is likely to be ongoing in coastal net fisheries (Pajuelo et al. 2018) and the IATTC tuna purse-seine fishery (Román et al. 2018), which both overlap with whale shark feeding and likely breeding areas in this region (Hearn et al. 2016; Ryan et al. 2017; Pajuelo et al. 2018; Román et al. 2018). While the previous target Western Pacific fisheries (i.e., in Taiwan and the Philippines) may also have affected whale shark abundance in the EP, and there may be ongoing issues with purse-seine bycatch (Clarke 2015) and potentially pelagic gillnet bycatch (Anderson et al. 2020) in the WCPFC region, the degree of connectivity across the vast Pacific Ocean is currently unclear. There have been no photo-identification matches across this distance (Norman et al.

2017). Tracking results are ambiguous, with two long-term movements of sharks apparently crossing the central Pacific (Eckert and Stewart 2001; Guzman et al. 2018), but it is difficult to confirm that either of these tags, which did not collect depth data, remained attached to the sharks throughout these unusually long (2–3 year) tracks. Other tracking results indicate a more regional population (Hearn et al. 2016; Ramírez-Macías et al. 2017). No significant genetic differences are known to exist across the Pacific (Yagishita et al. 2020). Given that a broad-scale decline in whale shark sightings across the Pacific, based on tuna fishery data, has not been explained by environmental variation (Neubauer et al. 2018), we take a precautionary approach here. We consider whale sharks to be Vulnerable (A2bd) in the EP. The population in this spatial unit is likely to have been affected by target and bycatch fisheries in the Western Pacific, bycatch in coastal South American fisheries and offshore tuna fisheries, and shipping mortality.

12.5.2.6 Ecological Functionality

A variety of parameters can be integrated to assess ecological functionality (Akçakaya et al. 2020). The ecosystem role of the whale shark has been provisionally discussed in Chapter 8. In brief, whale sharks are likely to be important transporters of nutrients from productive coastal waters (Rohner et al. 2018), and offshore frontal regions (Ramírez-Macías et al. 2017; Ryan et al. 2017), to nutrient-poor areas, such as most tropical oceanic habitats (Estes et al. 2016). Assessment of the whale sharks' contribution to ecosystem processes is at a very early stage, but they are thought to contribute to the resilience of tropical marine systems, as modeled for the Yucatan coast of Mexico (Ibarra-García et al. 2017). Whale sharks are also closely associated with tuna in many areas (Fox et al. 2013; Escalle et al. 2016b, 2018b; Fontes et al. 2020), which may represent a mutually beneficial interaction with these important oceanic predators.

Our limited understanding of the specific ecological role of whale sharks makes it simplest to use the species' pre-impact density as a proxy for full functionality, following Akçakaya et al. (2020). Here, we use the estimated current state of the whale shark population in a spatial unit, based on IUCN regional Red List criteria (discussed below), compared to their pre-impact state. For example, a Near Threatened (without continuing decline) designation here, based on approaching the A2 criteria (which implies a population reduction approaching 30% over ~75 years, for whale sharks), implies a remaining population that is at >70% of their pre-impact population density. This scenario would then receive a fine-resolution weight of 10 (functional in >70% of the spatial unit; Table 12.3).

12.5.3 Species Recovery Score Calculation

While significant data gaps remain, we can calculate a provisional SRS for the whale shark using the Green Score equation, the criteria in Table 12.3, and the information on spatial unit status outlined below (Table 12.4). Each spatial unit was broadly assessed against regional Red List criteria and assigned a corresponding score, as summarized in Table 12.3. The contemporary (likely) SRS for whale sharks is estimated to be 29%.

Assessors are encouraged to incorporate uncertainty in Green Score estimation. In this case, a pessimistic scenario, assuming that whale sharks are threatened (VU or EN) in all spatial units, matches the likely state. The minimum SRS is then 29%. An optimistic scenario assumes that both the Atlantic and EP populations have stabilized at a Near Threatened, without continuing decline, and therefore, the remaining population exceeds 70% of pre-impact numbers. Whale shark populations in these spatial units are then Functional rather than Present, providing a maximum value of 55% for the contemporary SRS. This 'confidence range' is a result of the relative lack of long-term sighting or abundance trend data from two large geographical regions, the Atlantic and EP, and the limited quantification of contemporary mortality from human threats. Further work in these two

Table 12.4   Contemporary Assessments of Regional States (and Their Associated Weights) Using Three Scenarios (Pessimistic, Likely, and Optimistic) to Account for Uncertainty; Weights Are Used to Calculate SRS for the Whale Shark

Spatial Unit[a]	Regional Red List Status[b]	Current State		
		Pessimistic	Likely	Optimistic
Atlantic Ocean	VU	Present (3.5)	Present (3.5)	Functional (10)
Western Indian Ocean	EN	Present (2.5)	Present (2.5)	Present (2.5)
Arabian Sea	EN	Present (2.5)	Present (2.5)	Present (2.5)
Southeast Asia+Australia	EN	Present (2.5)	Present (2.5)	Present (2.5)
Eastern Pacific Ocean	VU	Present (3.5)	Present (3.5)	Functional (10)
Species Recovery Score		29%	29%	55%

[a] As outlined in chapter text for 'Likely' scenario.
[b] As estimated formally (Atlantic and Arabia) in published assessments, or otherwise outlined in chapter text.

regions will be particularly important for refining the whale shark's current SRS, as is improved knowledge of the species' ecological role and contemporary functionality.

12.5.4 Conservation Impact Metrics

The Green Score can also be estimated across different timeframes to (a) retrospectively evaluate and (b) project the impact of future conservation actions. The differences between these estimated Green Scores and the contemporary SRS produce the four Conservation Impact Metrics that comprise the Green Status of Species (Figure 12.1).

12.5.4.1 Conservation Legacy

'Conservation Legacy' estimates the effect of past conservation interventions on species recovery. A high Legacy value indicates that prior conservation actions have substantially improved the species' status, while a low Legacy reflects a situation where conservation efforts have not been needed, been ineffective, or have not been attempted.

Here, we consider primarily the benefit of the closures of commercial fisheries, and subsequent protection efforts, in India, the Philippines, and Taiwan. We also cautiously include the mainland Chinese whale shark fishery here, due to the probable reduction of effort and the presence of legal protection in the country (at least on-paper), as well as the various national protections and RFMO initiatives now in place through the species' distribution.

In the absence of conservation, respective regional Red List assessments of Critically Endangered in the Arabian Sea (due primarily to the Indian fishery) and the Southeast Asian spatial unit (due to the other target fisheries) would have been likely. This scenario would produce a likely SRS of 25%. Therefore, the best estimate of Conservation Legacy (the difference from the current SRS) is 4%.

12.5.4.2 Conservation Dependence

'Conservation Dependence' indicates the reliance of the species on current conservation efforts to maintain the SRS over the next 10 years. A continuing reduction in catch off mainland China is likely to be the primary influence on this score, as most RFMOs have already made efforts to reduce bycatch and improve release practices in tuna fisheries. If the Chinese fishery is a continuing threat, particularly at the previous documented levels (Li et al. 2012), there is a chance that the Southeast

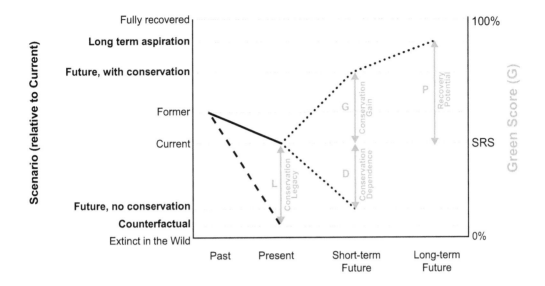

Figure 12.1   Green Status of Species conceptual diagram. The Green Score (G; right *y*-axis) is estimated at the time of assessment, giving the SRS. Green Scores can also be estimated for each of the bolded scenarios on the left *y*-axis (contextual reference points are also provided). The differences between the scenario-based Green Scores and the SRS provide the four Conservation Impact Metrics (green arrows and text). (Adapted from IUCN (in review) with permission.)

Asia and surrounding population will be regionally downgraded to Critically Endangered (a weight of 1.5). The best estimate of Conservation Dependence is therefore 2%.

12.5.4.3 Conservation Gain

'Conservation Gain' captures the improvement in status expected to occur in the next 10 years as a result of ongoing or planned conservation actions. While the speed of abundance increase in whale sharks is limited by their conservative life history (Chapter 2), we hope that the ongoing reduction of whale shark catch in China along with bycatch mitigation in gillnet and purse-seine fisheries will allow populations to stabilize, albeit at a low level. However, with such a reduced population in some spatial units, successful recovery may be dependent on mortality from other threats, such as shipping and plastic pollution, also being held to a low level. Even a small incidence of mortality from human threats in absolute terms could lead to continuing decline in small whale shark populations.

A possible best-case scenario for 2030 could see Atlantic whale sharks moving to Near Threatened, without ongoing decline (Viable; score 5.5), which – using the proxy metric for ecological function – would increase the state of the population to Functional (10). The EP population could also improve to Near Threatened, without ongoing decline (Functional; score 10). It is likely to take longer to demonstrate a broad recovery in the depleted Arabian Sea, WIO, and Southeast Asian whale shark populations. However, this scenario would still result in a SRS of 55% in 2030, a substantial Conservation Gain of 26% (best estimate).

12.5.4.4 Recovery Potential

Finally, 'Recovery Potential' represents the most ambitious recovery outcome that could be achieved within 100 years. This assumes that biological and ecological constraints are still present, but financial constraints are lifted (IUCN in review). In many cases, the best-case outcome for a species will be a figure of less than 100% (Figure 12.1), as it may be unrealistic for a species to re-establish its full prior range or functionality (such as when historical habitats have been irreversibly modified by human activities) over that period. It is also possible that the Recovery Potential will not be realized; it represents an upper bound on the available space in which conservation can work in the next 100 years, but real-world constraints and trade-offs will of course determine which actions are taken.

With current and ongoing conservation management, there appears to be little risk of whale sharks becoming regionally extinct in any individual spatial unit. For the moment, we are optimistic that climate shifts, and potential effects on plankton productivity, will not limit whale shark habitat availability. The species is, therefore, hoped to reach 100% in the next 100 years. Their relatively low numbers compared to baseline and projected slow rate of recovery, due to inherent biological constraints (Chapter 2), however, mean that successfully realizing this potential will clearly require concerted, long-term effort to reduce broad-scale threats.

12.6 CONSERVATION PLANNING

The publication of periodic Green Status reassessments will provide a dedicated, evolving tool for tracking the whale sharks' progress towards full recovery. How can we ensure that happens? Answering that question is the purpose of species conservation planning. The strategic aim of a species-level conservation plan is to provide SMART (specific, measurable, achievable, realistic, and time-bound) goals on what needs to be done, and how these actions can be accomplished (Roberts and Hamann 2016; Stephenson 2019).

The IUCN Shark Specialist Group has facilitated the production of species-level or group-level international or regional conservation strategies for some of the most threatened elasmobranchs over the past few years, including sawfishes (Harrison and Dulvy 2014), mobulid rays (Lawson et al. 2017), and angel sharks (Lawson et al. 2020). The intention of the inclusion of whale sharks within the Sharks MOU, and the Concerted Actions produced in conjunction with their listing on Appendix I of CMS, is also to support and promote a conservation plan. A broader IUCN framework has now been produced to provide detailed guidance on how best to approach the conservation planning process (IUCN/SSC 2017). Whale sharks have not been assessed using these guidelines, but we can use this document to build on the provisional Green Status assessment outlined above to provide a general outline of what the full process could involve. Figure 12.2 shows that the strategic planning cycle is split into two phases: the planning process itself, followed by practical implementation and adaptation. This creates the feedback loop required to periodically update the plan.

12.6.1 Preparing a Conservation Plan

First, why do whale sharks require conservation support? Hopefully, this far through the chapter and book, you are already well informed on this particular topic. In short, whale sharks are a globally Endangered species on the IUCN Red List, the species is highly distinctive in evolutionary terms (EDGE 2020 and Chapter 1), and whale shark conservation also benefits other threatened species and ecosystems, as well as creating economic opportunities for shark-watching tourism (Chapters 9 and 10). Our provisional Green Status assessment indicates that the full recovery and resumption of ecosystem function is possible for whale sharks. The purpose of creating a conservation plan, then, is to identify and assemble stakeholders, develop a shared vision for whale shark conservation and recovery, and jointly create a SMART plan to achieve these goals.

A status review involves the collation, analysis and reporting of relevant data on biology, ecology, and threatening processes. Many of these details are, conveniently, summarized in this book. The formal preparation of a Green Status assessment as a base metric for measuring future success would be an important concurrent output of this work. The purpose of vision and goals are to

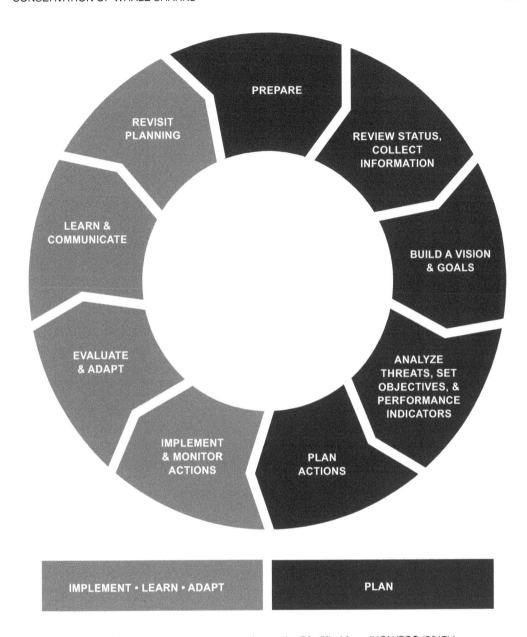

Figure 12.2 The SSC species planning conservation cycle. (Modified from IUCN/SSC (2017).)

articulate the desired future state for the species. The vision statement could usefully build on the Green Status goal of achieving full ecological recovery for whale sharks, while also integrating their future relationship with humans. In this case, the vision could be

> Whale sharks are restored to full ecological functionality, with all subpopulations either stable at a near-pristine level or increasing in numbers. Successful whale shark conservation provides an inspiration and justification for the ongoing management and improvement of key marine habitats, and enables the development of best-practice ecotourism ventures in the areas where these are appropriate.

The associated goals represent 'the vision redefined in operational terms' (IUCN/SSC 2017). These can include up to five practical, concrete steps that will contribute to achieving the desired vision. These should be time-bound, for instance, using a 3-year timescale and workplan, and be collaboratively formulated with all stakeholders involved. In this example, such goals could

1. Identify the most relevant geographical areas in which to manage and monitor whale sharks. This can include both ecological considerations, as summarized in Chapters 5 and 6, and the existing multilateral management frameworks such as RFMOs that facilitate implementation.
2. Develop best-practice long-term monitoring frameworks in each of these areas to track whale shark abundance over time.
3. Identify the priority threat/s in each area, with a focus on the specific habitats that are particularly important to whale sharks (i.e., transit points, feeding sites, reproductive aggregations), and mitigate or remove these.
4. Work with existing and developing ecotourism industries to ensure best practices are in place and followed, to create monitoring programs, and to foster sustainable development and marine conservation within communities.
5. Build awareness of whale shark conservation needs among governments, management agencies, and the public, to support the goals above.

12.6.1.1 Threat Identification and Prioritization

The next stage will normally take place in a workshop setting. The starting point for this stage is the identification of threats to the species (see Chapter 11 of this book). The production of a 'problem tree' to conceptualize threats, their causes (why these threats exist), and constraints (broader complexities), is a common approach to organizing this information. A number of examples are provided by the IUCN Conservation Planning Specialist Group (CPSG 2020). An example for whale sharks might be the threat of bycatch in artisanal coastal gillnet fisheries in the Western Indian Ocean, such as in Mozambique (i.e., Rohner et al. 2018). The causes of this could be the desire of fish for protein, the relative ease of this technique, and a lack of fishing-related equipment, such as boat motors, that might facilitate the use of more selective methods. The extreme poverty in the country, which leads to a lack of food security and a lack of alternative livelihoods available to coastal fishing communities, acts as a constraint. These factors may need to be addressed at a broader level, such as by working with governments and development agencies.

The most urgent and important threats are then prioritized to ensure that these are actively addressed within the timeframe of the plan. The purpose of this exercise is to focus on the most direct ways in which the species' conservation status can be improved. This can consider factors such as whether a given threat is expected to increase or decrease in the absence of conservation action, how much of the population is affected (and to what extent), and how reversible the threat is. For instance, the contemporary Red List assessment for whale sharks identified the Chinese mainland fishery as 'the largest single direct threat to whale shark recovery in the Indo-Pacific' (Pierce and Norman 2016). Obtaining more information on this threat, and implementing practical mitigation actions if required, will be a key objective.

The objectives are the specific, ideally quantitative statements of what must be done to meet the goals outlined above. For example, Goal 2 above identifies the need for long-term abundance monitoring of whale sharks in each geographical unit, so as to provide an independent metric of whale shark population trend. The associated objectives could then be summarized as

Objective 2.1. Publish clear details on how to set up a seasonal photo-identification data collection program for whale sharks that will enable (a) the calculation of abundance trend, referred to as lambda in mark-recapture studies, and (b) the minimum timeframe of monitoring activities that is likely to be required to achieve (a) at various levels of precision.

Objective 2.2. Identify and publish existing datasets that can provide a baseline for past and current whale shark abundance and trends.

Objective 2.3. Identify strategic locations within each geographical unit where long-term (>10 year) monitoring programs are already in progress, or could be developed, and support or create these if necessary.

Objective 2.4. Work with RFMOs to obtain ongoing access to tuna gillnet and purse-seine observer data on whale shark sightings. Use these to conduct broad-scale trend analyses across large oceanic areas on a periodic basis.

Actions are the building blocks of the conservation plan. Their implementation ensures that the objectives are met, thereby addressing the threats, and in doing so, achieving the goals. Setting actions involves the explicit consideration of what needs to be done and when, who the project team will be, and who is responsible for the team, what resources are needed, and the key metric used to track completion of the action. The results are the SMART performance indicators for assessing completion of the actions and progress towards meeting the objectives. For instance, using the example Objective 2.2 above, one of the actions could be to update the 2005–2011 analysis of whale shark sighting trends from Mozambique (Rohner et al. 2013) to provide long-term trend information from a key WIO aggregation, by the end of 2021. That would then require sourcing funding for the time required for an identified researcher to update this dataset and analyze the new results. The result, publication of these data as either a peer-reviewed paper or working document, would then constitute completion of that particular action. Ideally, though, each result will represent a meaningful conservation success – a new law, a new protected area, or demonstrated reduction in catch. In practical terms, that can help to ensure that media, funding agencies, and stakeholders remain supportive and engaged. The outcome of the planning process is a hierarchical list starting with (a) the agreed-upon long-term vision for the species, (b) up to five goals to achieve this vision, (c) specific objectives accompanying each goal, (d) the specific actions required to fulfil these objectives, and (e) results acting as a continual metric for progress.

12.6.2 Implementing the Plan

Aside from the IUCN Conservation Planning materials (IUCN/SSC 2017) this section is based on, it is also worth reading the 'Cat Conservation Compendium' (Breitenmoser et al. 2015), published by the IUCN Cat Specialist Group, for an applied example of a strategy document for a particular wildlife group. The authors pointed out that *un*successful conservation action plans typically have one or more of four shortcomings. The IUCN Conservation Planning framework is, of course, designed to avoid such failures, but they are certainly worth restating here:

1. The planning group had no political mandate and did not represent interest groups or other stakeholders affected by the planned actions.
2. The plan was too ambitious or impractical.
3. Specific responsibilities for the actions were not defined, or the actors were not ready to take responsibility.
4. Insufficient funding was available for implementation.

The actions identified during the planning phase are commonly grouped into a set of related projects for implementation purposes. Each project can then, hopefully, be supported by a single donor or donor suite, with a single implementation body or steering committee responsible for delivery. The project teams will typically comprise some of the planning team, but not necessarily all of them. It is important, in that case, to ensure that the full planning team is kept informed of progress.

A monitoring strategy is also needed to check that the actions, when implemented, are achieving the desired results, and that progress is being made towards meeting the goals. Monitoring may have to persist through multiple planning cycles, so considerations here include safe data storage, accessibility, and cost-effectiveness of the techniques used. The data used for progress tracking will ideally be the minimum needed for effective assessment to ensure the long-term sustainability of this monitoring work. A useful monitoring strategy enables a 'results-based management process.' If implementation of the actions is *not* leading to the desired results, or better outcomes could be obtained from a change in focus or resource availability, the existing scope of work may require modification. Scheduled periodic evaluations are required for this to work.

Effective conservation requires good communication. This encompasses both 'internal' communications, e.g., from project teams to local stakeholders, within and between project teams, and with the planning group and funders, as well as 'external' communications with media, other conservation practitioners, management authorities, and the interested public. 'Good' communication will, of course, be tailored to the audiences that need to be reached. For the whale shark, an important consideration is both the type of content produced – for instance, videos are often a more effective way to inform stakeholder groups than are large, formal text documents – and language, both in terms of the actual language spoken by the target audience, and that it is at a level that is accessible for the audience.

Full whale shark recovery to an ecologically functional state worldwide will take a long time. Unfortunately, donors normally provide funding for a maximum of 3–5 years per project. An effective whale shark conservation *strategy*, then, has to align with a realistic timeframe for actions to be completed and results to be reported. A time span of 3–10 years for a strategy is the maximum that is recommended by the IUCN, and the shorter end of that spectrum is likely to be the most practical. What happens after that? The completion of this first cycle is accomplished with an evaluation of monitoring results and a thorough, honest review of what worked, what did not, and the overall progress made towards meeting the objectives and goals. Given the short time span recommended above, there is a strong rationale for linking the desired results to human actions, for instance legislative change, rather than the whale shark's long-term biological response. That helps to demonstrate success early, and thereby enable the next planning cycle in an adaptive and iterative process. The plan is dead; long live the plan. Shorter timeframes do pose a challenge for effective measurement of success, in that 'real' whale shark population trends may take a long time to emerge. The strategy may also cover a shorter period than that used for Red List and Green Status updates. It is vital, though, that monitoring efforts are set up in such a way to meaningfully track the ultimate goal: saving the whale shark.

12.7 LOOKING FORWARD

Whale sharks are, finally, well-protected throughout most of their broad distribution, and many of their key habitats have also been protected. We are optimistic for the species' future. At the same time, we also need to be very clear-headed about the whale shark's current plight. Our best estimation of their current SRS is just 29%; it is likely to have been 100% within the last century. The global whale shark population has been halved since the 1980s, and the species remains at risk of extinction worldwide. Whale sharks can recover, but they will need our help.

12.7.1 Closing the Gaps in Existing Protections

Important management and conservation measures for whale sharks are now in place. Legal international trade is no longer an existential threat to whale sharks, based on the lack of NDFs or trade data submitted to CITES, although some illegal trade is likely to still be occurring. Where whale shark products are identified in trade, retrospective identification of the source population through forensic genetic techniques, as applied in some other shark species (Cardeñosa et al. 2020; Fields et al. 2020), would be useful for directing enforcement efforts.

Regulating the international movement of whale shark products has, however, led to a frustrating second-order effect. The various permissions required make it extremely difficult now to conduct research that requires even tiny quantities of whale shark tissue, such as global genetic studies, to be exported to other countries for processing. The serious (months, to several year) delays that result is a hindrance to the studies needed to identify, for example, spatial units for accurate Green Status assessment and regional Red List evaluations. This issue has been identified across multiple taxa since at least 1992 (Bowen 1992) and has been regularly debated at CITES meetings ever since. Reform of this system is needed to minimize the ongoing negative impact on non-harmful research that is needed to inform conservation efforts. We need a system that removes incentives for illegal trade among bad actors, while not restricting the important work of good actors.

The 2017 listing of the whale shark on Appendix I of CMS, coupled with their inclusion on the Sharks MOU, has the potential to increase management attention on the species in some important range states. If all CMS Parties do implement the obligatory conservation measures that are required by the whale shark's listing, the species will have effective protection throughout much of their range. However, there has been a clear gap between signatory countries acknowledging international conservation requirements, by accepting new species listings to CMS, and implementing actual conservation measures themselves (Lawson and Fordham 2018). A lack of capacity and resources in developing countries is a persistent barrier to fulfilling the listing goals of CMS (Lawson and Fordham 2018), and specific funding mechanisms and technical assistance to fulfil the required actions may be required. Private philanthropy and transnational organizations such as Organization of American States, the Global Environment Facility, and the World Bank may have important roles to play in helping these countries to meet their CMS commitments.

Some Parties to CMS, such as Gabon and Madagascar, have not yet formally protected whale sharks in their national waters. Significant whale shark numbers occur in both countries, so this is a useful short-term conservation goal, and efforts are underway in both countries at the time of writing (Sea Shepherd Legal 2019b; S. Diamant pers comm.). Some major fishing nations, including China and Oman, have not signed the CMS treaty. Targeted advocacy campaigns may be helpful to raise public awareness of whale shark conservation needs in these countries, and to drive political interest in their management. Oman has a growing dive tourism industry, which provides an economic incentive to maintain this natural resource.

12.7.2 'Big Data' for Habitat-Based Conservation

The reduction of directed fishing for whale sharks means that ship strikes and bycatch are likely to be the two most important ongoing threats across most of the species' range (Chapter 11). The well-documented tendency of whale sharks to repeatedly return to the same places makes area-based conservation a viable management tool to address these threats (Rigby et al. 2019b), at least at local scale. Whale sharks often show high site fidelity to seasonal feeding areas (Norman et al. 2017), use consistent movement corridors (Hearn et al. 2016; Rohner et al. 2018), and aggregate at oceanic islands as part of their reproductive ecology (Perry et al. 2020). While the species' high mobility and extensive use of oceanic habitats means that even the largest contemporary MPAs are unlikely to enclose the full life cycle of individual whale sharks, focusing on threat mitigation at specific locations could provide substantial benefits at a population level.

Whale shark mortalities from ship strikes can likely be reduced by implementing specific management measures at high-risk locations (Pirotta et al. 2019; Schoeman et al. 2020). Major shipping activity occurs near areas where large numbers of whale sharks routinely feed on the surface, such as the Al Shaheen area off Qatar (Robinson et al. 2013, 2017) and off the Yucatan coast of Mexico (de la Parra Venegas et al. 2011; Hueter et al. 2013). Real-time mitigation measures, such as temporary re-routing or speed reduction, could be implemented to reduce the danger to these whale sharks (see Chapter 11), as has been done successfully for some marine mammals (Pirotta et al. 2019; Schoeman et al. 2020). The International Maritime Organization (IMO) is the international body responsible for mitigating the environmental harm from shipping. A relevant area-based protection strategy for whale shark aggregations could be the designation of important sites as Particularly Sensitive Sea Areas (PSSAs) by the IMO (IMO 2020). Member governments of the IMO can nominate PSSAs based on areas being critical habitats for endangered species, such as feeding areas, that are susceptible to harm from shipping activities (IMO 2006). The creation of PSSAs allows for re-routing of ships and other measures deemed necessary to protect the area. Some whale shark habitats, such as Malpelo Island in Colombia, the Galapagos Archipelago in Ecuador, and Tubbataha Reefs Natural Park in the Philippines, have already been listed as PSSAs (IMO 2020), and a tangible conservation goal for whale sharks would be to extend such protection to other important constellation sites (Chapter 7).

Tracking data suggests that adult whale sharks, in particular, primarily live in oceanic habitats and probably forage near frontal zones (Ramírez-Macías et al. 2017; Ryan et al. 2017). These areas are also favored by many commercially targeted fish species, so there is a high overlap between whale shark movements in the open ocean and fisheries (Queiroz et al. 2019). Whale sharks use some well-known frontal zones, such as the Equatorial Front in the Eastern Pacific (Scales et al. 2014; Ryan et al. 2017), that are also hotspots for purse-seine bycatch of the species (Román et al. 2018). As Sequeira et al. (2019) discussed, identifying areas that are important to marine megafauna in international waters 'presents major challenges for conservation because these seascapes are often located over large areas remote from observation or potential enforcement.' To help address this, the UN has launched negotiations to develop a treaty that builds on UNCLOS to develop a mechanism for the creation of area-based management tools in international waters (Sequeira et al. 2019). Satellite oceanographic sensing has enabled automated, high-resolution identification of frontal zones right through the whale shark's oceanic distribution. This near-real-time mapping capability can inform the use of 'mobile' MPAs to protect megafauna that use these frontal zones (Game et al. 2009; Maxwell et al. 2015). The detailed purse-seine fishery risk assessment for whale sharks by Neubauer et al. (2018) has already identified areas where open-ocean MPAs could be particularly beneficial for whale shark conservation in the Pacific.

Whale shark research has been, and continues to be, a highly collaborative exercise across countries and research groups, which has enabled conservation successes like the CMS Appendix I listing. Tools for managing and distributing 'big data,' such as the Wildbook for Whale Sharks global photo-identification and sighting database (www.whaleshark.org), and the Global Shark Movement Project (www.globalsharkmovement.org) for storing satellite tracking datasets, allow for analyses of pooled data to address larger questions than possible for most individual research groups. Such initiatives will help us understand other global issues, such as impacts on the species from shipping, plastics, and climate change (Sequeira et al. 2019), which supports conservation planning.

12.7.3 Planning for Action

Effective conservation planning requires robust prioritization, the specific assignment of responsibilities, a sustainable funding mechanism, and a transparent reporting system. This chapter provides a template for that process. The Concerted Actions presented alongside the CMS Appendix I listing has already led to several positive steps, and production of a more formal, inclusive global species conservation plan for whale sharks can help accelerate the required activities. The high potential score for Conservation Gain in our provisional Green Status assessment emphasizes the benefits of fast action, and the whale shark's current Endangered status on the Red List demonstrates the high priority for this work.

A government signing a treaty, scientists publishing their research, or conservationists producing a strategic plan does not automatically result in management action. Somewhat ironically, plenty of published research is available to inform this statement on the 'research-practitioner divide' (Hays et al. 2019). As Hays et al. (2019) point out, 'early engagement between the data collectors and the stakeholders involved in policy development and implementation was often extremely important to help translate tracking data into conservation outcomes.' It is also important to reiterate the point we made earlier: a species' risk of extinction, even when well-documented and understood, is only one consideration in the minds of decision-makers. The conservation planning process emphasizes the importance of communication activities; conservation efforts are far more likely to be successful with community and stakeholder inclusion in planning and implementation, and with continued media interest and, to an extent, pressure.

Adaptive planning requires good monitoring data. In a retrospective on sea turtle conservation, Godley et al. (2020) emphasized that, 'long-term monitoring and protection of nesting sites, in some locations exceeding 50 years, have been central to recovery, understanding trends and determining the importance of previously underestimated aggregations.' Ningaloo Reef, the first whale shark aggregation 'discovered' by scientists, has monitoring data available back to the 1990s, unlocking detailed insights on individual residency, population abundance, and the changes occurring over time (Meekan et al. 2006; Bradshaw et al. 2008; Holmberg, et al. 2009; Sequeira et al. 2016; Norman et al. 2017). An increasing number of aggregations now have >10 yr datasets available, which will be of great assistance to conservation assessment going forward. While a regular feedback loop from scientists to assessors and managers is needed to track whale shark population trends, this can add considerably to the workload of the few experienced population modelers in this field. Following on from the notes in the preceding section on the usefulness of 'big data' automation, building a standardized data reporting template into the Wildbook for Whale Sharks system to provide a population monitoring 'dashboard' would provide significant benefits.

12.8 CONCLUSIONS

Whale sharks are increasingly well known, with aquariums and wild tourism producing hundreds of thousands of potential new advocates each year. The species has become the focus of a lot of research attention, much of which is synthesized in this book. Their iconic status makes them a useful flagship species in conservation campaigns, to the demonstrable benefit of the broader marine environment. Their high profile and feeding ecology make them a suitable flagship for countering large scale, increasing threats like plastic pollution. The innate charisma of whale sharks makes them an excellent candidate species to become an international marine conservation success story, like the humpback whale, but this will not happen without a lot of work.

While there have been promising advances in international and national policy development relevant to whale sharks, a significant gap still exists between 'paper protection' and practical conservation and management initiatives. Effective protection across their range, through both species conservation efforts and protected area management, remains a key goal to secure the future of whale sharks. People may have been the problem in the past, but people are now the solution. We hope that this chapter helps to prepare you with the knowledge base to save the earth's greatest fish.

REFERENCES

Abreo, N. A. S., D. Blatchley, and M. D. Superio. 2019. Stranded whale shark (*Rhincodon typus*) reveals vulnerability of filter-feeding elasmobranchs to marine litter in the Philippines. *Marine Pollution Bulletin* 141:79–83.

Acebes, J. M. V. 2013. Hunting 'big fish': A marine environmental history of a contested fishery in the Bohol Sea. PhD diss., Murdoch University.

Acuña-Marrero, D., J. Jiménez, F. Smith, et al. 2014. Whale shark (*Rhincodon typus*) seasonal presence, residence time and habitat use at Darwin Island, Galapagos Marine Reserve. *PLoS One* 9:e115946.

Afonso, P., N. McGinty, and M. Machete. 2014. Dynamics of whale shark occurrence at their fringe oceanic habitat. *PLoS One* 9:e102060.

Akçakaya, H. R., E. L. Bennett, T. M. Brooks, et al. 2018. Quantifying species recovery and conservation success to develop an IUCN Green List of Species. *Conservation Biology* 32: 1128–38.

Akçakaya, H. R., A. S. L. Rodrigues, D. A. Keith, et al. 2020. Assessing ecological function in the context of species recovery. *Conservation Biology* 34:561–71.

Akhilesh, K. V., C. P. R. Shanis, W. T. White, et al. 2013. Landings of whale sharks *Rhincodon typus* Smith, 1828 in Indian waters since protection in 2001 through the Indian Wildlife (Protection) Act, 1972. *Environmental Biology of Fishes* 96:713–22.

Alava, M. N. R., and E. R. Z. Dolumbalo. 2002. Fishery and trade of whale sharks and manta rays in the Bohol Sea, Philippines. In *Elasmobranch Biodiversity, Conservation and Management: Proceedings of the International Seminar and Workshop, Sabah, Malaysia, July 1997*, eds. S. L. Fowler, T. M. Reed, and F. A. Dipper, 132–48. Occasional Paper of the IUCN Species Survival Commission.

Anderson, R. C., M. Herrera, A. D. Ilangakoon, et al. 2020. Cetacean bycatch in Indian Ocean tuna gillnet fisheries. *Endangered Species Research* 41:39–53.

Andrzejaczek, S., J. Meeuwig, D. Rowat, et al. 2016. The ecological connectivity of whale shark aggregations in the Indian Ocean: A photo-identification approach. *Royal Society Open Science* 3:160455.

Araujo, G., A. Agustines, B. Tracey, et al. 2019. Photo-ID and telemetry highlight a global whale shark hotspot in Palawan, Philippines. *Scientific Reports* 9:17209.

Araujo, G., A. R. Ismail, C. McCann, et al. 2020. Getting the most out of citizen science for endangered species such as whale shark. *Journal of Fish Biology* 96: 864–67.

Barbosa-Filho, M. L. V., D. C. Tavares, S. Siciliano, et al. 2016. Interactions between whale sharks, *Rhincodon typus* Smith, 1928 (Orectolobiformes, Rhincodontidae), and Brazilian fisheries: The need for effective conservation measures. *Marine Policy* 73:210–15.

Bean, B. A. 1907. The history of the whale shark (*Rhinodon typicus* Smith). *Smithsonian Miscellaneous Collections* 48:139–48.

Boldrocchi, G., M. Omar, A. Azzola, and R. Bettinetti. 2020. The ecology of the whale shark in Djibouti. *Aquatic Ecology* 54: 535–51.

Borrell, A., A. Aguilar, M. Gazo, et al. 2011. Stable isotope profiles in whale shark (*Rhincodon typus*) suggest segregation and dissimilarities in the diet depending on sex and size. *Environmental Biology of Fishes* 92:559–67.

Bowen, B. 1992. Guest editorial: CITES and scientists – conservation in conflict. *Marine Turtle Newsletter* 58:5–6.

Bradshaw, C. J. A., B. M. Fitzpatrick, C. C. Steinberg, B. W. Brook, and M. G. Meekan. 2008. Decline in whale shark size and abundance at Ningaloo Reef over the past decade: The world's largest fish is getting smaller. *Biological Conservation* 141:1894–905.

Brand, L. E., and A. Compton. 2007. Long-term increase in *Karenia brevis* abundance along the southwest Florida coast. *Harmful Algae* 6:232–52.

Breitenmoser, U., T. Lanz, K. Vogt., and C. Breitenmoser-Würsten. 2015. How to save the cat – Cat conservation compendium, a practical guideline for strategic and project planning in cat conservation. *Cat News Special Issue* 9:36.

Brooks, K., D. Rowat, S. J. Pierce, et al. 2010. Seeing spots: Photo-identification as a regional tool for whale shark identification. *Western Indian Ocean Journal of Marine Science* 9:185–94.

Cagua, E. F., N. Collins, J. Hancock, and R. Rees. 2014. Whale shark economics: A valuation of wildlife tourism in South Ari Atoll, Maldives. *PeerJ* 2014:e515.

Campagna, C., F. T. Short, B. A. Polidoro, et al. 2011. Gulf of Mexico oil blowout increases risks to globally threatened species. *Bioscience* 61:393–97.

Capietto, A., L. Escalle, P. Chavance, et al. 2014. Mortality of marine megafauna induced by fisheries: Insights from the whale shark, the world's largest fish. *Biological Conservation* 174:147–51.

Cardeñosa, D., A. T. Fields, S. K. H. Shea, et al. 2020. Relative contribution to the shark fin trade of Indo-Pacific and Eastern Pacific pelagic thresher sharks. *Animal Conservation* 106. doi:10.1111/acv.12644.

CBD. 2010. Aichi biodiversity targets. COP 10 Decision X/2: Strategic Plan for Biodiversity, 2011–20. https://www.cbd.int/sp/targets/ (accessed November 17, 2020).

CBD. 2020. *Global Biodiversity Outlook 5*. Montreal: Secretariat of the Convention on Biological Diversity.

Chen, C. T., K.-M. Liu, and S.-J. Joung. 1997. Preliminary report on Taiwan's whale shark fishery. *TRAFFIC Bulletin* 17:53–57.

Clarke, S. 2015. Understanding and mitigating impacts to whale sharks in purse seine fisheries of the Western and Central Pacific Ocean. Western and Central Pacific Fisheries Commission. WCPFCSC11-2015/EB-WP-03 Rev. 1. Pohnpei, Federated States of Micronesia.

CMS. 2017a. Convention on Migratory Species: Proposal for inclusion of the whale shark (*Rhincodon typus*) on Appendix I of the Convention. UNEP/CMS/COP12/Doc.25.1.20.

CMS. 2017b. Concerted Action for the whale shark (*Rhincodon typus*). UNEP/CMS/Concerted Action 12.7. Convention on Migratory Species.

CMS. 2020a. Convention on the Conservation of Migratory Species of Wild Animals: Appendix I & II of CMS. https://www.cms.int/en/species/appendix-i-ii-cms (accessed November 17, 2020).

CMS. 2020b. Sharks: Memorandum of Understanding on the Conservation of Migratory Sharks. https://www. cms.int/sharks/en (accessed November 19, 2020).

Collette, B., S. K. Chang, A. Di Natale, et al. 2011. *Thunnus maccoyii*. The IUCN Red List of Threatened Species. https://www.iucnredlist.org/species/21858/9328286 (accessed November 17, 2020).

Colman, J. G. 1997. A review of the biology and ecology of the whale shark. *Journal of Fish Biology* 51:1219–34.

Courchamp, F., I. Jaric, C. Albert, Y. Meinard, W. J. Ripple, and G. Chapron. 2018. The paradoxical extinction of the most charismatic animals. *PLoS Biology* 16:e2003997.

CPSG. 2020. Threats analysis processes. http://www.cbsg.org/threats-analysis-processes (accessed November 18, 2020).

de la Parra Venegas, R., R. Hueter, J. González Cano, et al. 2011. An unprecedented aggregation of whale sharks, *Rhincodon typus*, in Mexican coastal waters of the Caribbean Sea. *PLoS One* 6:e18994.

Diamant, S., C. A. Rohner, J. J. Kiszka, et al. 2018. Movements and habitat use of satellite-tagged whale sharks off western Madagascar. *Endangered Species Research* 36:49–58.

Dulvy, N. K., D. P. Robinson, S. J. Pierce, et al. 2017. Whale shark *Rhincodon typus* Smith, 1828. In *The Conservation Status of Sharks, Rays, and Chimaeras in the Arabian Sea and Adjacent Waters*, edited by R. W. Jabado, P. M. Kyne, R. A. Pollom, et al., 2.1:236. Vancouver: Environment Agency – Abu Dhabi, UAE and IUCN Species Survival Commission Shark Specialist Group.

Eckert, S. A., and B. S. Stewart. 2001. Telemetry and satellite tracking of whale sharks, *Rhincodon typus*, in the Sea of Cortez, Mexico, and the North Pacific Ocean. *Environmental Biology of Fishes* 60:299–308.

Edgar, G. J., S. Banks, R. Bensted-Smith, et al. 2008. Conservation of threatened species in the Galapagos Marine Reserve through identification and protection of marine Key Biodiversity Areas. *Aquatic Conservation: Marine and Freshwater Ecosystems* 18:955–68.

EDGE. 2020. Top 50 EDGE sharks and rays. https://www.edgeofexistence.org/species/species-category/sharks-and-rays/ (accessed November 19, 2020).

Erisman, B., W. Heyman, S. Kobara, et al. 2017. Fish spawning aggregations: Where well-placed management actions can yield big benefits for fisheries and conservation. *Fish and Fisheries* 18:128–44.

Escalle, L., J. M. Amandé, J. D. Filmalter, et al. 2018a. Update on post-release survival of tagged whale shark encircled by tuna purse-seiner. *International Commission for the Conservation of Atlantic Tunas* 74:3671–78.

Escalle, L., D. Gaertner, P. Chavance, et al. 2018b. Catch and bycatch captured by tropical tuna purse-seine fishery in whale and whale shark associated sets: Comparison with free school and FAD sets. *Biodiversity and Conservation* 28:467–99.

Escalle, L., H. Murua, J. M. Amande, et al. 2016a. Post-capture survival of whale sharks encircled in tuna purse-seine nets: Tagging and safe release methods. *Aquatic Conservation: Marine and Freshwater Ecosystems* 26:782–89.

Escalle, L., M. G. Pennino, D. Gaertner, et al. 2016b. Environmental factors and megafauna spatio-temporal co-occurrence with purse-seine fisheries. *Fisheries Oceanography* 25:433–47.

Estes, J. A., M. Heithaus, D. J. McCauley, D. B. Rasher, and B. Worm. 2016. Megafaunal impacts on structure and function of ocean ecosystems. *Annual Review of Environment and Resources* 41:83–116.

FAO. 2016. FAO's input to the UN Secretary-General's comprehensive report for the 2016 resumed review conference on the UN Fish Stocks Agreement. http://www.un.org/Depts/los/2016_FAO_Overview.pdf (accessed November 17, 2020).

FAO. 2020a. Fishery statistical collections: Global capture production 1950-2018. http://www.fao.org/fishery/statistics/global-capture-production/en (accessed November 17, 2020).

FAO. 2020b. International Plan of Action for Conservation and Management of Sharks: National and Regional Plans of Action. http://www.fao.org/ipoa-sharks/national-and-regional-plans-of-action/en/ (accessed November 24, 2020).

Fields, A. T., G. A. Fischer, S. K. H. Shea, et al. 2020. DNA zip-coding: Identifying the source populations supplying the international trade of a critically endangered coastal shark. *Animal Conservation* 38. doi:10.1111/acv.12585.

Flewelling, L. J., D. H. Adams, J. P. Naar, et al. 2010. Brevetoxins in sharks and rays (Chondrichthyes, Elasmobranchii) from Florida coastal waters. *Marine Biology* 157:1937–53.

Fontes, J., N. McGinty, M. Machete, and P. Afonso. 2020. Whale shark-tuna associations, insights from a small pole-and-line fishery from the mid-North Atlantic. *Fisheries Research* 229:105598.

Fowler, S. 2016. Gap analysis of activities for the conservation of species listed in Annex 1 under relevant fisheries related bodies. Memorandum of Understanding on the Conservation of Migratory Sharks. http://www.cms.int/sharks/sites/default/files/document/CMS_Sharks_CWG1_Doc_2_1.pdf (accessed November 17, 2020).

Fox, S., I. Foisy, R. De La Parra Venegas, et al. 2013. Population structure and residency of whale sharks *Rhincodon typus* at Utila, Bay Islands, Honduras. *Journal of Fish Biology* 83:574–87.

Frias-Torres, S., and C. R. Bostater Jr. 2011. Potential impacts of the Deepwater Horizon oil spill on large pelagic fishes. In *Remote Sensing of the Ocean, Sea Ice, Coastal Waters, and Large Water Regions 2011*, 8175:81750F. International Society for Optics and Photonics.

Furby, K. 2018. A red tide ravaging Florida may have killed a whale shark for the first known time. *The Washington Post*, August 3, 2018. https://www.washingtonpost.com/news/speaking-of-science/wp/2018/08/03/a-red-tide-ravaging-florida-may-have-killed-a-whale-shark-for-the-first-known-time/ (accessed November 17, 2020).

Gallagher, A. J., and N. Hammerschlag. 2011. Global shark currency: The distribution, frequency, and economic value of shark ecotourism. *Current Issues in Tourism* 14:797–812.

Game, E. T., H. S. Grantham, A. J. Hobday, et al. 2009. Pelagic protected areas: The missing dimension in ocean conservation. *Trends in Ecology & Evolution* 24:360–69.

Germanov, E. S., A. D. Marshall, L. Bejder, et al. 2018. Microplastics: No small problem for filter-feeding megafauna. *Trends in Ecology & Evolution* 33:227–32.

Germanov, E. S., A. D. Marshall, I. G. Hendrawan, et al. 2019. Microplastics on the menu: Plastics pollute Indonesian manta ray and whale shark feeding grounds. *Frontiers in Marine Science* 6:679.

Godley, B. J., A. C. Broderick, L. P. Colman, et al. 2020. Reflections on sea turtle conservation. *Oryx* 54:287–89.

Grace, M., H. R. Akçakaya, E. Bennett, et al. 2019. Using historical and palaeoecological data to inform ambitious species recovery targets. *Philosophical Transactions of the Royal Society B* 374:20190297.

Grace, M., et al. In review. Testing a global standard for quantifying species recovery and assessing conservation impact *Conservation Biology*.

Graham, R. T. 2007. Whale sharks of the Western Caribbean: An overview of current research and conservation efforts and future needs for effective management of the species. *Gulf and Caribbean Research* 19:149–59.

Gudger, E. W. 1938. Four whale sharks rammed by steamers in the Red Sea region. *Copeia* 1938:170–73.

Gudger, E. W. 1941. The whale shark unafraid: The greatest of the sharks, *Rhineodon typus*, fears not shark, man nor ship. *The American Naturalist* 75:550–68.

Guzman, H. M., C. G. Gomez, A. Hearn, and S. A. Eckert. 2018. Longest recorded trans-Pacific migration of a whale shark (*Rhincodon typus*). *Marine Biodiversity Records* 11:8.

Haetrakul, T., S. Munanansup, N. Assawawongkasem, and N. Chansue. 2009. A case report: Stomach foreign object in whaleshark (*Rhincodon typus*) stranded in Thailand. *Proceedings of the 4th International Symposium on SEASTAR 2000 and Asian Bio-Logging Science (The 8th SEASTAR 2000 Workshop)*:83–85.

Hanfee, F. 2001. Gentle giants of the sea: India's whale shark fishery. TRAFFIC-India, WWF-India.

Haque, A. B., S. A. Das, and A. R. Biswas. 2019. DNA analysis of elasmobranch products originating from Bangladesh reveals unregulated elasmobranch fishery and trade on species of global conservation concern. *PloS One* 14:e0222273.

Harrison, L. R., and N. K. Dulvy. 2014. *Sawfish: A Global Strategy for Conservation*. IUCN Species Survival Commission's Shark Specialist Group.

Hays, G. C., H. Bailey, S. J. Bograd, et al. 2019. Translating marine animal tracking data into conservation policy and management. *Trends in Ecology & Evolution* 34:459–73.

Hearn, A. R., J. Green, M. H. Román, et al. 2016. Adult female whale sharks make long-distance movements past Darwin Island (Galapagos, Ecuador) in the Eastern Tropical Pacific. *Marine Biology* 163:214.

Heyman, W. D., L. M. Carr, and P. S. Lobel. 2010. Diver ecotourism and disturbance to reef fish spawning aggregations: It is better to be disturbed than to be dead. *Marine Ecology Progress Series* 419:201–10.

Heyman, W. D., R. T. Graham, B. Kjerfve, and R. E. Johannes. 2001. Whale sharks *Rhincodon typus* aggregate to feed on fish spawn in Belize. *Marine Ecology Progress Series* 215:275–82.

Holmberg, J., B. Norman, and Z. Arzoumanian. 2009. Estimating population size, structure, and residency time for whale sharks *Rhincodon typus* through collaborative photo-identification. *Endangered Species Research* 7:39–53.

Hsu, H. H., S. J. Joung, and K. M. Liu. 2012. Fisheries, management and conservation of the whale shark *Rhincodon typus* in Taiwan. *Journal of Fish Biology* 80:1595–1607.

Hueter, R. E., J. P. Tyminski, and R. de la Parra. 2013. Horizontal movements, migration patterns, and population structure of whale sharks in the Gulf of Mexico and northwestern Caribbean Sea. *PloS One* 8:e71883.

Ibarra-García, E. C., M. Ortiz, E. Ríos-Jara, A. L. Cupul-Magaña, Á. Hernández-Flores, and F. A. Rodríguez-Zaragoza. 2017. The functional trophic role of whale shark (*Rhincodon typus*) in the northern Mexican Caribbean: Network analysis and ecosystem development. *Hydrobiologia* 792: 121–35.

IMO. 2006. Revised guidelines for the identification and designation of Particularly Sensitive Sea Areas. Resolution A.982(24). International Maritime Organization.

IMO. 2020. Particularly sensitive sea areas. https://www.imo.org/en/OurWork/Environment/Pages/PSSAs.aspx (accessed November 2, 2020).

International Year of Biodiversity. 2010. National Biodiversity Strategies and Action Plans (NBSAPs). Convention on Biological Diversity. https://www.cbd.int/nbsap/ (accessed November 18, 2020).

IUCN. 2016. A Global Standard for the Identification of Key Biodiversity Areas, Version 1.0. First edition. Gland: IUCN.

IUCN. In review. IUCN Green Status of Species: A global standard for measuring species recovery and assessing conservation impact. Version 2.0. Gland: IUCN.

IUCN/SSC. 2017. Guidelines for Species Conservation Planning - Version 1.0. Gland: IUCN.

Key Biodiversity Areas. 2020. World database of Key Biodiversity Areas. http://www.keybiodiversityareas.org/home (accessed November 18, 2020).

LAMAVE. 2019. Report on the implementation of the Concerted Action for the whale shark (*Rhincodon typus*). UNEP/CMS/COP13/Doc.28.1.7.b. Convention on Migratory Species.

Lawson, J. M., and S. V. Fordham. 2018. Sharks ahead: Realizing the potential of the Convention on Migratory Species to conserve elasmobranchs. CMS/Sharks/MOS3/Inf 21:76.

Lawson, J. M., S. V. Fordham, M. P. O'Malley, et al. 2017. Sympathy for the devil: A conservation strategy for devil and manta rays. *PeerJ* 2017:e3027.

Lawson, J. M., R. A. Pollom, C. A. Gordon, et al. 2020. Extinction risk and conservation of critically endangered angel sharks in the Eastern Atlantic and Mediterranean Sea. *ICES Journal of Marine Science* 77:12–29.

Lee, S. 2019. Plastic bag causes death of whale shark in Sabah. https://www.thestar.com.my/news/nation/2019/02/08/plastic-bag-causes-death-of-whale-shark-in-sabah/ (accessed January 5, 2021).

Lester, E., M. G. Meekan, P. Barnes, et al. 2020. Multi-year patterns in scarring, survival and residency of whale sharks in Ningaloo Marine Park, Western Australia. *Marine Ecology Progress Series* 634: 115–25.

Li, W., Y. Wang, and B. Norman. 2012. A preliminary survey of whale shark *Rhincodon typus* catch and trade in China: An emerging crisis. *Journal of Fish Biology* 80:1608–18.

Matsumoto, R., M. Toda, Y. Matsumoto, et al. 2017. Notes on husbandry of whale sharks, *Rhincodon typus*, in aquaria. In *The Elasmobranch Husbandry* Manual II: Recent Advances in the Care of Sharks, Rays and Their Relatives, ed. Smith, M., D. Warmolts, D. Thoney, R. Hueter, M. Murray, and J. Ezcurra, 15–22. Columbus: Special Publication of the Ohio Biological Survey.

Mau, R. 2008. Managing for conservation and recreation: The Ningaloo whale shark experience. *Journal of Ecotourism* 7:213–25.

Maxwell, S. M., E. L. Hazen, R. L. Lewison, et al. 2015. Dynamic ocean management: Defining and conceptualizing real-time management of the ocean. *Marine Policy* 58:42–50.

Mazzoldi, C., G. Bearzi, C. Brito, et al. 2019. From sea monsters to charismatic megafauna: Changes in perception and use of large marine animals. *PLoS ONE* 14:e0226810.

McClain, C. R., M. A. Balk, M. C. Benfield, et al. 2015. Sizing ocean giants: Patterns of intraspecific size variation in marine megafauna. *PeerJ* 2015:e715.

McCoy, E., R. Burce, D. David, et al. 2018. Long-term photo-identification reveals the population dynamics and strong site fidelity of adult whale sharks to the coastal waters of Donsol, Philippines. *Frontiers in Marine Science* 5:271.

McGowan, J., L. J. Beaumont, R. J. Smith, et al. 2020. Conservation prioritization can resolve the flagship species conundrum. *Nature Communications* 11:994.

Meekan, M. G., C. J. A. Bradshaw, M. Press, et al. 2006. Population size and structure of whale sharks (*Rhincodon typus*) at Ningaloo Reef, Western Australia. *Marine Ecology Progress Series* 319:275–85.

Meyers, M. M., M. P. Francis, M. Erdmann, et al. 2020. Movement patterns of whale sharks in Cenderawasih Bay, Indonesia, revealed through long-term satellite tagging. *Pacific Conservation Biology*. doi:10.1071/PC19035.

Moazzam, M., and R. Nawaz. 2017. Major bycatch reduction of cetaceans and marine turtles by use of subsurface gillnets in Pakistan. IOTC-2017-WPEB13-19. Indian Ocean Tuna Commission.

Neubauer, P., Y. Richard, and S. Clarke. 2018. Risk to the Indo-Pacific Ocean whale shark population from interactions with Pacific Ocean purse-seine fisheries. WCPFC-SC14-2018/SA-WP-12 (rev. 2). Western and Central Pacific Fisheries Commission.

Norman, B. 2000. *Rhincodon typus*. The IUCN Red List of Threatened Species 2000: e.T19488A8912689.

Norman, B. 2005. *Rhincodon typus*. The IUCN Red List of Threatened Species 2005: e.T19488A8913502.

Norman, B. M., J. A. Holmberg, Z. Arzoumanian, et al. 2017. Undersea constellations: The global biology of an endangered marine megavertebrate further informed through citizen science. *Bioscience* 67:1029–43.

Oddenyo, R. M., E. Mueni, B. Kiilu, et al. 2018. Kenyan shark baseline assessment report for the National Plan of Action for the Conservation and Management of Sharks. IOTC-2019-WPEB15-11. Indian Ocean Tuna Commission.

O'Malley, M. P., K. A. Townsend, P. Hilton, et al. 2017. Characterization of the trade in manta and devil ray gill plates in China and South-East Asia through trader surveys. *Aquatic Conservation: Marine and Freshwater Ecosystems* 27:394–413.

Pajuelo, M., J. Alfaro-Shigueto, M. Romero, et al. 2018. Occurrence and bycatch of juvenile and neonate whale sharks (*Rhincodon typus*) in Peruvian waters. *Pacific Science* 72:463–73.

Parton, K. J., T. S. Galloway, and B. J. Godley. 2019. Global review of shark and ray entanglement in anthropogenic marine debris. *Endangered Species Research* 39:173–90.

Peñaherrera-Palma, C., I. van Putten, Y. V. Karpievitch, et al. 2018. Evaluating abundance trends of iconic species using local ecological knowledge. *Biological Conservation* 225:197–207.

Perry, C. T., E. Clingham, D. H. Webb, et al. 2020. St. Helena: An important reproductive habitat for whale sharks (*Rhincodon typus*) in the south central Atlantic. *Frontiers in Marine Science* 7: 899.

Pierce, S. J., and B. Norman. 2016. *Rhincodon typus*. The IUCN Red List of Threatened Species 2016:e. T19488A2365291.

Pirotta, V., A. Grech, I. D. Jonsen, W. F. Laurance, and R. G. Harcourt. 2019. Consequences of global shipping traffic for marine giants. *Frontiers in Ecology and the Environment* 17:39–47.

Pravin, P. 2000. Whale shark in the Indian coast - Need for conservation. *Current Science* 79:310–15.

Pravin, P., M. P. Ramesan, and K. K. Solanki. 1998. Commercial fishing of whale sharks (*Rhinodon typus* Smith) in Gujarat. *International Symposium on Large Marine Ecosystems*, 312–18.

Prebble, C. E. M., C. A. Rohner, S. J. Pierce, et al. 2018. Limited latitudinal ranging of juvenile whale sharks in the Western Indian Ocean suggests the existence of regional management units. *Marine Ecology Progress Series* 601:167–83.

Qatar Ministry of Environment. 2014. Qatar national biodiversity strategy and action plan 2015–2025. https://www.cbd.int/doc/world/qa/qa-nbsap-v2-en.pdf (accessed November 18, 2020).

Queiroz, N., N. E. Humphries, A. Couto, et al. 2019. Global spatial risk assessment of sharks under the footprint of fisheries. *Nature* 572:461–66.

Ramírez-Macías, D., N. Queiroz, S. J. Pierce, et al. 2017. Oceanic adults, coastal juveniles: Tracking the habitat use of whale sharks off the Pacific coast of Mexico. *PeerJ* 2017:e3271.

Razzaque, S. A., M. M. Khan, U. Shahid, et al. 2020. Safe handling & release for gillnet fisheries for whale shark, manta & devil rays and sea turtles. IOTC-2020-WPEB16-26_Rev1. Indian Ocean Tuna Commission.

Rigby, C. L., N. K. Dulvy, R. Barreto, et al. 2019a. *Sphyrna lewini*. The IUCN Red List of Threatened Species 2019:e.T39385A2918526.

Rigby, C. L., Simpfendorfer, C. A. and A. Cornish. 2019b. A practical guide to effective design and management of MPAs for sharks and rays. Gland: WWF.

Roberts, J. D., and M. Hamann. 2016. Testing a recipe for effective recovery plan design: A marine turtle case study. *Endangered Species Research* 31:147–61.

Robinson, D. P., M. Y. Jaidah, S. Bach, et al. 2016. Population structure, abundance and movement of whale sharks in the Arabian Gulf and the Gulf of Oman. *PLoS One* 11:e0158593.

Robinson, D. P., M. Y. Jaidah, S. Bach, et al. 2017. Some like it hot: Repeat migration and residency of whale sharks within an extreme natural environment. *PLoS One* 12:e0185360.

Robinson, D. P., M. Y. Jaidah, R. W. Jabado, et al. 2013. Whale sharks, *Rhincodon typus*, aggregate around offshore platforms in Qatari waters of the Arabian Gulf to feed on fish spawn. *PLoS One* 8:e58255.

Rohner, C. A., J. E. M. Cochran, E. F. Cagua, et al. 2020. No place like home? High residency and predictable seasonal movement of whale sharks off Tanzania. *Frontiers in Marine Science* 7: 423.

Rohner, C. A., S. J. Pierce, A. D. Marshall, et al. 2013. Trends in sightings and environmental influences on a coastal aggregation of manta rays and whale sharks. *Marine Ecology Progress Series* 482:153–68.

Rohner, C. A., A. J. Richardson, F. R. A. Jaine, et al. 2018. Satellite tagging highlights the importance of productive Mozambican coastal waters to the ecology and conservation of whale sharks. *PeerJ* 2018:e4161.

Roll, U., J. C. Mittermeier, G. I. Diaz, et al. 2016. Using Wikipedia page views to explore the cultural importance of global reptiles. *Biological Conservation* 204:42–50.

Román, M. H., A. Aires-Da-Silva, and N. W. Vogel. 2018. Whale shark interactions with the tuna purse-seine fishery in the Eastern Pacific Ocean: Summary and analysis of available data. BYC-08 INF-A. Inter-American Tropical Tuna Commission.

Ryan, J. P., J. R. Green, E. Espinoza, and A. R. Hearn. 2017. Association of whale sharks (*Rhincodon typus*) with thermo-biological frontal systems of the eastern tropical Pacific. *PLoS One* 12:1–22.

Sánchez, L., Y. Briceño, R. Tavares, et al. 2020. Decline of whale shark deaths documented by citizen scientist network along the Venezuelan Caribbean coast. *Oryx* 54:600–601.

Scales, K. L., P. I. Miller, L. A. Hawkes, et al. 2014. On the front line: Frontal zones as priority at-sea conservation areas for mobile marine vertebrates. *The Journal of Applied Ecology* 51:1575–83.

Schoeman, R. P., C. Patterson-Abrolat, and S. Plön. 2020. A global review of vessel collisions with marine animals. *Frontiers in Marine Science* 7:292.

Sea Shepherd Legal. 2019a. Legislative review and recommendations for implementation of the CMS Concerted Action for the whale shark (*Rhincodon typus*). UNEP/CMS/COP13/Inf.15. Convention on Migratory Species. https://www.cms.int/sites/default/files/document/cms_cop13_inf.15_whale-shark-ca-legislative-review-and-recommendation_e.pdf.

Sea Shepherd Legal. 2019b. Report on the implementation of the Concerted Action for the whale shark (*Rhincodon typus*). UNEP/CMS/COP13/Doc.28.1.7a. Convention on Migratory Species.

Sequeira, A. M. M., G. C. Hays, D. W. Sims, et al. 2019. Overhauling ocean spatial planning to improve marine megafauna conservation. *Frontiers in Marine Science* 6. doi:10.3389/fmars.2019.00639.

Sequeira, A. M. M., C. Mellin, D. Rowat, et al. 2012. Ocean-scale prediction of whale shark distribution. *Diversity and Distributions* 18:504–18.

Sequeira, A. M. M., C. Mellin, D. A. Fordham, et al. 2014. Predicting current and future global distributions of whale sharks. *Global Change Biology* 20:778–89.

Sequeira, A. M. M., M. Thums, K. Brooks, and M. G. Meekan. 2016. Error and bias in size estimates of whale sharks: Implications for understanding demography. *Royal Society Open Science* 3:150668.

Shahid, U., M. M. Khan, R. Nawaz, et al. 2015. A preliminary assessment of shark bycatch in tuna gillnet fisheries of Pakistan (Arabian Sea). IOTC-2015-WPEB11-46. Indian Ocean Tuna Commission.

Silas, E. G. 1986. The whale shark (*Rhiniodon typus*) in Indian coastal waters: Is the species endangered or vulnerable? *Marine Fisheries Information Service Technical and Extension Series* 66.

Smith, R. J., D. Veríssimo, N. J. B. Isaac, and K. E. Jones. 2012. Identifying Cinderella species: Uncovering mammals with conservation flagship appeal. *Conservation Letters* 5:205–12.

SPC-OFP. 2010. Summary information on whale shark and cetacean interactions in the tropical WCPFC purse seine fishery. Western and Central Pacific Fisheries Commission. https://www.wcpfc.int/system/files/WCPFC-2011-IP-01-Rev-1-Whale-shark-cetacean-interactions-(18-Jan-2012).pdf.

Speed, C. W., M. G. Meekan, D. Rowat, et al. 2008. Scarring patterns and relative mortality rates of Indian Ocean whale sharks. *Journal of Fish Biology* 72:1488–1503.

Steinke, D., A. M. Bernard, R. L. Horn, et al. 2017. DNA analysis of traded shark fins and mobulid gill plates reveals a high proportion of species of conservation concern. *Scientific Reports* 9505: 1–6.

Stephenson, P. J. 2019. The Holy Grail of biodiversity conservation management: Monitoring impact in projects and project portfolios. *Perspectives in Ecology and Conservation* 17:182–92.

Stephenson, P. J., C. Workman, M. K. Grace, and B. Long. 2020. Testing the IUCN Green List of Species. *Oryx* 54:10–11.

Temple, A. J., N. Wambiji, C. N. S. Poonian, et al. 2019. Marine megafauna catch in southwestern Indian Ocean small-scale fisheries from landings data. *Biological Conservation* 230:113–21.

UNESCO. 2020. World Heritage List. http://whc.unesco.org/en/list/ (accessed November 18, 2020).

United Nations. 2020a. United Nations Convention on the Law of the Sea of 10 December 1982: Overview and full text. https://www.un.org/Depts/los/convention_agreements/convention_overview_convention.htm (accessed November 24, 2020).

United Nations. 2020b. The United Nations Agreement for the Implementation of the Provisions of the United Nations Convention on the Law of the Sea of 10 December 1982 relating to the Conservation and Management of Straddling Fish Stocks and Highly Migratory Fish Stocks (in force as from 11 December 2001): Overview. https://www.un.org/Depts/los/convention_agreements/convention_overview_fish_stocks.htm (accessed November 24, 2020).

Veríssimo, D., H. A. Campbell, S. Tollington, et al. 2018. Why do people donate to conservation? Insights from a 'real world' campaign. *PLoS One* 13:e0191888.

Veríssimo, D., D. C. MacMillan, and R. J. Smith. 2011. Toward a systematic approach for identifying conservation flagships. *Conservation Letters* 4:1–8.

WCS. 2020. Wildlife. https://www.wcs.org/our-work/wildlife (accessed November 18, 2020).

White, E. R., M. C. Myers, J. M. Flemming, and J. K. Baum. 2015. Shifting elasmobranch community assemblage at Cocos Island – an isolated marine protected area. *Conservation Biology* 29:1186–97.

Wilcox, C., N. J. Mallos, G. H. Leonard, et al. 2016. Using expert elicitation to estimate the impacts of plastic pollution on marine wildlife. *Marine Policy* 65:107–14.

WildLifeRisk. 2014. Planet's biggest slaughter of whale sharks exposed in PuQi, Zhejiang Province, China. https://web.archive.org/web/20160503104157/http://wildliferisk.org/press-release/ChinaWhaleSharks-WLR-Report-ENG.pdf (accessed November 18, 2020).

Wu, J. 2016. *Shark Fin and Mobulid Ray Gill Plate Trade in Mainland China, Hong Kong and Taiwan.* Cambridge: TRAFFIC.

WWF. 2020a. Know your flagship, keystone, priority and indicator species. https://wwf.panda.org/our_work/our_focus/wildlife_practice/flagship_keystone_indicator_definition/ (accessed November 17, 2020).

WWF. 2020b. Priority species. https://wwf.panda.org/knowledge_hub/endangered_species/ (accessed November 18, 2020).

Wyatt, A. S. J., R. Matsumoto, Y. Chikaraishi, et al. 2019. Enhancing insights into foraging specialization in the world's largest fish using a multi-tissue, multi-isotope approach. *Ecological Monographs* 89:e01339.

Yagishita, N., S.-I. Ikeguchi, and R. Matsumoto. 2020. Re-estimation of genetic population structure and demographic history of the whale shark (*Rhincodon typus*) with additional Japanese samples, inferred from mitochondrial DNA sequences. *Pacific Science* 74:31–47.

Zacharias, M. A., and J. C. Roff. 2001. Use of focal species in marine conservation and management: A review and critique. *Aquatic Conservation: Marine and Freshwater Ecosystems* 11:59–76.

CHAPTER 13

Outstanding Questions in Whale Shark Research and Conservation

David Rowat
Marine Conservation Seychelles (ret.)

David P. Robinson
Sundive Research

Alistair D. M. Dove
Georgia Aquarium

Gonzalo Araujo
Large Marine Vertebrates (LAMAVE) Research Institute Philippines

Tonya Clauss and Christopher Coco
Georgia Aquarium

Philip Dearden
University of Victoria

Molly K. Grace
University of Oxford

Jonathan R. Green
Galapagos Whale Shark Project

Alex R. Hearn
Universidad San Francisco de Quito

Jason Holmberg
Wild Me (Wildbook)

Lisa Hoopes
Georgia Aquarium

Rui Matsumoto
Okinawa Churaumi Aquarium

Craig R. McClain
Louisiana Universities Marine Sciences Consortium

Mark G. Meekan
Australian Institute of Marine Science

Kiyomi Murakumo
Okinawa Churaumi Aquarium

Bradley M. Norman
ECOCEAN, Inc

Ryo Nozu
Okinawa Churaumi Aquarium

Sebástian A. Pardo
Dalhousie University

Emily E. Peele
University of North Carolina Wilmington

César Peñaherrera-Palma
Pontificia Universidad Católica del

Cameron Perry
Georgia Institute of Technology
and
Maldives Whale Shark Research Project

Clare E. M. Prebble
Marine Megafauna Foundation

Samantha Reynolds
The University of Queensland

Marlon Román
Inter-American Tropical Tuna Commission

Christoph A. Rohner
Marine Megafauna Foundation

Keiichi Sato
Okinawa Churaumi Aquarium

Jennifer V. Schmidt
Shark Research Institute

Christian Schreiber
Georgia Aquarium

Ana M. M. Sequeira
University of Western Australia

Freya Womersley
Marine Biological Association, UK

Makio Yanagisawa
Marine Palace Co. Ltd.

Kara E. Yopak
University of North Carolina Wilmington

Jackie Ziegler
University of Victoria

Simon J. Pierce
Marine Megafauna Foundation

CONTENTS

13.1 INTRODUCTION

In a recent review article on "The Origins and Rise of Shark Biology in the 20th Century," Jose Castro argued that the public perception of sharks has changed from being dangerous fish to being "totemic animals" and that, consequently, the study of shark biology has morphed into shark advocacy (Castro 2016). Whale sharks are bona fide members of the "charismatic megafauna" club (see Chapters 1 and 12). The same article pointed out that "the logistical difficulties of carrying out research on sharks are just as prevalent now as they have been in the past" and that much of the current research effort is targeted at only a few species such as the white shark *Carcharodon carcharias* and the whale shark. He concluded that, even after more than 30 years of research, knowledge concerning the whale shark is still fragmentary and lacks some essential information (Castro 2016). Notwithstanding the contents of this book, this is still largely true today.

The increase in global tourism since the 1980s, and particularly the popularization and growth of snorkeling, scuba diving, and digital underwater photography, has helped enable both a huge increase in the amount of scientific work done on whale sharks and also the contribution of "citizen scientists" to these efforts. This heralded the rise of the "modern" phase of whale shark research, particularly following Geoff Taylor's discovery of the whale shark constellation off Ningaloo Reef in Western Australia (Taylor 2019). A search on the global scientific literature database, Web of Science, for *Rhincodon typus* in December 2020 (using the Topic field) returned 391 results in the era of digital abstracting databases. While these will not all be unique peer-reviewed publications, it emphasizes the substantial body of research that has considered the species.

For such a well-known animal, a lot of mysteries still surround whale sharks. Through this book, we have summarized the state of knowledge on a range of topics, but each chapter also raises new questions on the biology and ecology of this iconic species. We invited all the chapter authors to contribute their most important question for the "next phase" of whale shark research, and outline those here. We hope that these will provide some direction and inspiration to those interested in the research and conservation of this awesome species.

13.2 WHAT IS THE GENOMIC BASIS OF GIGANTISM IN WHALE SHARKS, AND HOW DOES THAT RELATE TO PLANKTIVORY AND ENDOTHERMY?

The whale shark is likely the largest fish to have ever lived on this planet; longer even than the enigmatic and extinct filter-feeding teleost *Leedsichthys problematicus* and the Miocene mega-predatory shark popularly called "Megalodon" (Chapter 1). How the whale shark came to be such a large, pelagic filter feeder when its nearest relatives are demersal, predatory carpet sharks of unremarkable size is something of a mystery, although the slow metabolism of orectolobiforms may have aided them in their transition to planktivory in the nutrient poor tropical pelagic zone.

In considering why the whale shark isn't even larger, reaching the size of the great baleen whales, we speculated that there may be biomechanical limits on the bulk that can be supported by a cartilaginous skeleton, as there are with even the robust bony skeletons of mysticeti whales (Potvin et al. 2012). Or, that low food availability in the tropical pelagic zone may place an upper bound on size, past which it is not possible to harvest enough food energy to sustain larger body sizes (Goldbogen et al. 2011). Or, it could be both of these factors acting in concert, or there may be another explanation not yet identified.

In exploring these questions, science now has access to a potentially useful tool: the complete and annotated whale shark genome (Alam et al. 2014; Read et al. 2018). The whale shark is the first elasmobranch species to have that information available, although new shark genomes are now being published quite regularly. By comparing the whale shark genome with those of related species of demersal carpet sharks, we may be able to identify key genetic domains under positive selection that have allowed for such a divergent phenotype to emerge. Comparisons with other gigantic species across a wide range of marine and terrestrial taxa may identify a common genomic basis or bases that allows these species to so profoundly depart from the typical phenotype in their respective groups.

Two traits correlate with gigantism across a range of unrelated taxa in the ocean. First, gigantism in marine vertebrates is clearly correlated with planktivorous habits (Pimiento et al. 2019). The presence of the gigantism trait may therefore correlate with gene families that are associated with foraging for and eating a planktivorous diet. Second, several large pelagic sharks, fishes, and at least one turtle also show regional or partial endothermy, which has been facilitated by the evolution of several different adaptations in different lineages. There is no evidence yet that whale sharks have specific anatomical adaptations to partial endothermy, but they may display simple gigantothermy, in which metabolic heat or the onboarding of solar heat is retained by a favorable surface-area-to-volume ratio (Meekan et al. 2015). This likely allows them to act as effectively warm-blooded when they forage in cold waters at depth, until they lose body heat to the surrounding water and must return to the surface to warm up (Thums et al. 2013). So far, the direct evidence suggests that they have at very least great thermal stability due to their enormous size (Nakamura et al. 2020). Further exploration of how body size, body temperature regime, metabolic rate, and planktivory interplay, and are determined by genetics and environment, may provide new insights into the life of not only whale sharks, but many other giants in the pelagic zone.

13.3 HOW OFTEN DO WHALE SHARKS BREED?

Of all the unknown aspects of whale shark biology, questions about their reproduction are some of the most basic and long-standing. Almost 200 years after whale sharks were first described, we still have only a single record of a pregnant whale shark. That singular female, the "megamamma" shark harpooned off Taiwan in 1995 (Joung et al. 1996), contained around 300 fetal pups at various states of development. This huge litter is more than twice that reported from any other shark, so whale sharks appear to be divergent when it comes to reproduction, as they are with so many other aspects of their biology. We still don't know if this is the norm in whale sharks, but assuming it is, female sharks are clearly investing substantial energy in the potential next generation. That must incur a significant cost to the female, requiring more food energy and therefore more foraging time, which leads to our outstanding question: how often do whale sharks breed?

We investigated two different scenarios for this question while building a demographic model for whale sharks in Chapter 2. First, that adult females might breed every 2 years on average, perhaps introducing a resting year between cycles, similar to their relative the nurse shark *Ginglymostoma cirratum* (Castro 2000). Based on the little we know of whale shark reproductive biology, a biennial cycle is the shortest likely interval. However, sharks that have large ovarian follicles, as whale sharks do, often have a longer reproductive cycle; 3 years or more in the case of some other orectolobiform sharks, the wobbegongs *Orectolobus* spp. (Huveneers et al. 2007). With that in mind, the second scenario we introduced, as a probable upper bound, is that female whale sharks might have a 4–5-year reproductive cycle. That longer interval would give the sharks more time to regain body condition before their next pregnancy.

Which is closer to the truth, 2 years or 4–5? We do not know, but the answer has implications for both their ecology and conservation. If the reality is closer to 4 or 5 years, then whale sharks would have one of the "slowest" life histories of any shark, which emphasizes their susceptibility to human threats. If the sharks do swim to specific areas to mate and to give birth to their pups, as hypothesized in Chapter 7, the periodicity of such a migration would also have a major influence on the movements and habitat choices of adult female whale sharks.

In the absence of an opportunistic catch or stranding of another pregnant female, or successful captive mating and birth, this seemed like an intractable research question until recently. Now, however, the use of emerging research methods provides opportunities to study whale shark reproduction in the field. Waterproofed ultrasound units – which were in the realm of science fiction just a few years ago – now allow study of the internal development of reproductive organs (Figure 13.1), and the ability to collect blood samples from free-swimming sharks can allow their reproductive hormone levels to be evaluated. Answers to this enduring mystery may finally be forthcoming.

13.4 HOW DOES THE WHALE SHARK SENSE ITS ENVIRONMENT AND HOW ARE THESE SIGNALS PROCESSED IN THE BRAIN?

Our understanding of the neurobiology of whale sharks is in its infancy (Chapter 3). We know their brain is not fundamentally different in structure from that of other sharks, but we do not understand how signals are integrated there to elicit the appropriate behavioral responses to external stimuli. We know more about their sensory capabilities, but only slightly. Evidence suggests that they see without color but are likely to have very sensitive vision at depth because they have very blue-shifted visual pigment sensitivity (Hara et al. 2018). Their hearing capabilities are unknown. It

Figure 13.1   Rui Matsumoto assessing a large female whale shark with a waterproofed ultrasound unit in the Galapagos Islands. (Photo: Simon Pierce.)

remains possible that sound could play a part in their foraging behavior, so a better understanding of their frequency-specific sensitivity would be of great interest. This information can be obtained by evoked auditory potential studies, which measure brain activity in response to aural stimuli. This information would also find application in determining how whale sharks are affected by human activity. Toothed and baleen whales are profoundly affected by ocean noise from shipping, seismic surveys, and other man-made sources of sound, and it would be helpful for managers to understand how sound pollution affects whale sharks too. This is especially important because it may determine their ability to avoid boat and ship strike, which has been identified as one of the top two threats to whale shark populations globally (Chapter 11). Continuous exposure to anthropogenic noise may also act as a chronic stressor (Ito et al. 2017), which can suppress immune function and increase susceptibility to disease.

Among the most interesting questions about whale sharks is what role chemoreception (senses of smell and taste) plays in the foraging behavior of whale sharks. The importance of olfaction in large predatory sharks is widely understood by the general public – everyone has heard the apocryphal stories of sharks detecting a drop of blood from a mile away – but what role, and over what range, does scent play in the ability of whale sharks to find planktonic sources of food? While data are beginning to be gathered on this topic (e.g., Dove 2015), the truth is that we still do not really know.

With only a couple of brains available for examination so far and some extreme logistic challenges in conducting controlled experiments with this species, we expect that answers to these sorts of questions will come slowly. Aquarium animals may offer some opportunities for limited explorations of sensory biology (see Chapter 9 or, for example, Dove 2015), and it may be possible to do experiments with different scent compounds or sound sources to measure biologically relevant responses in the field, too. These sorts of studies will be difficult and hard to control (in the experimental design sense), but the benefits would be manifold.

13.5 WHAT IS THE NATURE OF THE RELATIONSHIPS OF DIFFERENT SPECIES OF MICROBES, PARASITES, AND FISHES THAT ASSOCIATE WITH WHALE SHARKS?

A variety of species associate with whale sharks in varying degrees of intimacy, from obligate parasites, to commensals, potential pathogens, and a number of other, more nebulous facultative relationships. Undoubtedly, parasites and pathogens have been drastically understudied in whale sharks. It is hard to redress this situation because sampling for these organisms is typically lethal to the host and thus presents a vexing ethical dilemma. Emerging eDNA methods may allow for some exploration of intestinal parasites from DNA traces in feces, but detailed morphological taxonomy will have to wait until a suitable, freshly dead, wild whale shark can be examined using the appropriate techniques. In the same way that much of what we know about whale shark reproduction derives from that one "megamamma supreme" in the Joung et al. (1996) paper, we foresee a similar individual necropsy in the future may cause a quantum leap forward in our understanding of parasites and pathogens infecting this species.

Included among the other "associates" of whale sharks are familiar and reliable species of fish, such as the remora *Remora remora*, live sharksucker *Echeneis naucrates*, and pilot fish, as well as more transient and nebulous associates like cobia *Rachycentron canadum* and various species of "baitfish." Many of these seem to use whale sharks for shelter, for energy efficient transport, or just as something to scratch themselves on. In most of these cases, the precise nature of the relationship needs much more exploration. A particularly good example is the remoras, which in our opinion do not remove parasites and are therefore not mutualists; they seem to use whale sharks for efficient long-distance transport in the pelagic zone and as a source of food via coprophagy (i.e. eating their feces).

Perhaps the most enigmatic of relationships is that between whale sharks and several species of tuna. In different contexts, whale sharks may prey on tuna eggs, tuna may prey on baitfish that seek shelter around whale sharks, or both whale sharks and tuna may feed cooperatively on the same patches of prey. Tunas and whale sharks often co-occur all around the world and can often be observed following each other; sometimes the pursuer, other times the pursued. Certainly, fishers have learned to exploit this relationship to increase tuna catches, sometimes to the detriment of whale sharks ensnared as bycatch (Chapter 12). These observations hint that "there is something important going on" that deserves some scientific attention. A better understanding of the dynamics of the whale shark/tuna dyad is a valuable goal because it has implications for deliberate or incidental mortality of whale sharks in tuna fisheries.

13.6 HOW CAN WE APPLY GENETIC AND GENOMIC METHODS TO DEFINITIVELY IDENTIFY POPULATION STRUCTURE AT A GLOBAL SCALE?

How genetic information transfer (i.e. gene flow) is mediated across the whale shark's broad distribution remains unclear. Five studies have shown that, at a genetic level, whale sharks across the Indo-Pacific are highly similar to one another and differentiated from Atlantic Ocean animals (Castro et al. 2007; Schmidt et al. 2009; Vignaud et al. 2014; Meekan et al. 2017; Yagishita et al. 2020). This may imply that at least some individuals have swum across and even between oceans. Long-term satellite-tracking (Chapter 6 and Section 13.7) and collaborative international photo-matching of individual sharks will help to identify what may be a small number of animals making long-range movements. Chapter 5 proposes that a critical time for inter-ocean movements might be upon reaching maturity, when sharks appear to leave many coastal feeding constellations (see also Chapter 7). Alternatively, it is also possible that sequential gene transfer may take place, as (hypothetical) regional populations interbreed at their edges. This scenario would imply a gradient of genetic differentiation across the species' range, which can be tested by including data from unstudied areas. Some studies have shown subtle differentiation between the eastern and western Indian Ocean, for instance (Vignaud et al. 2014). High-resolution markers from genomic studies may serve to illuminate any regional differences that exist. Similarly, sequential gene flow from the Atlantic to the Indian Ocean should move along the western coast of Africa, a region unstudied at the genetic level but known to host at least one globally significant constellation off Gabon (Escalle et al. 2016).

It is also important to understand the significance of the seemingly reduced genetic diversity identified in certain whale shark populations. While whale sharks globally show evidence of strong genetic diversity, Western Australian animals, as well as those in the Atlantic Ocean, have been reported to have reduced levels of genetic diversity (Vignaud et al. 2014). Tan et al. (2021) showed low levels of heterozygosity in the Taiwanese animal that was used for the whale shark genome sequencing effort described by Read et al. (2017), a finding that can be an indicator in individuals of low population genetic diversity. While low diversity does not necessarily lead to future problems for population-level health and survival, it can reduce the ability of these sharks to adapt to changing conditions. Human-induced population depletion can compound this issue by reducing the probability of migrants with new alleles. It is possible that a small number of historic founder animals may have given rise to the larger Atlantic population, providing an evolutionary explanation for the relatively low diversity in this large geographical region. Reduced diversity is more difficult to explain for the Australian whale sharks, who appear to be genetically part of the vast Indo-Pacific population (Vignaud et al. 2014). Increased sampling, and more robust genomic markers, will help to illustrate the magnitude and biogeography of these effects.

As we explore further in Section 13.8, habitat use by demographic groups other than juvenile and subadult male whale sharks is still relatively unknown. Studies on young female sharks, adult females and males, and newborn animals, were for many years limited by a lack of information on where these animals could be found. Two areas where large female whale sharks can be observed, in the Gulf of California and the Galapagos Islands, respectively, are now the topic of relatively long-term and ongoing studies (Ramírez-Macías et al 2012, 2017; Hearn et al. 2013, 2016; Ketchum et al 2013; Acuña-Marrero et al 2014; Ryan et al. 2017), although individual sharks are transient to both locations. Recently, progress has been made in identifying regions where both male and female adults may be found, such as Saint Peter and Saint Paul Archipelago, St. Helena Island, and the Arabian Gulf (Macena and Hazin 2016; Robinson et al. 2016; Perry et al. 2020). However, the habitat use of newborn whale sharks is still unknown. Individuals measuring less than 2 m in length are rarely seen even in constellations dominated by small 3–5 m juveniles, such as in Djibouti, the Red Sea, and Palawan in the Philippines (Rowat et al 2011; Cochran et al 2016; Araujo et al 2019). Recently -developed kinship genetic approaches may help in understanding newborn behavior patterns (Berger-Wolf et al 2007; Feutry et al 2017). Kinship studies on sharks at the constellations dominated by small juveniles will show if these individuals are more highly related than would be expected by chance, or than those present at other sites biased towards larger juveniles. Determining whether neonates from individual whale shark litters remain together as they move into juvenile foraging habitats, or whether regional breeding populations are likely to exist, would help illuminate the earliest life history of the species.

13.7 HOW CAN WE OBTAIN LONGER AND MORE DETAILED MOVEMENT TRACKS?

Since we cannot physically follow whale sharks on their oceanic journeys, which are only partially near the surface, attaching tags to the sharks to record and transmit their location, depth, and even aspects like acceleration is our best bet for tracking their movements. The longer we can keep these tags attached, the better information we get. That brings us to our major limitation: tag retention. These expensive devices fall off the sharks far too quickly.

Tags detach early for many and varied reasons. Predators attack the tags, the insertion point for tethered tags may not hold, and a unique problem with whale sharks is that they can, and do, dive deeper than the maximum depth the tag's inbuilt floatation can withstand (currently ~1,700 m). Other issues, such as hardware failure, can also occur. Fouling can interrupt the wet–dry sensor switch, leading the tag to conclude that it is constantly submerged and therefore stops transmitting. Anti-fouling paints have had some success in delaying, but not halting, this process. The result has been few longer tracks to work with, i.e. those exceeding 6 months. That makes it hard to assess seasonal movements, let alone habitat preferences, which likely vary between the sexes and change as the sharks mature. Longer tracks will be needed to answer important questions: Do whale sharks visit constellation sites yearly or at a longer interval? Where do they go in between? How do movement patterns and environmental preferences vary between demographic groups, such as females and males, and juveniles and adults in both? Where do transient whale sharks, which only occasionally visit constellations, spend the rest of their time? How deep do whale sharks actually dive, and what do they do at such great depths? Improved tag retention will allow us to answer some of these questions, but new methods also have to pass high ethical standards. After all, we track whale sharks to understand their lives and ultimately to improve their conservation status, and the last thing we want is to negatively impact their health.

There are several promising ways to improve overall tag retention. Attaching a near-real-time tag directly to the first dorsal fin, as has been done previously in other shark species (Hammerschlag et al. 2011), has enabled tracks exceeding a year in Indonesian whale sharks (Meyers et al. 2020). A further benefit of this fin-mounting technique is the low rates of injury or scarring observed on the tagged sharks; the solid attachment resists drag, which is a major consideration in a highly mobile species like the whale shark. The main challenge is the attachment, as it takes time to bolt the tags to the fin. This was overcome in Indonesia due to the sharks being routinely encircled in nets, where they can be quasi-restrained before release (Meyers et al. 2020) with minimal stress indicated by the concentration of stress indicators in blood (A. Dove, unpublished data.). However, it is also possible to attach tags to free-swimming sharks in the same manner when they are focused on feeding (R. de la Parra pers. comm.) – whale sharks definitely appreciate good plankton. A variation on this technique is to use fin-clamps, which can also increase retention rates over tethered tags (Norman et al. 2016). Clamps are easy and near instantaneous to deploy on free-swimming sharks, but care needs to be taken in their positioning as the clamp can cause injury if drag pushes it into the leading edge of the fin. A vertical position at the top of the fin (as shown in Figure 6.6) is likely to minimize the chance of tissue damage through drag, but re-sightings of tagged sharks will be needed to assess this over long timeframes. As there is no value to the tag's attachment outlasting its battery life, incorporating a programmable detachment time on fin-mounted tags will be an important consideration to ensure animal welfare.

A multitude of sensors are now available to complement the "traditional" data obtained by satellite-linked tags, i.e. location (or light-levels from which to derive location), depth (pressure), and water temperature. The inclusion of additional sensors, such as cameras, accelerometers, and gyroscopes, can provide novel data to better interpret the behavior of whale sharks during tracks (Gleiss et al. 2009). However, the increased data storage associated with these sensors means that the current bandwidth of satellite transmission is insufficient, so these tags currently have to be physically recovered to retrieve the data therein. Tag battery life also limits the duration of recording in these high-resolution sensors to just a few hours in some cases. While still useful, this obviously limits their use to document less common behaviors and those that may occur in less accessible habitats. Technological advancements, such as solar panels on tags to enhance battery life, or faster data transmission through "satellite internet," are likely to enable new possibilities and discoveries in the near future. Such advances will greatly increase our understanding of whale shark behavior away from coastal constellations, over longer migrations, and across their lives, especially if we can simultaneously solve the problem of tag retention.

13.8 WHAT DRIVES SEXUAL SEGREGATION IN CONSTELLATIONS?

Many of the preceding chapters have noted that our knowledge of whale shark ecology is highly biased towards juvenile and subadult males. This is a result of most constellations, whether located close to the coast or around offshore islands, being dominated by male whale sharks (more than

⅔ of the identified sharks in many locations; see Figure 7.2) between 3 and 9 m in length. At this point, a "predominantly male coastal feeding constellation" is practically a whale shark cliché. We have no field data from smaller juveniles or pups, but this sexual segregation clearly starts early in a whale shark's life; even the smallest size classes at most constellations (often 3–4 m) are biased towards males (Chapter 7). Often this discrepancy becomes even more pronounced among larger juvenile size-classes. But *why*? Female whale sharks of equivalent size will presumably have the same physical characteristics, i.e. in their feeding biomechanics and swimming ability, so why do they not frequent the same areas as the males? Particularly in locations where an energy-rich and predictable food source is available to them?

We discuss a few hypotheses in Chapter 7. If the predation risk to whale sharks is higher in coastal areas, females may choose a less "risky" offshore habitat, where they have a slower growth rate but a higher chance of survival (Meekan et al. 2020). Male whale sharks, on the other fin, maximize their odds for reproductive success by growing faster and attaining adulthood earlier, whereupon they can seek mating opportunities. Interestingly, one of the few exceptions to this male-dominated population structure is at Shib Habil reef in the Saudi Arabian Red Sea, where the same numbers of females and males are seen (Cochran et al. 2016). Some known and probable whale shark predators, including killer whales *Orcinus orca* and great white sharks *Carcharodon carcharias*, are rare in this area (Hardenstine et al. in review), which could hypothetically relax the predation pressure on this population. Further assessment of the different growth rates between sexes, and modeling the survival probability of males and females, is needed to test this idea among constellations.

Intertwined with the above explanation is the possibility that immature females are harassed by adult male whale sharks in constellations, to the extent that females prefer to avoid these areas. Such harassment has been occasionally observed in whale sharks, in both aquaria (Chapter 2) and the field (Chapter 7), and sexual conflict has been implicated as a driver of segregation in multiple other shark species (Mucientes et al. 2009; Jacoby et al. 2010; Wearmouth et al. 2012; Vandeperre et al. 2014). In whale sharks, however, a pronounced male bias is also present in constellations where mature males have not been observed over multiple years of research, such as Nosy Be in Madagascar (S. Diamant pers. comm.), the Gulf of Tadjoura in Djibouti (Rowat et al. 2011), and the Maldives (Perry et al. 2018).

Differing environmental preferences, such as juvenile females preferring cooler water, is another potential explanation based on data from other shark species (Vandeperre et al. 2014). Ultimately, in whale sharks, the resolution of this interesting ecological question will require more "big data" collaborative work across constellations, and longer-term tracking studies to compare and contrast fine-scale habitat preferences over time in both sexes. Advances in the biochemical assessment of long-term diet and habitat use (see Chapter 8) are also promising methods to obtain a population-level dataset that can complement such tracking results. Answering the question of why female and male whale sharks seem to prefer such different lives will be a significant step towards understanding whale shark movements and, most critically, helping the species to recover from human-induced declines (see Section 13.13).

13.9 WHAT DO NEONATES, ADULTS, FEMALES, AND TRANSIENT WHALE SHARKS FEED ON?

As whale sharks segregate by size and sex, our understanding of their feeding ecology is heavily biased towards the 3–9 m immature males that frequent coastal areas with high prey densities. Furthermore, we only see some of these juvenile males once, as constellations are generally comprised of some reliable "returning" sharks among many others that are transient at these areas. In addition to these transient animals, other demographic groups, such as neonates and small juveniles, large and mature sharks, and females are poorly known. To complete the picture, we need to examine their feeding ecology as well. Ontogenetic diet changes are common in predatory sharks: larger individuals can catch larger prey. For a filter feeder, it is not quite as simple, but it is possible that, for example, large whale sharks have a stronger suction feeding capacity, enabling them to catch larger, more mobile prey that could outswim the suction vortex of a smaller shark. Apart from prey type, we already know that feeding locations differ with size. Smaller individuals have to weigh up the higher risk of predation at inshore feeding sites versus the benefit of potentially more rewarding feeding opportunities. Mature adults may be able to access deeper water prey at cooler water temperatures, as well as the need to balance breeding opportunities with feeding. As such, direct observations of feeding might be difficult to achieve for most of the demographic groups of whale sharks because it likely happens in oceanic habitats, at depth or at night. Other indirect methods are therefore our best bet. First, biochemical analyses of stable isotopes and fatty acids hold some promise. The main benefit of these techniques is that they require only a small tissue or blood sample that can be collected from animals that are only seen once. For example, we can collect these samples from large, mature females off Darwin Island in the Galapagos, or from mature males in Qatar. Opportunistic samples from neonates might provide some information on the diet and habitat of very small whale sharks. Returning and transient individuals, determined retrospectively through photo-ID records or similar, can both be sampled to compare these groups. The main caveat with using biochemical methods is a glaring lack of understanding of basic biological processes in sharks, and whale sharks in particular, such as tissue turn-over rates, how dietary fatty acids are routed to the different tissues, or how periods of fasting affect their biochemistry. Only once this is improved can we exclude other, non-dietary influences on the stable isotope and fatty acid profiles in whale shark tissue. There are several ways to learn more about biochemistry in whale sharks. Strandings of whale sharks do occur and provide the best opportunity to collect samples of other tissues, such as liver and muscle, that are more indicative of diet than subdermal tissue. Strandings also allow the collection of vertebrae that accrete over time. The stable isotope profiles of different sections of a vertebra can reveal where the shark has fed at different stages of its life, including the developmental stage reflecting their mothers' movements and dietary preferences. The use of compound-specific stable isotope analysis will also go some way to separate the influence of diet and location. Application of this method in multi-tissue studies could deepen and clarify our understanding of whale shark feeding behavior over multiple time scales. New tags that can record a shark's three-dimensional movements and acceleration also allow for a detailed reconstruction of the shark's foraging and feeding behaviors. The main limitation with this technique is that physical recovery of the tag has been necessary to collect those data, rather than a transfer via satellite from a floating tag. Continued cost reduction and technological improvements, such as an increase in satellite data transmission bandwidth, could enable longer-term collection and remote transfer of these detailed data, which would increase our understanding of whale shark feeding behavior away from coastal feeding sites.

13.10 HOW CAN WE TAKE ADVANTAGE OF WHALE SHARKS IN AQUARIUMS TO FILL KNOWLEDGE GAPS?

The handful of public aquariums in the world that are able to house whale sharks have had to overcome a number of significant challenges in order to do so successfully (Dove et al. 2010a, Matsumoto et al. 2017, Schreiber and Coco 2017). Along the way, they have provided unique research opportunities that complement those available in the field, particularly longitudinal studies of growth, maturation, diet, energetics and behavior, and their dynamics in individuals over prolonged periods. They have also produced a number of methods in husbandry or veterinary care that then find applications in field research, blood draw being the most obvious example (Dove et al. 2010b). There are obvious limitations on manipulative research with priceless aquarium specimens, and caveats on interpreting the findings, but scientists should nonetheless consider what questions they may be able to answer by collaboration with accredited facilities that have whale sharks in their collection, because the potential benefits are many. Sensory ecology is one example of a field where aquariums may offer good opportunities (Dove 2015). But perhaps the best opportunities lie in reproductive biology (Matsumoto et al. 2019); while whale sharks have not yet bred in any aquarium, it seems likely to occur in the not too distant future, perhaps even before mating or pupping is observed in the field. This would provide a bonanza of life history information to fill in some of the more elusive gaps in our understanding of the whale shark life cycle, such as those described above in Section 13.3.

13.11 HOW CAN WE ENSURE THAT WHALE SHARK TOURISM DEVELOPS SUSTAINABLY?

Regular commercial whale shark tourism began at Ningaloo Reef in Western Australia and the Maldives in the 1980s and, particularly at Ningaloo, has been intensively and successfully managed to ensure the sustainability and longevity of the industry. Ningaloo whale shark tourism is typically regarded as setting international best practice. As we describe in Chapter 10, over 46 sites have now developed their own whale shark tourism industry, with substantial variation in standards of operation and interpretation. Although some aspects of the Ningaloo management measures may be difficult to replicate in less-resourced locations, a common theme is that engagement with the tourism industry and regulators by researchers has generally led to improved standards in practice, whether these be legally enforced or voluntarily applied. How, then, can we ensure that this limited expertise is equitably available to all operators?

Although almost every whale shark tourism site has distinct and unique operational considerations, a standard template for best practice, based on experimentation and experience (Table 10.6), has emerged across much of the world and is simple enough to be broadly applicable. The key limitation, then, is the distribution and awareness of this framework. One of the recommendations of the Concerted Action document produced in conjunction with the 2017 listing of the whale shark on Appendix I of the Convention on Migratory Species was to produce unified tourism guidelines. Such an initiative could be web-based, along the same lines as the Manta Trust's "How to swim with manta rays" website (https://swimwithmantas.org/), or the Green Fins Initiative for sustainable scuba diving (Hunt et al. 2013; https://greenfins.net/). Resources can then be made available to assist in both operator training and to provide interpretive materials for tourist education. This can then support the dual goals of building knowledge in new entrants to the industry, while maximizing the adoption of pro-conservation behaviors in tourists following their experience.

13.12 HOW PROBLEMATIC IS PLASTIC POLLUTION FOR WHALE SHARKS?

Marine pollution was not ranked as one of the primary threats to whale sharks during our prioritization exercise in Chapter 11. We are only aware of five documented records of whale shark mortalities associated with plastic ingestion: two from Japan (Matsumoto et al. 2017; R. Matsumoto unpubl. data), discussed below, and one each from Malaysia (Lee 2019, Sen Nathan pers. comm.), the Philippines (Abreo et al. 2019), and Thailand (Haetrakul et al. 2009), as well as an additional possible case in Brazil (Sampaio et al. 2018). These cases were all identified during necropsies to

Figure 13.2 A whale shark feeding near a plastic bag off Djibouti. (Photo: Thomas Peschak.)

identify the cause of death in stranded sharks or, in Japan, two sharks that died in aquaria from plastic ingested prior to capture. However, and despite the lack of data, we have chosen to highlight plastic pollution, and its effects on whale sharks, as an important research topic going forward. Why? Because surface filter-feeders are uniquely vulnerable to this emerging threat (Figure 13.2), which has the potential to affect whale sharks across much of their range (Germanov et al. 2018), especially since the global focus of marine plastic pollution (East and Southeast Asia, see Jambeck et al. 2015) is also home to a large portion of the Indo-Pacific whale shark population.

Whale sharks often spend a significant portion of their day, sometimes hours at a time, feeding on zooplankton at the surface (Chapter 8). In some areas, a substantial quantity of floating plastic and "microplastics" (often defined as <5 mm pieces: Fossi et al. 2017; Germanov et al. 2019) is found at the surface too, concentrated by the same oceanographic factors that lead to the accumulation of high zooplankton densities. Larger plastic pieces can cause physical damage or digestive tract blockage in whale sharks, as noted in the cases reported above. Two cases of delayed mortality from plastic ingestion have been recorded in Japan (R. Matsumoto unpubl. data). One whale shark, 533 cm TL at death, was caught in a set net in 1984 and transported to an aquarium where it was held for 297 days. One week prior to its death, the shark stopped feeding, and a necropsy revealed a plastic plate obstructing the pylorus (Matsumoto et al. 2017). Plating of this type was not present in the aquaria, so it is assumed that ingestion occurred prior to capture. In the second case, a whale shark caught in a set net was held in aquaria for 201 days, apparently healthy, but suddenly stopped feeding 4 days prior to death. Dissection found a plastic comb in the stomach, again apparently ingested prior to capture (Nishida 2001). This emphasizes the cryptic threat to whale sharks posed by marine debris.

The ingestion of microplastic could also lead to sublethal chronic health problems in affected sharks. Marine plastic pollution often contains various toxic pollutants, either purposely added during manufacture or adsorbed from the environment (Germanov et al. 2018). Trace levels of plastic additives and persistent organic pollutants, which could potentially be incorporated into tissue via plastic ingestion, have been identified from skin samples obtained from live whale sharks in the Gulf of California, Mexico (Fossi et al. 2017). The scale of potential ingestion in some areas, with a staggering 137 pieces per hour estimated for whale sharks feeding off Java in Indonesia (Germanov et al. 2019), emphasizes the importance of quantifying these risks.

While even "low-risk" sharks may only need to ingest a single piece of plastic for a terminal result, a probabilistic approach to threat assessment is possible and useful. Models have been developed to estimate the distribution of surface plastics, which have provisionally identified two important whale shark habitats – the Gulf of Mexico and the Coral Triangle (a biodiverse area in the Southeast Asian and western Pacific region) – as high-priority areas for further investigation (Germanov et al. 2018). Such work has also been suggested as a research need for ecologically similar mobulid rays (Germanov et al. 2018; Stewart et al. 2018). The potential also exists for population-level health assessments of whale sharks, through skin biopsies (Fossi et al. 2017) and blood samples (Chapter 2), in constellations that may be affected. Whale sharks may also be a suitable "flagship species" for raising awareness of the broad threat to marine life from plastic pollution (see Chapter 12; Germanov et al. 2018).

13.13 HOW CAN THE WHALE SHARK'S RECOVERY POTENTIAL BE IMPROVED?

To both arrest the species' decline, and to minimize the whale shark's potential recovery time, we need to maximize the survival of large juvenile and adult female sharks. These sharks have survived through their early years, when they face a high risk from natural predators, and are either close to joining, or already part of the breeding population that can help to "restock" areas where whale sharks have been depleted. Any additive mortality among these females from human threats can be particularly damaging to the population as a whole, but the effective protection of this breeding and pre-breeding segment of the population can pay disproportionate dividends.

Our lack of knowledge on the ecology of these females (see Chapter 7) and the threats they face (see Chapter 11) represent a major hindrance to species-level conservation planning. Understanding the changing ecology of whale sharks throughout their lives is certainly the long-term goal, but minimizing their threats can proceed independently, and immediately. We have identified shipping and fisheries as the two largest contemporary threats to whale sharks (Chapter 11). Areas where shipping corridors overlap with or run close to feeding constellations, where whale sharks of both sexes spend a disproportionate time at the surface, are clearly sites where the implementation of go-slow zones or seasonal re-routing of ships can minimize the chance of ship strikes. As female whale sharks appear to primarily live offshore, they may be especially vulnerable to pelagic gillnet fisheries. Initiatives to assess and mitigate whale shark bycatch in these fisheries, which are both large and poorly monitored (Anderson et al. 2020), are critical.

The whale shark fishery from China was flagged as "likely to be the largest single direct threat to whale shark recovery in the Indo-Pacific" during the contemporary IUCN Red List assessment for the species (Pierce and Norman 2016). Little further information has emerged since, apart from occasional media reports of catches. It is possible that whale shark catches may have reduced, as part of a broader decrease in shark catches (Wu 2016), but the potential scope of this fishery – which was estimated to land around 1000 whale sharks per year in 2010 (Li, Wang, and Norman 2012) – demands that its current status be proven, not just inferred.

13.14 ARE WHALE SHARKS STILL A MYSTERIOUS SPECIES?

In short, yes. Despite a huge amount of dedicated research on this species, and having just completed this textbook, there remain fundamental and pressing questions about whale shark biology and ecology that are still unanswered, unclear, or understudied. The fact that so many fundamental questions remain starkly underscores the myriad difficulties of conducting research on extremely large-bodied and highly mobile marine species that regularly undertake staggering horizontal and vertical movements, the majority of which are beyond our current technological ability to document fully.

Despite these challenges, new and emerging technologies, together with the world-wide and internet-connected power of engaged citizens, may yet provide some of these answers, prompting what we hope will be quantum leaps in our knowledge of this species and how it relates to the rest of the great web of ocean biodiversity. With any luck, these advances, when combined with shifting public perceptions of whale sharks as bona fide charismatic megafauna, will galvanize our efforts to protect them so that they may continue to inspire future generations, just as we have been inspired to conduct the works described in this book. We certainly hope so, because it will be the very next generation that has the greatest opportunity to reverse population trends and conserve this species in perpetuity. This is important because, as we aimed to illustrate throughout this book, the ocean, and indeed the world, is a far more interesting, gratifying, and beautiful place with whale sharks in it.

REFERENCES

Abreo, N. A. S., D. Blatchley, and M. D. Superio. 2019. Stranded whale shark (*Rhincodon typus*) reveals vulnerability of filter-feeding elasmobranchs to marine litter in the Philippines. *Marine Pollution Bulletin* 141:79–83.

Acuña-Marrero, D., J. Jiménez, F. Smith, et al. 2014. Whale shark (*Rhincodon typus*) seasonal presence, residence time and habitat use at Darwin Island, Galapagos Marine Reserve. *PLoS One* 9:e115946.

Alam, M. T., Petit, R. A., Read, T. D., and A. D. M. Dove. 2014. The complete mitochondrial genome sequence of the world's largest fish, the whale shark (*Rhincodon typus*), and its comparison with those of related shark species. *Gene* 539(1):44–49.

Anderson, R. C., M. Herrera, A. D. Ilangakoon, et al. 2020. Cetacean bycatch in Indian Ocean tuna gillnet fisheries. *Endangered Species Research* 41:39–53.

Araujo, G., A. Agustines, B. Tracey, et al. 2019. Photo-ID and telemetry highlight a global whale shark hotspot in Palawan, Philippines. *Scientific Reports* 9:1–12.

Berger-Wolf, T. Y., S. I. Sheikh, B. DasGupta, et al. 2007. Reconstructing sibling relationships in wild populations. *Bioinformatics* 23:i49–56.

Castro, J. I. 2000. The biology of the nurse shark, *Ginglymostoma cirratum*, off the Florida east coast and the Bahama Islands. *Environmental Biology of Fishes* 58:1–22.

Castro, J. I. 2016. The origins and rise of shark biology in the 20th century. *Marine Fisheries Review* 78:14–33.

Castro, A. L. F., B. S. Stewart, S. G. Wilson, et al. 2007. Population genetic structure of Earth's largest fish, the whale shark (*Rhincodon typus*). *Molecular Ecology* 16:5183–92.

Cochran, J. E. M., R. S. Hardenstine, C. D. Braun, et al. 2016. Population structure of a whale shark *Rhincodon typus* aggregation in the Red Sea. *Journal of Fish Biology* 89:1570–82.

Dove, A. D. M., J. Arnold and T. M. Clauss. 2010b. Blood cells and serum chemistry in the world's largest fish: The whale shark *Rhincodon typus*. *Aquatic Biology* 9(2):177–83.

Dove, A. D. M., C. Coco, T. Binder, et al. 2010a. Care of whale sharks in a public aquarium setting. *Proceedings of the 2nd US/Russian Bilateral Conference on Aquatic Animal Health*. East Lansing, Michigan State University Press.

Dove, A. D. M. 2015. Foraging and ingestive behaviors of whale sharks, *Rhincodon typus*, in response to chemical stimulus cues. *The Biological Bulletin* 228(1):65–74.

Escalle, L., M. G. Pennino, D. Gaertner, et al. 2016. Environmental factors and megafauna spatio-temporal co-occurrence with purse-seine fisheries. *Fisheries Oceanography* 25:433–47.

Feutry, P., O. Berry, P. M. Kyne, et al. 2017. Inferring contemporary and historical genetic connectivity from juveniles. *Molecular Ecology* 26:444–56.

Fossi, M. C., M. Baini, C. Panti, et al. 2017. Are whale sharks exposed to persistent organic pollutants and plastic pollution in the Gulf of California (Mexico)? First ecotoxicological investigation using skin biopsies. *Comparative Biochemistry and Physiology Part C: Toxicology & Pharmacology* 199:48–58.

Germanov, E. S., A. D. Marshall, L. Bejder, et al. 2018. Microplastics: No small problem for filter-feeding megafauna. *Trends in Ecology & Evolution* 33:227–32.

Germanov, E. S., A. D. Marshall, I. G. Hendrawan, et al. 2019. Microplastics on the menu: Plastics pollute Indonesian manta ray and whale shark feeding grounds. *Frontiers in Marine Science* 6:679.

Gleiss, A. C., B. Norman, N. Liebsch, et al. 2009. A new prospect for tagging large free-swimming sharks with motion-sensitive data-loggers. *Fisheries Research* 97:11–16.

Goldbogen, J. A., Calambokidis, J., Oleson, E., et al. 2011. Mechanics, hydrodynamics and energetics of blue whale lunge feeding: Efficiency dependence on krill density. *Journal of Experimental Biology* 214:131–46.

Haetrakul, T., S. Munanansup, N. Assawawongkasem, and N. Chansue. 2009. A case report: Stomach foreign object in whaleshark (*Rhincodon typus*) stranded in Thailand. *Proceedings of the 4th International Symposium on SEASTAR 2000 and Asian Bio-Logging Science (The 8th SEASTAR 2000 Workshop)*:83–85.

Hammerschlag, N., A. J. Gallagher, and D. M. Lazarre. 2011. A review of shark satellite tagging studies. *Journal of Experimental Marine Biology and Ecology* 398:1–8.

Hara, Y., Yamaguchi, K., Onimaru, K., Kadota, M., Koyanagi, M., Keeley, S.D., Tatsumi, K., Tanaka, K., Motone, F., Kageyama, Y. and Nozu, R., 2018. Shark genomes provide insights into elasmobranch evolution and the origin of vertebrates. *Nature Ecology & Evolution* 2(11):1761–1771.

Hardenstine, R. S., S. He, J. E. M. Cochran, et al. In review. Pieces in a global puzzle: Population genetics at two whale shark aggregations in the western Indian Ocean. *Ecology and Evolution*.

Hearn, A. R., J. R. Green, E. Espinoza, et al. 2013. Simple criteria to determine detachment point of towed satellite tags provide first evidence of return migrations of whale sharks (*Rhincodon typus*) at the Galapagos Islands, Ecuador. *Animal Biotelemetry* 1:11.

Hearn, A. R., J. R. Green, M. H. Román, et al. 2016. Adult female whale sharks make long-distance movements past Darwin Island (Galapagos, Ecuador) in the Eastern Tropical Pacific. *Marine Biology* 163:214.

Hunt, C. V., J. J. Harvey, A. Miller, et al. 2013. The Green Fins approach for monitoring and promoting environmentally sustainable scuba diving operations in South East Asia. *Ocean & Coastal Management* 78:35–44.

Huveneers, C., T. I. Walker, N. M. Otway, and R. G. Harcourt. 2007. Reproductive synchrony of three sympatric species of wobbegong shark (genus *Orectolobus*) in New South Wales, Australia: Reproductive parameter estimates necessary for population modelling. *Marine and Freshwater Research* 58:765–77.

Ito, T., K. Onda, and K. Nishida. 2017. Effects of noise and vibration on the behavior and feeding activity of whale sharks, *Rhincodon typus* (Smith, 1828), in Osaka Aquarium Kaiyukan. In *The Elasmobranch Husbandry Manual II: Recent Advances in the Care of Sharks, Rays and Their Relatives*, eds. Smith, M., D. Warmolts, D. Thoney, R. Hueter, M. Murray, and J. Ezcurra, 159–67. Columbus: Special Publication of the Ohio Biological Survey.

Jacoby, D. M. P., D. S. Busawon, and D. W. Sims. 2010. Sex and social networking: The influence of male presence on social structure of female shark groups. *Behavioral Ecology: Official Journal of the International Society for Behavioral Ecology* 21:808–18.

Jambeck, J. R., R. Geyer, C. Wilcox, et al. 2015. Plastic waste inputs from land into the ocean. *Science* 347(6223):768–771.

Joung, S.-J., C.-T. Chen, E. Clark, et al. 1996. The whale shark, *Rhincodon typus*, is a livebearer: 300 embryos found in one 'megamamma' supreme. *Environmental Biology of Fishes* 46:219–23.

Ketchum, J. T., F. Galván-Magaña, and A. P. Klimley. 2013. Segregation and foraging ecology of whale sharks, *Rhincodon typus*, in the southwestern Gulf of California. *Environmental Biology of Fishes* 96:779–95.

Lee, S. 2019. Plastic bag causes death of whale shark in Sabah. https://www.thestar.com.my/news/nation/2019/02/08/plastic-bag-causes-death-of-whale-shark-in-sabah/ (accessed January 5, 2021).

Li, W., Y. Wang, and B. Norman. 2012. A preliminary survey of whale shark *Rhincodon typus* catch and trade in China: An emerging crisis. *Journal of Fish Biology* 80:1608–18.

Macena, B. C. L., and F. H. V. Hazin. 2016. Whale shark (*Rhincodon typus*) seasonal occurrence, abundance and demographic structure in the mid-equatorial Atlantic Ocean. *PLoS One* 11:e0164440.

Matsumoto, R., Matsumoto, Y., Ueda, K., et al. 2019. Sexual maturation in a male whale shark (*Rhincodon typus*) based on observations made over 20 years of captivity. *Fishery Bulletin* 117:78–86.

Matsumoto, R., M. Toda, Y. Matsumoto, et al. 2017. Notes on husbandry of whale sharks, *Rhincodon typus*, in aquaria. In *The Elasmobranch Husbandry Manual II: Recent Advances in the Care of Sharks, Rays and Their Relatives*, ed. Smith, M., D. Warmolts, D. Thoney, R. Hueter, M. Murray, and J. Ezcurra, 15–22. Columbus: Special Publication of the Ohio Biological Survey.

Meekan, M. G., Fuiman, L. A., Davis, R., et al. 2015. Swimming strategy and body plan of the world's largest fish: Implications for foraging efficiency and thermoregulation. *Frontiers in Marine Science* 2:64.

Meekan, M., C. M. Austin, M. H. Tan, et al. 2017. iDNA at sea: Recovery of whale shark (*Rhincodon typus*) mitochondrial DNA sequences from the whale shark copepod (*Pandarus rhincodonicus*) confirms global population structure. *Frontiers in Marine Science* 4:420.

Meekan, M.G., Taylor, B.M., Lester, E., Ferreira, L.C., Sequeira, A.M., Dove, A.D., Birt, M.J., Aspinall, A., Brooks, K. and Thums, M., 2020. Asymptotic growth of whale sharks suggests sex-specific life-history strategies. *Frontiers in Marine Science* 7:774.

Meyers, M. M., M. P. Francis, M. Erdmann, et al. 2020. Movement patterns of whale sharks in Cenderawasih Bay, Indonesia, revealed through long-term satellite tagging. *Pacific Conservation Biology*. doi:10.1071/PC19035.

Mucientes, G. R., N. Queiroz, L. L. Sousa, et al. 2009. Sexual segregation of pelagic sharks and the potential threat from fisheries. *Biology Letters* 5:156–59.

Nakamura, I., Matsumoto, R., and K. Sato. 2020. Body temperature stability in the whale shark, the world's largest fish. *Journal of Experimental Biology* 223. doi:10.1242/jeb.210286.

Nishida, K. 2001. Whale shark - the world's largest fish. In *Fishes of the Kuroshio Current, Japan*, ed. T. Nakabo, Y. Machida, K. Yamaoka, and K. Nishida, 20–26. Osaka: OsakaAquarium Kaiyukan.

Norman, B. M., S. Reynolds, and D. L. Morgan. 2016. Does the whale shark aggregate along the Western Australian coastline beyond Ningaloo Reef? *Pacific Conservation Biology* 22:72–80.

Perry, C. T., J. Figueiredo, J. J. Vaudo, et al. 2018. Comparing length-measurement methods and estimating growth parameters of free-swimming whale sharks (*Rhincodon typus*) near the South Ari Atoll, Maldives. *Marine and Freshwater Research* 69:1487–95.

Perry, C. T., E. Clingham, D. H. Webb, et al. 2020. St. Helena: An Important reproductive habitat for whale sharks (*Rhincodon typus*) in the central south Atlantic. *Frontiers in Marine Science* 7:899.

Pimiento, C., Cantalapiedra, J. L., Shimada, K., et al. 2019. Evolutionary pathways toward gigantism in sharks and rays. *Evolution* 73:588–99.

Potvin, J., Goldbogen, J. A., and R. E. Shadwick. 2012. Metabolic expenditures of lunge feeding rorquals across scale: Implications for the evolution of filter feeding and the limits to maximum body size. *PLoS ONE* 7:e44854.

Pierce, S. J., and B. Norman. 2016. *Rhincodon typus*. The IUCN Red List of Threatened Species 2016:e. T19488A2365291.

Ramírez-Macías, D., A. Vázquez-Haikin, and R. Vázquez-Juárez. 2012. Whale shark *Rhincodon typus* populations along the west coast of the Gulf of California and implications for management. *Endangered Species Research* 18:115–28.

Ramírez-Macías, D., N. Queiroz, S. J. Pierce, et al. 2017. Oceanic adults, coastal juveniles: Tracking the habitat use of whale sharks off the Pacific coast of Mexico. *PeerJ* 5:e3271.

Read, T. D., Petit, R. A, Joseph, S. J., et al. 2017. Draft sequencing and assembly of the genome of the world's largest fish, the whale shark: *Rhincodon typus* Smith 1828. *BMC Genomics* 18:1–10.

Robinson, D. P., M. Y. Jaidah, S. Bach, et al. 2016. Population structure, abundance and movement of whale sharks in the Arabian Gulf and the Gulf of Oman. *PLoS One* 11:e0158593.

Rowat, D., K. Brooks, A. March, et al. 2011. Long-term membership of whale sharks (*Rhincodon typus*) in coastal aggregations in Seychelles and Djibouti. *Marine and Freshwater Research* 62:621–27.

Ryan, J. P., J. R. Green, E. Espinoza, and A. R. Hearn. 2017. Association of whale sharks (*Rhincodon typus*) with thermo-biological frontal systems of the eastern tropical Pacific. *PLoS One* 12:e0182599.

Sampaio, C. L. S., L. Leite, J. A. Reis-Filho, et al. 2018. New insights into whale shark *Rhincodon typus* diet in Brazil: An observation of ram filter-feeding on crab larvae and analysis of stomach contents from the first stranding in Bahia state. *Environmental Biology of Fishes* 101:1285–93.

Schmidt, J. V., C. L. Schmidt, F. Ozer, et al. 2009. Low genetic differentiation across three major ocean populations of the whale shark, *Rhincodon typus*. *PloS One* 4:e4988.

Schreiber, C., and C. Coco. 2017. Husbandry of whale sharks. In *The Elasmobranch Husbandry Manual II: Recent Advances in the Care of Sharks, Rays and Their Relatives*, ed. Smith, M., D. Warmolts, D. Thoney, R. Hueter, M. Murray, and J. Ezcurra, 87–98. Columbus: Special Publication of the Ohio Biological Survey.

Stewart, J. D., F. R. A. Jaine, A. J. Armstrong, et al. 2018. Research priorities to support effective manta and devil ray conservation. *Frontiers in Marine Science* 5:314.

Tan, M., A. K. Redmond, H. Dooley, et al. 2021. The whale shark genome reveals patterns of vertebrate gene family evolution. *eLife* (in review)

Taylor, G. 2019. *Whale Sharks: The Giants of Ningaloo Reef.* 2nd Revised Edition. Perth: Scott Print.

Thums, M., Meekan, M., Stevens, J., et al. 2013. Evidence for behavioral thermoregulation by the world's largest fish. *Journal of the Royal Society Interface* 10:20120477.

Vandeperre, F., A. Aires-da-Silva, J. Fontes, et al. 2014. Movements of blue sharks (*Prionace glauca*) across their life history. *PLoS One* 9:103538.

Vignaud, T. M., J. A. Maynard, R. Leblois, et al. 2014. Genetic structure of populations of whale sharks among ocean basins and evidence for their historic rise and recent decline. *Molecular Ecology* 23:2590–601.

Wearmouth, V. J., E. J. Southall, D. Morritt, et al. 2012. Year-round sexual harassment as a behavioral mediator of vertebrate population dynamics. *Ecological Monographs* 82:351–66.

Wu, J. 2016. *Shark Fin and Mobulid Ray Gill Plate Trade in Mainland China, Hong Kong and Taiwan.* Cambridge: TRAFFIC.

Yagishita, N., S-I. Ikeguchi, and R. Matsumoto. 2020. Re-estimation of genetic population structure and demographic history of the whale shark (*Rhincodon typus*) with additional Japanese samples, inferred from mitochondrial DNA sequences. *Pacific Science* 74:31–47.

Index

Note: **Bold** page numbers refer to tables and *italic* page numbers refer to figures.